R. C. Maurya, J. M. Mir

Molecular Symmetry and Group Theory

Also of Interest

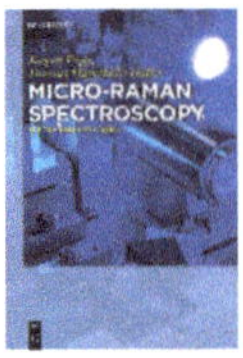

Micro-Raman Spectroscopy.
Theory and Application
Popp, Mayerhöfer, 2019
ISBN 978-3-11-051479-7, e-ISBN 978-3-11-051531-2

Elastic Light Scattering Spectrometry.
Huang, Ling, Wang, 2019
ISBN 978-3-11-057310-7, e-ISBN 978-3-11-057313-8

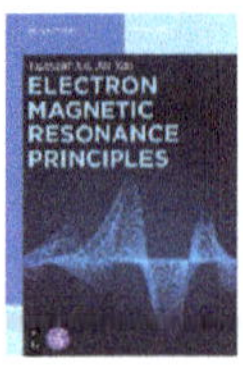

Electron Magnetic Resonance Principles.
Xu, Yao, 2019
ISBN 978-3-11-052800-8, e-ISBN 978-3-11-062010-8

Theoretical Mass Spectrometry.
Tracing Ions with Classical Trajectories
Song, Spezia, 2018
ISBN 978-3-11-044200-7, e-ISBN 978-3-11-043489-7

Elemental Analysis.
An Introduction to Modern Spectrometric Techniques
Schlemmer, Balcaen, Todolí, Hinds, 2019
ISBN 978-3-11-050107-0, e-ISBN 978-3-11-050108-7

R. C. Maurya, J. M. Mir

Molecular Symmetry and Group Theory

Approaches in Spectroscopy and Chemical Reactions

DE GRUYTER

Authors
Prof. Dr. R.C. Maurya
D. Sc, CChem FRSC (UK)
Professor of Inorganic Chemistry
Former Head, Department of Chemistry and Pharmacy, and
Dean, Faculty of Science
Rani Durgavati University
Jabalpur-482001 (M. P.)
India

Dr. J.M. Mir
Assistant professor
Department of Chemistry
Islamic University of Science and Technology,
Srinagar (J&K), India

ISBN 978-3-11-063496-9
e-ISBN (PDF) 978-3-11-063503-4
e-ISBN (EPUB) 978-3-11-063512-6

Library of Congress Control Number: 2019932032

Bibliographic information published by the Deutsche Nationalbibliothek
The Deutsche Nationalbibliothek lists this publication in the Deutsche Nationalbibliografie; detailed bibliographic data are available on the Internet at http://dnb.dnb.de.

Typesetting: Integra Software Services pvt. Ltd.
Printing and binding: CPI books GmbH, Leck
Cover image: PRImageFactory / iStock / Getty Images Plus

www.degruyter.com

Dedicated to

Mrs. Usha Rani Maurya,
who has always been a source of inspiration for Dr. Maurya
throughout his growth

Rani Durgavati University, Jabalpur (M.P.), India

Professor Kapil Deo Mishra
Vice-Chancellor

Saraswati Vihar, Pachperi,
Jabalpur-482001 (M.P.), India
Phone: +91-761- 2600568

Foreword

The concept and application of Symmetry and Group Theory leading to the understanding of molecular structures, chemical bonding, spectral properties of molecules and symmetry allowing chemical processes have become vital for students as well as faculty members of science, especially chemistry and physics. A very well presented and lucid text on the theme with student-friendly approach, especially targeting university and college students, is in demand. In this regard, the present book is a treasure trove because the author of the text has been Professor and Head of Chemistry and Pharmacy Department, and Dean Faculty of Science of our University with a long and wide teaching experience enabling him with rich inputs on the students' requirement.

The present book with the subject matter being carefully distributed over four chapters will serve the purpose for which it has been designed. This must have been a tediously hard work of Professor R. C. Maurya, D.Sc., CChem FRSC (UK), for which he deserves all appreciations. I am very sure that this book would be

https://doi.org/10.1515/9783110635034-201

very much liked by the teaching community and students of chemistry as well physics too.

September, 2018 **(Kapil Deo Mishra)**

Preface

Molecular symmetry and Group theory is a powerful theoretical tool to study and predict chemical bonding, and spectral properties of molecules and symmetry allowing chemical processes. It is only the last three decades or so that chemists are making use of this tool to understand the fundamental concepts involved in bonding, structures of organic and inorganic compounds, spectral properties and occurrence of chemical reactions.

Group theory and its applications have been part of the curriculum of postgraduate courses, in number of universities for the last several years. In our university, this is recently included in M.Phil. and Ph.D. course work also as a part of the syllabus.

In the initial stages, the problem faced by faculty members and students was the lack of sufficient reference materials in this area. Since then, a limited number of textbooks have been published. Several inorganic and physical chemistry textbooks included a chapter on Group theory, but it was found that most of them confined to the basic principles or to certain specific applications and hence inadequate. Now, we have so many books available in our library on Group theory and its applications but detailed mathematical calculations are missing in these books. That is why these books are not of very much liking by students and faculty members.

This book is the outcome of RCM's teaching of the subject for more than two and half decades to several batches of Masters, M.Phil. and Ph.D. students at the Department of Chemistry, Rani Durgavati University, Jabalpur. He has benefitted enormously from the response, questions and criticisms of my students. Extensive use of standard books/research papers on group theory has been made and that are listed under Bibliography.

Looking over the problems of students in learning this subject, we have tried my best to present the subject with student-friendly approach, that is, expressing it in an interactive manner and in simple language with many illustrative examples. Moreover, the mathematical part wherever required is given in details to make the subject easily understandable. Exercises are given at the end of each chapter. Therefore, we hope that the book will serve as text for M.Sc., M.Phil. and Ph.D. students of chemistry. We wish the readers a pleasant journey through this book and trust it will inspire them to delve deeper into the exciting field of Group theory.

In spite of serious attempts to keep the text free of errors, it would be presumptuous to hope that no error has crept in. We will be grateful to all those who may care to send their criticism and suggestions for the improvement of the book on my e-mail ID (rcmaurya1@gmail.com).

https://doi.org/10.1515/9783110635034-202

Final touch to this book was given in San Jose, California, during RCM's stay with his son, Animesh Maurya, for which he deserves thanks. Finally, RCM is thankful to his wife Mrs. Usha Rani Maurya for her encouragement and sacrifices.

September 20, 2018

R. C. Maurya
J. M. Mir

Contents

1 Symmetry elements and symmetry operations: molecular symmetry

1.1 Introduction

Symmetry is a very important and fascinating property of molecules. The concept of molecular symmetry is important from the view point of their applications to chemical problems, viz., molecular orbitals in polyatomic molecules, and thus their structure and bonding, crystal field and molecular orbital theory of complex compounds, electronic and vibrational spectra of simple and complex compounds and so on.

In order to use the symmetry properties of molecules in solving chemical problems as mentioned above, it is necessary to have some acquaintance with the branch of mathematics known as the *Group Theory*. It can be applied to any set of elements, which obey the necessary conditions to be called a Group.

A *Group* in the mathematical sense is a collection of elements having certain properties in common, which enable a wide variety of algebraic manipulations to be carried out. The elements could be numbers, matrices, symmetry operations or symmetry elements. All the symmetry operations present in a molecule forms a Group. Moreover, each symmetry operation in a molecule can be represented by a matrix. *Hence, the symmetry operations/symmetry elements of a molecule constitute what is known as a mathematical group.*

Hence, the formal treatment of the concept of molecular symmetry is the subject matter of this chapter.

1.2 Molecular symmetry: in non-mathematical and geometrical sense

The word Symmetry is derived from the Greek word *Summetria* meaning "similar measure." It implies that each part of an object is in harmony with each other and is well balanced. Hence, the term Symmetry is almost synonymous to beauty because the nature has made most of its creation symmetrical. The Sun, the planets, the human beings, animals and plants are all symmetrical. Thus, in non-mathematical (physical) sense, symmetry is a familiar term in our world, and is best described as

> "A thing of beauty is a joy forever."

In geometrical sense, when we say that a molecule has symmetry, we *mean that certain parts of it can be interchanged with other without altering the identity or the orientation of the molecule.* In other words, an object or a molecule is said to be

https://doi.org/10.1515/9783110635034-001

symmetrical if it can take more than one equivalent (or indistinguishable) orientation. It may not necessarily be identical with the original configuration because only some equivalent parts may have been interchanged. For example,

Let us consider H_2 (Fig. 1.1) molecule in orientation (a) H and H′ are same. It can be rotated by 180° to the orientation (b). (b) It cannot be distinguished from (a), that is, they are equivalent.

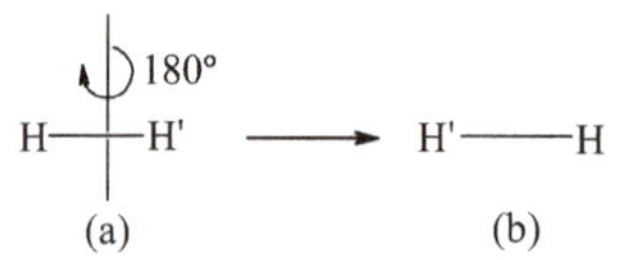

Fig. 1.1: Rotation of H_2 molecule through 180 in orientation (a) to get orientation (b).

Consider another example of PF_5 molecule (Fig. 1.2). It is a trigonal bipyramidal molecule. The three equatorial bonds, P–F_1, P–F_2 and P–F_3 are equivalent. Similarly, the axial P–F bonds, P–F_4 and P–F_5 are equivalent. However, axial and equatorial bonds are of different types, and if one of each was to be interchanged, the molecule will alter its identity.

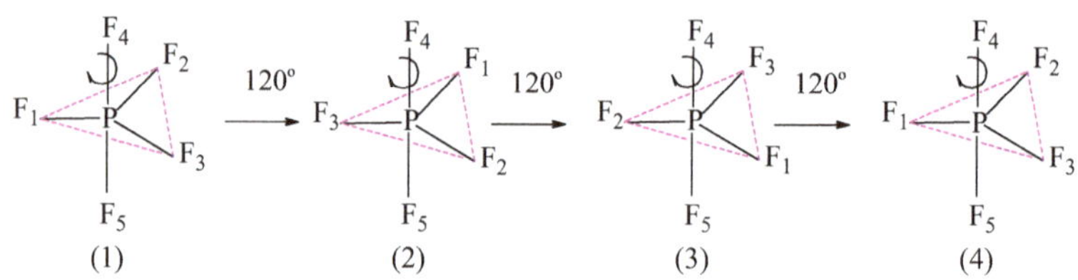

Fig. 1.2: Equivalence of equatorial and axial P-F bonds in PF_5 molecule.

1.3 Symmetry operations and symmetry elements

In order to make the idea of molecular symmetry more useful or for the systematic and detailed consideration of symmetry, certain formal tools are needed. The first set of tools is a set of symmetry operations generated by the symmetry elements. Thus, a fundamental concept of group theory is the symmetry operations.

The two things symmetry elements and symmetry operations are so closely related that they sometime confuse beginners. However, they are quite different. Hence, it is important to have clear understanding of the difference between them.

Symmetry operations

It is possible to classify symmetry of an object or a molecule in terms of symmetry operations defined as follows:

"A symmetry operation, in general, is defined as a movement of an object such that it brings the object into an equivalent configuration. Such a configuration is indistinguishable from the original one. It may not be necessarily being identical with the original configuration because some equivalent parts only may have been exchanged."

Symmetry elements

A symmetry element is defined as a geometrical entity such as a line, plane or a point with respect to which one or more symmetry operations may be carried out.

Symmetry elements and symmetry operations are closely interrelated as the symmetry operation can be defined only with respect to symmetry element, and at the same time the existence of a symmetry element can be demonstrated only by showing the appropriate symmetry operation exists.

Five types of symmetry elements and symmetry operations need be considered in treating the molecular symmetry (Table 1.1).

Table 1.1: Various symmetry elements, symmetry operation and their Schoenflies symbols.

S. No.	Symmetry element	Symmetry operation	Schoenflies symbol
1.	Axis of symmetry/ *n*-fold symmetry axis/ proper axis	Rotation by $2\pi/n$ or $360°/n$	C_n
2.	Mirror plane/ plane of symmetry	Reflection in the plane	σ
3.	Center of symmetry/ Center of inversion	Inversion of all atoms through the centre	i
4.	Improper axis/ *n*-fold axis of improper rotation	Rotation through $360°/n$ followed by reflection in a plane $\perp$r to the rotation axis	S_n
5.	–	Identity (*Trivial operation leaving the molecule unchanged or in its original position)	E or I

* In identity operation, the symmetry element can be thought of as the whole space.

1.4 Naming systems of notation for symmetry operations/elements

There are two types of naming systems commonly used for describing symmetry notation or designation:

Schoenflies notation
This type of notation is commonly used for molecules by spectroscopists. Such notations for axis of symmetry, plane of symmetry, center of symmetry, improper axis of symmetry and identity are C_n, σ, i, S_n and E, respectively. The point groups of molecules are designated as C_{2v}, C_{3v}, T_d, O_h, D_{4h} and so on.

Hermann–Mauguin notation
This is an international notation for molecular crystals preferred by crystallographers. Some such notations are: 1, 2, 3, 4, 6, m, 1(bar), 3(bar), 4 (bar), 6 (bar) and 22 more, for different crystal systems.

1.5 Proper axis of symmetry

It is an imaginary axis passing through the molecule, over which, rotation by an angle θ can be carried out to take the molecule from one orientation to another equivalent orientation. It may be possible to carry out several operations on a rotation axis. If a molecule can have n different orientations about this axis, the axis is said to be order n.

The general symbol for a proper axis of rotation is C_n, where the subscript n denotes the order of the axis, and the symbol C stands for circular rotation around an axis. The meaning of order n of an axis is to say that it is the number of times that the smallest rotation is capable of giving an equivalent configuration and it must be repeated in order to give a configuration not merely equivalent to the original but also identical to it in a circular rotation around the imaginary axis. Hence,

$$n = 2\pi \text{ radians}/\theta \text{ or } 360^\circ/\theta$$

$$\theta = 360^\circ/n$$

Examples
H_2O: In this molecule, the axis passing through O atom and in between the two H atoms has order 2, or is an axis of twofold symmetry. The molecule has to be rotated by 2π radians/2 or 360°/2 = 180° to get equivalent orientation (Fig. 1.3)

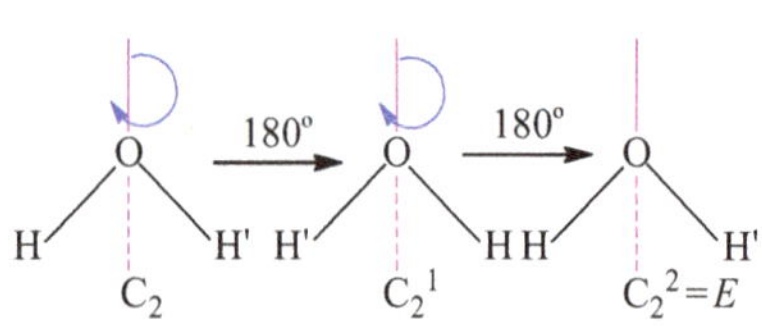

Fig. 1.3: Twofold axis of symmetry in H_2O molecule.

NH_3, PH_3: In case of NH_3, the axis passing through *N* and the center of the triangle containing 3Hs is threefold axis. The molecule has to be rotated by 360/3 = 120° (Fig. 1.4).

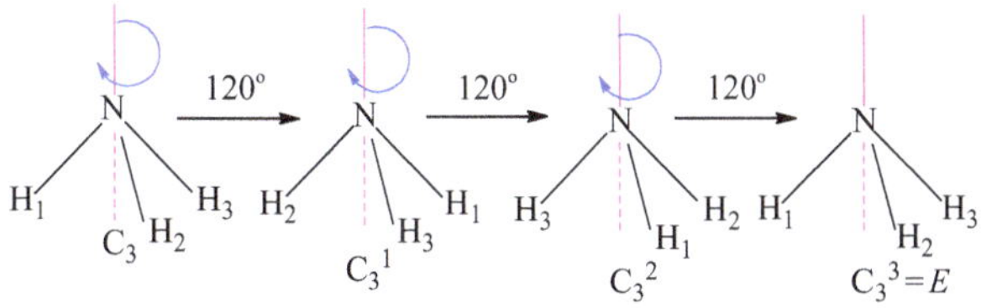

Fig. 1.4: Threefold axis of symmetry in NH_3 molecule.

BF_3: In the triangular planar BF_3 molecule, the axis passing through B and ⊥r to the triangle is the axis of threefold symmetry. The molecule has to be rotated by 360/3 = 120° (Fig. 1.5).

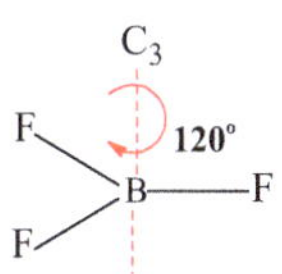

Fig. 1.5: Axis of threefold symmetry in BF_3 molecule.

$[Ni(CN)_4]^{2-}$: In square planar $[Ni(CN)_4]^{2-}$, the axis passing through Ni and ⊥r to the plane is an axis of fourfold symmetry (Fig. 1.6).

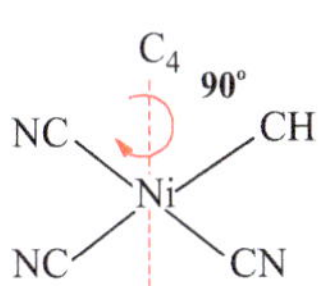

Fig. 1.6: Axis of fourfold symmetry in $[Ni(CN)_4]^{2-}$ molecule.

CH_4: In this case, there are $4C_3$. The axis contains each of the C–H bonds and center of the triangle containing three H atoms (Fig. 1.7). In fact, each of the four C_3 axes passes through one apex and center of the opposite face of the cube.

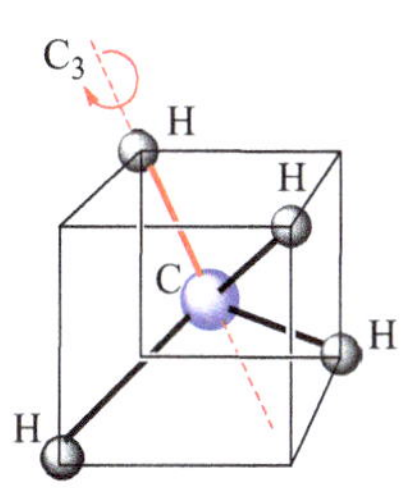

Fig. 1.7: C_3 axis of symmetry passing through one of the C-H bonds in CH_4.

Molecules with more axes

C_2H_4: This is an example of a molecule containing three twofold axes (Fig. 1.8). Here, one C_2 is collinear with C-C axis. A second C_2 is ⊥r to the plane of the molecule and bisects the C-C line. The third is ⊥r to the first two and intersects both at the midpoint of the C-C line.

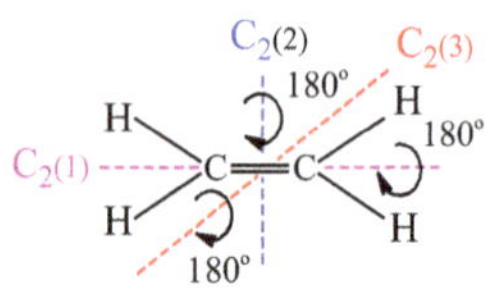

Fig. 1.8: C_2H_4 molecule containing three twofold axes of symmetry.

Allene ($CH_2 = C = CH_2$): Similar to C_2H_4, allene molecule also contains three twofold axes. The axis passing to the three carbons is a C_2. The second C_2 is ⊥r to the plane of the molecule and passing through mid carbon. The third is ⊥r to the first two and intersects both at the mid carbon (Fig. 1.9).

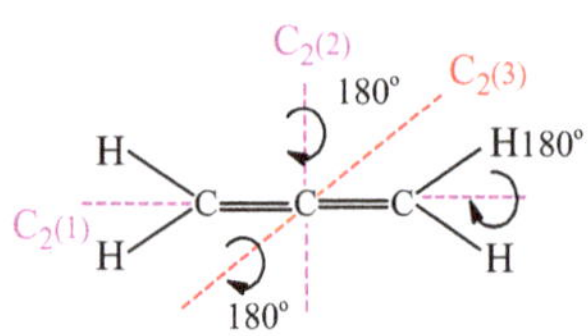

Fig. 1.9: Allene molecule containing three twofold axes of symmetry.

BF_3: In BF_3, besides the C_3, there are three C_2s, passing through B and each of the F atom, and are in the plane of the molecule (Fig. 1.10).

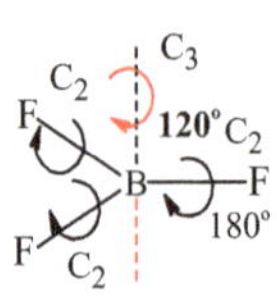

Fig. 1.10: BF_3 molecule containing C_3 and $3C_2$s passing through B.

$[Ni(CN)_4]^{2-}$: In **$[Ni(CN)_4]^{2-}$**, besides the C_4, there are $4C_2$s, two passing through Ni and trans CN^- and two passing through Ni and the centers of two opposite edges of square plane (Fig. 1.11). These are in the plane of the molecule.

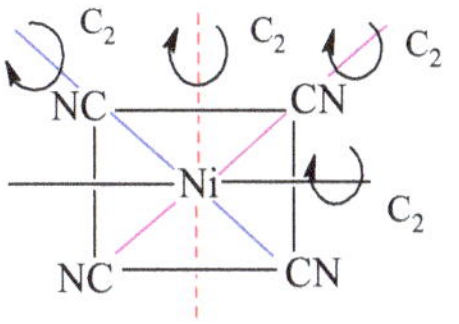

Fig. 1.11: $[Ni(CN)_4]^{2-}$ containing C_4 and $4C_2$s passing through Ni.

Benzene: In C_6H_6, besides the C_6, there are $6C_2$s, **three** passing through the center of benzene and two opposite C atoms, and three passing through the center of benzene and center of the two opposite edges (Fig. 1.12). C_3 is also present in benzene.

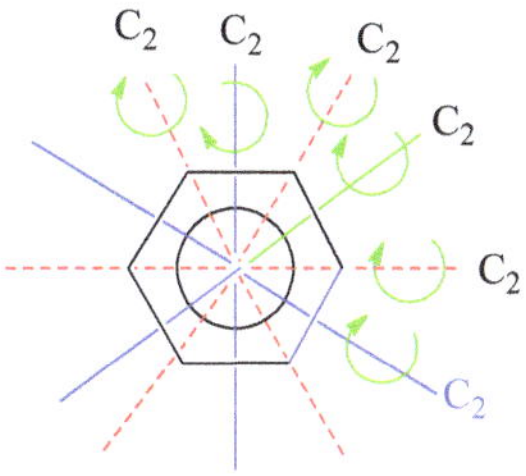

Fig. 1.12: C_6 and $6C_2$s in benzene passing through its centre.

CH_4: In CH_4, besides four C_3, there are $3C_2$s as shown in Fig. 1.13. In fact, each of the Cartesian axes, *x*, *y* and *z* bisecting two HCH bond angles, is the C_2 axis.

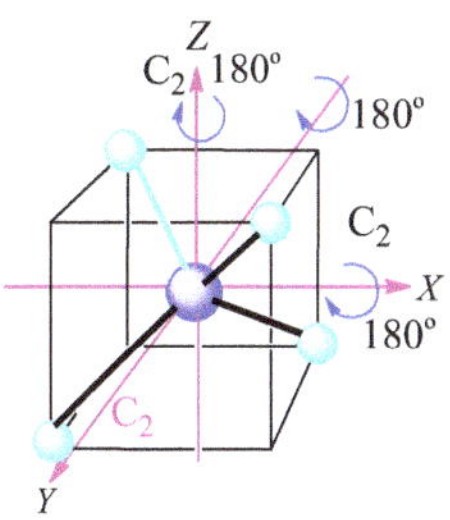

Fig. 1.13: CH_4 molecule containing $4C_3$s and $3C_2$s.

In cases of molecules with more than one axis of symmetry, the axis with highest fold symmetry is called the *principal axis*. For examples in BF_3, $Ni(CN)_4^{2-}$, CH_4 and C_6H_6, C_3, C_4, C_3 and C_6 are the principal axes, respectively. In each case there are subsidiary axes of twofold symmetry.

C_∞ axis: For linear molecules such as H_2, CO_2, HCN and HCl, the molecular axis is an infinite fold axis of rotation [Fig. 1.14(a)]. This is because rotation by any fraction of angle along the molecular axis will give infinite equivalent orientations. Symmetrical linear molecules such as H_2 and CO_2 possess, in addition, an infinite number of C_2 axes passing through the center of gravity of the molecule [Fig. 1.14(b)].

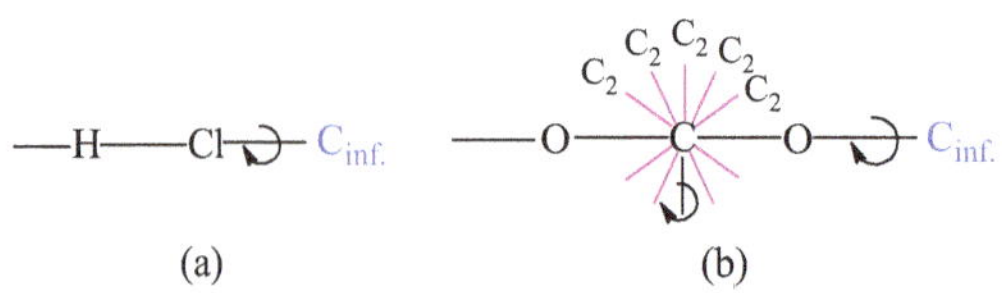

Fig. 1.14: (a) C ∞ in linear molecules and (b) ∞C_2s in symmetrical linear molecules.

Some important points

The process of rotation along the axis of symmetry is a symmetry operation. The possible number of symmetry operations on a C_n axis is n–1. The nth rotation will bring the molecule to the original orientation, that is, identity (E).

On C_2 the only operation possible is C_2^1, that is, rotation by 180°. C_2^2, that is, rotation by 360° will bring back the molecule to the original orientation. Over C_3, two operations C_3^1 (120°) and C_3^2 (240°) are possible. Over, C_4, three operations are possible, that is, C_4^1 (90°), C_4^2 (180°) and C_4^3 (270°). Over C_6 five operations, C_6^1 (60°), C_6^2 (120°), C_6^3 (180°), C_6^4 (240°) and C_6^5(300°).

It can be seen that C_2^1 (180°), = C_4^2 (180°) = C_6^3 (180°). These are equivalent operations because in each case there is rotation of 180°. Similarly, C_3^1 (120°), = C_6^2 (120°) and C_3^2 (240°) = C_6^4 (240).

The above rotations are all clockwise. One can argue that there can be anticlockwise rotation and they are also symmetry operations. For example, a rotation of 120° in the anticlockwise rotation along C_3, that is, C_3^{-1} should also give equivalent orientation. However, this gives the same orientation obtained by clockwise rotation of 240°. Thus, $C_3^{-1} \equiv C_3^2$, and hence, need not be considered as a separate symmetry operation (Fig. 1.15).

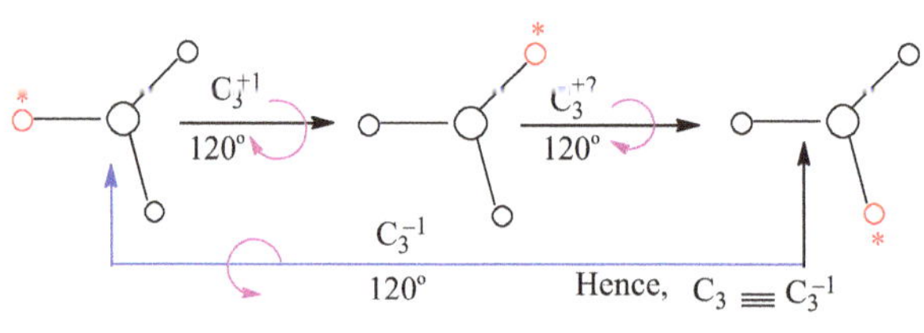

Fig. 1.15: Orientation obtained by clockwise and anticlockwise rotations.

Other examples: $[C_7H_7]^+$: C_7^- axis; $(C_8H_8)_2U$ (Uranocene):C_8, $Cr(C_6H_6)_2$:C_6 (⊥r to the plane), 6C_2s on the plane, $(C_5H_5)_2Fe$ (Ferrocene): C_5

1.6 Plane of symmetry

It is defined as an imaginary plane within the molecule that acts as a mirror and divides it into two parts, which are mirror images of each other. If a reflection of any atom of the molecule is carried out on the plane "σ" it must result in coincidence with another equivalent atom. In other words if a ⊥r is drawn from any atom on the plane of symmetry and extended on the other side up to the same length it must meet another atom of the same type (Fig. 1.16).

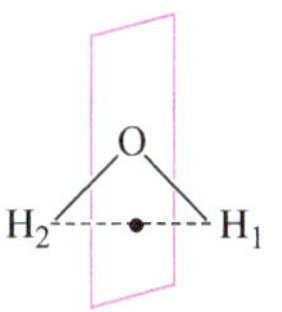

Fig. 1.16: ⊥r drawn from H_1 on the plane and extending other side meeting H_2 in hydrogen molecule.

If the reflection is carried out over the plane of the symmetry and the new positions of the atom noted, the new orientation will be equivalent to the original orientation. For example in H_2O molecule, the pane passing through C_2 axis and ⊥r to the molecular plane is a plane of symmetry (Fig. 1.17).

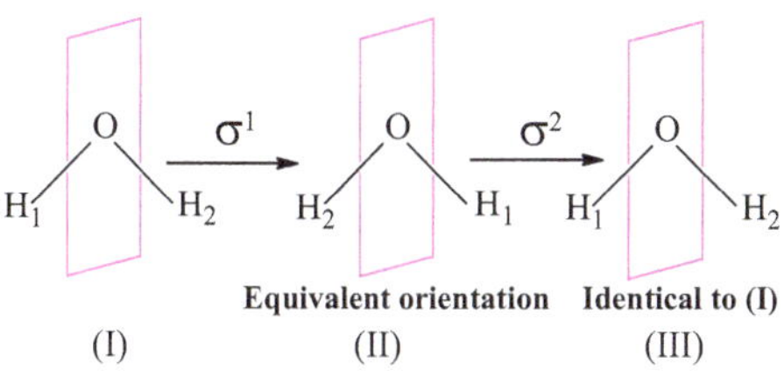

Fig. 1.17: Reflection over the plane of symmetry generating equivalent and identical orientation in H_2O.

Thus, the plane is the element of symmetry and the process of reflection is a symmetry operation. Only one operation is possible on one plane. If the second reflection is carried out it gives back the original orientation.

The atom(s) on the surface of the plane of symmetry remains unshifted as a result of reflection, for example, O atom in case of water molecule. All the atoms, other than those on the plane of the molecule, should be in pairs. If the molecule has more than one plane, the unpaired atom must occur at the intersection of the planes of symmetry. The planes of symmetry and reflection plane on it are represented as σ.

Types of planes

They can be classified into three types depending on their relationship with the principal axis.

(i) The plane passing through the principal axis and one of the subsidiary axes (if present) is called vertical plane. It is represented by σ_v.

(ii) The plane ⊥r to the principle axis is called horizontal plane, represented as σ_h.

(iii) The plane passing through the principal axis (i.e., vertical plane) and bisecting the angle between the two C_2 axes in a molecule is called dihedral plane, represented as σ_d.

Examples

Human and most of the animals have a symmetry plane.

H_2O: It has two reflection planes: (i) a molecular plane and (ii) a ⊥r plane bisecting the HOH angle, that is, in between the two H atoms. Both these planes contain the C_2 axis of rotation, which is also the principal axis. Taking C_2 as the z axis, the two symmetry planes may be denoted as $\sigma_{v(xz)}$ and $\sigma_{v(yz)}$. These are often denoted as σ_v and σ_v'. There is no σ_h in H_2O (Fig. 1.18).

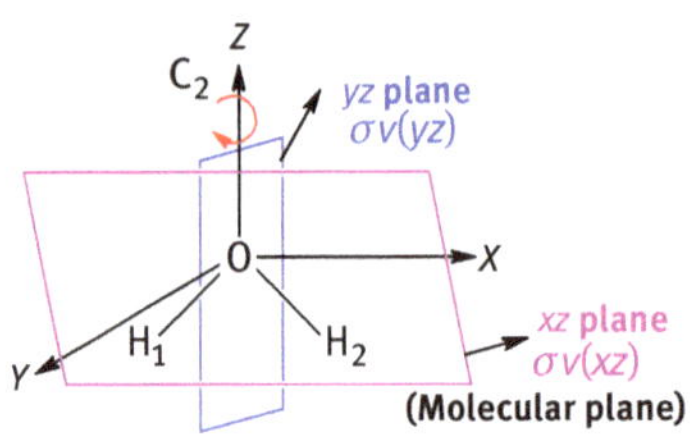

Fig. 1.18: Two vertical symmetry planes [$\sigma_{v(xz)}$ and $\sigma_{v(yz)}$] in H_2O molecule.

NH_3, PH_3: In NH_3 molecule, there are three σ, each passing through N and one of the H atoms. As they contain the principal axis C_3, they are represented as $\sigma_{v(a)}$, $\sigma_{v(b)}$ and $\sigma_{v(c)}$ (Fig. 1.19).

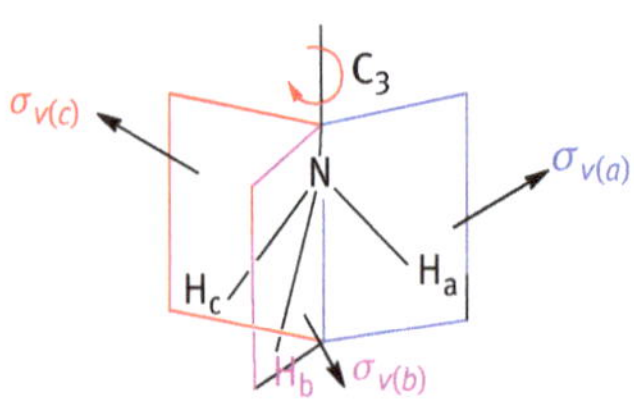

Fig. 1.19: Three vertical symmetry planes in NH_3 molecule.

BF_3: In case of trigonal planar BF_3 molecule, there are three σ_v, each passing through C_3 and one of the C_2s, that is, through one of the B–F bonds. As these planes bisect the angle between the two C_2 axes, they are dihedral plane (σ_d). Being planar molecule, the plane of the molecule is also a plane of the symmetry. This is $\perp$r to C_3 and is σ_h (Fig. 1.20). Similar to BF_3, NO_3^-, CO_3^{2-} and SO_3 also contain C_3, $3C_2$s, $3\sigma_d$ and σ_h (Fig. 1.21).

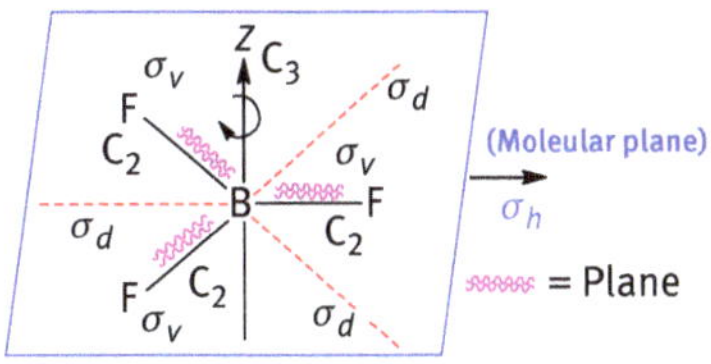

Fig. 1.20: Three dihedral planes and one horizontal plane in BF_3 molecule.

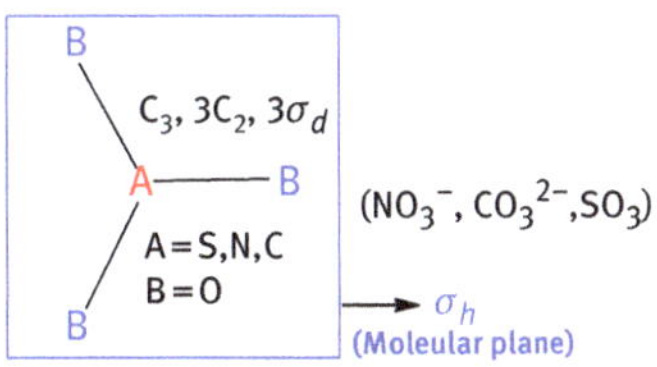

Fig. 1.21: Three dihedral planes and one horizontal plane in NO_3^-, CO_3^{2-} and SO_3 species.

PF_5: In PF_5 with trigonal bipyramidal structure, there are three σ_v, passing through P and three F atoms F_1 F_2 F_3, F_1 F_2 F_4 and F_1 F_2F_5. They contain the principal axis C_3. Unlike BF_3, C_2 is not present in the molecule and hence no σ_d. This molecule also contains one σ_h as shown by the pink dotted triangle, including axial planar F atoms, $F_3F_4F_5$ (Fig. 1.22).

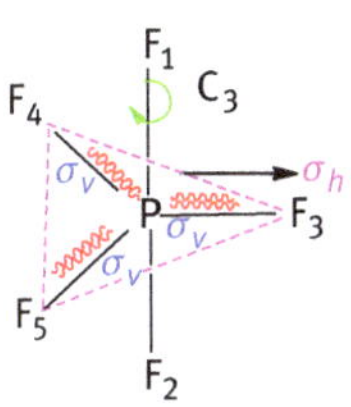

Fig. 1.22: Three vertical planes and one horizontal plane in PF_5 molecule.

$[Ni(CN)_4]^{2-}$, XeF_4: The square planar complex ion has $4\sigma_v$. Each one contains C_4 and one of the C_2s, and hence passes through central Ni (Fig. 1.23). Moreover, two σ_v passing through minimum number of atoms (only Ni, as shown in the Fig. 1.23) and bisecting the angle between two C_2 axes, are σ_d planes. Being a planar molecule, it also contains a σ_h. Similarly, XeF_4 also contains $2\sigma_v$, $2\sigma_d$ and a σ_h (Fig. 1.24).

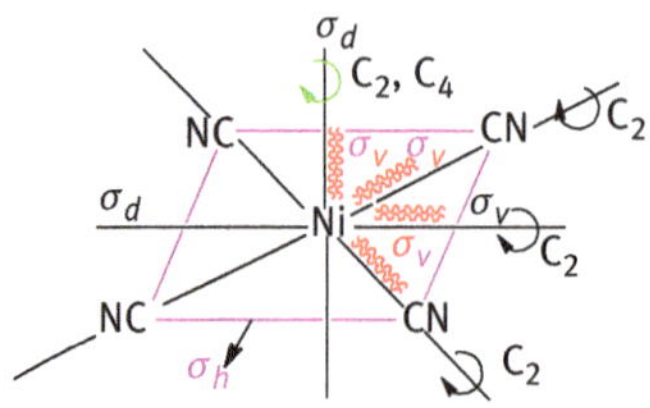

Fig. 1.23: $2\sigma_v$, $2\sigma_d$ and a σ_h planes in $[Ni(CN)_4]^{2-}$.

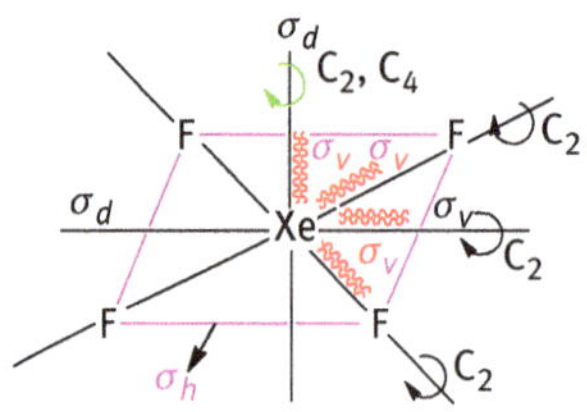

Fig. 1.24: $2\sigma_v$, $2\sigma_d$ and a σ_h planes in XeF_4.

C_6H_6: This planar molecule contains six σ_v. Each one passes through C_6 and one of the C_2s. Three pass through opposite C atoms, and three pass through the centers of the opposite edges. Out of six σ_v, $3\sigma_v$ passing through centers of opposite edges, are σ_d. The plane passing through six atoms is $\perp$r to C_6, and is σ_h (Fig. 1.25).

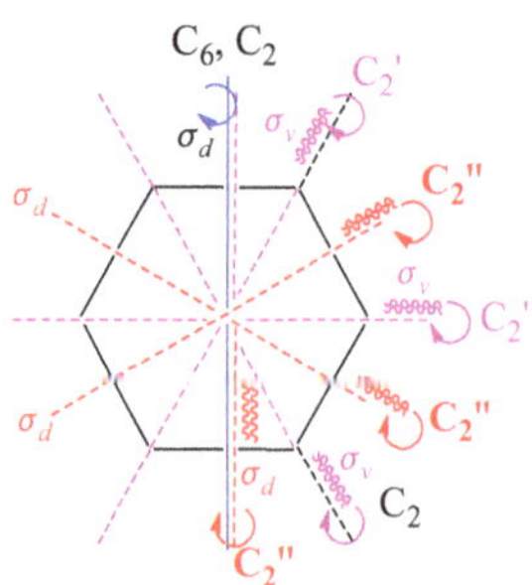

Fig. 1.25: $3\sigma_v$, $3\sigma_d$ and a σ_h planes in benzene molecule.

Allene ($H_2C = C = CH_2$): This molecule has three C_2 axes: one passing through the molecular axis, the other two C_2' and C_2'' are $\perp$r to it. The first plane passes through the molecular axis and H_1H_2, that is, PQRS, and the second passes through the molecular axis and H_3H_4, that is, ABCD. As these planes are passing in between the two subsidiary axes [the plane PQRS passes in between C_2 and C_2'; the plane ABCD passes in between C_2' and C_2''], these are dihedral planes (σ_ds) (Fig. 1.26).

AB_4: A regular tetrahedral molecule possesses six planes of symmetry. Using the numbering system (Fig. 1.27), we may specify these symmetry planes by stating

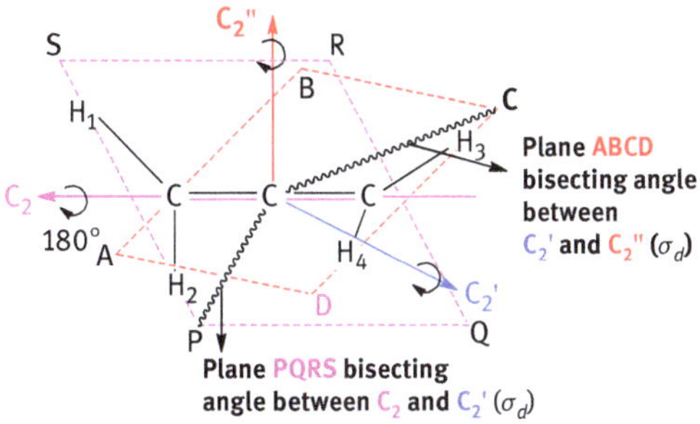

Fig. 1.26: Two dihedral planes in Allene molecule.

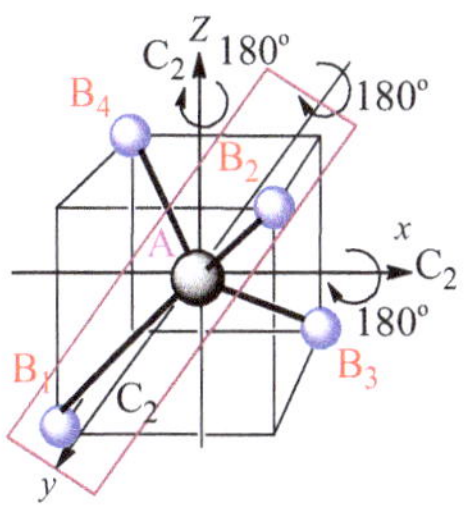

Fig. 1.27: Six dihedral planes (AB_1B_2, AB_1B_3, AB_1B_4, AB_2B_3, AB_2B_4 and AB_3B_4) in AB_4 tetrahedral molecule.

the atoms they contain: $\mathbf{AB_1B_2}$, AB_1B_3, AB_1B_4, AB_2B_3, AB_2B_4 and AB_3B_4. These planes are dihedral planes (σ_ds) as they pass in between two C_2's.

$\mathbf{AB_6}$**:** A regular octahedral possesses a total of nine symmetry planes. Referring to Fig. 1.28, there is first three of the same type, namely, those including the following sets of atoms: $\mathbf{AB_1B_2B_3B_4}$, $\mathbf{AB_1B_5B_3B_6}$ and $\mathbf{AB_2B_5B_4B_6}$ (σ_h, *perpendicular to principal axis*). There are *three more* of the second type, including atoms, viz., AB_5B_6, AB_1B_3 and AB_2B_4 (σ_d). The rest three planes (σ_d) are those which pass through the centers of the opposite edges, including "**A**." Three such planes (〰〰) are shown in Fig. 1.29.

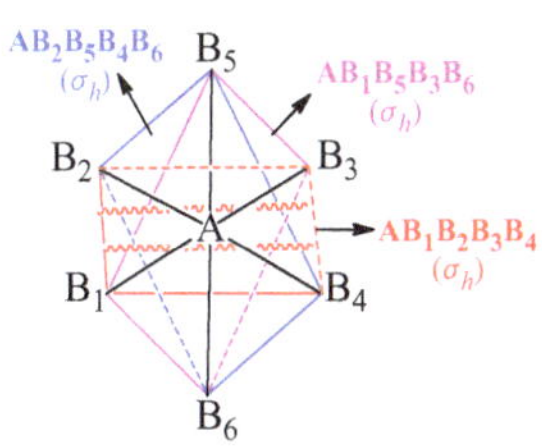

Fig. 1.28: Three horizontal planes ($AB_1B_2B_3B_4$, $AB_1B_5B_3B_6$ and $AB_2B_5B_4B_6$) and three dihedral planes (AB_5B_6, AB_1B_3 and AB_2B_4) in a regular octahedral AB_6 molecule.

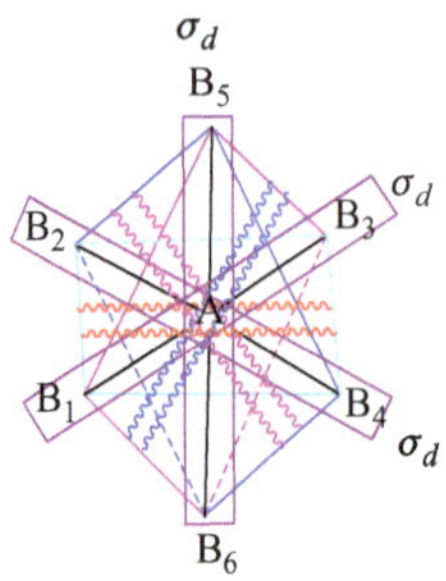

Fig. 1.29: Three σ_d planes passing through centers of two opposite edges including A in AB_6 octahedral molecule.

$Re_2Cl_8^{2-}$: It has shape of a square parallelepiped or right square prism. This is a more complex example, in which all four types of symmetry operations and symmetry elements are present (Fig. 1.30).

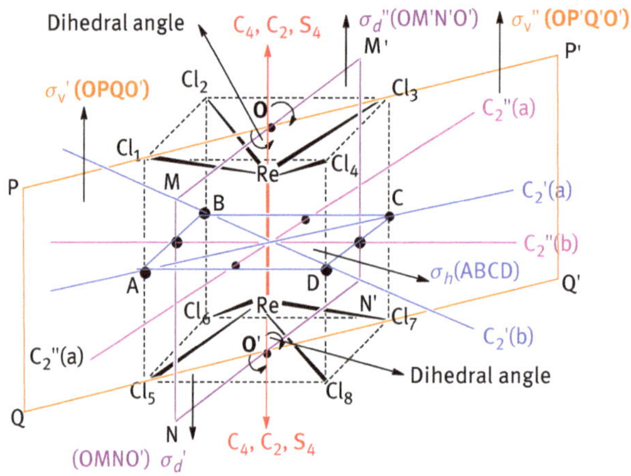

Fig. 1.30: All four types of symmetry operations present in $Re_2Cl_8^{2-}$ anion.

Proper rotation axis: This ion has six axes of proper rotation, of four different kinds. First, the Re-Re line is an axis of fourfold proper rotation, and four operations, C_4^1, C_4^2, C_4^3, $C_4^4 = E$ may be carried out. This same line is also C_2 axis, generating C_2 operation. It is notable here that the C_4^2 operation means $2 \times 2\pi/4$, which is equivalent to rotation by $2\pi/2$ that is, C_2 operation. Thus, C_2 axis and C_2 operation are not independent of C_4 axis. There are, however, two other types of C_2 axis that exist independently. There are two of the type that pass through the centers of the opposite vertical edges of the prism, that is, C_2'(a) and C_2'(b), and the two more that pass through the centers of the opposite vertical faces of the prism, namely, C_2'' (a) and C_2'' (b).

Symmetry planes: This ion has three different kinds of symmetry planes. There is one, ABCD, which bisects the Re-Re bond and all the vertical edges of the prism. This is σ_h, as it is ⊥r to the C_4 axis (Principal axis). There are then two types of vertical symmetry planes, namely, the two [OPQO′ and OP′Q′O′] that contain *opposite vertical edges*, and the two others [OMNO′ and OM′N′O′] that cut the centers of two *opposite vertical faces*. The first two planes, OPQO′ and OP′Q′O′ may be designated as $\sigma_{v'}$ and $\sigma_{v''}$ as they contain the principal axis C_4. Since the second set of planes, OMNO′ and OM′N′O′ bisect the dihedral angles between those of the first set, they are designated as $\sigma_{d'}$ and $\sigma_{d''}$.

$POBrCl_2$: There can be molecules with no axis of symmetry but have plane of symmetry. For example, $POBrCl_2$ has no centre of symmetry, but the plane containing POBr is a plane of symmetry (Fig. 1.31).

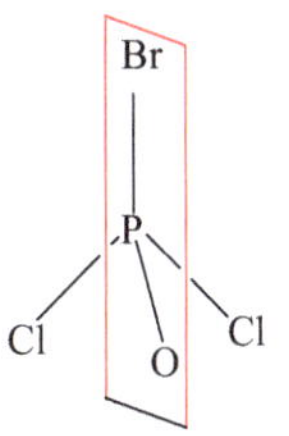

Fig. 1.31: Plane of symmetry present in $POBrCl_2$ molecule.

H_2 and CO_2: In symmetrical linear molecules, such as, CO_2 and H_2, there are infinite numbers of C_2 axes ⊥r to C_∞. The plane passing through C_∞ and one of the C_2 axes will make infinite number of vertical planes (Fig. 1.32).

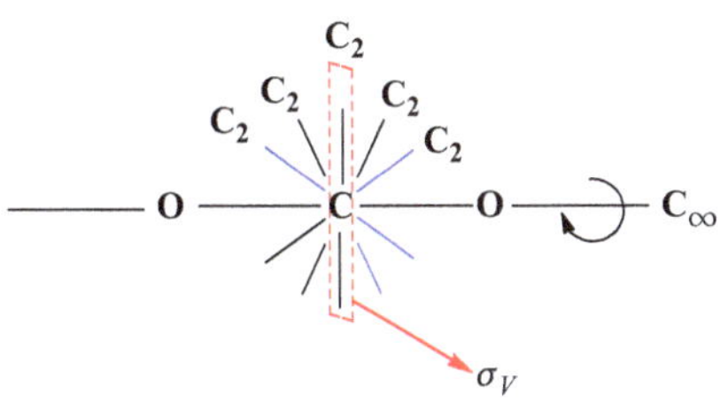

Fig. 1.32: C∞ and ∞C_2s present on symmetrical linear molecule CO.

1.7 Center of symmetry/inversion center

This is an imaginary point in the center of the molecule, through which the reflection of each atom can be carried out to result in its coincidence with an equivalent atom. In other words, if any atom in the molecule is connected with the center of symmetry and extended equally on the other side, it meets another equivalent atom. The following examples (Fig. 1.33) make it clear.

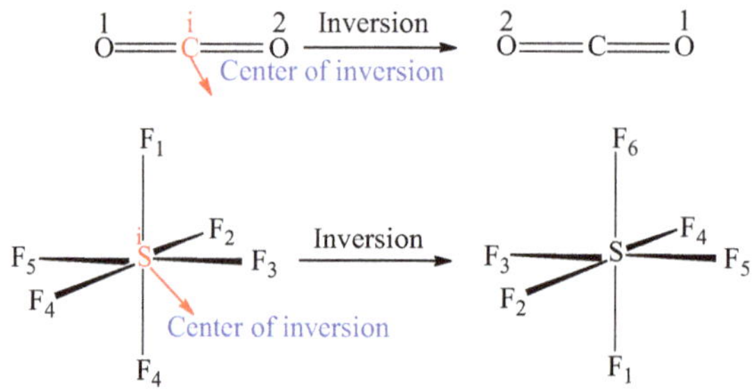

Fig. 1.33: Inversion operation and inversion center in CO_2 and SF_6.

The center of inversion of CO_2 lies at the **C** nucleus and that of SF_6 lies at the nucleus of **S** atom.

There need not be an atom at the center of inversion. For example, C_2H_4 or N_2 molecule (Fig. 1.34) has a center of inversion midway between the two carbon/nitrogen nuclei.

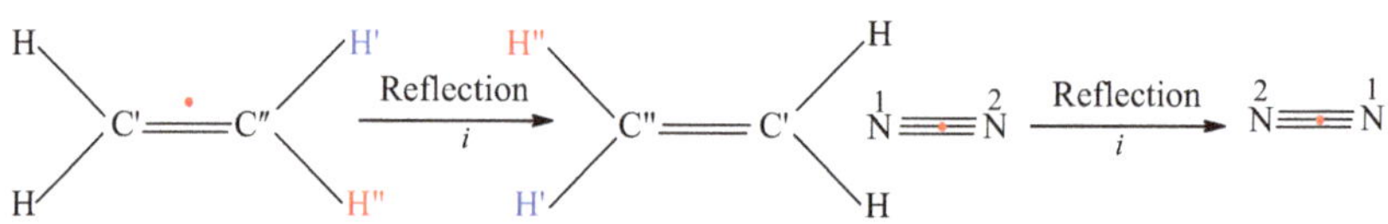

Fig. 1.34: C_2H_4 and N_2 molecules having inversion centre midway between two carbon/nitrogen nuclei, but no atom at the inversion centre.

The center of symmetry is the symmetry element, and the process of reflection through it leading to an equivalent orientation, is a symmetry operation.

As a result of this operation each molecule gets completely inverted to an equivalent orientation. Hence, the operation is termed as inversion (i) and the center of symmetry is also called inversion center. Only one inversion operation is possible. Second reflection gives back the original orientation.

Other examples

(i) This can be illustrated by the molecules, such as, C_2H_4, N_2O_2, square planar $[Ni(CN)_4]^{2-}$, octahedral $[Co(NH_3)_6]^{3+}$, $[(gly)_2Co(OH)_2(gly)_2]$ {isomers (a, b and c only)} and so on (Fig. 1.35).

(ii) *Trans*-dichloroethylene has a center of symmetry, but *cis*-dichloroethane does not. Benzene and *p*-substituted benzene have center of symmetry, but mono and *o*, *m*- disubstituted benzene have no center of symmetry (Fig. 1.36).

(iii) An H_2O molecule does not possess a center of inversion. No tetrahedral AB_4 molecule has a center of inversion.

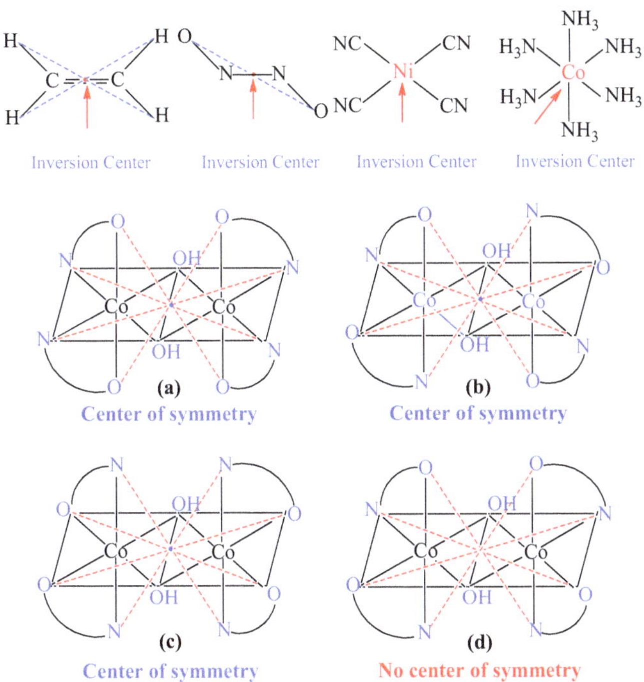

Fig. 1.35: Centre of symmetry in some molecules and three isomers of $[(gly)_2Co(OH)_2(gly)_2]$.

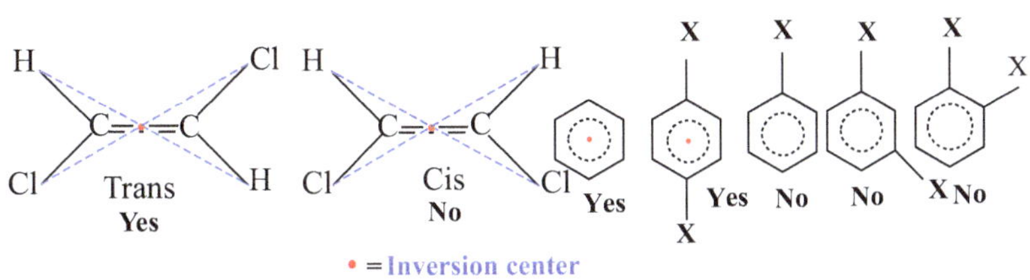

Fig. 1.36: Molecules having centre of symmetry and no centre of symmetry.

1.8 Rotation–reflection axis or axis of improper rotations

It is a composite operation, and one of the most difficult symmetry operations to identify in a molecule. It consists of a proper rotation of the molecule through a certain angle $2\pi/n$ around an imaginary axis followed by reflection through a plane perpendicular to that rotation axis to attain an equivalent/indistinguishable orientation. The axis about which this occurs is called an axis of improper rotation or, more briefly, an improper axis. It is denoted by the symbol S_n, where n indicates the order of the axis. The operation of improper rotation is also denoted by the symbol S_n.

The rotation operations (C_n) are sometimes termed as proper rotations, to distinguish them from the improper rotations.

Examples

(i) ***Trans*-dichloroethylene**

The axis of improper rotations can be illustrated by considering the example of *trans*-dichloroethylene. Simple rotation by 180° on the molecular axis does not give equivalent orientation. Rotation followed by reflection in a plane perpendicular to the rotation axis gives equivalent orientation. This is an axis of improper rotation of order 2, that is, S_2.Only one rotation operation S_2^1 is possible. S_2^2 gives the original orientation (Fig. 1.37). Thus, on an axis S_n, the number of possible operations are n–1. It is notable here that S_2^1 operation is same as inversion.

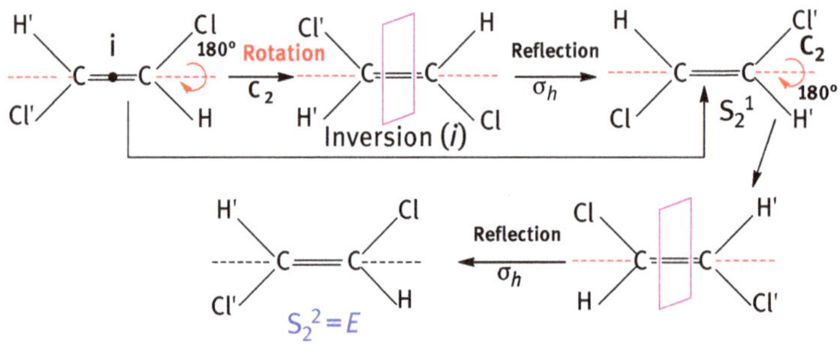

Fig. 1.37: *Trans*-dichloroethylene showing improper axis of symmetry (S_2).

(ii) ***Trans*-tetrafluorohydrazine (N_2F_4)**

In the *trans*-configuration of N_2F_2, if we perform a C_2 operation along the molecular axis followed by a σh operation, we will have a successful S_2^1 operation. S_2^2 gives the original orientation (Fig. 1.38). Again S_2^1 operation is same as inversion.

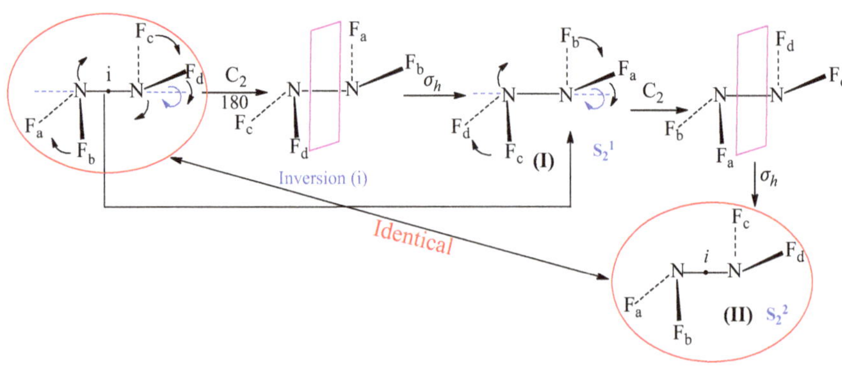

Fig. 1.38: *Trans*-tetrafluorohydrazine showing axis of improper rotation (S_2).

(iii) **BF$_3$**

This is an example of three fold improper axis (S$_3$) (Fig. 1.39).

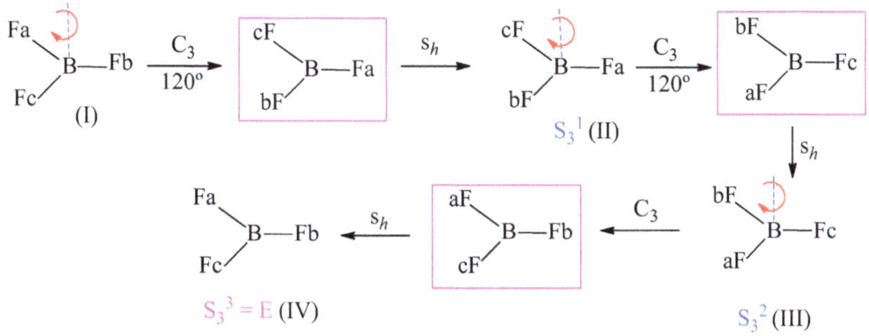

Fig. 1.39: BF$_3$ molecule showing three fold improper axis (S$_3$).

Thus, $S_3 = \sigma_h.C_3$. In the product $\sigma h.C_3$, the first operation carried out is C$_3$ and σh is the next operation. In many cases, the order of symmetry operation is important, but in S$_3$, order is immaterial, that is, $S_3 = \sigma_h C_3 = C_3 \sigma_h$.

(iv) **Staggered form of ethane (C$_2$H$_6$)**

In the staggered form of ethane the C–C line is a C$_3$ axis, but certainly not a C$_6$ axis. However, C$_3$ is the S$_6$ axis as Fig. 1.40 shows.

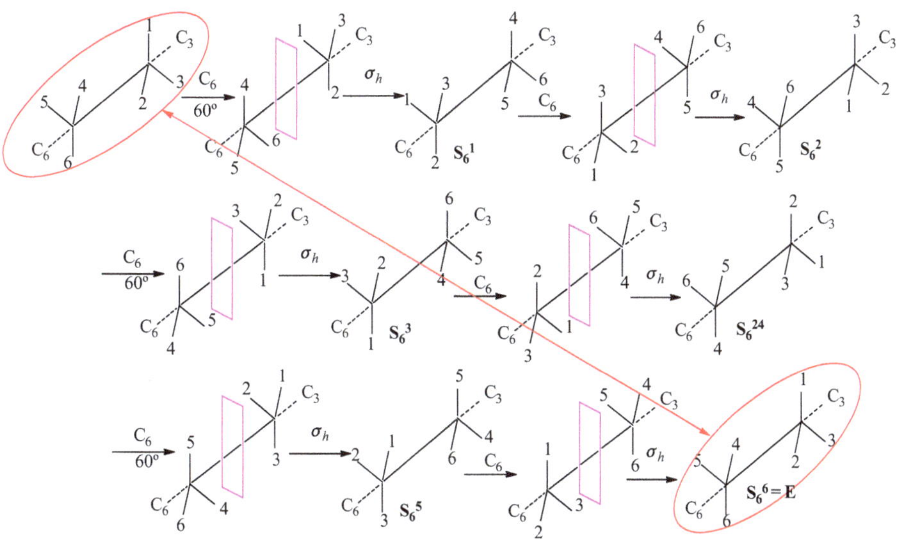

Fig. 1.40: Staggered form of ethane exhibiting six fold improper axis of symmetry (S$_6$).

It is seen that a simple rotation by 60° on the C_3 (molecular axis) does not give equivalent orientation. Rotation followed by reflection in a plane perpendicular to the rotation axis gives equivalent orientation. Hence, C_3 is an axis of improper rotation of order 6, that is, S_6. It is also clear that $S_6^6 = E$.

The very interesting point in the staggered C_2H_6 molecule is that it has S_6, but it does not have C_6 or σ_h. On the other hand, BF_3 has a C_3 and a σ_h. Since $S_n = \sigma_h C_n$, the presence of σ_h and a C_n necessitates the presence of a S_n. But the presence of a S_n does not necessitate the presence of a C_n and a σ_h.

(v) **Tetrahedral AB_4 type molecules (CH_4, CCl_4, SiF_4, $TiCl_4$)**

Tetrahedral AB_4 type molecules have no C_4, though they have $3C_2$s. Each of the Cartesian axes, *x*, *y* and *z* bisecting two HCH bond angles, is the C_2 axis. In fact, each of these C_2 axes is an S_4 axis as clear from Fig. 1.41. It is seen that neither the C_2 nor the σ_h is the true symmetry element in tetrahedral AB_4 type molecules. Combination of these two operations, however, results in a new orientation (III), which is indistinguishable to the original orientation (I). Hence, C_2 is an axis of improper rotation of order 4, that is, S_4.

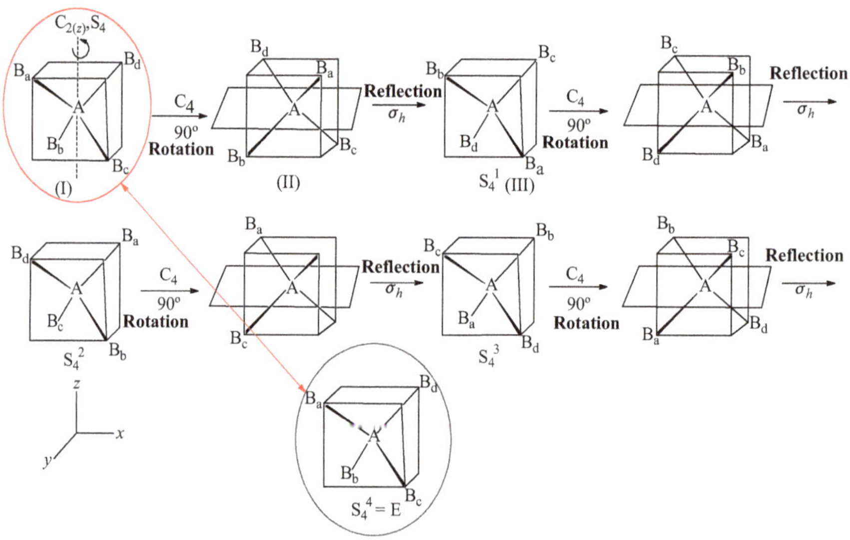

Fig. 1.41: Tetrahedral AB_4 type molecules showing fourfold axis of improper rotation (S_4).

(vi) **Octahedral complexes/simple molecules of type AB_6^{n+} ($[Ni(NH_3)_6]^{2+}$)/AB_6 (SF_6)**

Sometimes an axis of proper rotation can also be considered an axis of improper rotation of the same order, that is, $C_n \equiv S_n$. In the octahedral complex, $[Ni(NH_3)_6]^{2+}$ in a C_4 axis is also a S_4 axis (Fig. 1.42). Here, the axis of improper rotation is said to be coincident with the axis of proper rotation.

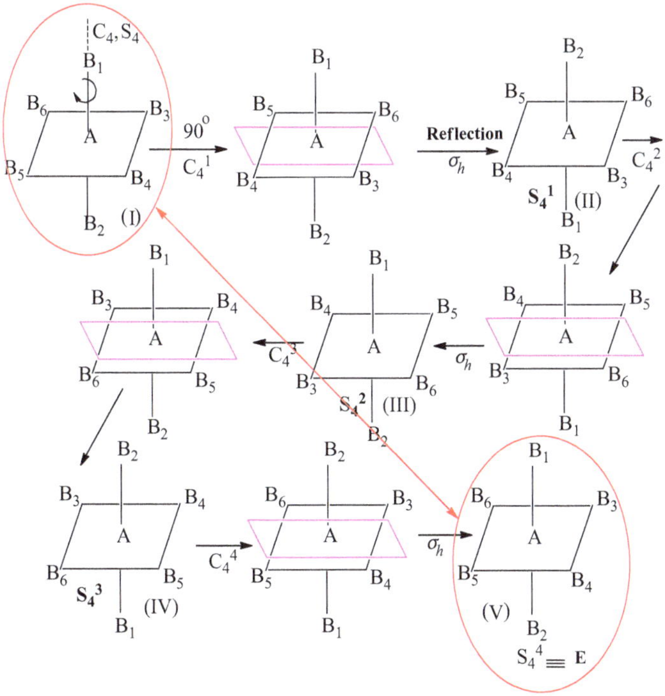

Fig. 1.42: C_4 coincident with S_4 axis in octahedral AB_6 type molecules.

(vii) **Distorted octahedral *trans* complex ML_4X_2**

In a distorted octahedral *trans* complex ML_4X_2, the axis passing through X–M–X is C_4. The same axis is also S_4, that is, C_4 is coincident with S_4. On S_4 three operations are possible and these are S_4^1, S_4^2 and S_4^3 as shown in Fig. 1.43.

Other examples

S_1 axis: Every molecule with only a plane of symmetry has an S_1 axis perpendicular to the plane of symmetry. In monochloroethylene (Fig. 1.44) the plane of symmetry is the molecular plane (*xy*-plane) and the *z*-axis is then S_1 axis. The rotation of 360° around *z*-axis and the reflection in the *xy*-plane result in the original configuration.

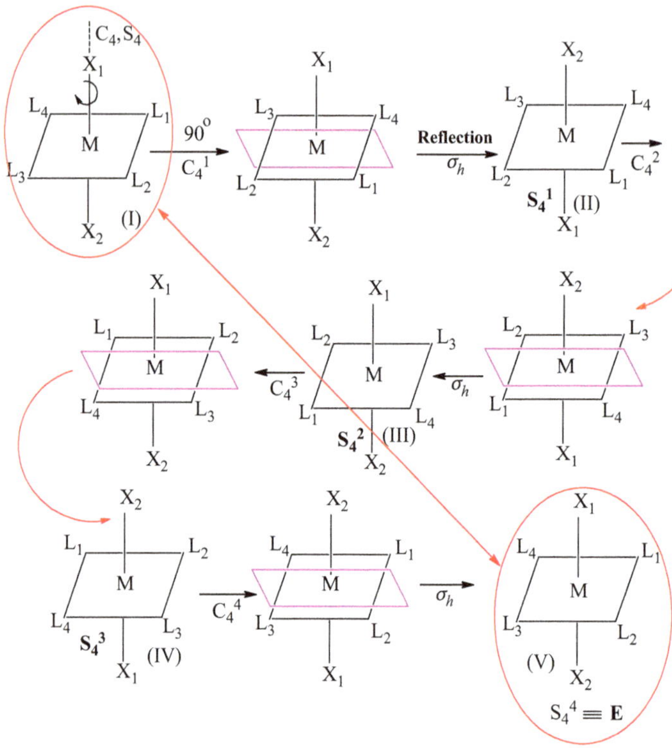

Fig. 1.43: C_4 coincident with S_4 axis in distorted octahedral *trans* ML_4X_2 complex.

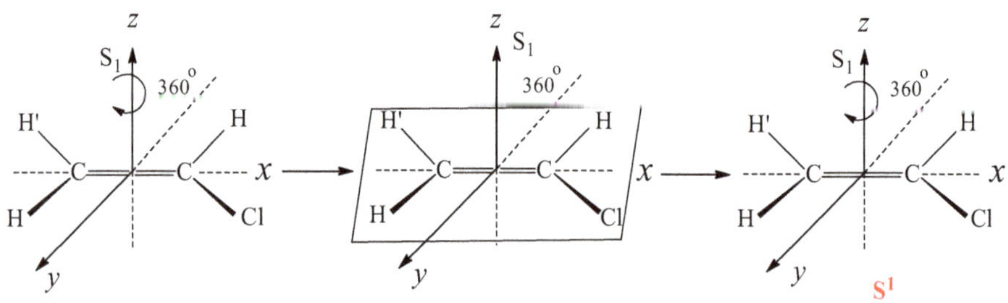

Fig. 1.44: S_1 axis in monochloroethylene with molecular plane as the horizontal plane.

S_1, S_2, S_3, S_4, S_5, S_6, S_7, S_8 and S_{10} axes: Compound showing these improper axes of symmetry are given in Fig. 1.45.

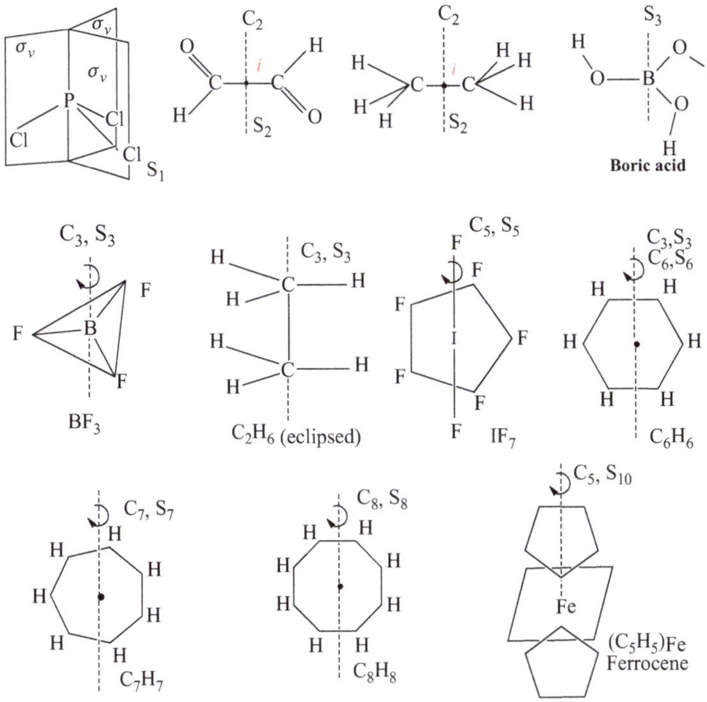

Fig. 1.45: Compounds showing S_1, S_2, S_3, S_4, S_5, S_6, S_7, S_8 and S_{10} improper axis of symmetry.

Some important features of improper axis

(i) We have $S_n = C_n.\sigma_h$. Obviously, S_1 is equivalent to σ_h because $C_1 = E$, and therefore the second step, reflection, is σ_h.

(ii) S_2^1 is equivalent to inversion operation *i*, as seen in the examples (Figs. 1.37 and 1.38). If a molecule has a center of symmetry *i*, it must also have an S_2 axis. Thus, *i* implies S_2 and *vice versa*.

(iii) In general, an S_n axis does not necessarily imply the existence of a separate C_n and σ_h. This becomes clear by considering the example of staggered C_2H_6 (Fig. 1.40). The molecule in this conformation contains an S_6 axis. But it does not have either a C_6 axis or a σ_h plane.

(iv) Like the ordinary rotation axis, C_n, and the rotation–reflection axis, S_n, gives rise to more than one distinct symmetry operations, because successive rotation–reflections (improper rotations) about the axis do not give identical results. The number of operations generated by a rotation–reflection axis depends on whether *n* in S_n is even or odd. Moreover, some features of S_n are different when *n* is even and odd. So, these two cases will be considered here separately.

If n is even in S_n:

(i) For n even, it can be proved that the operation $S_n^n = E$. The S_n^n means $C_n^n . \sigma_h^n$. But $C_n^n = E$, and also for even n, σ_h^n is also equal to E. Hence, $S_n^n = E$.

(ii) The existence of an S_n axis of even order always requires the existence of a $C_{n/2}$ axis. Further, the element S_n with n even generates "n" operations. The following examples make these two features clear.

Let us consider the operations generated by S_4 and S_6 axes:

$$S_4^1, S_4^2, S_4^3, S_4^4; \qquad S_6^1, S_6^2, S_6^3, S_6^4, S_6^5, S_6^6$$

In terms of C_n and σ_h operations, these two sets of operations may be written as:

$$C_4^1\sigma_h^1, C_4^2.\sigma_h^2, C_4^3\sigma_h^3, C_4^4\sigma_h^4; \qquad C_6^1\sigma_h^1, C_6^2\sigma_h^2, C_6^3\sigma_h^3, C_6^4\sigma_h^4, C_6^5\sigma_h^5, C_6^5\sigma_h^5$$

This gives operations,

$$S_4^1, C_2^1, S_4^3, E; \qquad S_6^1, C_3^1, i, (C_2^1\sigma_h^1 = S_2 = i), C_3^2, S_6^5, E$$

Thus, the element S_4 generates four, and the element S_6 generates six distinct operations. Hence, in general, S_n with n even generates "n" operations. It is notable that in case of element S_4, ($C_2^1$and E) are just the operations generated by C_2, and in case of element S_6, (C_3^1, C_3^2 and E) are just the operations generated by C_3. Hence, the existence of an S_n axis of even order always requires the existence of a $C_n/2$ axis.

If n-odd in S_n:

(i) For n is odd, it can be proved that the operation $S_n^n = \sigma_h$. The S_n^n means $C_n^n.\sigma_h^n$. But $C_n^n = E$, and for odd n, σ_h^n is equal to σ_h. Hence, $S_n^n = \sigma_h$.

(ii) An odd element S_n requires that C_n and a σ perpendicular to it, that is, σ_h must exist independently. It can be easily proved by the argument given below, and also by step (iii).

(iii) We have just seen that $S_n^n = \sigma_h$. It means that the element S_n generates a symmetry operation σ_h. But if the symmetry operation σ_h exists, the plane to which it is referred must be a symmetry element in its own right. Now, the operation S_n requires to reflect the object in the plane σ_h, thus carrying an orientation **I** into another orientation **II**, and then to rotate by $2\pi/n$, thus carrying out orientation **II** into an orientation **III**. Since, S_n is a symmetry operation; **I** and **III** must be equivalent configuration. However, when n is odd, σ_h is itself a symmetry operation, so that **II** is equivalent to **I**. Then, **II** is also equivalent to **III.** Thus, the operation C_n is also a symmetry operation in its own right. This proves that for n odd, C_n and a σ_h must exist independently.

(iv) In general, the element S_n with n odd generates $2n$ operations. It can also be easily shown by taking example of S_3 operation in PF_5 molecule generating six operations, S_3^1, S_3^2, S_3^3, S_3^4, S_3^5, S_3^6 (Fig. 1.46).

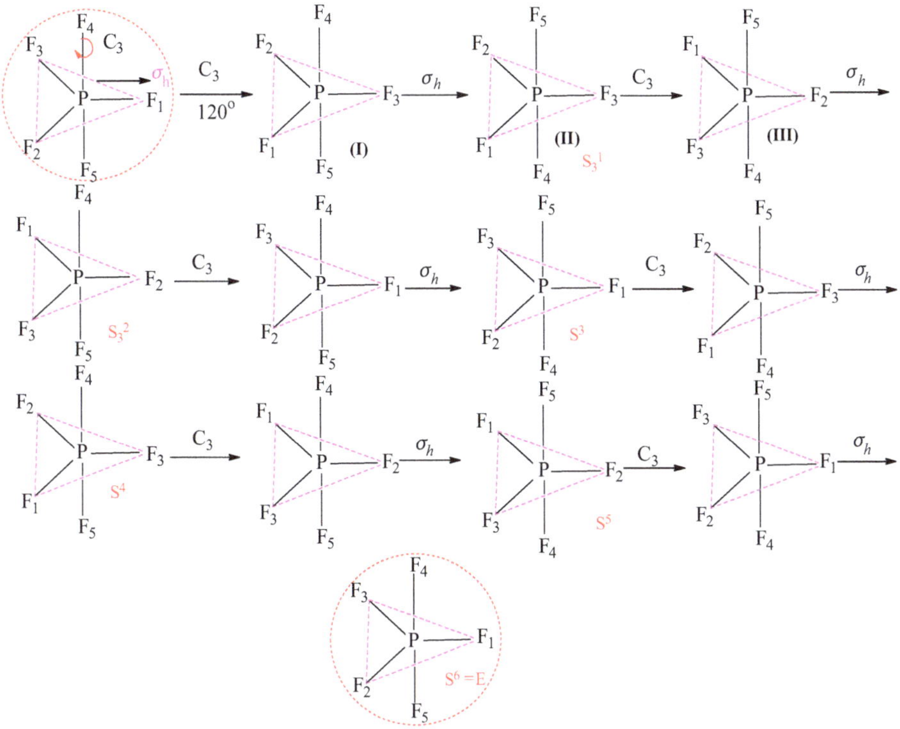

Fig. 1.46: PF_5 molecule generating six improper operations S_3^1, S_3^2, S_3^3, S_3^4, S_3^5, S_3^6.

Let us consider the operations, S_3^1, S_3^2, S_3^3, S_3^4, S_3^5, S_3^6 generated by S_3 axes. In terms of C_n and σ_h operations, we can write:

$C_3^1\sigma_h^1$, $C_3^2\sigma_h^2$, $C_3^3\sigma_h^3$, $C_3^4\sigma_h^4$, $C_3^5\sigma_h^5$, $C_3^6\sigma_h^6$, that is, S_3^1, C_3^2, σ_h, C_3^1, S_3^5, E = 6 distinct operations, that is, $2n$ operations [($C_3^3 = E$, and $\sigma_h^3 = \sigma_h^2.\ \sigma_h = E.\ \sigma_h$, hence $C_3^3\sigma_h^3 = E.\ E.\ \sigma_h = \sigma_h$) and $C_3^6\sigma_h^6 = E.\ E = E$].

The operations, C_3^1, C_3^2, E (or C_3^3) are just the operations generated by C_3 axis. This again shows that the element S_n requires that C_n and a σ_h must exist independently.

The same conclusion, that is, S_n requires C_n and a σ_h independently, may be derived by considering the operations generated by S_5 axes,

$$S_5^1, S_5^2, S_5^3, S_5^4, S_5^5, S_5^6, S_5^7, S_5^8, S_5^9, S_5^{10}$$

In terms of C_n and σ_h operations, we can write:

$C_5^1\sigma_h^1$, $C_5^2\sigma_h^2$, $C_5^3\sigma_h^3$, $C_5^4\sigma_h^4$, $C_5^5\sigma_h^5$, $C_5^6\sigma_h^6$, $C_5^7\sigma_h^7$, $C_5^8\sigma_h^8$, $C_5^9\sigma_h^9$, $C_5^{10}\sigma_h^{10}$, that is, S_5^1, C_5^2, S_5^3, C_5^4, σ_h, C_5^1, S_5^7, C_5^3, S_5^9, E = 10 distinct operations, that is, $2n$ operations.

We see that for S_3 through S_3^6, and for S_5 through S_5^{10} (in general S_n through S_n^{2n}), the operations are all different ones, but commencing with S_n^{2n+1}, repetition of the sequence begins, for example, $S_3^7 = C_3^7\sigma_h^7 = C_3^1\sigma_h^1 = S_3^1$ and $S_5^{11} = C_5^{11}\sigma_h^{11} = C_5^1\sigma_h^1 = S_5^1$.

Again $C_5^1, C_5^2, C_5^3, C_5^4, E\ (=C_5^5)$ means C_5 axis, and σ_h is already there in S_5 axis. Hence, this again shows that, in general, S_n requires the independent existence of C_n and σ_h.

1.9 Identity

If a sequence of symmetry operations brings the molecule back to its original configuration, the net operation is called identity operation, E. For example, two successive rotations about the C_2 axis in water are equivalent to identity. This fact is expressed as $C_2^2 = E$. Similarly, $\sigma_{v(xz)}^2 = E$ and $\sigma_{v\ (yz)}^2 = E$. In BF_3, $C_3^3 = E$, $C_2^2 = E$, $C_2'^2 = E$, $C_2''^2 = E$, $\sigma_d^2 = E$ and so on.

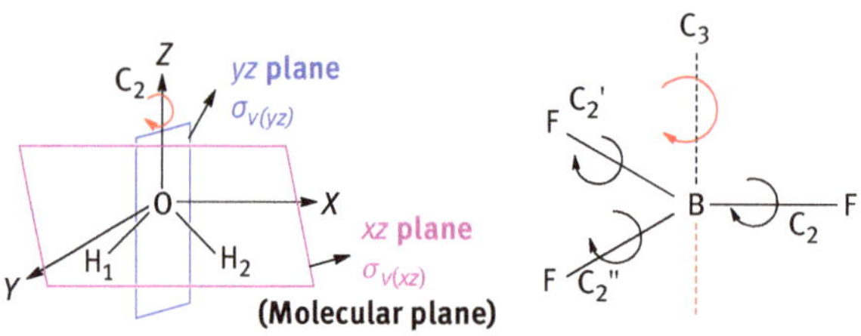

The identity operation leaves the molecule unchanged and it appears as if no operation has been performed. So, this operation sounds trivial (having little value) but its importance lies in considering the molecules as a group and to apply the group theory to molecules.

Exercises

Multiple choice questions

1. Symmetry properties of molecules are used in solving:
   (a) Physical problems (b) Chemical problems
   (c) Both (a) and (b) (d) None of these

2. A symmetry element is a entity called:
   (a) Geometrical (b) Non-geometrical
   (c) Any one of (a) and (b) (d) None of these

3. For getting equivalent orientations, ammonia molecule has to be rotated by:
   (a) 60° (b) 90° (c) 120° (d) 180°

4. H_2O molecule contains planes known as:
   (a) Vertical (b) Horizontal
   (c) Dihedral (d) Both vertical and horizontal

5. Which one of the molecules contain both C_3 and S_3
   (a) PCl_5 (b) NH_3 (c) BF_3 (d) SO_3

6. In general, S^n with n odd generates operations in number
   (a) n (b) $2n$ (c) $n + 2$ (d) $2n + 2$

7. The number and types of symmetry planes in a regular octahedral molecule is:
   (a) $3\sigma_h$ and $6\sigma_v$ (b) $6\sigma_h$ and $3\sigma_v$ (c) $6\sigma_d$ and $3\sigma_v$ (d) $3\sigma_h$ and $6\sigma_d$

8. The possible number of symmetry elements in a regular tetrahedral molecule is
   (a) 12 (b) 24 (c) 36 (d) 48

Short answer type questions

1. Differentiate between symmetry elements and symmetry operations.
2. How many symmetry planes do you know? Define.
3. Name the various symmetry operations in NH_3 molecules.
4. Give the name of three molecules having horizontal plane.
5. Define center of symmetry with three examples.
6. Taking a suitable example explain improper axis of symmetry
7. How many proper axis of symmetry a benzene molecule has?
8. Give some examples of molecule having infinite fold axis of symmetry

Long answer type questions

1. Differentiate between proper and improper axis of symmetry with two suitable examples of each.
2. Write down the different symmetry operations generated by C_5, C_6, S_5 and S_6.
3. Justify that if S_n exists (a) with even value of n, there exists a $C_{n/2}$ and (b) with odd value of n, there exists a C_n and a σ_h.
4. Write an explanatory note of plane of symmetry with suitable examples.
5. Taking suitable examples explain identity operation. In what way it is an important symmetry operation?
6. What is center of symmetry and why it is called inversion center? Explain with at least four suitable examples.

2 Application of group theory to electronic spectroscopy

2.1 Introduction

Spectroscopy deals with the transitions induced in a chemical species by its interaction with the photons of electromagnetic radiation. Thus, spectroscopy is defined as the interaction of electromagnetic radiation with matter. Group Theory can be applied to predict the probability of transitions, number of bands in the spectrum and their intensities, which transition is allowed or forbidden, and so on. Atomic and molecular spectra provide detailed information of electronic distributions, bond lengths, bond angles, molecular symmetry, and so on. Presently, spectroscopy is the one of the important tools to elucidate the structure of organic and inorganic compounds. In this chapter, discussion will be centered on electronic spectroscopy only.

2.2 Electronic spectroscopy

The basic principle in spectroscopy is that all kinds of spectra arise due to transition of electrons from lower energy to high energy quantized levels by the absorption of electromagnetic radiations. In electronic spectroscopy (UV-visible spectroscopy), the transition takes place between electronic energy levels by the absorption of UV or visible light. The molecular orbital theory is very helpful to understand the electronic spectra of molecules. Accordingly, the transition takes place from lower filled molecular orbital (HOMO) to the higher vacant orbital (LUMO) by the absorption of light of energy equal to the difference in the energies of HOMO and LUMO. Depending upon the separation between the two molecular orbitals, the energy is absorbed in visible or ultraviolet region of the electromagnetic radiation.

2.2.1 Electronic spectra of organic compounds

Electronic transitions in organic compounds can be discussed with the help of molecular orbital (MO) theory, wherein the energy of MOs of a molecule are in the order $\sigma < \pi < n < \pi^\star < \sigma^\star$ as shown in the Fig. 2.1.

As we know that the electronic transition takes place from highest filled (HOMO) to lowest vacant molecular orbital (HUMO), the possible transitions for different compounds could be $\sigma \rightarrow \sigma^*$, $\pi \rightarrow \pi^*$, $n \rightarrow \sigma^*$ and $n \rightarrow \pi^*$. Consequently, corresponding to each transition, there will appear a band in the absorption spectrum of

https://doi.org/10.1515/9783110635034-002

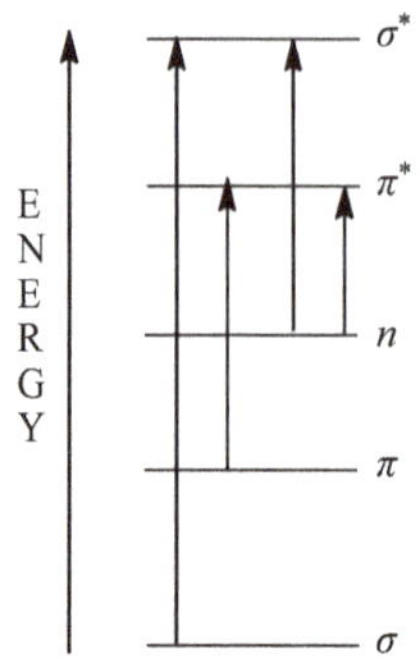

Fig. 2.1: Molecular orbital energy diagram of an organic molecule.

the compound. Thus, each compound is characterized by a specific number of bands occurring in specific region of the electromagnetic radiation.

As $\sigma \rightarrow \sigma^*$ is the highest energy transition for saturated organic compounds like CH_4, this transition takes place in the far UV region (λ_{max} = 135 nm), whereas in unsaturated compounds, in addition to $\sigma \rightarrow \sigma^*$ transition, there will be $\pi \rightarrow \pi^*$ transition also which will occur in the lower energy UV region. For example, in ethylene, $\pi \rightarrow \pi^*$ transition is observed at λ_{max} = 171 nm. Compounds such as $CHCl_3$ and CH_3OH also possess nonbonding electrons on hetero atoms, and hence $n \rightarrow \sigma^*$transition is also possible in such compounds in far UV region. In aldehydes and ketones, $n \rightarrow \pi^*$ transition also occur along with $\pi \rightarrow \pi^*$ transition in still lower energy region. For example, in formaldehyde (HCHO), all five types of molecular orbitals (σ, π, n, π^* and σ^*) are there, and hence four transitions, $\sigma \rightarrow \sigma^*$, $\pi \rightarrow \pi^*$, $n \rightarrow \sigma^*$ and $n \rightarrow \pi^*$ are possible.

2.2.2 Allowed and forbidden transition: prediction through group theory

It can be predicted by the use of Group Theory that of the four transitions mentioned above which transitions are allowed and which are forbidden. In other words, selection rules for electronic transitions can be derived.

For a transition to occur and appearance of a band (its intensity) in UV-Vis. spectrum of a compound, the following two conditions are to be fulfilled.

(i) There must be change in the dipole moment of the molecule as a result of transition. It means the dipole moment of the molecule should be different in the ground and excited states. This condition is required due to the fact that absorption of light by the electron takes place only when there is matching (resonance) of the electrical moment of the light radiation with the electrical component of dipole moment of the molecule in which the electronic transition is to occur. Also, as a result of transition, if the magnitude of the change in dipole moment is large, then intensity of absorption band will also be more.

The intensity of an absorption band or probability of an electronic transition is expressed by the equation:

$$I\,(\text{intensity}) \propto |\int_{-\infty}^{+\infty} \psi^{*}\,\mu M\,\psi\,d\tau|^{2}$$

Here, the integral is termed as "transition moment integral", ψ and ψ^{*} are the wave functions of the ground and excited states, respectively. μ_M is the electric dipole moment operator, and dτ is the volume element (i.e., $d\tau = d_x.\, d_y.\, d_z$). It is well clear from the above equation that if the value of transition moment integral is zero, the intensity also becomes zero. Hence, such electronic transition will not occur and the transition is called forbidden. But if the integral has nonzero value, the transition is called an allowed transition.

In order to determine whether the transition moment integral is zero or nonzero, we have to obtain the direct product representation of the wave functions of ground and excited states. Since the dipole moment μ_M is a vector quantity, hence it has three components along x, y and z axes. So, the transition moment integral has three components as given below:

$$\mathrm{e}\int \psi^{*}\,\mu x\,\psi\,d\tau$$

$$\mathrm{e}\int \psi^{*}\,\mu y\,\psi\,d\tau \quad \text{and}$$

$$\mathrm{e}\int \psi^{*}\,\mu z\,\psi\,d\tau$$

where μ_x, μ_y and μ_z are the change in dipole moment in the x, y and z directions, respectively, and e is the electronic charge.

If out of the three integrals any one is nonzero (i.e., the dipole moment change occurs in at least one direction), then electronic transition is allowed.

Now, since each transition moment component is a number, hence transition moment component will remain unchanged by all the symmetry operations of the point group to which the molecule belongs. This should be true for the total transition moment integral $\int_{-\infty}^{+\infty} \psi^{*}\,\mu M\,\psi\,d\tau$, that is, this integral should not change by the symmetry operations of the point group. Therefore,

$$\int_{-\infty}^{+\infty} \psi^{*}\,\mu M\,\psi\,d\tau \xrightarrow{\mathrm{R}} \int_{-\infty}^{+\infty} \psi^{*}\,\mu M\,\psi\,d\tau$$

The above condition is possible only if the integral is either zero or it transforms as the totally symmetric representation of the group. This is because only totally symmetric representation of the group remain unchanged under all symmetry operations of the group (character of each symmetry operation of the group is +1 in totally symmetric representation). In other words, the above condition can be fulfilled if the integral transforms as

$$\int_{-\infty}^{+\infty} \psi^{*}\ \mu M\ \psi\ d\tau\ \rightarrow \mathrm{A_{1g}}$$

Then one important selection rule we got here is that:

"Any electronic transition is allowed, if in a molecule having center of symmetry, the product of the representation of the ground state and the excited state wave functions with the particular dipole moment vector component should be A_{1g}.".

Now, if we look at the integral, the dipole moment (μM is a vector quantity and has a direction, so it is essentially transformed as a representation which will not be symmetric to the center of symmetry (i.e., it transform as u, noncentrosymmetric representation).

So, to get total product $\left[\int_{-\infty}^{+\infty} \psi^{*}\ \mu M\ \psi\ d\tau\right]$ as centrosymmetric (g), the product of the representation of ψ^{*} and ψ should be transformed as noncentrosymmetric (u) representation such that $u \times u = g$. Now, the product $\psi^{*}\psi$ can be equal to u only if one of the wave functions is u and the other is g, so that $u \times g = u$.

Therefore, for an electronic transition to occur in UV or visible region, the ground and excited state wave functions cannot be both g and both u. This condition derived from group theoretical consideration gives an important selection rule as the following:

"In centrosymmetric molecules, transitions between electronic states of same symmetry parity (g or u), that is, g→g and u→u are forbidden. This is called Laporte selection rule."

Let us consider the case of ethylene ($CH_2{=}CH_2$) molecule to verify the *Laporte selection* **rule.** Its molecular orbital energy level diagram is qualitatively shown in Fig. 2.2 along with symmetry of the σ- and π- molecular orbitals.

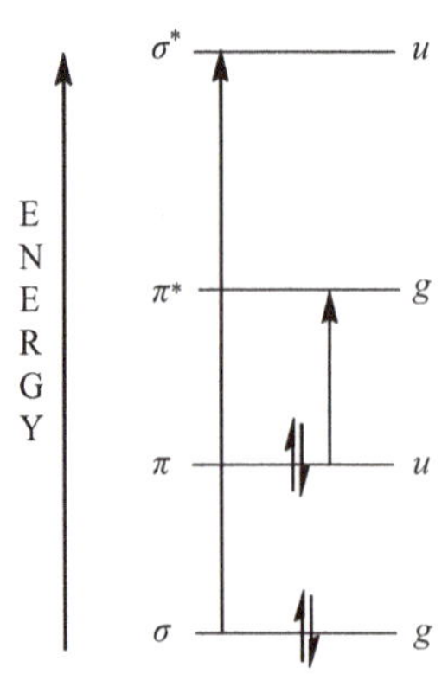

Fig. 2.2: Molecular orbital energy diagram of ethylene molecule.

As per Laporte selection rule, only two transitions, $\sigma_{(g)}\to\sigma^*_{(u)}$ and $\pi_{(u)}\to\pi^*_{(g)}$, are allowed. In fact, the $\pi_{(u)}\to\pi^*_{(g)}$ transition in ethylene occurs at 171 nm (near UV region) and $\sigma_{(g)}\to\sigma^*_{(u)}$ transition occurs in the far UV region. The remaining two transitions, $\sigma_{(g)}\to\pi^*_{(g)}$ and $\pi_{(u)}\to\sigma^*_{(u)}$, are forbidden and, therefore, should not occur. However, forbidden transitions are also observed in many compounds with low intensity. The reason for these partial permissible forbidden transitions is termed as "vibronic coupling".

2.2.3 Vibronic coupling

The vibronic coupling arises due to the fact that each electronic energy level in a molecule is composed of vibrational levels and each vibrational level has rotational energy levels also. So, when an electron is excited from the ground electronic state to the excited/higher electronic state, this transition is followed by changes in vibrational and rotational energy levels. The electron may go from ground vibrational level (v = 0) in the ground electronic state to higher vibrational levels (v = 1, 2, 3,. . . . . ., n) in excited electronic state. This is shown in Fig. 2.3, where ground and excited electronic states are represented by Morse's potential curves and transition like $v_0\to v_0^1$, $v_0\to v_1^1$ or $v_0\to v_2^1$ are possible.

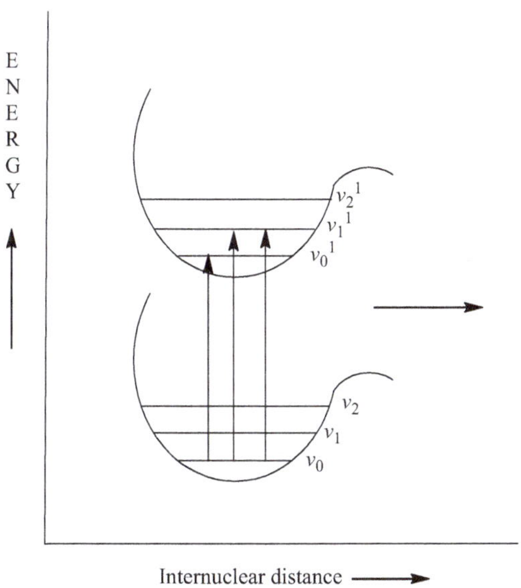

Fig. 2.3: Representation of ground and excited electronic states in a compound by Morse's potential curve.

These transitions are of different energies. Hence, due to vibrational changes involved in the electronic transition, the broad bands (or splitting) are observed in the electronic spectrum of compound corresponding to such vibrationally coupled electronic transitions.

(ii) According to Frank-Condon principle, during the electronic transition, the internuclear distance remain unchanged (because electronic transition takes place very fast as compared to the vibration of the molecule). Therefore, vibrational changes associated with electronic transitions will not affect the dipole moment. So, we can apply the above selection rule for vibrationally coupled electronic transitions also.

However, in polyatomic molecules, an odd vibration can remove the center of symmetry for a time being and so $g \rightarrow g$ electronic transition can bring some change in dipole moment of the molecule due to associated vibrational change, and, therefore, in such cases forbidden transition may become partially allowed (vibronic coupling) .

It is, therefore, well understood that forbidden transition may occur due to coupling of vibrational and electronic wave functions. In such cases, the intensity of transition is expressed as

$$\mathrm{I} = \left| \int \psi\, elect.\, \mu M.\ \psi^{*}\ elect.\, d\tau.\ \int \psi vib.\, \mu M\, \psi^{*}\ vib.\ d\tau \right|^{2}$$

So, the changes in electronic and vibrational wave functions both contribute to the intensity of transition. However, such transitions and their corresponding bands will have low intensity.

The selection rules derived above are based on g and u symmetries of ground and excited electronic states, but what happens for nonlinear polyatomic (noncentrosymmetric) molecules is worth considering. In fact, in such cases, we need to determine the symmetry species of the ground and electronic states, and then to see how the integral $\int \psi\, ground.\, \mu M\, \psi^{*}\, excited.\, d\tau$ transform as A_{1g} or A_1. Thereafter, conclusion is derived for the transition to be allowed or forbidden.

The conclusion derived from the above discussion can be understood by some illustrative examples given below.

(a) Formaldehyde (HCHO) molecule

This molecule belongs to C_{2v} point group having symmetry operations E, C_2, $\sigma_{v(xz)}$ and $\sigma_{v(yz)}$. The carbon atom in carbonyl group is sp^2 hybridized. Out of the three sp^2 hybrid orbitals, two form σ-bonds with two hydrogen atoms by overlapping with s-orbital of H atoms and the third sp^2 orbital of carbon overlap with p_z-orbital of oxygen to form a σ-bond.

The p_x-orbital of carbon and p_x-orbital of oxygen combine to form π and π^* molecular orbitals. The remaining p_y-orbital on oxygen will be a nonbonding MO (actually it is atomic orbital), and it will be perpendicular to the direction of σ and π-orbitals. The MO diagram of the molecule is shown in Fig. 2.4.

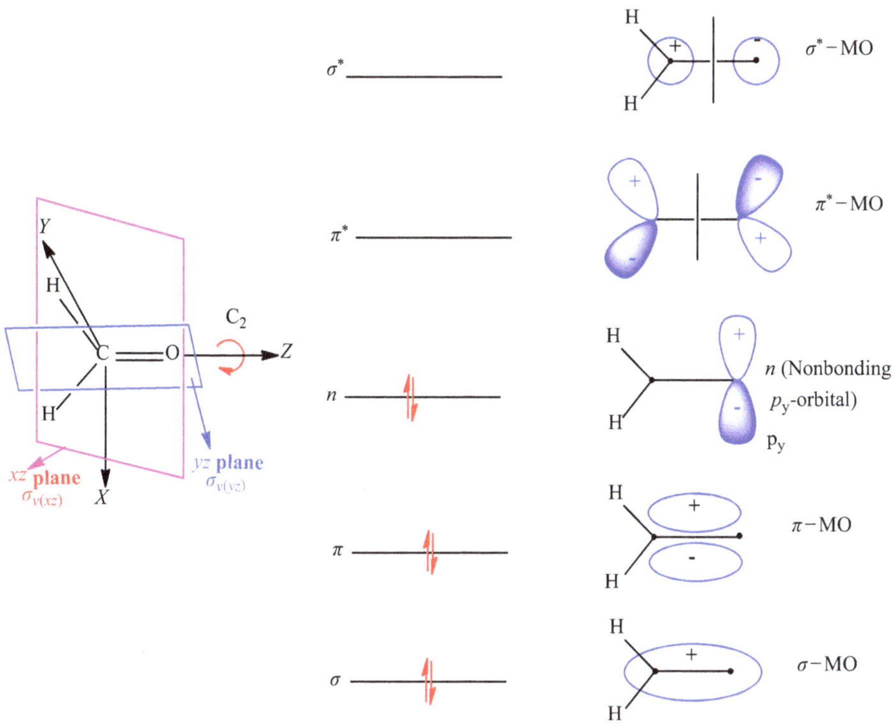

Fig. 2.4: Molecular orbital diagram of >C=O group.

The electronic configuration of >C=O group in ground state is as follows:

$$(\sigma_{CO})^2(\pi_{CO})^2(n)^2(\pi^*_{CO})^0(\sigma^*_{CO})^0$$

As per electronic configuration and also the Mo diagram shown above, the possible transitions are from $(\sigma_{CO})^2(\pi_{CO})^2$ $(n)^2$ MOs to higher energy vacant MOs. Thus, the transitions are confined to carbonyl group only. Hence, we are required to know the symmetries of the molecular orbitals of the >C=O group only, the shapes of which group are also given in Fig. 2.4. In order to know the symmetries of these MOs, it is required to perform the symmetry operations, E, C_2, $\sigma_{v(xz)}$ and $\sigma_{v(yz)}$ of C_{2V} point group on these MOs one by one and then determine the way in which these MOs transform to the corresponding irreducible representation of this point group.

On performing symmetry operations, the MO transformation can be observed as given in Table 2.1.

Table 2.1: MO transformation on performing symmetry operations.

MOs	E	C_2	$\sigma_{v(xz)}$	$\sigma_{v(yz)}$	Corresponding irreducible representation of C_{2v}
σ	+1	+1	+1	+1	A_1
π	+1	−1	+1	−1	B_1
n	+1	−1	−1	+1	B_2
π^*	+1	−1	+1	−1	B_1
σ^*	+1	+1	+1	+1	A_1

It is notable here that if the symmetry operation performed on MO does not change the direction of MO, the character for that operation is taken as +1. On the other hand, if on performing symmetry operation the direction of the MOs is reversed, the character for that operation is taken as −1.

$n \rightarrow \pi^*$ transition

The ground and excited state electronic configuration for $n \rightarrow \pi^*$ transition is shown in the Fig. 2.5.

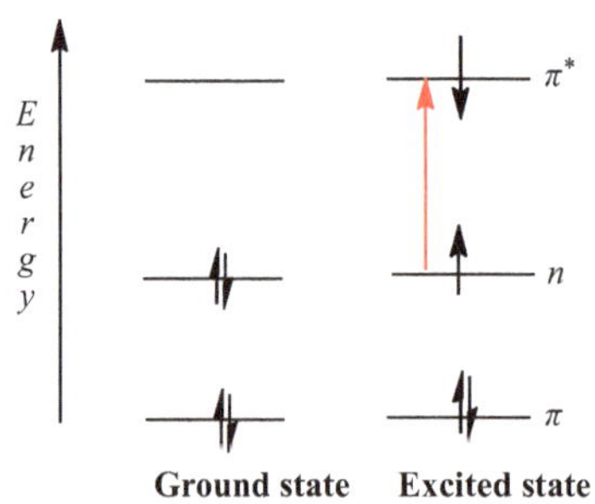

Fig. 2.5: The ground and excited state electronic configuration for $n \rightarrow \pi^*$ transition.

Let us now see whether $n \rightarrow \pi^*$ is allowed or not. This can be known through Group Theory.

In fact, the transition $n \rightarrow \pi^*$ is electronically allowed if the direct product representation of π_{ground} and π_{excited} has the same symmetry as one of the Cartesian coordinates x, y and z.

The direct product representation of ψ_{ground} and ψ_{excited} can be determined from the direct product of the representations of the electronic stat. In the present case, the ground state has both the electrons in nonbonding orbital, $(n)^2$, and the symmetry of the n-orbital is B_2. So, each electron has B_2 symmetry. Hence, ground electronic state or representation of ground state wave function ψ_{ground} will be

$$\Gamma_{\text{ground}} = \Gamma^2{}_{\pi} \times \Gamma^2 \text{n} \tag{2.1}$$

From the Table 2.1, it is notable that π and n-MOs belong to B_1 and B_2 representation respectively. Hence,

$$\Gamma_{\text{ground}} = (B_1)^2 \times (B_2)^2 \tag{2.2}$$

Squaring the character of B_1 and B_2 representations of C_{2v} point group, we get

$$(B_1)^2 = (+1)^2 \;\; (-1)^2 \;\; (+1)^2 \;\; (-1)^2 = +1 \;\; +1 \;\; +1 \;\; +1 = A_1$$

$$(B_2)^2 = (+1)^2 \;\; (-1)^2 \;\; (-1)^2 \;\; (+1)^2 = +1 \;\; +1 \;\; +1 \;\; +1 = A_1$$

Putting the values of $(B_1)^2$ and $(B_2)^2$ in eq. (2.2), we get

$$\Gamma_{\text{ground}} = A_1 \times A_1 = A_1$$

The representation for excited state wave functions will be

$$\begin{aligned} \Gamma_{\text{excited}} &= \Gamma^2{}_{\pi} \times \Gamma^1{}_{\text{n}} \times \Gamma^1{}_{\pi^*} \\ &= (B_1)^2 \times (B_2)^1 \times (B_1)^1 \end{aligned} \tag{2.3}$$

We have observed above that $(B_1)^2 = A_1$, and

$$(B_2)^1 \times (B_1)^1 = B_2 \times B_1 = (+1 \times +1) \;\; (-1 \times -1) \;\; (-1 \times +1) \;\; (+1 \times -1)$$

$$= 1 \quad 1 \quad -1 \quad -1 = A_2 \text{ (from the character Table)}$$

So, $$\Gamma_{\text{ground}} \times \Gamma_{\text{excited}} = A_1 \times A_2 = (+1 \times +1) \;\; (+1 \times +1) \;\; (+1 \times -1) \;\; (+1 \times -1)$$

$$= \;\; 1 \quad 1 \quad -1 \quad -1$$

$$\text{or } \Gamma_{\text{ground}} \times \Gamma_{\text{excited}} = A_2$$

Looking over the character table of C_{2v} point group (shown below, Table 2.2), it is crystal clear that A_2 representation of C_{2v} point group does not correspond to the representation of any one of the Cartesian coordinates (x, y or z). *This concludes that n→π* is not allowed electronically.*

Table 2.2: Character Table of C_{2v} point group.

C_{2v}	E	C_2	$\sigma_{v(xz)}$	$\sigma_{v(yz)}$		
A_1	1	1	1	1	z	x^2, y^2, z^2
A_2	1	1	−1	−1	R_z	xy
B_1	1	−1	1	−1	x, R_y	xz
B_2	1	−1	−1	1	y, R_x	yz

$\pi \rightarrow \pi^*$ transition

In order to know whether $\pi \rightarrow \pi^*$ is allowed or not, we again find the direct product representation for ground and excited state wave functions and then conclude. The ground and excited state configurations are as follows:

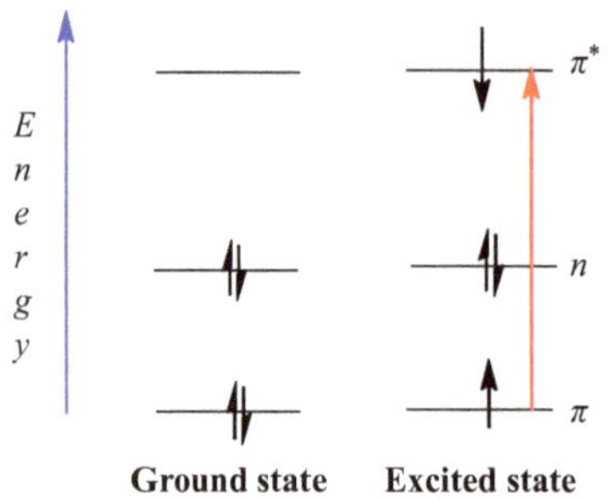

$$\psi_{ground} = (\pi)^2$$
$$\psi_{excited} = (\pi)^1(\pi^*)^1$$

Now, the representation for ground and excited state wave function will be

$$\Gamma_{ground} = (B_1)^2 = B_1 \times B_1 = A_1 \text{ and}$$

$$\Gamma_{excited} = (B_1)^1(B_1)^1 = B_1 \times B_1 = A_1$$

So, the direct product representation of ψ_{ground} and $\psi_{excited}$ can be written as

$$\Gamma_{ground} \times \Gamma_{excited} = A_1 \times A_1 = A_1$$

From the character table of C_{2v} point group, it is apparent that A_1 representation corresponds to the symmetry species of z-coordinate or z-vector. Therefore, $\pi \rightarrow \pi^*$ transition is an electronically allowed transition.

The $\pi \rightarrow \pi^*$ transition can be represented as ${}^1A_1 \leftarrow {}^1A_1$, the superscript indicates spin multiplicity, that is, it is a singlet to singlet transition. This ($\pi \rightarrow \pi^*$) transition will bring change in the dipole moment in the z-direction and $\mu_M(z) \rightarrow A_1$. Hence, the total integral will be equal to $A_1 \times A_1 \times A_1 = A_1$. Hence, it is an allowed transition.

$\sigma \rightarrow \sigma^*$ transition

This transition is also an electronically allowed transition because both ground and excited states have same symmetry as shown below:

$$\psi_{ground} = (\sigma)^2$$
$$\psi_{excited} = (\sigma)^1(\sigma^*)^1$$

Now, the representation for ground and excited state wave function will be

$$\Gamma_{\text{ground}} = (A_1)^2 = A_1 \times A_1 = A_1 \qquad \text{and}$$
$$\Gamma_{\text{excited}} = (A_1)^1 (A_1)^1 = A_1 \times A_1 = A_1$$

So, the direct product representation of ψ_{ground} and ψ_{excited} can be written as

$$\Gamma_{\text{ground}} \times \Gamma_{\text{excited}} = A_1 \times A_1 = A_1$$

$n \rightarrow \sigma^*$ transition

Similarly, $n \rightarrow \sigma^*$ transition is an allowed transition in HCHO. This is well evident from the following.

$$\psi_{\text{ground}} = (n)^2$$
$$\psi_{\text{excited}} = (n)^1 (\sigma^*)^1$$
$$\Gamma_{\text{ground}} = (B_2)^2 = B_2 \times B_2 = A_1 \qquad \text{and}$$
$$\Gamma_{\text{excited}} = (B_2)^1 (A_1)^1 = B_2 \times A_1 = B_2$$
$$\Gamma_{\text{ground}} \times \Gamma_{\text{excited}} = A_1 \times B_2 = B_2$$

Looking over the character table of C_{2v} point group, it is noticeable that B_2 representation corresponds to the symmetry species of *y*-coordinate or *y*-vector. Therefore, $n \rightarrow \sigma^*$ transition is an electronically allowed transition.

The UV-spectrum of formaldehyde exhibits three bands. The high intensity band at 180 nm and 160 nm are due to $\pi \rightarrow \pi^*$ and $n \rightarrow \sigma^*$ transitions, respectively. The low intensity band at 280 nm is due to forbidden $n \rightarrow \pi^*$ transition. This transition becomes partially allowed (low intensity) due to vibronic coupling as discussed above. There also appears a very low intensity band at 350 nm which is due to fully forbidden $n \rightarrow \pi^*$ ($^3A_2 \leftarrow {}^1A_1$) transition.

A transition is fully allowed if it is allowed orbitally (i.e., symmetry requirement as discussed above) as well as spin allowed. A spin-allowed transition is one that takes place between the states of the same multiplicity.

Now, in $n \rightarrow \pi^*$ transition, we can see two types of situation.

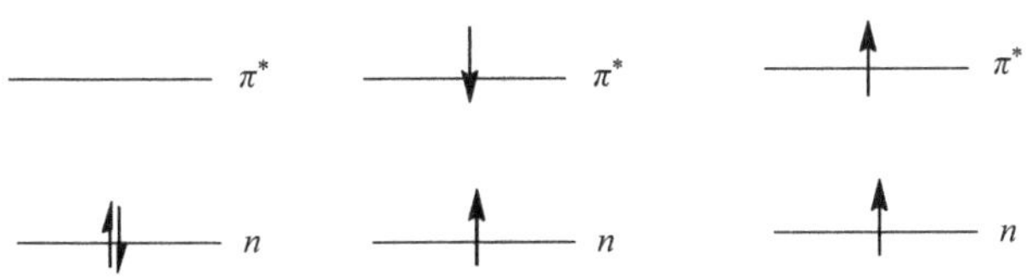

	S=+1/2 – 1/2=0	S=+1/2 – 1/2=0	S=+1/2 + /2=1
Spin multiplicity	2S+1=0+1=1	2S+1=0+1=1	2S+1=2+1=3
	(Singlet)	(Singlet)	(Triplet)

Based on the above, we can represent the $n \rightarrow \pi^*$ transition in two ways as:

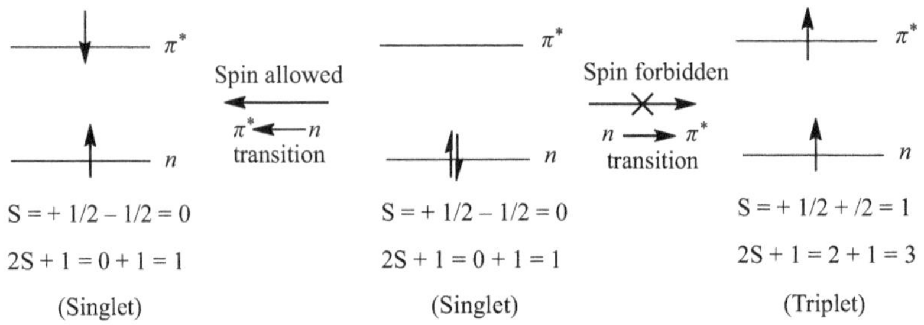

The singlet to singlet $n \rightarrow \pi^*$ transition is symmetrically or orbitally forbidden but spin allowed. On the other hand, singlet to triplet $n \rightarrow \pi^*$ transition is fully forbidden, orbitally as well spin forbidden. That is why the band corresponding to ${}^3A_2 \leftarrow {}^1A_1$ transition is of very low intensity.

The complete wave function for an electronic state is expressed as the product of orbital and spin wave functions, that is,

$$\psi_{\text{electronic}} = \psi_{\text{orbital}} \times \psi_{\text{spin}}$$

Since the dipole moment operators (μ_i) do not interact with the spin of electron, they can be separated out from transition moment integral (M_i) as expressed by:

$$M_i = \int \psi orbital. \mu i. \psi * orbital. dT \int \psi spin. \psi * spin dT$$

The integral of spin wave function determines the spin selection rules. Since the spin functions form an orthogonal set, that is,

$$\int \psi spin\,(i).\ \psi * spin\,(j) = \delta_{ij}$$

$$\text{where } \delta_{ij} = 0 (\text{if } i \neq j) \text{ and } \delta_{ij} = 1\ (\text{if } i = j)$$

So, the transition moment integrals become zero if spin multiplicities of ground and excited states are not same, and the whole transition moment integral M_i will become zero if the spin wave functions are orthogonal.

On the basis of the above discussion, the transition selection rules are summarized as follows:

"*Transitions between the states of different spin multiplicities are forbidden*", that is, the spin-allowed transitions are singlet to singlet, doublet to doublet, triplet to triplet, and so on, whereas spin-forbidden transitions are singlet to triplet, doublet to quartet, and so on

(b) Benzene (C_6H_6) molecule

Benzene molecule belongs to D_{6h} point group. We are aware that the six π-molecular orbitals of benzene have A_{2u}, E_{1g}, E_{2u} and B_{2g} symmetry. Accordingly, the ground and excited state configuration of benzene is shown in Fig. 2.6.

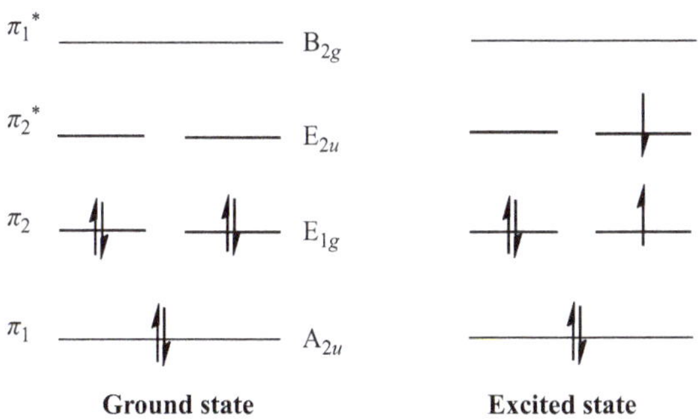

Fig. 2.6: Ground and excited state of electronic configurations of benzene.

Similar to the case of HCHO, the direct product representation for ground state wave function can be determined by the configuration as

$$\Gamma_{\text{groundstate}} = (\pi_1)^2(\pi_2)^2 = (\pi_2)^2$$

An electron present in an orbital of A_{2u} or E_{1g} symmetry has the same symmetry as the orbital itself. But if there is pair of electrons, then symmetry of this arrangement of electrons is determined by squaring the characters of A_{2u} or E_{1g} representations of the point group D_{6h}. So, this will result into

$$\begin{aligned}\Gamma_{\text{groundstate}} &= (\pi_1)^2(\pi_2)^2 = (\pi_2)^2\\ &= (A_{2u})^2(E_{1g})^2 = (E_{1g})^2\\ &= (A_{1g})(A_{1g})\\ &= A_{1g}\end{aligned}$$

This shows that when electrons are filled in pairs, symmetry of the ground state is A_{1g}.

When an electron is excited from E_{1g} to E_{2u} orbital by absorption of UV radiation, the first excited state is obtained as given in Fig. 2.6. As both the MOs (E_{1g} and E_{2u}) are doubly degenerate, the excited state will belong to different symmetry. The symmetry of the excited state can be determined by finding that the product of the characters of the E_{1g} and E_{2u} representations occupied one unpaired electron in each. All other electrons are present in pairs, and so belong to A_{1g} symmetry. It is notable here that closed-shell configurations belong to totally symmetric representation of the group, and, hence, it will not affect the total symmetry of the excited state. So, the symmetry of excited state will depend only on the unpaired electrons.

As stated above, in order to get symmetry of the excited states, we multiply the characters of E_{1g} and E_{2u} representation of D_{6h} point group, and, consequently, we get

D_{6h}	E	$2C_6$	$2C_3$	C_2	$3C_2'$	$3C_2''$	i	$2S_3$	$2S_6$	σh	$3\sigma_d$	$3\sigma_v$
$\chi(E_{1g})\cdot\chi(E_{2u})$	4	−1	1	−4	0	0	−4	1	−1	4	0	0

This reducible representation can be reduced by applying the reduction formula and using the character table of D_{6h} point group as follows:

$$N_i = \frac{1}{h}\sum_R \chi(R)\cdot n\cdot x_i(R)$$

where Ni is the number of times the ith irreducible representation occurs in a reducible representation, h is the order of group (number of symmetry operations), $\chi(R)$ is the character of a particular operation in the reducible representation, n is the number of operation of that type and $\chi i(R)$ is the character of the same operation in the irreducible representation.

D_{6h}	E	$2C_6$	$2C_3$	C_2	$3C_2'$	$3C_2''$	i	$2S_3$	$2S_6$	σh	$3\sigma_d$	$3\sigma_v$
$\chi(E_{1g})\cdot\chi(E_{2u})$	4	−1	1	−4	0	0	−4	1	−1	4	0	0

D_{6h}	E	$2C_6$	$2C_3$	C_2	$3C_2'$	$3C_2''$	i	$2S_3$	$2S_6$	σh	$3\sigma_d$	$3\sigma_v$		
A_{1g}	1	1	1	1	1	1	1	1	1	1	1	1		$x^2+y^2+z^2$
A_{2g}	1	1	1	1	−1	−1	1	1	1	1	−1	−1	R_z	
B_{1g}	1	−1	1	−1	1	−1	1	−1	1	−1	1	−1		
B_{2g}	1	−1	1	−1	−1	1	1	−1	1	−1	−1	1		
E_{1g}	2	1	−1	−2	0	0	2	1	−1	−2	0	0	(R_x, R_y)	(xz, yz)
E_{2g}	2	−1	−1	2	0	0	2	−1	−1	2	0	0		(x^2-y^2, xy)
A_{1u}	1	1	1	1	1	1	−1	−1	−1	−1	−1	−1		
A_{2u}	1	1	1	1	−1	−1	−1	−1	−1	−1	1	1	z	
B_{1u}	1	−1	1	−1	1	−1	−1	1	−1	1	−1	1		
B_{2u}	1	−1	1	−1	−1	1	−1	1	−1	1	1	−1		
E_{1u}	2	1	−1	−2	0	0	−2	−1	1	2	0	0	(x, y)	
E_{2u}	2	−1	−1	2	0	0	−2	1	1	−2	0	0		

$$\begin{aligned} NA_{1g} &= 1/24[(4.1.1) + (-1.2.1) + (1.2.1) + (-4.1.1) + 0 + 0 + (-4.1.1) + (1.2.1) \\ &\quad + (-1.2.1) + (4.1.1) + 0 + 0] \\ &= 1/24[(4 - 2 + 2 - 4 + 0 + 0 - 4 + 2 - 2 + 4 + 0 + 0] \\ &= 1/24[0] = 0 \end{aligned}$$

$$
\begin{aligned}
NA_{2g} &= 1/24[(4.1.1) + (-1.2.1) + (1.2.1) + (-4.1.1) + 0 + 0 + (-4.1.1) + (1.2.1) \\
&\quad + (-1.2.1) + (4.1.1) + 0 + 0] \\
&= 1/24[(4 - 2 + 2 - 4 + 0 + 0 - 4 + 2 - 2 + 4 + 0 + 0] \\
&= 1/24[0] = 0
\end{aligned}
$$

$$
\begin{aligned}
NB_{1g} &= 1/24[(4.1.1) + (-1.2. - 1) + (1.2.1) + (-4.1. - 1) + 0 + 0 + (-4.1.1) \\
&\quad + (1.2. - 1) + (-1.2.1) + (4.1. - 1) + 0 + 0] \\
&= 1/24[(4 + 2 + 2 + 4 + 0 + 0 - 4 - 2 - 2 - 4 + 0 + 0] \\
&= 1/24[0] = 0
\end{aligned}
$$

$$
\begin{aligned}
NB_{2g} &= 1/24[(4.1.1) + (-1.2. - 1) + (1.2.1) + (4.1. - 1) + 0 + 0 + (-4.1.1) \\
&\quad + (1.2. - 1) + (-1.2.1) + (4.1. - 1) + 0 + 0] \\
&= 1/24[(4 + 2 + 2 + 4 + 0 + 0 - 4 - 2 - 2 - 4 + 0 + 0] \\
&= 1/24[0] = 0
\end{aligned}
$$

$$
\begin{aligned}
NE_{1g} &= 1/24[(4.1.2) + (-1.2.1) + (1.2. - 1) + (4.1. - 2) + 0 + 0 + (-4.1.2) \\
&\quad + (1.2.1) + (-1.2. - 1) + (4.1. - 2) + 0 + 0] \\
&= 1/24[(8 - 2 - 2 + 8 + 0 + 0 - 8 + 2 + 2 - 8 + 0 + 0] \\
&= 1/24[0] = 0
\end{aligned}
$$

$$
\begin{aligned}
NE_{2g} &= 1/24[(4.1.2) + (-1.2. - 1) + (1.2. - 1) + (-4.1.2) + 0 + 0 + (-4.1.2) \\
&\quad + (1.2. - 1) + (-1.2. - 1) + (4.1.2) + 0 + 0] \\
&= 1/24[(8 + 2 - 2 - 8 + 0 + 0 - 8 - 2 + 2 + 8 + 0 + 0] \\
&= 1/24[0] = 0
\end{aligned}
$$

$$
\begin{aligned}
NA_{1u} &= 1/24[(4.1.1) + (-1.2.1) + (1.2.1) + (-4.1.1) + 0 + 0 + (-4.1. - 1) \\
&\quad + (1.2. - 1) + (-1.2. - 1) + (4.1. - 1) + 0 + 0] \\
&= 1/24[(4 - 2 + 2 - 4 + 0 + 0 + 4 - 2 + 2 - 4 + 0 + 0] \\
&= 1/24[0] = 0
\end{aligned}
$$

$$
\begin{aligned}
NA_{2u} &= 1/24[(4.1.1) + (-1.2.1) + (1.2.1) + (-4.1.1) + 0 + 0 + (-4.1. - 1) \\
&\quad + (1.2. - 1) + (-1.2. - 1) + (4.1. - 1) + 0 + 0] \\
&= 1/24[(4 - 2 + 2 - 4 + 0 + 0 + 4 - 2 + 2 - 4 + 0 + 0] \\
&= 1/24[0] = 0
\end{aligned}
$$

$$
\begin{aligned}
NB_{1u} &= 1/24[(4.1.1) + (-1.2. - 1) + (1.2.1) + (-4.1. - 1) + 0 + 0 + (-4.1. - 1) \\
&\quad + (1.2.1) + (-1.2. - 1) + (4.1.1) + 0 + 0] \\
&= 1/24[(4 + 2 + 2 + 4 + 0 + 0 + 4 + 2 + 2 + 4 + 0 + 0] \\
&= 1/24[24] = 1
\end{aligned}
$$

$$
\begin{aligned}
NB_{2u} &= 1/24[(4.1.1) + (-1.2. - 1) + (1.2.1) + (-4.1. - 1) + 0 + 0 + (-4.1. - 1) \\
&\quad + (1.2.1) + (-1.2. - 1) + (4.1.1) + 0 + 0] \\
&= 1/24[(4 + 2 + 2 + 4 + 0 + 0 + 4 + 2 + 2 + 4 + 0 + 0] \\
&= 1/24[24] = 1
\end{aligned}
$$

$$
\begin{aligned}
NE_{1u} &= 1/24[(4.1.2) + (-1.2.1) + (1.2. - 1) + (-4.1. - 2) + 0 + 0 + (-4.1. - 2) \\
&\quad + (1.2. - 1) + (-1.2.1) + (4.1.2) + 0 + 0] \\
&= 1/24[(8 - 2 - 2 + 8 + 0 + 0 + 8 - 2 - 2 + 8 + 0 + 0] \\
&= 1/24[24] = 1
\end{aligned}
$$

$$
\begin{aligned}
NE_{2u} &= 1/24[(4.1.2) + (-1.2. - 1) + (1.2. - 1) + (-4.1.2) + 0 + 0 + (-4.1. - 2) \\
&\quad + (1.2.1) + (-1.2.1) + (4.1. - 2) + 0 + 0] \\
&= 1/24[(8 + 2 - 2 - 8 + 0 + 0 + 8 + 2 - 2 - 8 + 0 + 0] \\
&= 1/24[0] = 0
\end{aligned}
$$

The above calculations suggest that the the reducible representation reduced to:

$$\Gamma = B_{1u} + B_{2u} + E_{1u}$$

This result idicates that the lowest energy excited state of the benzene molecule consists of two nondegenerate B_{1u} and B_{2u} states and one doubly degenerate E_{1u} state. Hence, the possible transitions are

$$
\begin{aligned}
&A_{1g} \rightarrow B_{1u} \\
&A_{1g} \rightarrow B_{2u} \quad \text{and} \\
&A_{1g} \rightarrow E_{1u}
\end{aligned}
$$

For a transition to be allowed, as per selection rule, direct product representation for ground and excited states wave functions should correspond to totally symmetric representation. We know that ground state is A_{1g}, and then μM × excited state should also be A_{1g}. Only then we get

$$\int \psi \mu M \, \psi^{*} \, d\tau = A_{1g}$$

The above is possible only when the excited representation belongs to the representation of one of the dipole moment vector *x*, *y* or *z*. Form the character table of D_{6h} point group, one can see that the E_{1u} representation belongs to *x* and y. So, $A_{1g} \rightarrow E_{1u}$ transition is only an allowed transition. This occurs in the UV spectrum of benzene at 180 nm, and it is the most intense band.

Since B_{1u} and B_{2u} do not correspond to any dipole moment vectors *x*, *y* and *z*, the transitions $A_{1g} \rightarrow B_{1u}$ and $A_{1g} \rightarrow B_{2u}$ are forbidden but occur with low intensity at 210 and 250 nm, repectively. These two transitions are partially allowed due to vibronic coupling. A band with very low intensity is also observed at 330 nm. This is due to spin-forbidden transition ${}^1A_{1g} \rightarrow {}^3B_{1u}$. The spin-forbidden transitions also become partially allowed when spin-orbit coupling is present and the singlet state may also have the same total angular momentuum as the triplet state and the two states can interact to give rise to such transition.

Summary

The overall discussion concludes that the electronic transitions are governed by three selection rules and these are as follows:

(i) Transition between the states of same parity (*g* or *u*) is forbidden. This selection rule is called Laporte rule, parity rule and orbital or symmetry selection rule. This rule is for centrosymmetric molecules. For noncentrosymmetric molecules, the direct product representation for ground and excited state wave functions should be totally symmetric. For atomic orbitals, this rule suggests that if $\Delta l = 0$ or ± 2, transitions are forbidden. But for $\Delta l = \pm 1$, transitions will be allowed. Hence, $s \rightarrow s$, $p \rightarrow p$, $d \rightarrow d$, $f \rightarrow f$, $s \rightarrow d$ and $p \rightarrow f$ are forbidden transition but $s \rightarrow p$, $p \rightarrow d$, $d \rightarrow$f transition are allowed.

(ii) It is notable here that the dipole moment operator (μ_M) is a one-electron operator. Hence, an electronic transition (electric-dipole transition) can only involve one electron excitation. So, any transition involving two or more electrons is forbidden. Simultaneous excitation of more than one electron from ground to excited states is forbidden. For polyatomic atoms, this selection rule may be expressed as $\Delta L = \pm 1$, where L is the total angular momentum quantum number.

(iii) Transitions between states of same multiplicities are allowed.

2.2.4 Charge transfer spectra in simple and coordination compounds

Let us consider the formation of a molecule AB formed by the overlapping of atomic orbitals of A and B. According to MO theory, if there is difference in the energy of atomic orbitals A and B, then the bonding MO has greater character of A and antibonding MO has more character of higher energy orbital of B as shown in Fig. 2.7.

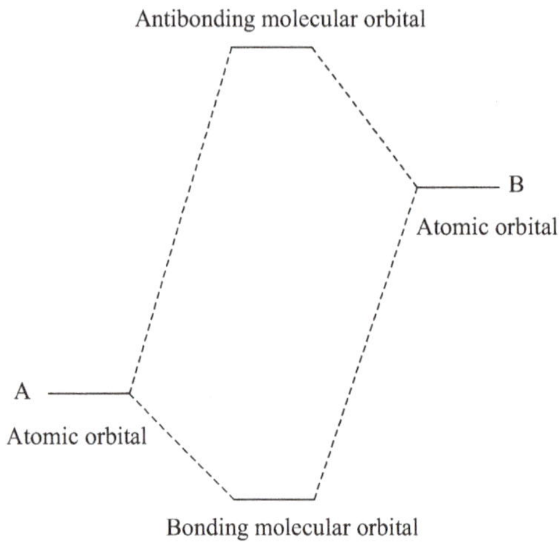

Fig. 2.7: MO diagram of AB molecule.

In such compounds, the transition of electrons from bonding MO to anti-bonding MO is considered as charge transfer from atom A to atom B, because in bonding MO electron charge density is near atom A, while in anti-bonding MO electron charge density occurs more near atom B. The band corresponding to such a transition in the electronic (UV- visible) spectrum of a compound is called charge transfer band.

The charge transfer is always followed by a change in the dipole moment, and, therefore, such transitions are allowed and they appear with high intensity.

The charge transfer may be due to $n \rightarrow \pi^*$, $\pi \rightarrow \pi^*$, $\sigma \rightarrow \sigma^*$ or $n \rightarrow \sigma^*$ transitions. Mostly, charge transfer transitions occur in higher energy region, but, in some cases, they appear in visible region also giving color to the compound.

(a) Compounds showing charge transfer transition in visible region

$KMnO_4$ and $K_2Cr_2O_7$ are purple and yellow in color, respectively. The color in these two compounds is due to charge transfer transitions by absorption of visible light. The metal ions in $KMnO_4$ and K_2CrO_4 are Mn^{+7} and Cr^{6+}containing no d-electrons, and, therefore, charge transfer takes place from O^{2-} to M^{n+} (metal ion). In fact, in $KMnO_4$ (purple) and K_2CrO_4 (yellow), the lowest energy L→M charge transfer is due to transition of a nonbonding 2p oxygen electron to the unoccupied molecular orbital level of d^0 tetrahedral compound. As the positive charge on metal ion increases, its electronegativity also increases, and, therefore, transfer of charge (electrons) from O^{2-} to M^{n+} ion may take place in lower energy region (visible region). Solutions of $TcO_4^-$and ReO_4^- are colorless because charge transfer bands in these compounds occur at high energy in UV region. Similarly, $[HgI_4]^{2-}$, $[HgBr_4]^{2-}$and $[HgCl_4]^{2-}$are colorless because charge transfer transitions in these complexes occur at 31,000 cm^{-1} (322 nm), 40,000 cm^{-1} (250 nm) and 43,700 cm^{-1} (229 nm), respectively.

In the process of charge transfer, the electron density decreases over O^{2-} and increases at M^{n+}. So, we can say that O^{2-} is oxidized and M^{n+} is reduced. This is why charge transfer spectra are also called redox spectra. In charge transfer process, the oxidation of $O^{2-}$and reduction of metal ion may be understood by taking the example of $KMnO_4/MnO_4^-$ in aqueous medium.

$$\begin{array}{ccc} O^{2-} & \longrightarrow & O^- + e^- \text{ (oxidation)} \\ (O = -2) & & (O = -1) \\ MnO_4^- + e^- & \longrightarrow & MnO_4^{2-} \text{ (Reduction)} \\ (Mn = +7) & & (Mn = +6) \\ \hline \end{array}$$

$$O^{2-} \text{ (Reducing agent)} + MnO_4^- \text{ (oxidising agent)} \rightarrow O^- + MnO_4^{2-}$$

Charge transfer also takes place in metal halides. For instance, in AgI (d^{10}), the charge transfer takes place from less acidic I^- to Ag^+. It occurs in visible region so that the color of AgI is yellow. The colors of cadmium sulphide (CdS) (yellow), arsenic sulphide (As_2S_3) (bright yellow) and antimony sulphide (Sb_2S_3) (orange) are also due to the same phenomena.

The charge transfer transition is also observed in organic compounds. If the charge transfer transitions are fully allowed, that is, Laporte allowed as well as spin allowed (have higher intensity), then they are observed in high energy region (UV).

(b) Metal complexes exhibiting charge transfer transitions

Since a charge transfer transition originates from the redox character of the metal and ligand, it is of two distinct types, viz., (i) M→L and (ii) L→M. In transition metal complexes, charge transfer bands appear in their electronic spectra at very high energy side. In M→L type, the metal ion must have some reducing character. Such a transition is possible when the ligand possesses low lying empty anti-bonding orbitals (π_L*) and the metal ion has filled orbitals which are above the highest filled ligand orbitals (Fig. 2.8).

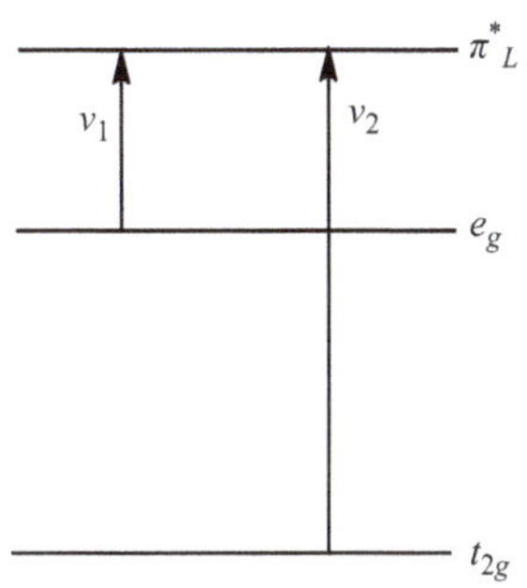

Fig. 2.8: Partial MO diagram depicting M→L charge transfer transitions.

The charge transfer bands have 100–1,000 times greater intensities than the band due to ligand field d-d transitions because the d-d transitions are Laporte-forbidden transitions. The charge transfer bands have been observed in many complexes. For example, $Cr(CO)_6$ exhibits a charge transfer band at 36,000 cm^{-1} due to transition of an electron from filled t_{2g} orbital of Cr(0) to the unoccupied π^* MO of CO by the absorption of UV light (M→L type).

The L→M charge transfer transition is more common. In this type of transition, the ligand has a pronounced reducing ability and the metal ion has oxidizing properties. The charge transfer that actually takes place is from a low-lying filled molecular orbital having mostly ligand character to a higher energy empty molecular orbital having mostly metal character (Fig. 2.9)

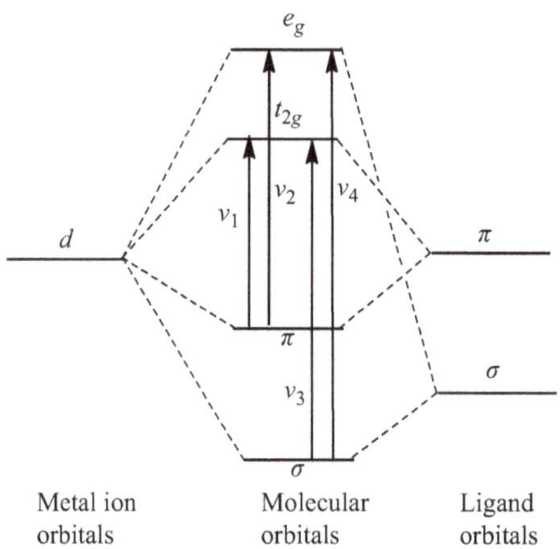

Fig. 2.9: Molecular orbital diagram depicting L→M charge transfer transitions.

The metal complexes, such as, $[MCl_6]^{n-}$, $[MBr_6]^{n-}$ and $[MI_6]^{n-}$ (halo complexes) and $[Co(NH_3)_5X]^{2+}$($X = Cl^-$, Br^- and I^-) (amino complexes) are examples of complexes exhibiting L→M charge transfer transitions. The charge transfer bands were observed at 37,100 cm^{-1} in $[Co(NH_3)_5Cl]^{2+}$ and at 32,000 cm^{-1} in $[Co(NH_3)_5Br]^{2+}$.

The position of the charge transfer bands depends on the oxidizing or reducing power of metal ion and ligand, and the intensity of the charge transfer bands depends on the extent of overlapping of the orbitals between which the electron transfer occurs. However, in the complexes of the second- and third-row transition series metals, where Δ is high, there may be overlapping of the d-d transition and charge transfer transitions.

2.2.5 Electronic spectra of transition metal complexes

In transition metal complexes, metal atom or ion may contain more than one electron; the spectroscopic transitions can be well understood by using the *Term symbols* for the *ground* and *excited states.* A term is nothing but the energy level of an atom or ion specified by an electronic configuration. Electrons in such atoms have the orbital angular momentum (μ_l) and spin angular momentum (μ_s) and μ_l and μ_s can couple. There are two ways of spin-orbit coupling, and these are:

(i) L-S or Russell–Saunders coupling

This is the coupling of the total orbital angular momentum L with the total spin angular momentum S, to give a total angular momentum J. For light atoms, this coupling first involves the independent coupling of individual electronic orbital angular momenta to give a total resultant orbital angular momentum (L), and the independent coupling of individual spin-orbital momenta to give a total resultant spin angular momentum (S). In other words, L = Σl_i (l_i = orbital angular momentum of an electron *i*) and S = Σs_i (s_i = spin angular momentum of an electron i). These two resultant angular momenta, L and S, couple to give J. Thus,

$$J = (L+S),\ (L+S-1),\ (L+S-2)..........|(L-S)| \text{ or } (L+S)\text{to}(L-S)$$

(ii) j-j Coupling

This is the second type of spin-orbit coupling that may be operating in the case of atoms belonging to heavy elements of second- and third-transition series. In this coupling scheme, for each electron (*i*), the orbital and spin momenta l_i and s_i are combined to give total angular momentum j_i, that is,

$$l_1 + s_1 = j_1,\ \ l_2 + s_2 = j_2,\ \ l_3 + s_3 = j_3 \text{ and so on}$$

In heavy metal atoms or ions, the nuclear charge is high enough to force the coupling of *l* and *s* of individual electrons to give the resultant *j*. In such cases, the orbital angular momentum of a given electron interacts more strongly with its own spin angular momentum than with the orbital angular momentum of the other electrons.

Now, the individual electron's total angular momenta j_1, j_2, j_3...... couple to produce the total angular momentum J, for all the electrons, that is,

$$J = j_1 + j_2 + j_3 + \ .... + J_i \ \ (\text{J is always positive})$$

This type of coupling is known as j-j coupling scheme.

2.2.5.1 Determination of Ttrms or term symbols

Now, after having acquaintance with spin-orbital coupling, the next step is to know the meaning and method of the determination of terms or term symbols.

A *Term* here means *energy level/energy state/electronic state* of an atom specified by an electronic configuration and arises due to inter-electronic repulsion of the order of ~10,000 cm^{-1} magnitude. Each energy level/state is indicated by a term symbol. A term symbol for any atomic energy state for which L, S and J are known is expressed as

$$^{(2S+1)}\mathbf{L}_J$$

Here, (2S + 1) = spin multiplicity of the state, L = total orbital angular momentum and J = total angular momentum.

In many electron atoms, due to electron-electron repulsion, several energy states or electronic states are possible for an electronic configuration. For instance, carbon atom has the electronic configuration $1s^2\ 2s^2\ 2p^2$. This can be expressed in the following ways:

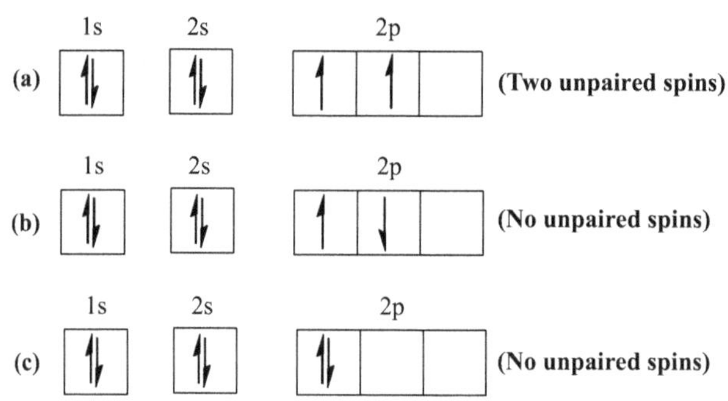

Now, it is well clear that the three arrangements (a), (b) and (c) of electrons in the orbitals have different energies. Each of these arrangements has the same electronic configuration, $1s^2\ 2s^2\ 2p^2$. So, in case of carbon, we have three electronic states for one electronic configuration.

Now, the question arises, how many number of states are possible for an electronic configuration and what will be the term symbol for each state. This can be worked out in a simple way as:

First determine the L, S and J values, then for

L =	0	1	2	3	4	5	6	7	8.....
Term/electronic state ⟶	S	P	D	F	G	H	I	K	L......

It is notable that J is not included as term here because we have already assumed J as total angular momentum.

2.2.5.2 Assignment of term symbols of different atoms

(i) Hydrogen atom

(a) Ground state

Electronic configuration: $1s^1$

$$L = 0 \text{ (because of s-orbital)};\ S = 1/2$$
$$J = 0 + 1/2 = 1/2;\ 2S + 1 = 2(1/2) + 1 = 2.$$

Hence, the term symbol for the ground state of H-atom is $^{(2S+1)}L_j$ = $^2S_{1/2}$ (read as double S one half)

(b) Excited state

For excited state of H-atom with $2s^1$ electronic configuration, we have

$$L = 0 \ \text{(because of s-orbital)};\ S = 1/2$$
$$J = 0 + 1/2 = 1/2;\ 2S + 1 = 2(1/2) + 1 = 2.$$

Hence, the term symbol for the excited state of H-atom is $^{(2S+1)}L_j$ = $^2S_{1/2.}$

Thus, both ground and excited sates have the same term symbol $^2S_{1/2}$ but ground and excited state term have different energy, so these states are written as $^2S_{1/2}$ (1s) and $^2S_{1/2}$ (2s) to differentiate them.

For excited state of H-atom with $2p^1$ electronic configuration, we have

$$L = 1 \ \text{(because of p-orbital)};\ S = 1/2$$
$$J = 1 + 1/2, 1 + 1/2 - 1, 1 - 1/2$$
$$= 3/2, 1/2$$
$$2S + 1 = 2(1/2) + 1 = 2.$$

So, we will have two terms with symbols $^2P_{3/2}$ and $^2P_{1/2}$. It is notablehere that two electronic states (energy levels) are produced in one electronic configuration. Here, the two different states are due to spin-orbit coupling (L-S coupling) because L and S can combine in two different manner, that is, L ± S to get the J values as 3/2 and 1/2. The difference of energy between two energy states is small, that is, spin-orbit coupling is small for H-atom. For atoms having large atomic numbers, spin-orbit coupling is important.

The L-S coupling scheme can be applied to the elements having atomic number up to 30. After that the extent of spin-orbit coupling increases, and therefore, for heavy atoms the j-j coupling scheme is employed.

(ii) Boron atom

Electronic configuration: $1s^22s^22p^1$ (closed shells do not contribute to term symbol). Hence, only $2p^1$ electron will be considered.

$$L = 1 \text{ (because of p – orbital); } S = 1/2; (2S + 1) = (2 \times 1/2 + 1) = 2$$

$$J = 1 + 1/2 = 3/2, 1 + 1/2 - 1 = 1/2 \text{ or } 1 - 1/2 = 1/2$$

Hence, the term symbol is $^{(2S+1)}L_j = {}^2P_{1/2}, {}^2P_{3/2}$

(iii) Helium atom

Electronic configuration: $1s^2$ = ↑↓

$$L = 0 \text{ (because of s – orbital); } S = 0 \text{ (both the electrons are paired).}$$

$$J = L + S = 0; 2S + 1 = 2(0) + 1 = 1.$$

Hence, the term symbol is $^{(2S+1)}L_j = {}^1S_0$

(iv) Empty subshells and fully filled subshells

Empty subshell configurations, viz., p^0, d^0, f^0 and so on, and fully filled subshell configurations, such as, p^6, d^{10}, f^{14} and so on, have always the term symbol $^1S_0$ since the resultant spin (S) and angular momenta (L) are equal to zero.

All the inert gases have term symbols for their ground state $^1S_0$ because S = 0 and L = 0. Similarly, all alkali metals reduce to one electron (ns^1) problems since closed-shell core contributes nothing to L, S and J. Their ground state term symbol is $^2S_{1/2}$ (S = 1/2 and L = 0).

2.2.5.3 Terms for atoms having more than one electron

(a) Equivalent electrons

In case of an atom/metal ion having more than one electron, the terms for possible state of a given electronic configuration can be worked out as follows:

The total orbital angular momentum (L) and total spin angular momentum (S) can have quantized value for their components as

$$M_L = m_{l1} + m_{l2} + m_{l3} + \dots\dots\dots m_{ln}$$

or $M_L = \Sigma m_{li}$ (M_L is the vector sum of the m_l of the individual electrons)

$$\text{or} \quad \begin{aligned} M_L &= L, L-1, L-2, L-3, \dots\dots 0, \dots\dots -L \\ &= (2L+1) \text{values} \end{aligned}$$

and $M_S = m_{s1} + m_{s2} + m_{s3} + \quad \dots\dots\dots m_{sn}$
or $M_S = \Sigma m_{si}$ (M_S is the vector sum of the m_s of the individual electrons)

$$\text{or} \quad \begin{aligned} M_S &= S, S-1, S-2, S-3, \dots\dots 0, \dots\dots -S \\ &= (2S+1) \text{ values} \end{aligned}$$

Since, the term is characterized by a particular L and a particular S, so the term will have

Orbital degeneracy = (2L + 1)
and Spin degeneracy = (2S + 1).

Hence, total degeneracy of the term = (2L + 1)(2S + 1).

Now, we can understand that each possible pair of m_l and $m_s$values will produce a microstate of the configuration. For example, in nd^1 configuration, that is, a single electron in a degenerate set of 5*d*-orbitals, the electron can have any value for m_l = +2, +1, 0, −1, −2 and for m_s = ±1/2. So, there will be 10 possible combinations [5 (orbital)×2 (spin) = 10] of m_l and m_s values. That is, there will be 10 ways of arranging the electron with a particular spin value and a particular orbital's m_l value as shown below:

These 10 arrangements of electrons are actually called microstates of the d^1-configuration. It is remarkable to note that all microstates are associated with the ground state configuration.

With increase in the number of electrons, the number of microstates increases noticeably, and then some microstates will be associated with different energy states of the configuration. In general, for any allowed number of equivalent electrons in a set of degenerate orbitals, the number of possible microstates is given by

$$\text{No. of microstates (Total deneracy of the term)} = \frac{\mathbf{n!}}{\mathbf{r!(n-r)!}}$$

Where *n* = twice the No. of orbitals, r = No. of electrons in an orbital, *n*! = is factorial *n* and r! is factorial r.

Thus, for a p^3 case, there are three orbitals, so *n* = 6 and three electrons, so r = 3. So, the number of microstates equals to:

$$\frac{\mathbf{6!}}{\mathbf{3!(6-3)!}} = \frac{6\times5\times4\times\cancel{3}\times\cancel{2}\times\cancel{1}}{(3\times2\times1)(\cancel{3}\times\cancel{2}\times\cancel{1})} = \frac{120}{6} = 20\,\text{microstates}$$

The number of microstates for various electron arrangements is given in Table 2.3:

Table 2.3: Number of microstates for various electron arrangements in p- and d-orbitals.

p^n	p^1	p^2	p^3	p^4	p^5	p^6
microstates	6	15	20	15	6	1

d^n	d^1	d^2	d^3	d^4	d^5	d^6	d^7	d^8	d^9	d^{10}
microstates	10	45	120	210	252	210	120	45	10	1

(b) Nonequivalent electrons

For electronic configuration involving inequivalent electrons i.e., electrons may have same value of *n* but different *l* values or same value of *l* but different *n* values or both n and l are different, that is, configurations, np^1nd^1,$(n\text{-}1)p^1np^1$ or $(n\text{-}1)d^1np^1$ microstates are obtained by the formula given below:

If both the subshells contain one electron in each (np^1nd^1), then the number of microstates is given as:

$(2l_1+1)(2s_1+1)(2l_2+1)(2s_2+1)$, where l_1, s_1 for p-electron and l_2, s_2 for d-electron
Therefore, $(2l_1+1)(2s_1+1)2l_2+1)(2s_2+1) = (2\times1+1)(2\times1/2+1)(2\times2+1)(2\times1/2+1)$
$= 3\times2\times5\times2 = 60$

For $(n-1)p^1np^1 = (2\times1+1)(2\times1/2+1)(2\times1+1)(2\times1/2+1) = 3\times2\times3\times2 = 36$
For $(n-1)d^1np^1 = (2\times2+1)(2\times1/2+1)(2\times1+1)(2\times1/2+1) = 5\times2\times3\times2 = 60$

If number of electrons in both shells is more than one and these electrons may or may not be equal, then the formula used is:

$$\text{No. of microstates} = \left[\frac{n!}{r!(n-r)!}\right]_p \cdot \left[\frac{n!}{r!(n-r)!}\right]_d$$

For np^2nd^2 configuration, the number of microstates is calculated as follows:

$$= \left[\frac{(2\times3)!}{2!\,(2\times3-2)!}\right]\left[\frac{(2\times5)!}{2!\,(2\times5-2)!}\right]$$
$$= \left[\frac{6!}{2!\times4!}\right]\left[\frac{10!}{2!\times8!}\right]$$
$$= \left[\frac{6\times5\times4!}{2!\times4!}\right]\left[\frac{10\times9\times8!}{2!\times8!}\right]$$
$$= 15\times45 = 675$$

(c) Pigeon hole method
Term symbols and microstates for carbon atom

The electronic configuration for C-atom is $1s^22s^22p^2$. In this case, we need to consider only $2p^2$ configuration because for computing L, closed shell and subshell ($1s^2$, $2s^2$) can be ignored since their contribution to L is zero. Here m_l values for 2p electrons are +1, 0 –1.

A very interesting method called pigeon hole method is described here to know the possible terms in this case, and for the other cases also. In this method, the pigeon hole diagram is constructed by using the following points:

(i) Use one horizontal row for each value of m_l.
(ii) In a given column, put first electron with spin up in the lowest hole.
(iii) Fill other electrons systematically in higher holes one after another above the first electron till all possibilities are completed. For pairing the electrons, pairing should be started from the lowest row so that duplication is avoided.
(iv) Pauli's exclusion rule is obeyed in pairing the electrons in a hole.

Table 2.4: Pigeon hole diagram for p^2-configuration.

↓m_l									
−1		↑	↑		↓	↓			↑↓
0	↑		↑	↓		↑		↑↓	
+1	↑	↑		↑	↑		↑↓		
M_L→	+1	0	−1	+1	0	−1	+2	0	−2
S→	1	1	1	0	0	0	0	0	0

Based on the above points, pigeon hole is constructed for all possible arrangements of two ($2p^2$) electrons as shown in the Table 2.4.

Thus, there are nine possible arrangements (allowed according to Pauli's exclusion principle) of two electrons in ^2p configuration.

As there are two electrons in $2p^2$, there can be two possible values of total spin S = 0 or 1. Now for S = 0, we have two combinations which are nonidentical, that is,

$$M_L = 0\ (L = 0) \quad \text{and}$$

$$M_L = +2, +1, 0, -1, -2 \quad (L = 2)$$

So, the term corresponding to S = 0 and L = 0 will be ^1S, and for S = 0 and L = 2, we have another term or state ^1D.

For S = 1, and M_L = +1, 0, −1 (L = 1), the term will have the symbol ^3P.

Thus, we have three terms, ^1S, ^3P and ^1D. These three terms give 15 microstates as shown below:

$$\begin{array}{lcl} ^1\text{S gives} & \rightarrow & 1\times 1 = 1\,\text{microstate} \\ ^3\text{P gives} & \rightarrow & 3\times 3 = 9\,\text{microstate} \\ ^1\text{D gives} & \rightarrow & 1\times 5 = 5\,\text{microstate} \\ \hline & \text{Total microstates} = 15 & \end{array}$$

Term symbols and microstates for d^2 configuration

In the present case, two electrons are in d-orbital with l = 2. So, we have m_l = +2,+1, 0, −1, −2. So, for each value of m_l, we can draw a horizontal row, then construct a pigeon hole diagram as shown in Table 2.5.

There are in all 25 possible arrangements following Pauli's exclusion principle of 2 electrons in $3d^2$ configuration.

Now, first we pick up M_L values, which have same S values 1. So, M_L values having S = 1 are +3, +2, +1, 0, −1, −2, −3 (L = 3). Here spin multiplicity is 2S + 1 = 2 × 1 + 1 = 3. So, the term corresponding to S = 1 is ^3F.

Table 2.5: Pigeon hole diagram for d^2-configuration.

m_l																									
−2				↑			↑		↑	↑				↓			↓		↓	↓					↑↓
−1			↑			↑		↑		↑			↓			↓		↓		↑				↑↓	
0		↑			↑			↑	↑			↓			↓			↑	↑				↑↓		
+1	↑				↑	↑	↑				↓				↑	↑	↑					↑↓			
+2	↑	↑	↑	↑							↑	↑	↑	↑							↑↓				
M_L	+3	+2	+1	0	+1	0	−1	−1	−2	−3	+3	+2	+1	0	+1	0	−1	−1	−2	−3	+4	+2	0	−2	−4
S	1	1	1	1	1	1	1	1	1	1	0	0	0	0	0	0	0	0	0	0	0	0	0	0	0

From the Table 2.5, we also have for S = 1, another set of M_L values, +1, 0, −1 (L = 1). The corresponding term symbol will be ^3P.

For S = 0, we have

(a) M_L values, +4, +3, +2, +1, 0, −1, −2, −3, −4 (L = 4). So, term symbol will be ^1G.
(b) M_L values, +2, +1, 0, −1, −2 (L = 2). Hence, term symbol is ^1D.
(c) M_L value = 0 (L = 0); term symbol ^1S.

Therefore, all possible terms or states are ^3F, ^3P, ^1G, ^1D and ^1S. These five terms give 45 microstates as shown below:

$$
\begin{array}{lcl}
{}^3\text{F gives} & \rightarrow & 3\times7=21 \text{ microstate} \\
{}^3\text{P gives} & \rightarrow & 3\times3=9 \text{ microstate} \\
{}^1\text{G gives} & \rightarrow & 1\times9=9 \text{ microstate} \\
{}^1\text{D gives} & \rightarrow & 1\times5=5 \text{ microstate} \\
{}^1\text{S gives} & \rightarrow & 1\times1=1 \text{ microstate} \\
\hline
\end{array}
$$

$$\text{Total microstates} = 45$$

2.2.6 Hund's Rules: determination of ground state terms for many electron atoms/ions

We have so far worked out the terms for p^2/d^2 electronic configuration. Now the question arises, out of these terms which one will be ground state term? This can be known by Hund's rule. He proposed three rules in this regard, and these are given below:

(i) *Hund's first rule* is that of maximum *multiplicity. Accordingly, among the states arising from an electronic configuration, the term with the maximum spin multiplicity always has the lowest energy*. Thus, the triplet spin states are lower in energy than the singlet spin states; quartets are lower in energy than doublets and so on. This is well clear from the energy states of p^2-configuration (Fig. 2.10).

   A system with *maximum number of parallel spins* will be stabilized by the exchange energy resulting from their more favurable spatial distribution when compared with that of paired electron.

(ii) *The second rule states* that among the states of same multiplicity, the state with larger *L value* lies lower in energy. Thus, the *^1D state (L = 2)* lies lower in energy than the *^1S state (L = 0)*. The greater stability of states in which the electrons are coupled to produce maximum angular momentum is also related to the special distribution and movement of electrons.

   Applying the (i) and (ii) rules, the above derived energy states/terms for p^2 and d^2 configurations can be arranged in the relative order of increase in energy as

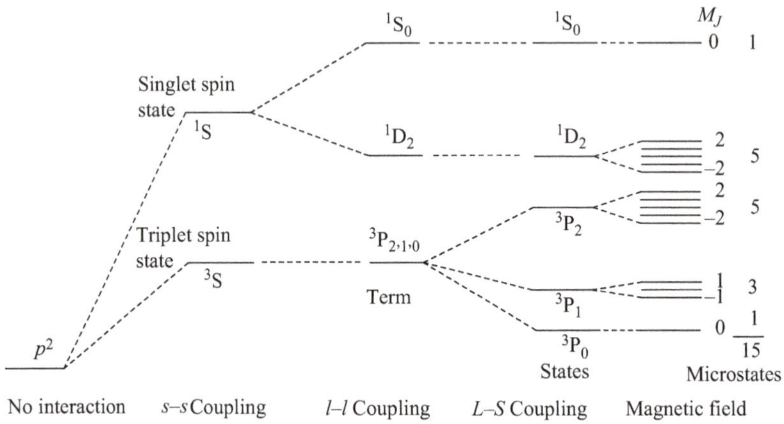

Fig. 2.10: Splitting of energy states arising from p^2 configuration.

$$^3P < {}^1D < {}^1S \qquad (p^2\text{configuration})$$

$$^3F < {}^3P < {}^1G < {}^1D < {}^1S \qquad (d^2\text{configuration})$$

(iii) If there is still uncertainty, for the given values of L and S in a subshell, the *Hund's 3rd rule* decides the state of lowest energy. This rule states that for a given configuration whose subshell is *less than half-full, the states with lower J are lower in energy*. On the other hand, for a subshell that *is more than half-full, states with higher J values lie lower in energy.*

Thus, the ground state term symbol for oxygen (p^4) is $^3P_2$ and for carbon (p^2), it is ^3P_O among $^3P_{O}$, $^3P_1$ and $^3P_2$ states (L = 1 and S = 1 for each). The splitting of energy states arising from p^2 configuration with $3P^0$ ground state is shown below (Fig. 2.11).

Similarly, the ground state term for d^2 configuration (less than half full) is $^3F_2$ (L = 3, S = 1; J = 4, 3, 2) among $^3F_4$, $^3F_3$, $^3F_2$ (Fig. 2.12).

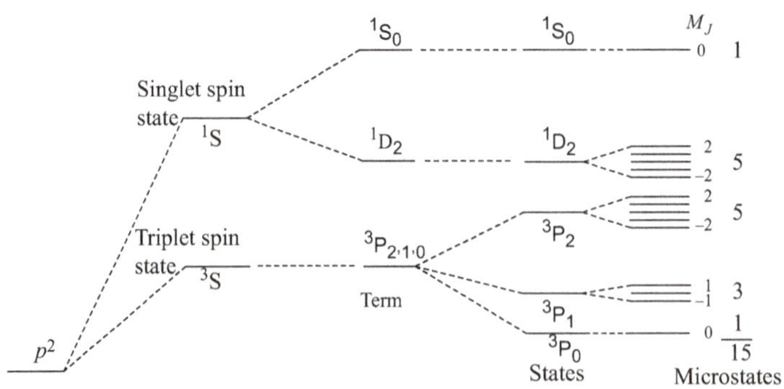

Fig. 2.11: Splitting of energy states arising from p^2 configuration.

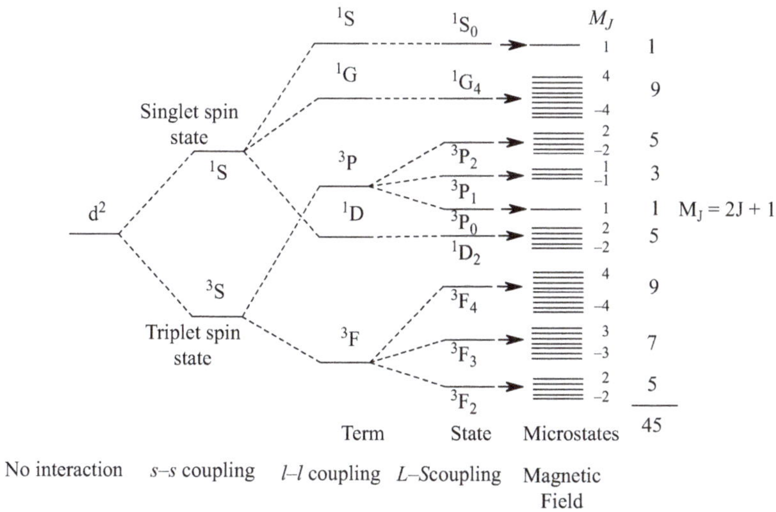

Fig. 2.12: Splitting of energy states arising from d^2 configuration.

Ground state term for nonequivalent electron

(a) Let us consider the case of p^2 configuration when the two electrons are present in *p*-orbitals of different principal quantum numbers, viz., $2p^1$ $3p^1$. In this case, Pauli's exclusion principle is not relevant as the two electrons are not equivalent. Now, we have $l_1 = 1$ and $l_2 = 1$, $L = \Sigma m_l = 1 + 1 = 2$. So, the possible L and S will be

$$L = 2, 1, 0 \quad \text{and}$$
$$S = 1, 0$$

For L = 2 and S = 1, J will be 3, 2, 1; Term/State $= {}^3D_3, {}^3D_2, {}^3D_1$
For L = 1 and S = 1, J will be 2, 1, 0; Term/State $= {}^3P_2, {}^3P_1, {}^3P_0$
For L = 0 and S = 1, J will be 1; Term/State $= {}^3S_1$
For L = 2 and S = 0, J will be 2; Term/State $= {}^1D_2$,
For L = 1 and S = 0, J will be 1; Term/State $= {}^1P_1$
For L = 0 and S = 0, J will be 0; Term/State $= {}^1S_0$,

On applying Hund's rule, it is concluded that $^3D_1$ (with J) or ^3D (without J) is the lowest energy ground state term.

(b) Let us consider the case of another nonequivalent electron, such as, p^1d^1 configuration.

In the present case, $l_1 = 1$ and $l_2 = 2$; $L = l_1 + l_2, l_1 + l_2 - 1, l_1 + l_2 - 2, \ldots\ldots$ *to* $l_1 - l_2$ *or* $l_2 - l_1$, = 3, 2, 1 and S = 1, 0

Now, coupling L and S gives two possibilities.

(i) $$S=1, L=3; J=4,3,2 \rightarrow {}^3F_J \text{ states (that is, } {}^3F_4, {}^3F_3, {}^3F_2)$$
$$S=1, L=2; J=3,2,1 \rightarrow {}^3D_J \text{ states (that is, } {}^3D_3, {}^3D_2, {}^3D_1)$$
$$S=1, L=1; J=2,1,0 \rightarrow {}^3P_J \text{ states (that is, } {}^3P_2, {}^3P_1, {}^3P_0)$$

(ii) $$S=0, L=3; J=3 \rightarrow {}^1F_J \text{ states (that is, } {}^1F_3)$$
$$S=0, L=2; J=2 \rightarrow {}^1D_J \text{ states (that is, } {}^1D_2)$$
$$S=0, L=1; J=1 \rightarrow {}^1P_J \text{ states (that is, } {}^1P_1)$$

On applying Hund's rule, it is concluded that ${}^3F_2$ (with J) or ${}^3F$ (without J) is the lowest energy ground state term.

2.2.7 Hole formulation: term symbols for p^n and p^{6-n} and d^n and d^{10-n} configurations

When a subshell (*p* or *d*) is more than half full, it is simpler and more convenient to work out the terms by considering the holes, that is, the vacancies in the various orbitals rather than the larger number of electrons actually present.

The terms derived in this way for the ground state of oxygen, which has a p^4 configuration, and hence two "holes" are the same as for carbon with a p^2 configuration, that is, ${}^3P$, ${}^1D$ and ${}^1S$. As the oxygen has a more than half-full subshell, according to Hund's third rule, the energy of the triplet *P* states for oxygen are ${}^3P_0 > {}^3P_1 > {}^3P_2$, making ${}^3P_2$ the ground state.

In a similar manner, by considering "holes", the pairs of atoms with p^n and p^{6-n} configurations, and also d^n and d^{10-n}, give rise to identical terms as given in Table 2.6.

2.2.8 Symmetry species of terms

We describe the symmetry of the atomic orbitals using atomic wave functions as basis for irreducible representations of the point groups. Terms are also basic function (many electrons atomic function) for the irreducible representation of the point groups. So, we can obtain character for a rotational group for any L value of a term using the equation given below:

$$\chi(C_n) = \sin\,(L + 1/2)\alpha / \sin\alpha/2 \ (\alpha \text{ is the angle of rotation) and}$$
$$\chi(E) = (2L+1)$$

Here, L is used in place of *l* for term. In CFT, *l* (azimuthal quantum number) has been used for determining the symmetry species of atomic orbitals (d-orbitals).

Table 2.6: Free ion Terms for s^n, p^n and d^n configurations.

Configuration	Term (Bold face: Ground state)
s^1	**^{2}S** (1)
s^0,s^2	**^{1}S** (1)
p^1, p^5	**^{2}P** (1)
p^2, p^4	**^{3}P**, ^{1}D, ^{1}S (3)
p^3	**^{4}S**, ^{2}P, ^{2}D (3)
p^6	**^{1}S (1)**
d^1, d^9	**^{2}D (1)**
d^2, d^8	**^{3}F**, ^{3}P, ^{1}G, ^{1}D, ^{1}S (5)
d^3, d^7	**^{4}F**, ^{4}P, ^{2}H, ^{2}G, ^{2}F, 2 × ^{2}D, ^{2}P (8)
d^4, d^6	**^{5}D**, ^{3}H, ^{3}G, 2 × ^{3}F, ^{3}D, 2 × ^{3}P, ^{1}I, 2 × ^{1}G, ^{1}F, 2 × ^{1}D, 2 × ^{1}S (16)
d^5	**^{6}S**, ^{4}G, ^{4}F, ^{4}D, ^{4}P, ^{2}I, ^{2}H,, 2 × ^{2}G, 2 × ^{2}F, 3 × ^{2}D, ^{2}P, ^{2}S (16)
d^{10}	**^{1}S (1)**

Finally, the irreducible representations are same for a term as for the corresponding orbital set. The orbital set is designated by lower case letters, while terms are designated by capital letters.

In fact, some rules are to be known to assign *g* and *u* subscript for irreducible representations obtained from the terms and these are:

(i) If center of symmetry is absent, then *g* and *u* subscripts are not used.

(ii) If the group has center of symmetry, then *g* and *u* subscripts are assigned as follows:

  (a) Terms derived from orbitals having center of symmetry (even parity) have *g* label. Such orbitals are *s, d, g,* and so on. For example, for d^2 configuration, the terms are S, P, D, F, G, and when metal ion having this configuration is surrounded by octahedral ligand field, the above terms will split into irreducible representations having label *g*.

  (b) If terms derived from those orbitals (*p, f*) which do not have center of symmetry (odd parity), then *g* label is given for even number of electrons, and *u* label is given for odd number of electrons in a configuration. For example, for p^n configuration of the type p^1, p^5, we have ^{2}P term, and when this term is split in octahedral ligand field, then the corresponding irreducible representation will be given *u* label ($^2T_{1u}$) because *n*, the number of electron in *p*-orbital, is odd.

  (c) The spin multiplicity for the irreducible representations will be same as those for the parent terms. Also the spin multiplicity of the term is retained when it splits in any ligand environment.

2.2.9 Splitting of terms: step to arrive to Orgel diagrams

In order to interpret the electronic spectra of transition metal complexes, it is necessary to know how the energies of the various states which arise from the d^n configuration change with the crystal field splitting (or ligand field splitting). So, our main task is to obtain the correlation diagram between the free metal ion terms and terms in the crystal field environment.

In crystal field theory, we know how atomic orbitals split in different ligand fields of different symmetry (using Group Theory). The interesting point is that the terms of free metal ions also split in the same manner as the atomic orbitals split in different environment. For atomic orbitals, we use lower-case symbols, whereas for terms we use capital letters. If we see the character tables of O_h and T_d point groups, and find the irreducible representation which belongs to atomic orbital corresponding to spectral term, then we can easily identify the splitting of terms in these two fields as given in Table 2.7. These split terms are shown in Mulliken Symbols.

Table 2.7: Splitting of free ion terms in O_h and T_d fields.

Spectroscopic Terms of free ion	Degeneracy	Group theoretical Terms (States) in cubic fields (in Mulliken Symbols)	
		Octahedral (O_h)	Tetrahedral (T_d)
S	1	A_{1g}	A_1
P	3	T_{1g}	T_1
D	5	$T_{2g} + E_g$	$T_2 + E$
F	7	$A_{2g} + T_{1g} + T_{2g}$	$A_2 + T_1 + T_2$
G	9	$A_{1g} + E_g + T_{1g} + T_{2g}$	$A_1 + E + T_1 + T_2$

Now, for d^1 configuration of metal ion, ground state term is ^{2}D, and for f^1 configuration, the ground state term is ^{3}F. So, from the above table, we can represent their splitting in O_h field as shown in the Fig. 2.13.

It is notable from the above splitting diagrams that the spin multiplicity of the terms does not change due to splitting in crystal field because the field does not interact with the spins of electrons.

2.2.10 How to decide the ground state in group theoretical terms as Mulliken symbols in cubic field?

Because of splitting of a term in various fields, we can understand the nature of electronic transition, for example, for octahedral complexes with d^1 metal ion configuration (^{2}D spectroscopic term). This term splits into T_{2g} and E_g in octahedral

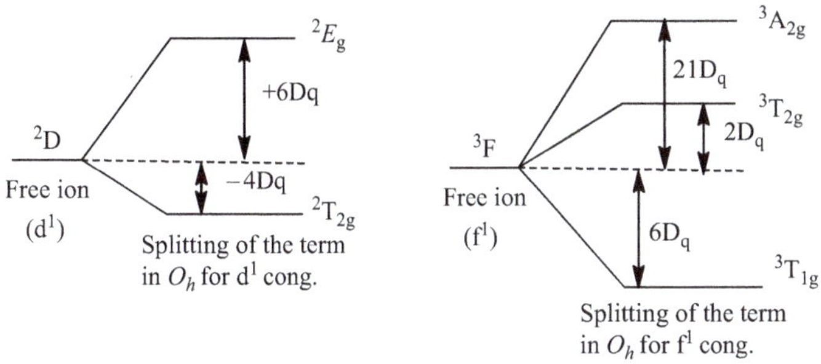

Fig. 2.13: Splitting of ^{2}D and ^{3}F terms in octahedral field.

ligand field. Now, the question arises as to how we will decide that which state is lower in energy. The electronic arrangements (No. of microstates) will be a deciding factor in lower or higher energy of the group theoretical terms because the spin multiplicity is same.

In T_{2g} term, we can see that there are three electronic arrangements with the same energy for $t_{2g}{}^1e_g{}^0$configuration, that is, $d_{xy}{}^1d_{yz}{}^0d_{xz}{}^0$, $d_{xy}{}^0d_{yz}{}^1d_{xz}{}^0$ and $d_{xy}{}^0d_{yz}{}^0d_{xz}{}^1$, making the state triply degenerate. In E_g term, an electron can have only two electronic arrangements for $t_{2g}{}^0e_g{}^1$ configuration, that is, $d_{x2\text{-}y2}{}^1d_{z2}{}^0$ and $d_{x2\text{-}y2}{}^0d_{z2}{}^1$, making the state doubly degenerate. Hence, T_{2g} will be lower state and E_g is a higher state. So, the transition can be designated as

$$t_{2g}{}^1e_g{}^0 \xrightarrow{hv} t_{2g}{}^0e_g{}^1$$

or in term symbols, this can be represented as transfer of electron from T_{2g} to E_g state

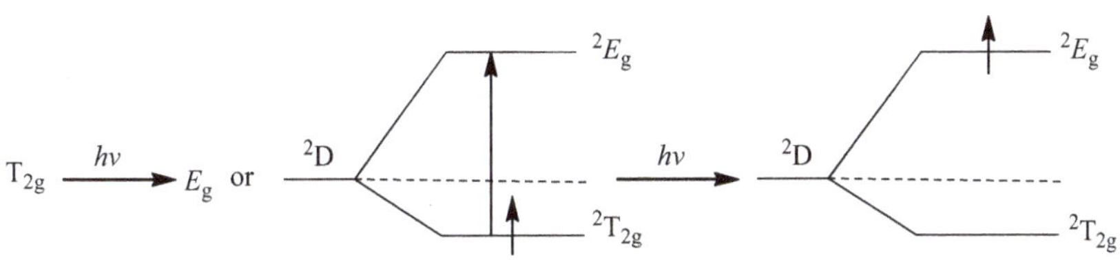

Let us now see what will be the splitting pattern of the term (state) for octahedral complex having d^9 configuration. This can be understood through d^9 configuration given below with m_l values.

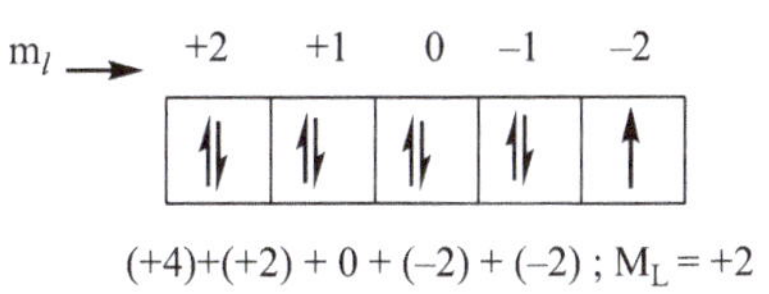

As the highest value of $M_L = 2$ and so $L = 2$. Also there is one unpaired electron in the configuration, so the term symbol of free metal ion with d^9 configuration is ^2D similar to d^1 configuration. Here in d^9 configuration, there are 10 possible electronic arrangements (microstates). The ground state electronic configuration of d^9in octahedral field is $(t_{2g})^6(e_g)^3$, which has one unpaired electron in e_g orbital. The symmetry of the state is determined by the unpaired electron only (paired electrons belong to symmetric irreducible representation). Hence, the ground state is ^2E_g in which two electronic arrangements are of the same energy, and these are: $t_{2g}{}^6d_{x2-y2}{}^0d_{z2}{}^1$ and $t_{2g}{}^6d_{x2-y2}{}^1d_{z2}{}^0$.

Let us work out the symmetry of the excited state. The excited state is obtained by promotion of an electron from $(t_{2g})^6$ to $(e_g)^3$, giving the electronic configuration $(t_{2g})^5(e_g)^4$. As in the excited state, the unpaired electron is in the t_{2g} orbital; the term for excited state will be $^2T_{2g}$. This has three electronic arrangements, which are $d_{xy}{}^2d_{yz}{}^2d_{xz}{}^1e_g{}^4$, $d_{xy}{}^2d_{yz}{}^1d_{xz}{}^2e_g{}^4$ and $d_{xy}{}^1d_{yz}{}^2d_{xz}{}^2e_g{}^4$, hence triply degenerate. As transition occurs always from ground state to excited state, the electronic transition for d^9 can be represented as

$$^2E_g \xrightarrow{hv} {}^2T_{2g}$$

This is just opposite to the transition of d^1 because the splitting of d^9 case is just opposite to that of d^1 case. We can represent these two splitting in O_h for d^1 and d^9 configuration as shown in Fig. 2.14.

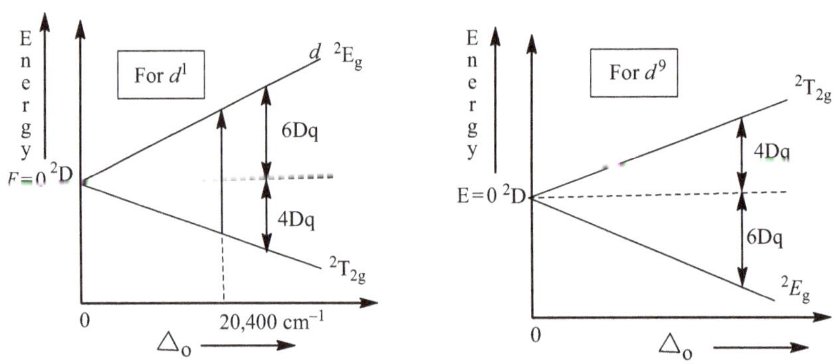

Fig. 2.14: Splitting of ^2D term for d^1 and d^9 configurations in octahedral field.

For example, an absorption band appears at 20,300 cm^{-1} (Δ_o = 235 kJ mol^{-1}) for the octahedral complex ion $[Ti(H_2O)_3]^{3+}$ (d^1case) (Fig. 2.15). This band corresponds to $^2T_{2g} \rightarrow {}^2E_g$, transition which causes violet color to the aqueous solution of Ti(III) ion. Similarly, this transition occurs at 13,000 cm^{-1} in $[TiCl_4]^{3-}$, 18,900 cm^{-1} in $[TiF_4]^{3-}$ and 22,300 cm^{-1} in $[Ti(CN)_4]^{3-}$.

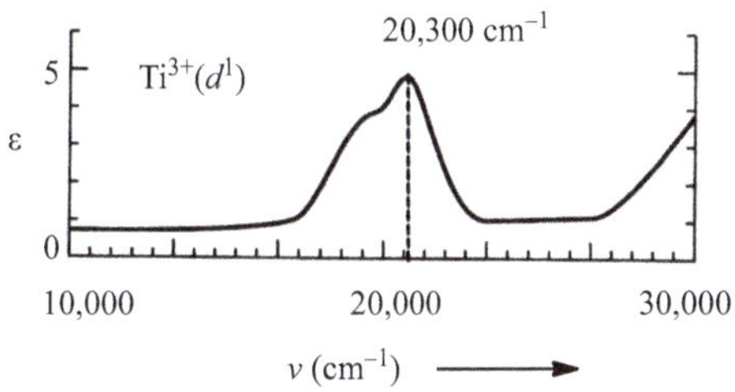

Fig. 2.15: Electronic spectrum of $[Ti(H_2O)_3]^3$.

The above interpretation can also be arrived by splitting of d-orbitals in octahedral crystal field and the band for the Ti(III) (d^1) complexes can be interpreted as due to d-d transition, that is, t_{2g}→e_g. But for multi-electron ions, the crystal field splitting diagrams are not sufficient for predicting the electronic spectra of complexes. We actually need term diagram which represents relative energy of the various terms (that arises due to the crystal field of a particular symmetry point group) with respect to the change in field strength. For d^1 and d^9 octahedral complexes, the combined crystal field terms diagram can be represented as given in the Fig. 2.16. Such a diagram is also known as *Orgel diagram.*

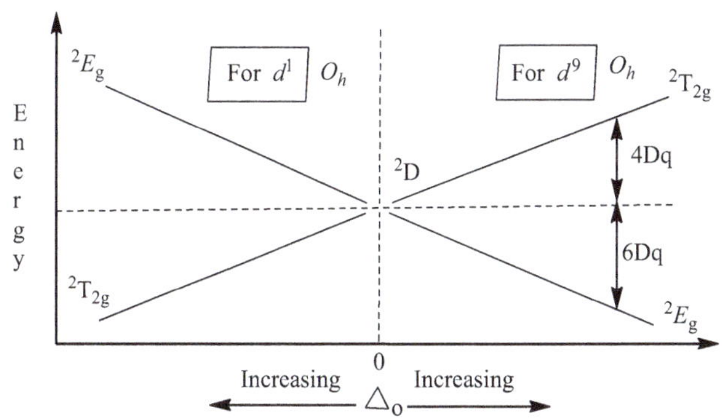

Fig. 2.16: Orgel diagram for d^1 and d^9 configurations in O_h field.

E. H. Orgel suggested such term diagrams for weak field complexes. He plotted the splitting of the ground state terms as a function of crystal field splitting energy (Δ)

2.2.10.1 Tetrahedral complexes with d^1 and d^9 electronic configurations

It is well known that the d-orbital splitting in tetrahedral ligand field is reverse to that of octahedral ligand field, that is, two doubly degenerate orbitals (*e*) are of lower energy and three triply degenerate orbitals (t_2) are of higher energy. Hence, the splitting of ground state term of d^1 configuration for tetrahedral is similar to that of ground state term of d^9configuration for octahedral. Also basic difference between Δ_t and Δ_o is $\Delta_o = 4/9\Delta_t$. The splitting of ground state term of d^9 configuration for tetrahedral will be reverse of splitting of d^1configuration ground state term for tetrahedral. Here also the ground state term for d^1 and d^9 configuration is ^{2}D. Therefore, splitting of energy levels (terms) for d_1 and d^9 configuration in tetrahedral ligand field can be represented as shown in Fig. 2.17.

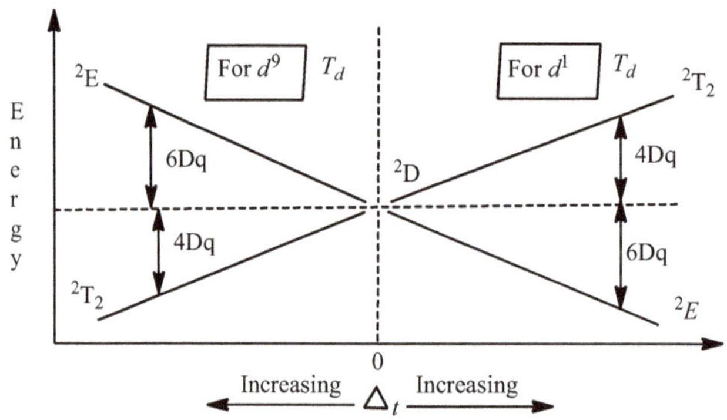

Fig. 2.17: Orgel diagram for d^1 and d^9 configuration in T_d field.

2.2.10.2 Octahedral complexes with d^4 and d^6 electronic configurations

The L value for the ground state is same in case of d^1 and d^6 configurations, but the spin multiplicities are different.

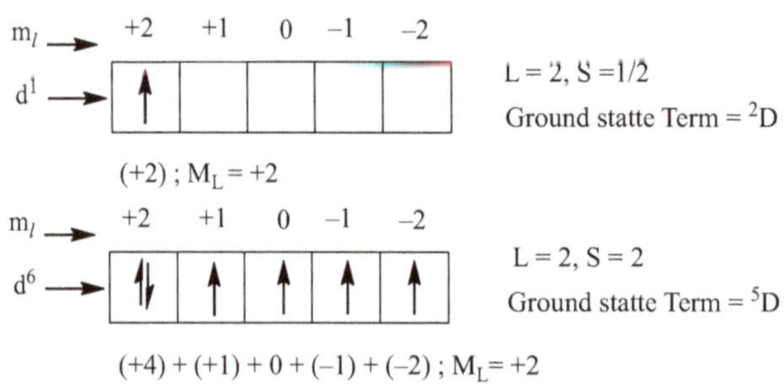

In an octahedral field, the ground state term ^{5}D of the metal ion for d^6 electronic configuration splits into $^5T_{2g}$ of lower energy ($t_{2g}^4 e_g^2 \rightarrow d_{xy}^2 d_{yz}^1 d_{xz}^1 e_g^2$, $d_{xy}^1 d_{yz}^2 d_{xz}^1 e_g^2$ and $d_{xy}^1 d_{yz}^1 d_{xz}^2 e_g^2$) (No. of microstates = 3) and ^5E_g of higher energy ($t_{2g}^3 e_g^3 \rightarrow t_{2g}^3 d_{x2-y2}^2 d_{z2}^1$ and $t_{2g}^3 d_{x2-y2}^1 d_{z2}^2$) (No. microstates = 2). So, the single d-d transition can be represented as shown in Fig. 2.18.

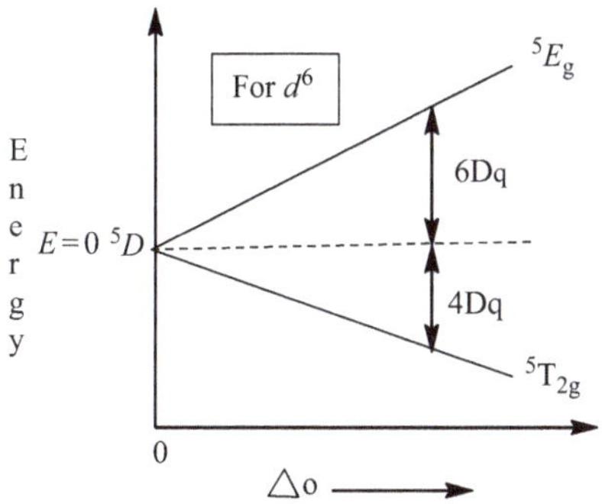

Fig. 2.18: Splitting of the ground state term ^{5}D of the metal ion for d^6 electronic configuration in Oh field.

$$^5T_{2g} \rightarrow {^5E_g}$$

This is similar to the case of octahedral d^1 configuration of metal ion except multiplicity of the states.

Similarly, electronic transition in d^4 octahedral and d^9 (^{2}D) octahedral will also be similar because in d^4 configuration one electron is less than in half-filled configuration. The ground state has L = 2 and S = 2, so ground state term will be ^{5}D and it splits into lower energy ^5E_g state ($t_{2g}^4e_g^0 \rightarrow d_{xy}^2d_{yz}^1d_{xz}^1e_g^0$, $d_{xy}^1d_{yz}^2d_{xz}^1e_g^0$ and $d_{xy}^1d_{yz}^1d_{xz}^2e_g^0$) (No. of microstates = 3) and higher energy excited state $^5T_{2g}$ ($t_{2g}^3e_g^1 \rightarrow t_{2g}^3d_{x2-y2}^1d_{z2}^0$and t_{2g}^3 d_{x2-y2}^0 d_{z2}^1) (No. microstates = 2) as shown in Fig. 2.19.

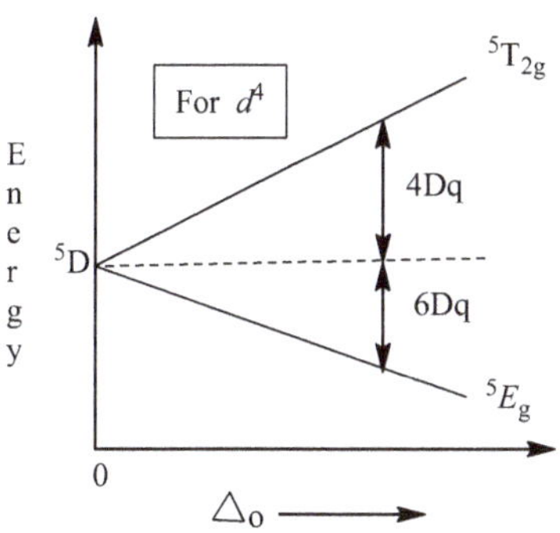

Fig. 2.19: Splitting of the ground state term ^{5}D of the metal ion for d^4 electronic configuration in O_h field.

From the above discussion it is concluded that transition in metal ions with a *difference* of *five (5) electrons* in their electronic configurations is similar for the given O_h or T_d field.

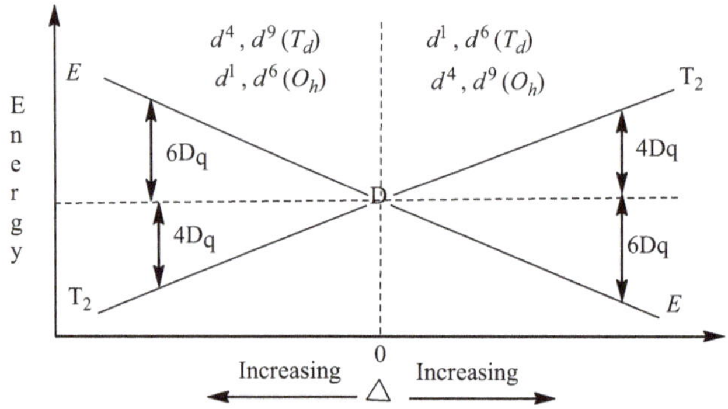

Fig. 2.20: Orgel diagram for d^1, d^4, d^6 and d^9octahedral and tetrahedral complexes. Octahedral will bear g subscript with T_2 and E both.

The above-discussed results can be combined in a single-term energy level diagram known as Orgel diagram shown below in Fig. 2.20.

From the above diagram, it is well clear that d^1, d^4, d^6 and d^9 metal ions will give only one broad band corresponding to a single d-d transition because there will occur only single d-d transition possible either from $T_2 \rightarrow E$ or $E \rightarrow T_2$. It is notable here that transition in case of octahedral will bear *g* subscript with T_2 and *E* both.

It is remarkable to note here that the d-d transitions in O_h complexes with center of symmetry are Laporte forbidden because it is *g*→*g* transition. This is due to the fact that the *g*→*g* transition will not cause any change in electric dipole moment, that is $|\int \psi \mu M\ \psi^* dT|^2 = 0$ However, this selection rule is not strictly followed and the forbidden transition do occur because of the three reasons given below:

(a) Due to vibration of the complex molecule, it may be distorted from regular octahedral geometry. Consequently, it loses the center of symmetry. Therefore, due to coupling of the electronic and vibrational wave functions, that is, the vibronic coupling, the d-d transition becomes partially allowed. Hence, low intensity broad bands are observed corresponding to d-d transitions.
(b) During the vibration of a regular O_h molecule, certain molecules may be distorted, and during distortion the regular octahedral symmetry is lost temporarily. This causes some mixing of *d* and *p* orbitals, and, hence, the $t_{2g} \rightarrow e_g$ transition is not purely d→d transition. In fact, they become dp→ dp transitions and so are partially allowed, resulting in a low intensity broad band.
(c) It has been observed that in some complexes the intensity of the d → d transition is much greater than expected by the above two reasons. This may be explained on the basis of overlapping of the d-orbital with the ligand orbitals. Consequently, character of the pure d- orbitals is lost and, hence, d → d transition becomes allowed. It has been experimentally verified by electron spin resonance and other spectroscopic methods. Moreover, the reasonably good intensity band due to d → d transition is self-explanatory of metal-ligand overlap in metal complex.

2.2.10.3 Octahedral complexes with d^2, d^8 and T_d complexes with d^2 electronic configurations

Table 2.6 shows that the terms arising for a d^2 configuration are ^{3}F, ^{3}P, ^{1}G, ^{1}D and ^{1}S, and ^{3}F is the ground state term. Splitting of these terms in octahedral field is:

$$^3\mathrm{F}\rightarrow{}^3\mathrm{A}_{2g}+{}^3\mathrm{T}_{2g}+{}^3\mathrm{T}_{1g}$$
$$^3\mathrm{P}\rightarrow{}^3\mathrm{T}_{1g}(\text{no splitting})$$
$$^1\mathrm{G}\rightarrow{}^1\mathrm{A}_{1g}+{}^1\mathrm{T}_{2g}+{}^1\mathrm{T}_{1g}+{}^1\mathrm{E}_{g}$$
$$^1\mathrm{D}\rightarrow{}^1\mathrm{T}_{2g}+{}^1\mathrm{E}_{g}$$
$$^1\mathrm{S}\rightarrow{}^1\mathrm{A}_{1g}(\text{no splitting})$$

Our main concern here will be with the splitting of the ground state term ^{3}F and other singlet terms, ^{1}G, ^{1}D and ^{1}D can be ignored because the transitions from ^{3}F to any of these singlet states is spin forbidden and will be very weak to be observed. Hence, we will consider only ^{3}F and ^{3}P states because they can have spin-permitted transitions. As the *p*-orbitals of metal ions are directed along three Cartesian axes, they will be raised in energy equally by ligand point charges situated along the axes in cubic geometry. So, they remain degenerate. Therefore, fee ion P term will remain unsplit in cubic crystal field and transferred to $^3T_{1g}$ (in O_h) or $^3T_1$ (in T_d).

As given above, the ^{3}F term splits into $^3A_{2g}$ + $^3T_{2g}$ + $^3T_{1g}$ in O_h field. Out of these three states, the lowest energy state (term) may be $^3T_{1g}$ because lowest energy configuration d^2 in O_h will be $t_{2g}{}^2e_g{}^0$, where the three degenerate possibilities of putting two electrons are as follows: $d_{xy}{}^1d_{yz}{}^1d_{xz}{}^0$, $d_{xy}{}^0d_{yz}{}^1d_{xz}{}^1$and $d_{xy}{}^1d_{yz}{}^0d_{xz}{}^1$. So, the lowest energy term is triply degenerate. But, there are two triply degenerate $^3T_{1g}$ and $^3T_{2g}$ in the present case. Out of the two, the first one $^3T_{1g}$ has subscript 1 indicating that it is symmetric with respect to rotation axis other than the principal axis, whereas in $^3T_{2g}$ the subscript 2 suggests that it is antisymmetric with respect to other rotation axis. Therefore, $^3T_{1g}$ is the lowest energy state and the next higher will obviously be $^3T_{2g}$. The next higher to $^3T_{2g}$ is $^3A_{2g}$. The Orgel diagram can be represented for d^2 configuration in O_h field as shown in Fig. 2.21.

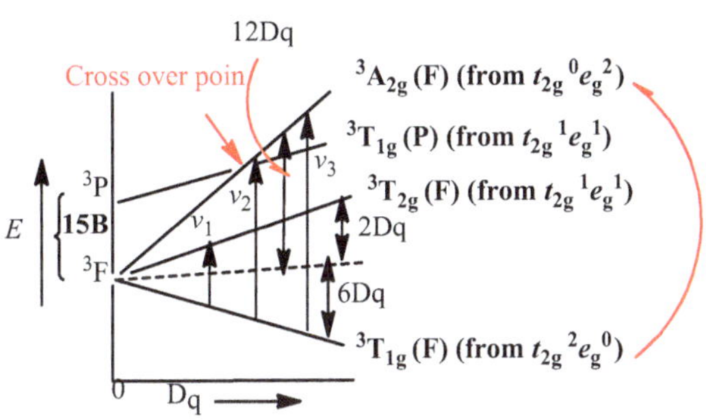

Fig. 2.21: Orgel diagram for d^2 O_h complexes.

Hence, for octahedral complexes of d^2 ion, such as $[V(H_2O)_6]^{3+}$, three spin-allowed transitions expected in order of increasing energy (as shown by three arrows in Fig. 2.21) are as follows:

$$^3T_{1g}(F) \rightarrow {}^3T_{2g}(F)(\nu_1)\ (17,400\ cm^{-1})$$

$$^3T_{1g} \rightarrow {}^3T_{1g}(P)(\nu_2)\ (25,200\ cm^{-1})\ \text{and}$$

$$^3T_{1g}(F) \rightarrow {}^3A_{2g}(F)(\nu_3)\ (34,500\ cm^{-1})$$

But, actually only two absorption bands located at 17,200 cm^{-1} (ν_1) and 25,600 cm^{-1} (ν_2) are observed in weak field $[V(H_2O)_6]^{3+}$ (Fig. 2.22) and alike complexes.

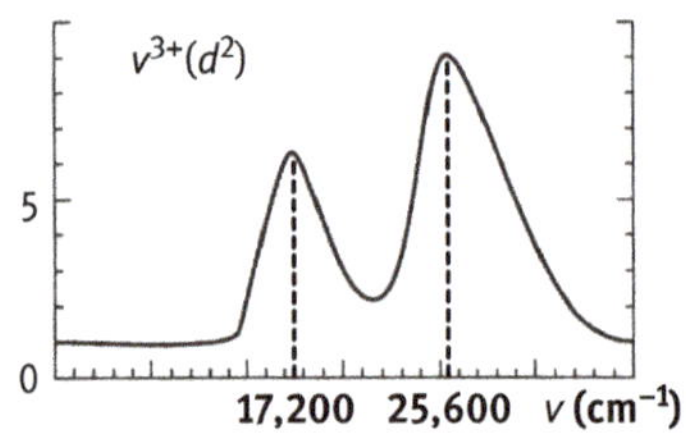

Fig. 2.22: Electronic spectrum of $[V(H_2O)_6]^{3+}$.

In the strong field configuration, the $^3T_{1g}$ (F) $\rightarrow {}^3A_{2g}$ (F) (ν_3) band corresponds to a two-electron excitation ($t_{2g}{}^2 \rightarrow e_g{}^2$) and hence is ought to be forbidden, besides being forbidden by the usual parity ($g \rightarrow g$ transition) restriction. It is thus relatively less probable. It is not observed owing to the intense charge transfer band occurring in this region of the spectrum and also probably due to its low intensity.

It is notable here that only two peaks should occur in the spectrum if the crystal field strength of the ligand results in a transition close to the crossover point between $^3T_{1g}$ (P) and $^3A_{2g}$ F).

The spectral data for some vanadium(III) octahedral complexes are given in the Table 2.8.

Table 2.8: Electronic Spectral data of Some Complexes.

Compound	$^3T_{1g}(F) \rightarrow {}^3T_{2g}(F)$ (ν_1, cm^{-1})	$^3T_{1g}(F) \rightarrow {}^3T_{1g}(P)$ (ν_2, cm^{-1})
$[V(H_2O)_6]^{3+}$	17, 200	25, 600
$[VF_6]^{3-}$	14, 800	23, 000
$[V(urea)_6]^{3+}$	16, 200	24, 200

In T_d complexes with d^2 configuration of metal ion, the Orgel diagram will have the same states (terms) as for d^2, but in the reverse order as shown in the Fig. 2.23.

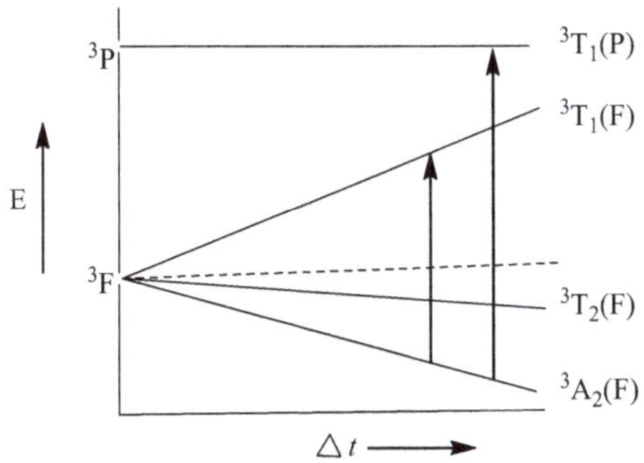

Fig. 2.23: Orgel diagram for d^2 T_d complexes.

Some tetrahedral complexes of V(III) are known to show two bands corresponding to the transition $^3A_2(F) \rightarrow {}^3T_1(F)$ at ~9000 cm^{-1} (near IR) and $^3A_2(F) \rightarrow {}^3T_1(P)$ at ~15,000 cm^{-1}. The transition $^3A_2(F) \rightarrow {}^3T_1(F)$ is of very low energy observed in near IR region. So, in the electronic spectra of tetrahedral d^2 V(III) complexes, one d→d band is observed in UV-visible region.

In d^8 octahedral complexes, there are two holes in the e_g level. Hence, promoting an electron is equivalent to transferring a hole from e_g to t_{2g} level. This is the inverse of the d^2 case. So, the energy level diagram for d^8 configuration in octahedral field will be reverse of d^2 O_h complexes and may be represented as given in Fig. 2.24. Thus, it is similar to d^2 tetrahedral complexes.

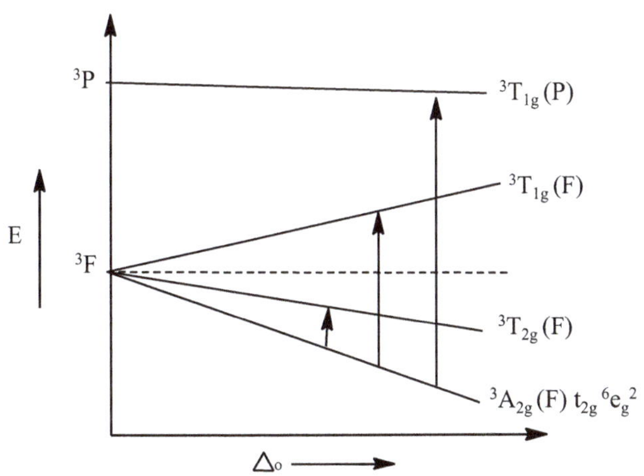

Fig. 2.24: Orgel diagram for d^8 O_h Complexes.

The octahedral ground state electronic configuration is $[t_{2g}{}^6e_g{}^2]$, which is nondegenerate, that is, $[(t_{2g}{}^6)(d_{x^2-y^2})^1(d_{z^2})^1]$, and, therefore, it is $^3A_{2g}$(F). When one electron is excited, the configuration of excited state will be $[t_{2g}{}^5e_g{}^3]$. Hence, there will be six possible electron arrangements corresponding to $^3T_{2g}$(F) and $^3T_{1g}$(P). Excitation of two electrons results in the electronic configuration $[t_{2g}^4 e_g^4]$. This state is triply degenerate, that is, $d_{xy}^2 d_{yz}^1 d_{xz}^1 e_g^4$, $d_{xy}^1 d_{yz}^2 d_{xz}^1 e_g^4$ and $d_{xy}^1 d_{yz}^1 d_{xz}^2 e_g^4$. This triply degenerate state is designated as $^3T_{1g}$(F).

Thus, in d^8 octahedral complexes (similar to d^2 O_h), three transitions are expected: two corresponding to one electron transition $t_{2g}^6 e_g^2 \rightarrow t_{2g}^5 e_g^3$, that is, $^3A_{2g}$ (F)→$^3T_{2g}$(F) and $^3A_{2g}$(F)→$^3T_{1g}$(P), and one corresponding to two electron transition $t_{2g}^6 e_g^2 \rightarrow t_{2g}^4 e_g^4$, that is, $^3A_{2g}$(F)→$^3T_{1g}$(F).

The d^8 octahedral nickel(II) complexes are most common and have been extensively studied. For example, aqueous solutions of nickel(II) salts containing the $[Ni(H_2O)_6]^{2+}$ ion have light green color, which is due to the presence of weak bands in the *red* and *blue* portions of the visible spectrum. The three absorption bands exhibited by this complex ion are at 8,700 (ε = 1.6, f = 1.8 × 10^{-5}) 14,500 (ε = 2.0, f = 3.0 × 10^{-5}) (red) and 25,300 cm^{-1} (ε = 4.6, f = 7.0 × 10^{-5}) (blue). There are also very weak bands (f ~10^{-6}) in the region 15,000–18,000 cm^{-1} which are attributed to spin-forbidden triplet $^3A_{2g}$(F)→singlet (^{1}G, ^{1}D, ^{1}S) transitions (Fig. 2.25)

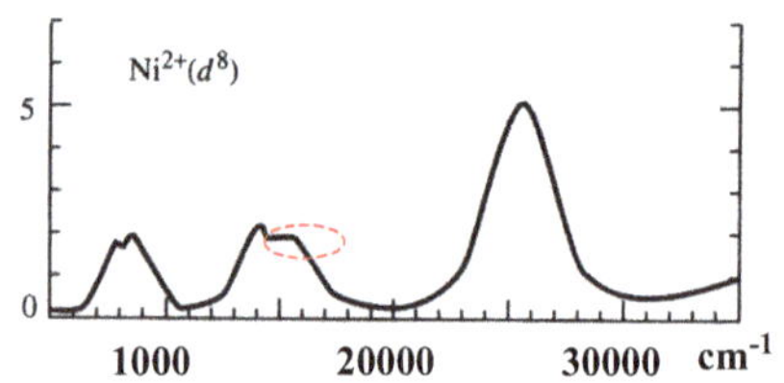

Fig. 2.25: Electronic spectrum of $[Ni(H_2O)_6]^{2+}$.

The three bands in the electronic spectrum of $[Ni(NH_3)_6]^{2+}$ at 10,750, 17,500, 28,200 cm^{-1} are identified with $^3A_{2g}$(F)→ $^3T_{2g}$(F), $^3A_{2g}$(F)→ $^3T_{1g}$(F) and $^3A_{2g}$(F)→ $^3T_{1g}$ (P) transitions, respectively. In some nickel(II) octahedral complexes, the band corresponding to $^3A_{2g}$(F)→ $^3T_{1g}$(F) transition has a shoulder on a band because it is a two-electron transition, which is generally not allowed. The combined Orgel diagram for d^2 and d^8 electron configuration is shown in the Fig. 2.26.

It is notable from the above Orgel diagram that the two T_{1g} states, one from the P state and other from the F state, are slightly curved lines because they have the same symmetry and they interact with each other. This effect of inter-electronic repulsion lowers the energy of the lower state $^3T_{1g}$ (F) and increases the energy of the higher $^3T_{1g}$ (P) state. This effect is more pronounced on the left side of the diagram because the two levels are close in energy. If the lines are not curved and become straight, they would cross each other, that is, at the cross-point the two electrons in an atom may have the same symmetry and the same energy. This would be impossible, and is prohibited by the noncrossing rule which says that the states of the same symmetry cannot cross each other.

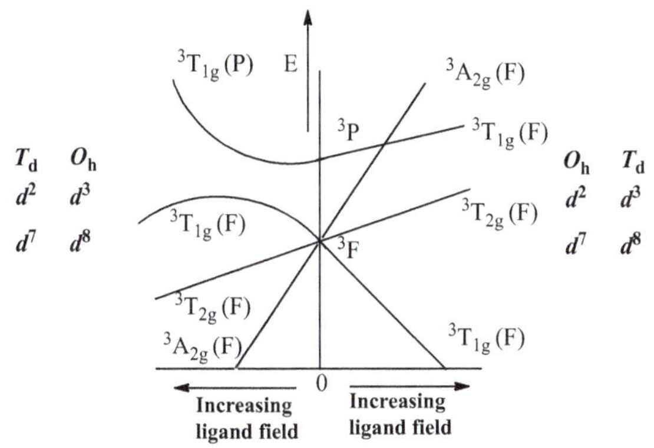

Fig. 2.26: Orgel diagram for d^2, d^3, d^7, d^8 octahedral and tetrahedral complexes. Spin multiplicity will be changed with electronic configuration. Superscript *g* and *u* will be dropped in tetrahedral cases.

2.2.10.4 Octahedral complexes with d^3, d^7 and T_d complexes with d^2, d^3, d^7, d^8 electronic configurations

In the free ion with d^3 configuration, the ground state term is ^4F corresponding to $t_{2g}{}^3e_g{}^0$ configuration. Here also, another state of the same multiplicity is ^4P. This case is similar to d^8 with two holes. The splitting pattern for d^3 is also same as for d^8. In d^3, the ground state is $^4A_{2g}$ and three possible transitions are as given in the Orgel diagram (Fig. 2.27).

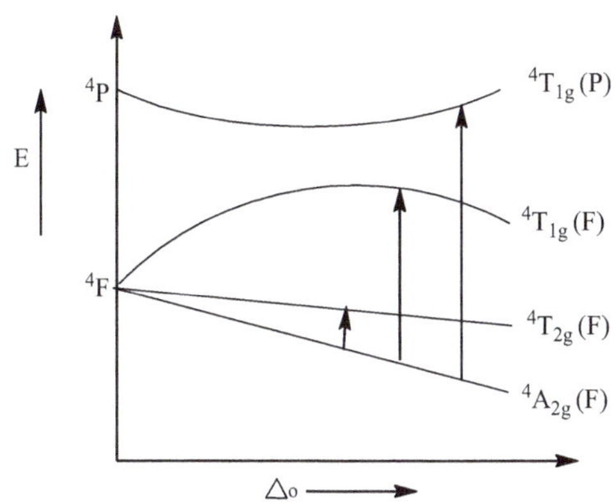

Fig. 2.27: Orgel diagram for d^3 O_h Complexes.

For example, in $[Cr(H_2O)_6]^{3+}$, these transitions correspond to the absorption bands at 17,400 cm^{-1}, 24,400 cm^{-1} and 38,600 cm^{-1}. These transitions are:

${}^4A_{2g}(F) \rightarrow {}^4T_{2g}(F)$, ${}^4A_{2g}(F) \rightarrow {}^4T_{1g}(F)$ and ${}^4A_{2g}(F) \rightarrow {}^4T_{1g}(P)$. Here also, ${}^4T_{1g}(F)$ and ${}^4T_{1g}(P)$ are slightly curved, because they interact with each other due to same symmetry and cannot cross each other. The electronic spectrum of $[Cr(H_2O)_6]^{3+}$ is shown in Fig. 2.28.

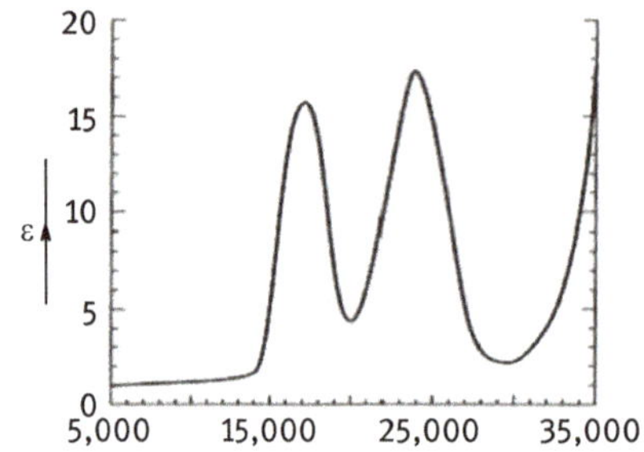

Fig. 2.28: Electronic spectrum of $[Cr(H_2O)_6]^{3+}$.

In tetrahedral d^2 complexes, the splitting pattern is same as in octahedral because as per Group Theory, splitting of the F and P Terms is same in tetrahedral and octahedral fields, that is,

$$F \rightarrow A_2 + T_2 + T_1 \quad \text{and}$$
$$P \rightarrow T_1$$

But the order of energy of the spectral states in tetrahedral is reverse to that of d^2 octahedral field. This can be understood as in the O_h field, the ground state configuration of d^2 is $e^2t_2^0$ and this configuration gives nondegenerate state A_2.

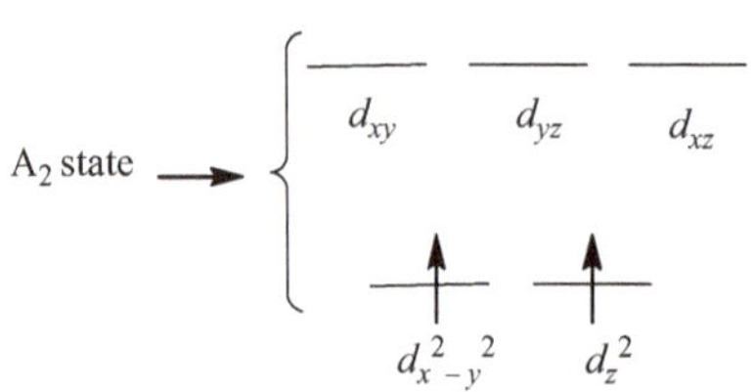

On excitation of one electron, the arrangement of electrons will become $e^1t_2^1$, which corresponds to two triply generate states $T_2(F)$ and $T_1(P)$.

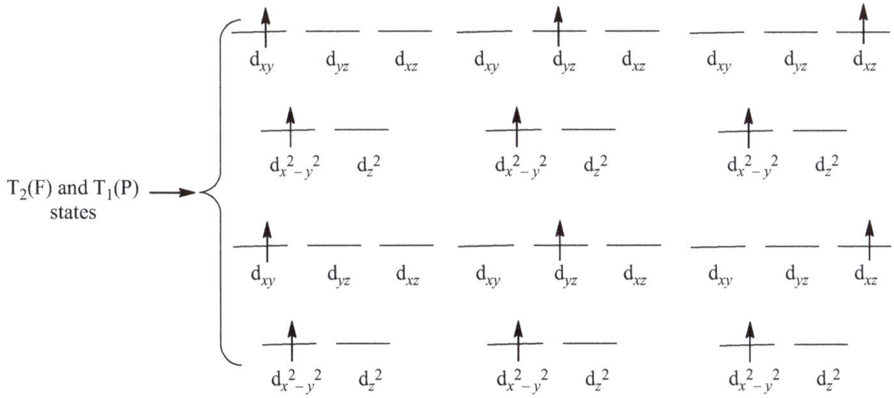

If two electrons are excited, then the electron arrangement is $e^0t_2^2$ and the corresponding triplet state is $T_1(F)$.

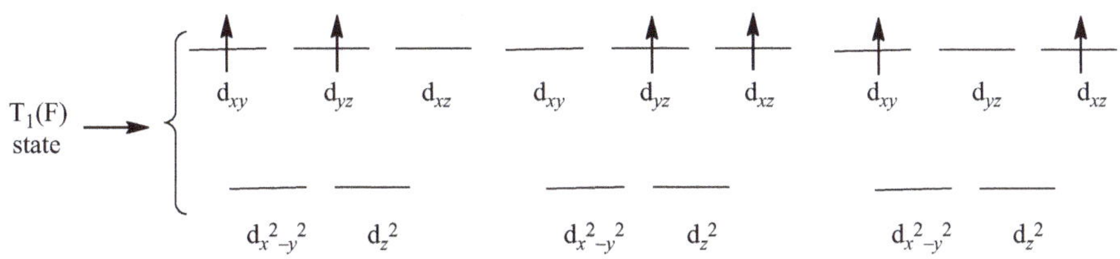

For tetrahedral d^2 complexes, transitions can be identified by the Orgel diagram given in Figure 2.29 (*left side*) as $^3A_2 \rightarrow ^3T_2$ (F), $^3A_2 \rightarrow ^3T_1$ (F) and $^3A_2 \rightarrow ^3T_1$ (P).

Similarly, the splitting pattern for d^3, d^7 and d^8 configuration of metal ion in tetrahedral field can be observed and the pattern is opposite to octahedral field. Thus, for d^3 tetrahedral complexes, the transition can be recognized by the Orgel diagram given in Fig. 2.29 (right side) as $^4T_1$ (F)$\rightarrow ^4T_2$ (F), $^4T_1$ (F)$\rightarrow ^4T_1$ (P) and $^4T_1$ (F)$\rightarrow ^4A_2$ (F).

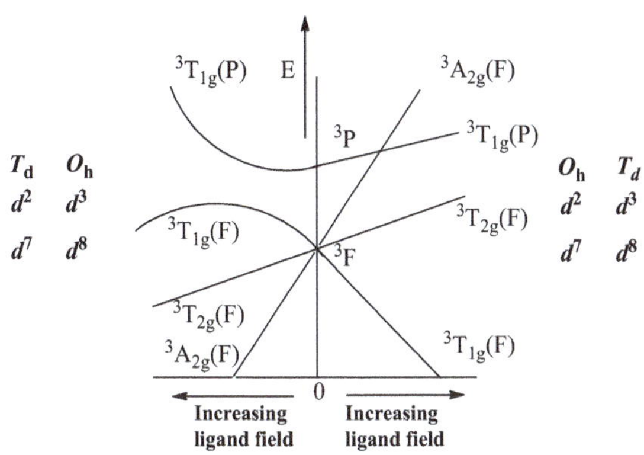

Fig. 2.29: Orgel diagram for d^2, d^3, d^7, d^8 octahedral and tetrahedral complexes. Spin multiplicity will be changed with electronic configuration. Superscript *g* and *u* will be dropped in tetrahedral cases.

In case of d^7 octahedral complexes, the electronic configuration of the metal ion is $t_{2g}^5e_g^2$, for which ^4F is the ground state term. This splits into $^4T_{1g}$ (F) + $^4T_{2g}$ (F) + $^4A_{2g}$ (F). The other term of the same multiplicity ^4P does not split but transforms to $^4T_{1g}$ (P) in octahedral field as shown below in the Orgel diagram (Fig. 2.30).

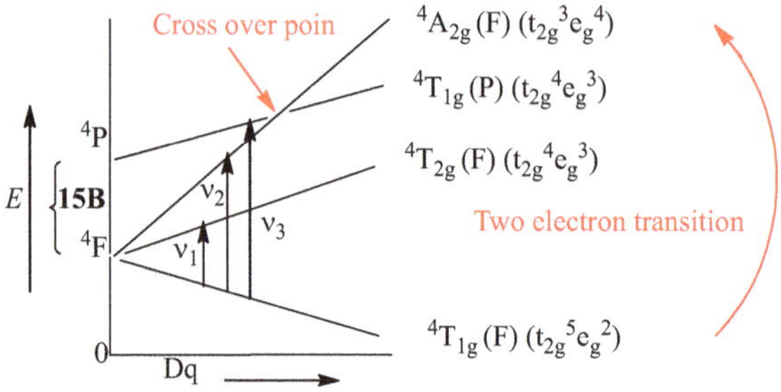

Fig. 2.30: Orgel diagram of d^7 metal ion in octahedral environment.

The six-coordinate high-spin Co(II) is a commonly encountered $3d^7$ system. The aqueous solutions of cobaltous (Co^{2+}) salts containing $[Co(H_2O)_6]^{2+}$ complex ion are pale pink. The electronic spectrum of $[Co(H_2O)_6]^{2+}$ is given in Fig. 2.31. A weak, well-resolved absorption band is observed around 8000 cm^{-1}, and a multiple absorption band comprising of three overlapping peaks at about 20,000 cm^{-1}.

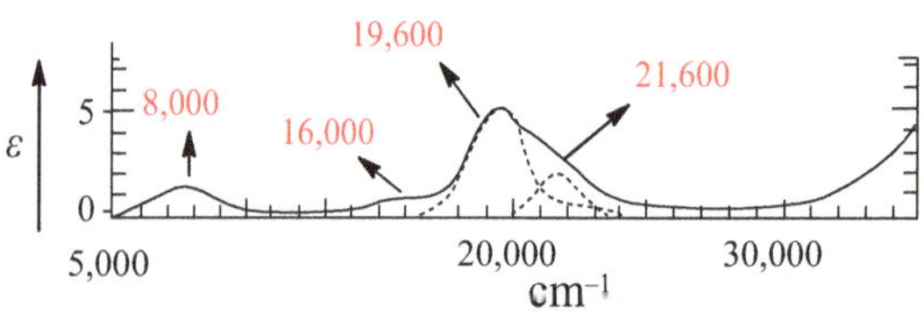

Fig. 2.31: Electronic spectrum of $[Co(H_2O)_6]^{2+}$.

The lowest energy band at 8,000 cm^{-1} is assigned to the $^4T_{1g}$ (F)→$^4T_{2g}$ (F)(ν_1) transition. The multiple bands have three peaks at about 16, 000, 19, 4000 and 21, 600 cm^{-1}. Two of these low energy peaks at 16, 000, 19, 4,000 cm^{-1} are generally accepted as due to $^4T_{1g}$(F)→$^4A_{2g}$(F) (ν_2) and $^4T_{1g}$(F)→$^4T_{1g}$(P) (ν_3), respectively. Since, these two peaks are close together, it indicates that these two transitions are close to the crossover point between the $^4A_{2g}$(F) and $^4T_{1g}$(P) on the energy diagram. The fourth band at 21, 600 cm^{-1} is attributed either to spin-orbit coupling effects or to transitions to doublet states, ^2H, ^2G, ^2F, ^2D and ^2P.

It has frequently been found in the high-spin octahedral cobalt(II) complexes that the ν_3 band involving the one-electron transition is stronger than the ν_2 band involving the two-electron transition, the electronic configurations of the ${}^4T_{1g}(F)$, ${}^4A_{2g}(F)$ and ${}^4T_{1g}(P)$ levels being $t_{2g}{}^5e_g{}^2$, $t_{2g}{}^3e_g{}^4$ and $t_{2g}{}^4e_g{}^3$, respectively (*vide* Fig. 2.30). The ν_2 band, therefore, usually appears as a shoulder or often doesn't appear at all.

In case of d^7 tetrahedral complexes having the electronic configuration of the metal ion $e^4t_2{}^3$, the Orgel diagram is same as for the d^3octahedral complexes, which is shown in Fig. 2.32.

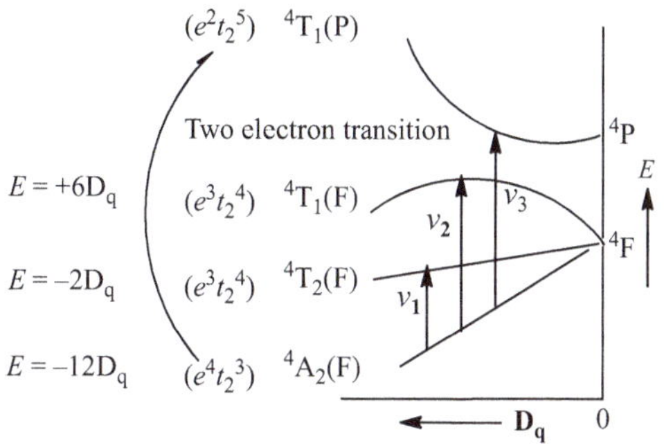

Fig. 2.32: Orgel diagram of d^7 metal ion in tetrahedral environment.

A classic example of a tetrahedral Co(II) complex (d^7) is the intensely blue-colored $[CoCl_4]^{2-}$. It shows two absorption bands (Fig. 2.33) at 5,800 cm^{-1} (in the near IR region ($\varepsilon = 64$) and at 15,000 cm^{-1} (in the visible region, $\varepsilon = 520$). These two bands are assigned as ${}^4A_2(F) \rightarrow {}^4T_1(F)$ (ν_2) and ${}^4A_2(F) \rightarrow {}^4T_1(P)$ (ν_3), respectively

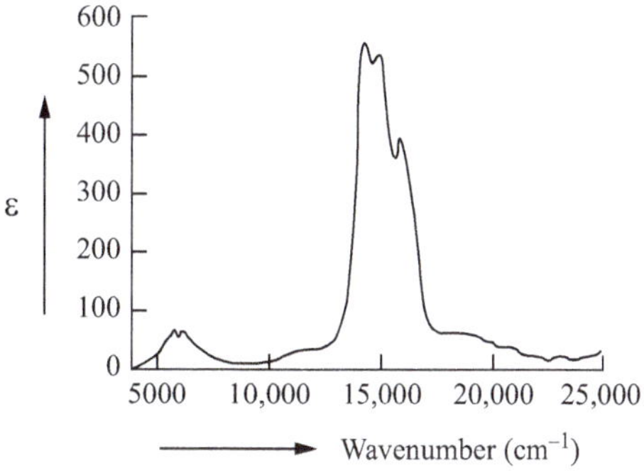

Fig. 2.33: Electronic spectrum of $[CoCl_4]^{2-}$.

The absorption band of lowest energy, that is, due to transition ${}^4A_2(F) \rightarrow {}^4T_2(F)$ is expected at 3,500 cm^{-1}, in the near IR region and is, therefore, out of range of common UV-Vis. spectrophotometer, and is often overlapped by ligand bands due to vibrational transitions. We can calculate 10 Dq from 5,800 cm^{-1} absorption band. Thus, 6Dq-(−12 Dq) = 18 Dq = 5,800 cm^{-1}. Hence, 10 Dq = 5800 ´ 10/18 = 3220 cm^{-1}. The energy gap between $^4A_2(F)$ and $^4T_2(F)$ is 10 Dq. Therefore, first transition should also appear at about 3,200 cm^{-1}. This transition is orbitally forbidden in regular tetrahedral cobalt(II) complexes, although vibronically allowed. It is, therefore, expected to be very weak and is rarely found.

For d^8 tetrahedral complexes, the right half the Orgel diagram (Fig. 2.29) after writing appropriate spin multiplicities on the terms and splitting (Fig. 2.34) can be used to interpret the electronic spectra of tetrahedral complexes. From this diagram, three spin-allowed transitions, $^3T_1(F) \rightarrow {}^3T_2(F)$ (ν_1), $^3T_1(F) \rightarrow {}^3A_2(F)$ (ν_2) and $^3T_1(F) \rightarrow {}^3T_1(P)$ (ν_3), are expected for the tetrahedral complexes.

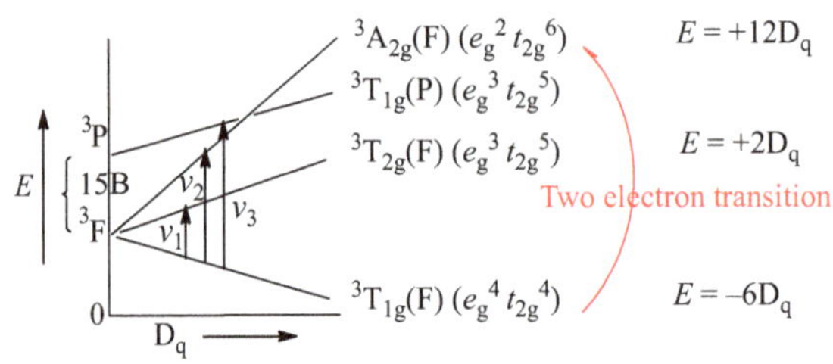

Fig. 2.34: Orgel diagram of d^8 metal ion in tetrahedral environment.

The solution of $[Ni(OAsPh_3)_2Cl_2]$ in benzene exhibits three absorption bands (Fig. 2.35) at 7,800 cm^{-1} (ε = 12), 14,600 cm^{-1} (ε = 70) and 16,400 cm^{-1} (ε = 71) cm^{-1}. The complex has a distorted octahedral structure.

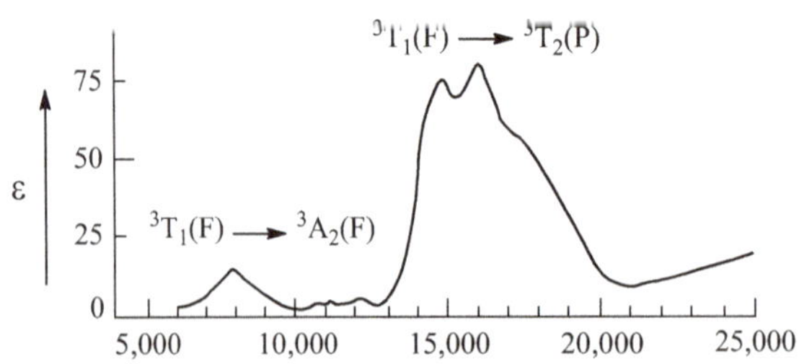

Fig. 2.35: Electronic spectrum of $[Ni(OAsPh_3)_2Cl_2]$.

The first absorption band at 7,800 cm^{-1} in $[Ni(OAsPh_3)_2Cl_2]$ has been assigned as $^3T_1(F) \rightarrow {}^3A_2(F)$ (ν_2) and the two other bands at ~ 15,000 cm^{-1} must represent a single transition $^3T_1(F) \rightarrow {}^3T_1(P)$ (ν_3), split by low symmetry ligand field causing splitting of

$^3T_1(P)$ term. Vibronic or spin-orbit coupling may also be a mechanism of band splitting. The center of gravity of the band is at 16,000 cm^{-1}.

From the electronic spectrum, it is clear that the energy gap corresponding to the transition $^3T_1(F) \rightarrow ^3A_2(F)$ (ν_2) is 12Dq-(−6Dq) = 18 Dq = 7,800 cm^{-1}. Hence, 10 Dq = 7800 × 10/18 = 4,300 cm^{-1}. The ν_2 is a two-electron transition in the strong field limit, and, hence, it is expected to be weaker than others. The band for the transition $^3T_1(F) \rightarrow ^3T_2(F)$ (ν_1) is predicted to occur at ~ 3,500 cm^{-1}, but is not observed as the frequency of absorption of this band falls in the IR region.

The first complex of Ni(II) which was shown to be tetrahedrally coordinated was the blue $[Ni(Cl)_2(PPh_3)_2]$. The electronic spectrum of this compound exhibits three absorption maxima at 10,800, 16,300 and 17,600 cm^{-1}.

Tetrahedral Ni(II) complexes with phosphine oxide, arsine oxide and various halides are known. Ni(II) halide complexes, $[NiX_4]^{2-}$ (X = Cl, Br or I), are of particular interest. By using bulky organic cations, a number of compounds containing $[NiX_4]^{2-}$ tetrahedral group have been synthesized and characterized. In $[NiX_4]^{2-}$, the bands are more intense due to lack of center of symmetry. Orbital mixing in non-centrosymmetric molecules also enhances the intensity.

2.2.10.5 Octahedral complexes with d^5 electronic configuration

The commonly encountered d^5 metal ions are Mn^{2+} and Fe^{3+}. In high-spin octahedral complexes formed by these metal ions involving weak field ligands, viz., $[MnF_6]^{4-}$, $[Mn(H_2O)_6]^{2+}$, $[FeF_6]^{3-}$and so on, there are five unpaired electrons with parallel spins.

Any electronic transition within the d-levels must involve a reversal of spins. Similar to all other spin-forbidden transitions, any absorption band in these complexes will be extremely weak. This is why most of the Mn(II) salts are pale pink, whereas Fe(III) alum is pale violet.

The ground term for the d^5 metal ion is ^6S. The other 11 excited state terms are ^4G, ^4F, ^4D, ^4P, ^2I, ^2H, ^2G, ^2F, ^2D, ^2P and ^2S. Since there are no excited state terms of sextet spin multiplicity, no spin-allowed d-d transitions are possible. Moreover, none of the 11 excited states can be attained without reversing the sign of an electron, and, hence, probability of such transition is extremely low. Of the 11 excited states, the four quartets, ^4G, ^4F, ^4D and ^4P involve the reversal of electron spin only one time. The other seven excited states which are doublets, are doubly spin forbidden, and are unlikely to participate in electronic transitions. In an O_h field, these four quartet states split into 10 states and, hence, up to 10 extremely weak absorption bands may be observed. The spectrum of $[Mn(H_2O)_6]^{2+}$ along with the Orgel diagram for d^5 O_h configuration including only quartet terms are shown in Fig. 2.36.

It is notable from the Orgel diagram that

(i) The ground state term ^6S is not split, and transforms to the $^6A_{1g}$ state which is drawn along the horizontal axis.

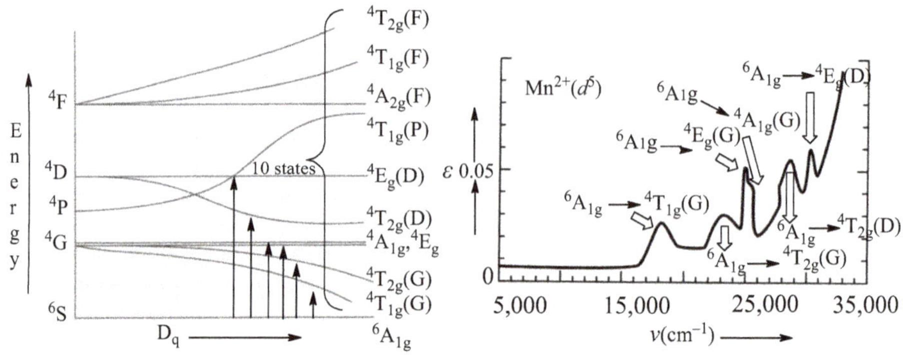

Fig. 2.36: Orgel diagram and electronic spectrum of $[Mn(H_2O)_6]^{2+}$.

(ii) The excited states $^4E_g(G)$, $^4A_{1g}(G)$, $^4E_g(D)$ and $^4A_{2g}(F)$ are also horizontal lines on diagram, and so their energies are independent of the crystal field.

The broadening of the spectral peaks is related to the slop of the exited states on the Orgel diagram. As the slope of the ground state term $^6A_{1g}$ is zero, and the slopes of the $^4E_g(G)$, $^4A_{1g}(G)$, $^4E_g(D)$ and $^4A_{2g}(F)$ terms are also zero, transitions from the ground state to these states should give rise to sharp peaks. By the same reasoning, the transitions from the ground state term $^6A_{1g}$ to states such as $^4T_{1g}(G)$ and $^4T_{2g}(G)$ whose slopes are appreciable, give broader bands.

The electronic spectral bands in $[Mn(H_2O)_6]^{2+}$ are assigned as follows:

$$\begin{aligned}^6A_{1g}\rightarrow &{}^4T_{1g}(G);18,870\ \mathrm{cm}^{-1}\ \text{(Broad)}\\ \rightarrow &{}^4T_{2g}(G);23,120\ \mathrm{cm}^{-1}\ \text{(Broad)}\\ \rightarrow &{}^4E_g(G);24,960\ \mathrm{cm}^{-1}\ \text{(Sharp)}\\ \rightarrow &{}^4A_{1g}(G);25,275\ \mathrm{cm}^{-1}\ \text{(Sharp)}\\ \rightarrow &{}^4T_{2g}(D);27,980\ \mathrm{cm}^{-1}\ \text{(Sharp)}\\ \rightarrow &{}^4E_g(D);29,750\ \mathrm{cm}^{-1}\ \text{(Sharp)}\end{aligned}$$

The same diagram applies to tetrahedral d^5 complexes with omission of g subscripts.

2.3 Effect of Jahn-Teller distortion on electronic spectra of complexes

Jahn-Teller theorem [37] states that for a nonlinear molecule in electronically degenerate electronic state, distortion must occur to lower the symmetry, remove the degeneracy and lower the energy. Such distortions create additional broadening features in the electronic spectra of transition metal complexes because splitting pattern of metal *d*-orbitals changes on account of change in the field symmetry from O_h to D_{4h} due to this distortion. This can be understood as follows how distortion in a nonlinear molecule takes place.

Consider an O_h molecule (nonlinear) in which the ligands on the *z*-axis have moved out as a distortion. As a result, it interacts less with orbitals having *z* components,that is, d_{z^2}, d_{xz} and d_{yz}, and these orbitals are stabilized, and, as a result of center of gravity, the orbitals without z components, that is, $d_{x^2-y^2}$ and d_{xy}, are raised a corresponding amount.

This theorem *per se* (itself) does not predict which type of distortion (*z* ligands out or *z* ligands in) will take place other than that center of gravity will remain.

In the case of "*z*-ligands in" distortion, the splitting is similar but secondary splitting within e_g/E_g and t_{2g}/T_{2g} levels are inverted.

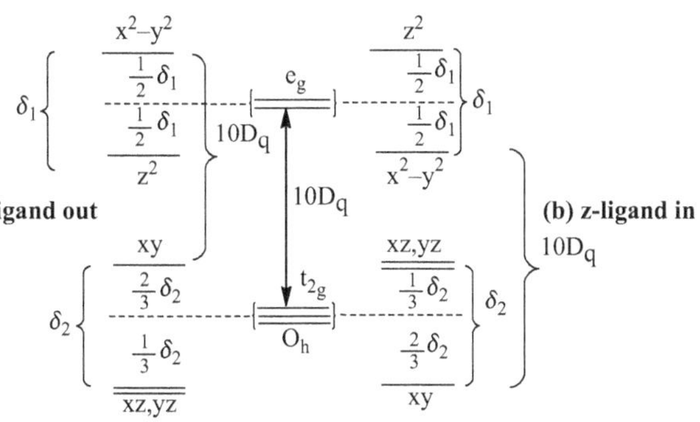

In O_h $[Ti(H_2O)_6]^{3+}$ (d^1), it is easy to see that stabilization of odd electron (1e) in the "z-in" distortion is twice that of "z-out" distortion. Hence, one would expect a distorted octahedron in compound, compressed along the z-axis. The CFSE energy will be $2/3\delta_2$ larger compared to a regular octahedron.

For example

(i) d^1-Metal ion in octahedral environment

Octahedral $[Ti(H_2O)_6]^{3+}$ belongs to d^1 system and it has only one energy term ^2D. The Ti(III) in O_h field has the configuration t_{2g}^1. As the t_{2g} level is triply degenerate electronically in this case, Jahn-Teller theorem forbids/prevents it to be occupied by single electron without undergoing distortion. It is notable here that in Cr(III) ion having d^3 configuration no degeneracy occurs due to three equivalent orbital electronic arrangements d_{xy}^1, d_{xz}^1 andd_{yz}^1.

Although the upper e_g levels are split as well, there is no energetic effect since they are unoccupied in the ground state. The excited state, $t_{2g}^0e_g^1$, will also be subject of distortion because of the electronically degenerate electronic state. The spectrum of

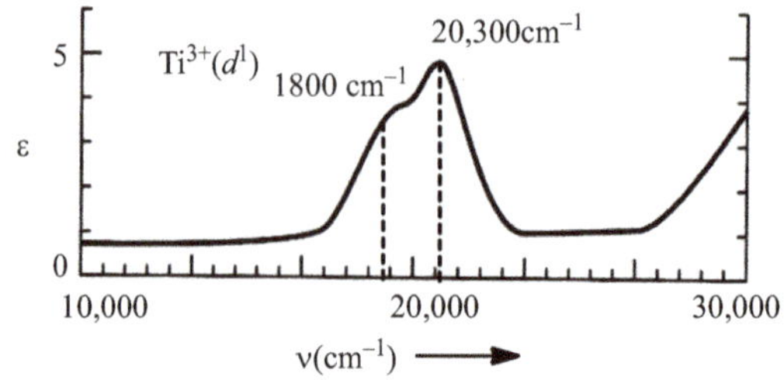

Fig. 2.37: Electronic spectrum of $[Ti(H_2O)_6]^{3+}$ with an absorption peak showing a shoulder.

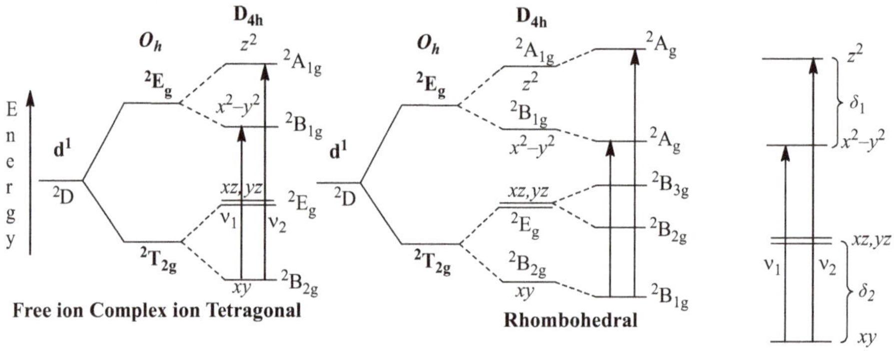

Fig. 2.38: Splitting of T_{2g} and E_g levels due to Jahn- Teller distortion in of $[Ti(H_2O)_6]^{3+}$ resulting two transitions one from the ground state to the lower E_g level $d_{x^2-y^2}$ and the other to the upper d_{z^2} level.

$[Ti(H_2O)_6]^{3+}$ shows the result of this e_g splitting. Instead of a single Gaussian curve, the absorption peak shows a shoulder at a value of some 1,800 cm^{-1} (Fig. 2.37) resulting from the superimposition of two peaks, one from the ground state to the lower e_g level $d_{x^2-y^2}$ and the other to the upper d_{z^2} level as shown in Fig. 2.38. Thus, Jahn-Teller distortion creates additional broadening features in the electronic spectrum of the complex because splitting pattern of metal *d* orbitals changes on account of change in the field symmetry from O_h to D_{4h} either through tetragonal or rhombohedral distortion.

The magnitude of the splitting $\Delta o/\Delta E$ depends on the nature of ligands, and, hence, the frequency of maximum absorption in the spectrum. Thus, main peak occurs at 13,000 cm^{-1} in $[TiCl_6]^{3-}$, 18,900 cm^{-1} in $[TiF_6]^{3-}$, 20,300 cm^{-1} in $[Ti(H_2O)_6]^{3+}$ and 22,300 cm^{-1} in $[Ti(CN)_6]^{3-}$. Similar to $[Ti(H_2O)_6]^{3+}$, spectrum in each case reveals a shoulder band.

Most of the vanadium(IV) complexes contain oxovanadium(IV) group (vanadyl cation, VO^{2+}). Six-coordinate oxovanadium(IV) complexes have compressed tetragonal structure and belong to C_{4v} point group symmetry. The $p\pi$ of oxide oxygen is involved in strong π-bonding with the d_{xz} and d_{yz} orbitals of vanadium(IV). The ordering of the d-orbital splitting for the tetragonally compressed $[VO(H_2O)_5]^{2+}$ ion is shown in Fig. 2.39.

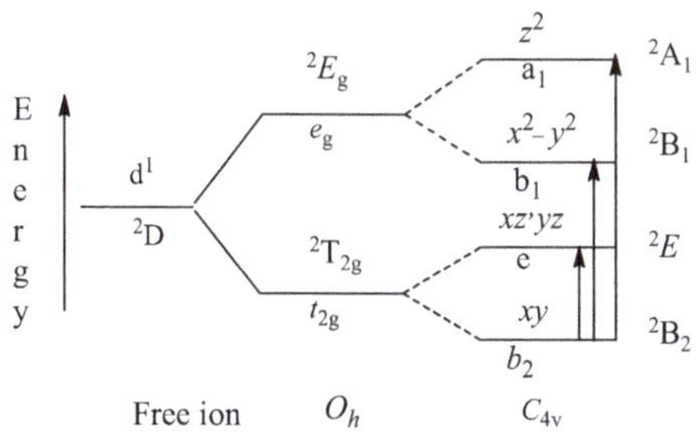

Fig. 2.39: The splitting of d-orbitals in oxovanadium(IV) complexes. Energy states are shown by capital letters.

Typically, three bands are observed in the spectra of oxovanadium(IV) compounds, viz., $[VO(H_2O)_5]^{2+}$,$[VO(NCS)_5]^{3-}$, $[VO(oxalate)_2]^{2-}$ and $[VO(acac)_2]$ (acacH = acetyl acetone) as given in Table 2.9.

Table 2.9: Electronic spectral data of some oxovanadium (IV) complexes.

Compound	$^2B_2 \to {}^2E$ (cm^{-1})	$^2B_2 \to {}^2B_1$ (cm^{-1})	$^2B_2 \to {}^2A_1$ (cm^{-1})
$[VO(H_2O)_5]^{2+}$	13,100	16,000	–
$[VO(NCS)_5]^{3-}$	13,500	17,200	23,900
$[VO(oxalate)_2]^{2-}$	12,600	16,500	29,400
$[VO(acac)_2]$	13,000	17,450	25,630

Ballhausen and Gray, in a molecular orbital study of the $[VO(H_2O)_5]^{2+}$ ion, suggested the *d*-orbital energy order as $b_2 < e_\pi^* < b_1^* < a_1^*$ (Fig. 2.40).

Considering the data given in Table 2.9, the three transitions, $^2B_{2g} \to {}^2E_g$, $^2B_{2g} \to {}^2B_{1g}$ and $^2B_{2g} \to {}^2A_{1g}$, correspond to $b_2 \to e_\pi^*$, $b_2 \to b_1^*$ and $b_2 \to a_1^*$, respectively, as shown by arrows in the molecular orbital diagram given in Fig. 2.40.

The distortion from C_{4v} symmetry could occur especially when L-L is a bidentate ligand in common complexes of the type $[VO(L\text{-}L)_2]$. In such a case, the highest possible symmetry is C_{2v} and the degeneracy of e_π^* orbitals is lifted. Hence, an additional transition may be seen.

(ii) d^6-Metal ion in octahedral environment

(a) Most of the iron (II) complexes are spin-free type. A somewhat strong field is required to cause the formation of low-spin complexes. Octahedral iron(II) complexes usually exhibit a broad band which is usually split into two components. For example, the electronic spectrum of the pale green aqueous $[Fe(H_2O)_6]^{2+}$ complex ion shows a double peak, a band at 10,400 cm^{-1} ($\varepsilon = 1.1$, f = 1.6×10^{-5}) and a shoulder at 8,300 cm^{-1} (Fig. 2.41c).

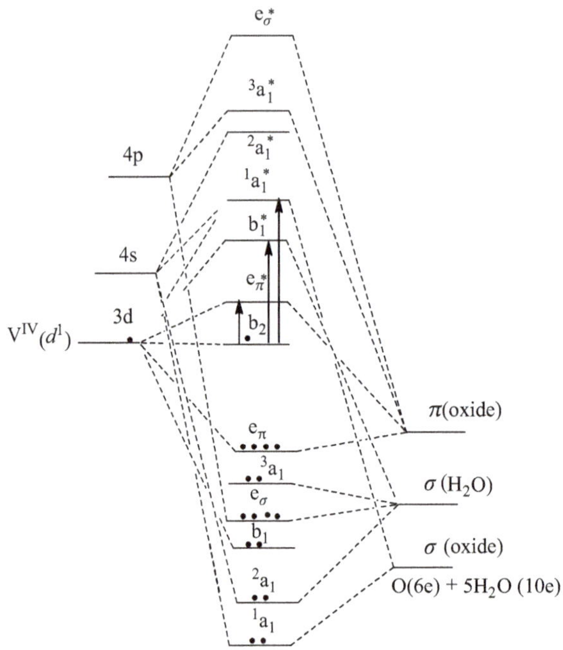

Fig. 2.40: M.O. diagram of tetragonally compressed $[VO(H_2O)_5]^{2+}$ ion (17e) of C_{4v} symmetry (paramagnetic w. r. t one unpaired electron).

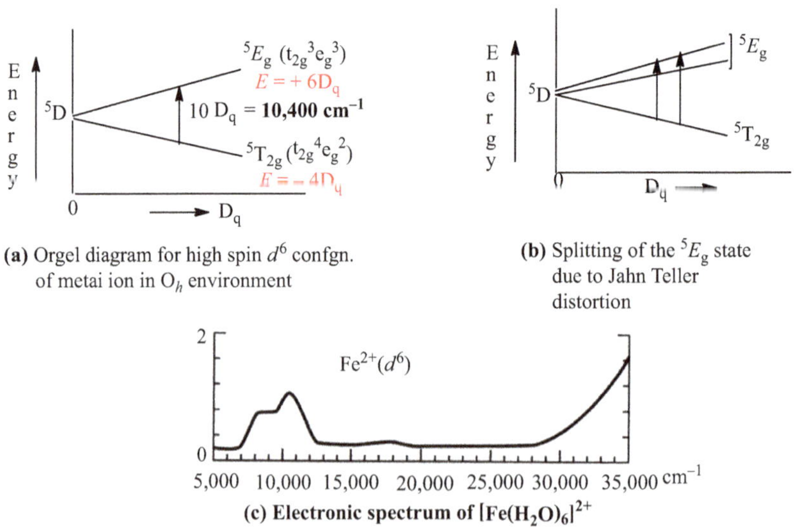

Fig. 2.41: **(a)** Orgel diagram for high spin d^6 confgn. of metai ion in O_h environment **(b)** Splitting of the ^5E_g state due to Jahn Teller distortion **(c) Electronic spectrum of $[Fe(H_2O)_6]^{2+}$.**

The spin multiplicity of the free ion ground term and of the energy levels T_{2g} and E_g obtained from the splitting of free ion term is 5 (Fig. 2.41a). The absorption band observed is, therefore, assigned to $^5T_{2g} \rightarrow {}^5E_g$, which corresponds to 10Dq. Hence, 10Dq for $[Fe(H_2O)_6]^{2+}$ complex ion is 10,400 cm^{-1}. The appearance of a composite band (double peak) in the electronic spectrum (Fig. 2.41c) of the complex ion is due to $^5T_{2g} \rightarrow {}^5E_g$ transition split by ~ 2,000 cm^{-1} owing to Jahn–Teller distortion. This splits up the excited state energy level ^5E_g into two nondegenerate energy levels (Fig. 2.41b).

In complexes which don't have six identical ligands around the metal ion, for example, $[Fe(Cl)_2(H_2O)_4]$, the ^5E_g level undergoes a more pronounced splitting.

(b) The Co^{3+} ion having d^6 configuration is isoelectronic with divalent iron (Fe^{2+}). The well-investigated high-spin octahedral cobalt(III) complexes are blue $[CoF_6]^{3-}$ and $[CoF_3(H_2O)_3]$ species. Their absorption spectra reveal a broad band at around 1,300 cm^{-1} assignable to $^5T_{2g} \rightarrow {}^5E_g$ transition. The absorption band of $[CoF_6]^{3-}$ is split into two distinct peak due to Jahn-Teller effecting the excited ^5E_g state (Fig. 2.42)

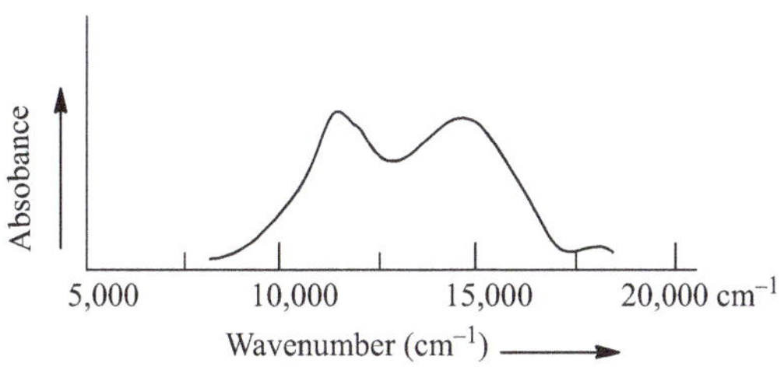

Fig. 2.42: Electronic spectrum of $[CoF_6]^{3-}$.

(iii) d^9-Metal ion in octahedral environment

The complex ion $[Cu(H_2O)_6]^{2+}$ (d^9-configuaration) in aqueous solution absorbs in the visible region around 12,000 cm^{-1} and gives blue-colored solution. The octahedral complexes of ions with a d^9-configuaration can be described in a similar way to the Ti^{3+} octahedral complexes with a d^1 arrangement.

In a d^1 case there is a single electron in the lower t_{2g} /T_{2g} level, while in the d^9 case there is single hole in the upper e_g/E_g level. Thus, the transition in d^1 case is promoting electron from T_{2g} level to E_g level, while in d^9 case it is simpler to consider the promotion of an electron as the transfer of a hole from E_g to T_{2g}. The energy level diagram for d^9 is, therefore, other way round, that is, the inverse of d^1 case (Fig. 2.43).

The absorption spectrum of $[Cu(H_2O)_6]^{2+}$ (Fig. 2.43b) consists of a weak broad band in the visible region at ~12,000 cm^{-1}. It is attributed to the transition $^2E_g \rightarrow {}^2T_{2g}$. The band is asymmetric on account of splitting of a low symmetry ligand field

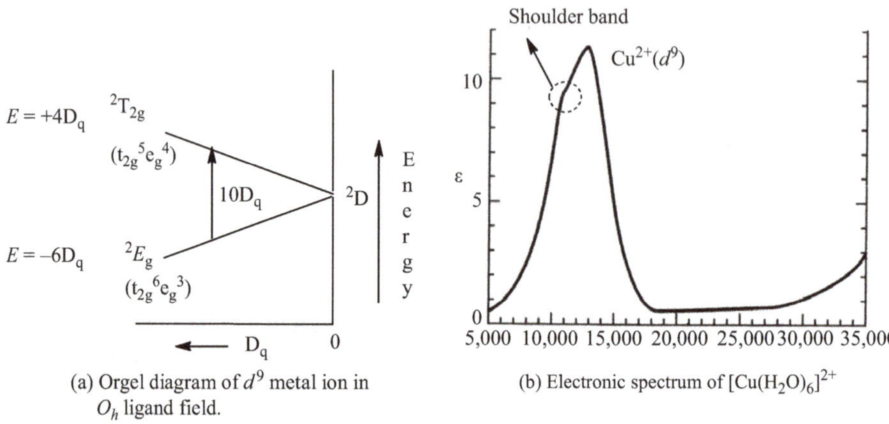

Fig.2.43: (a) Orgel diagram of d^9 metal ion in O_h ligand field. (b) Electronic spectrum of $[Cu(H_2O)_6]^{2+}$.

component ^2E_g owing to Jahn-Teller distortion. It is concluded from the spectrum that, for $[Cu(H_2O)_6]^{2+}$ ion, 10 Dq = ~12,000cm^{-1}.

Copper(II) complexes are known in a wide variety of structures. A very broad band with a peak maximum near 15,000 cm^{-1} is observed for most geometry. Thus, the electronic spectrum of copper(II) is often of little value in structural assignment.

Tetragonal distortion is usually assumed to be the most common example of Cu^{2+} coordination. The energy levels of an axially elongated octahedron are shown in Fig. 2.44.

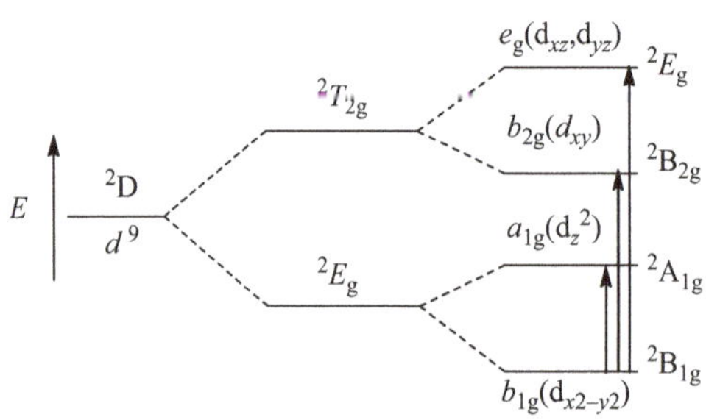

Fig. 2.44: Energy level for axially elongated octahedral Cu(II) complexes of D_{4h} symmetry.

The ground state of Cu(II) ion in an axially elongated tetragonally distorted octahedral crystal field of D_{4h} symmetry may be described as a single electron in $d_{x^2-y^2}$ (b_{1g}) orbital or a $^2B_{1g}$ spectroscopic state. In a tetragonal field, expected transitions are $^2B_{1g} \rightarrow {}^2A_{1g}$, $^2B_{1g} \rightarrow {}^2B_{2g}$ and $^2B_{1g} \rightarrow {}^2E_g$, and they occur in the range 12,000–17,1000, 15,500–18,000, 17,000–20,000 cm^{-1}, respectively.

The absorption spectrum of solid $CuSO_4.5H_2O$ can also be interpreted in terms of D_{4h} symmetry. The crystal structure of $CuSO_4.5H_2O$ has shown that each copper (II) ion is surrounded by four water molecules arranged in square planar geometry, with two sulfate oxygens at a slightly greater distance from the metal ion in *trans*-position as shown in Fig. 2.45.

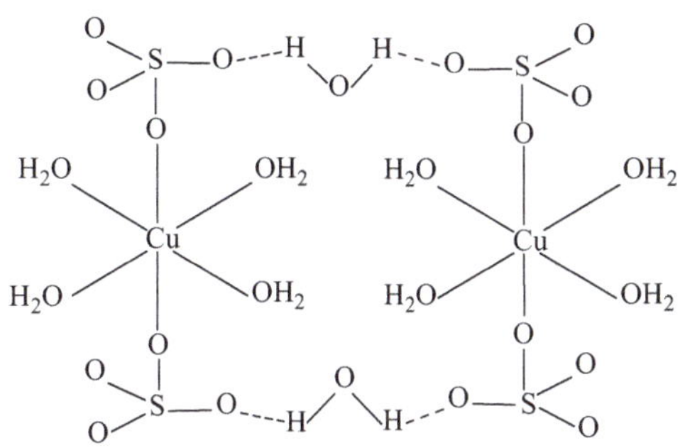

Fig. 2.45: Octahedral environment of Cu^{2+} in $CuSO_4 \cdot 5H_2O$.

The crystal field is then of D_{4h} symmetry. The analysis of the polarized absorption spectrum shows three bands at 10,500, 13,000 and 14,500 cm^{-1}, as expected. The blue color of the salt is due to it absorption at 13,000 cm^{-1}, assignable to $^2B_{1g} \rightarrow ^2B_{2g}$ transition (10 Dq = 13, 000 cm^{-1}). When $CuSO_4.5H_2O$ is heated, it gives $CuSO_4.H_2O$. The monohydrate is no more octahedrally surrounded in lack of four water molecules. Consequently, the splitting of d- orbitals doesn't occur in case of $CuSO_4.H_2O$. The energy levels due to the splitting of d orbitals will also vanish, and, hence, there is no question of occurrence of crystal field electronic transition in the visible region in $CuSO_4.H_2O$. This is why $CuSO_4.H_2O$ is colorless.

It is worthwhile to mention here that one more sulfate of copper, namely, Cu_2SO_4 is colorless. Obviously, it is so because copper(I) in the present case has a d^{10} configuration. Hence, no transition of d-d origin will occur in its electronic spectrum.

Although two types of distorted octahedral complexes of Cu(II) may be the outcome of Jahn Teller distortion, one axially elongated and other axially compressed. So, far, only one example of axially compressed octahedron namely, K_2CuF_4 is known (2Fs at 1.95 Å and 4Fs at 2.08 Å. The crystal structure of this compound shows that each Cu^{2+} ion has six F^- ions as immediate neighbors, as shown in Fig. 2.46.

Axially elongated octahedral complexes are numerous in number.

(iv) d^4-Metal ion in octahedral environment

Typical metal ions of high-spin $3d^4$ configuration are Cr(II) and Mn(III). The spectral measurements on Cr(II) complexes are difficult due to ease of their oxidation to Cr(III).

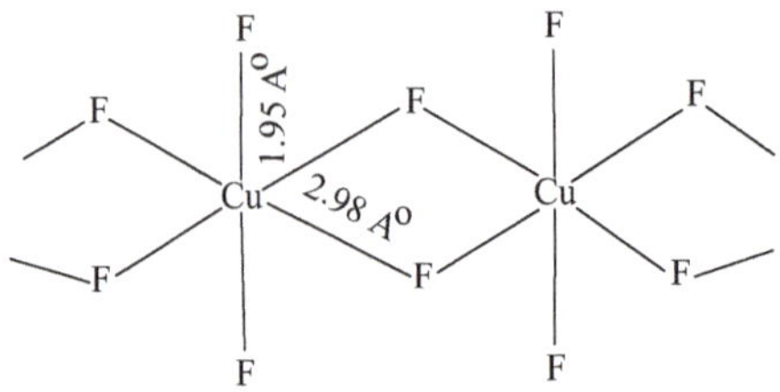

Fig. 2.46: Crystal structure of $[CuF_4]^{2-}$ in solid state.

Aqueous solutions of Cr^{2+} ion containing $[Cr(H_2O)_6]^{2+}$ is pale blue in color. The absorption spectrum (Fig. 2.47) of such a solution shows a broad weak band near 14,000 cm^{-1} ($\varepsilon = 5.2$, $f = 1 \times 10^{-4}$).

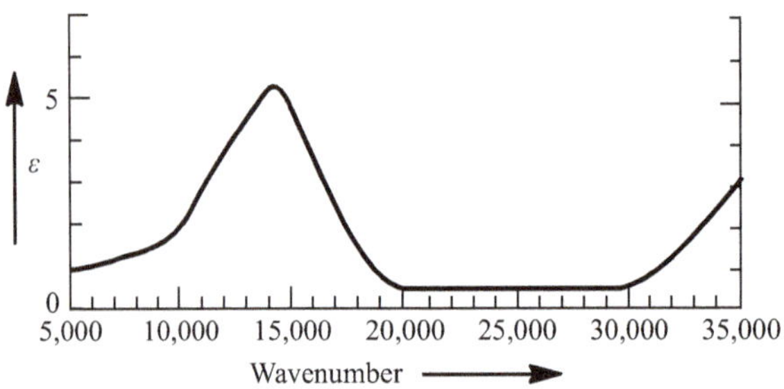

Fig. 2.47: Electronic spectrum of $[Cr(H_2O)_6]^{2+}$.

O_h complexes of d^4 configuration in a high-spin arrangement may be considered as having one hole in the upper e_g level, and, thus, they are analogous to the d^9 configuration. The Orgel diagram (Fig. 2.48) after writing down a spin multiplicity of 5 over each term/energy level, can be used for assigning absorption bands in octahedral complexes of Cr(II) as well as Mn(III). The absorption band at 14,000 cm^{-1} in the spectrum of $[Cr(H_2O)_6]^{2+}$, therefore, appears due to $^5E_g \rightarrow {}^5T_{2g}$ transition.

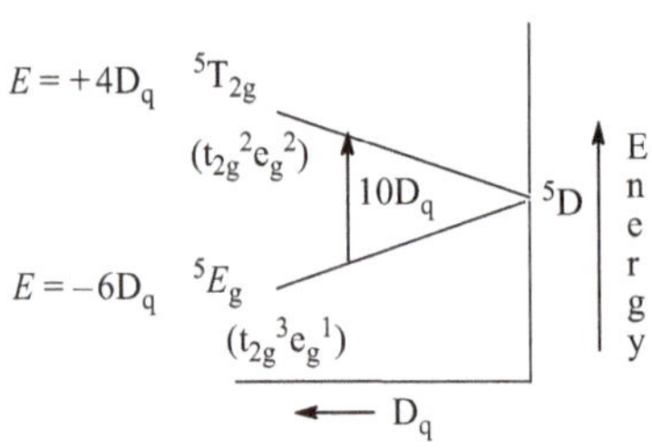

Fig. 2.48: Orgel diagram for d^4 configuration of metal ion in O_h ligand field.

The high-spin octahedral complexes of Cr(II) and Mn(III) of d^4 configuration usually show appreciable splitting of absorption band as a result of strong Jahn-Teller splitting of ^5E_g state resembling those of Cu(II) O_h complexes.

The crystal structures of CrF_2 and MnF_3 for example, have shown highly tetragonally distorted octahedral geometry and these are interpreted owing to Jahn-Teller distortion. The crystal structure [Fig. 2.49, T. Chatterji and T. C. Hansen, J. Phys.: Condensd matter, 23 (27) (2011)] of CrF_2 shows that four Cr–F bonds have length ~2.0 Å and the other two Cr–F bonds have length 2.43 Å. A Jahn-Teller splitting of the ^5E_g state is calculated to be 6,000 cm^{-1} in CrF_2. Evidence for Jahn-Teller distortion is also observed in the tetragonal distortion of MnF_3 octahedron.

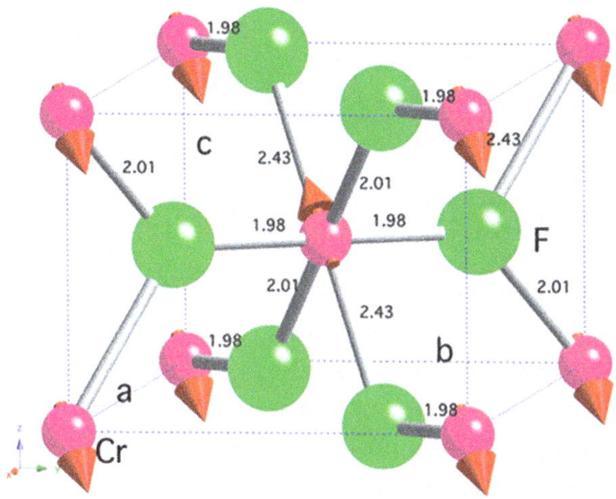

Fig. 2.49: Crystal structure of CrF_2.

The splitting of the ^5D term under tetragonal distortion (D_{4h}) of the O_h symmetry is shown in Fig. 2.50.

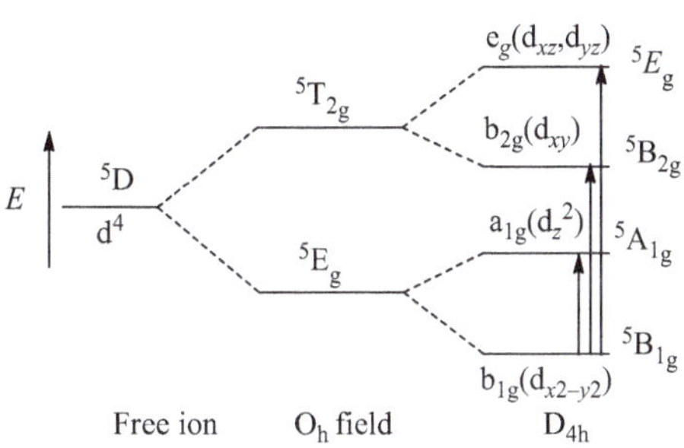

Fig. 2.50: Effect of Jahn-Teller distortion on d^4 metal ion.

Three bands arising from the transitions, $^5B_{1g}{\rightarrow}^5A_{1g}$ (ν_1), $^5B_{1g}{\rightarrow}^5B_{2g}$ (ν_2) and $^5B_{1g}{\rightarrow}^5E_g$ (ν_3) may be observed under favorable cases. The spectral data of some high-spin chromium(II) and manganese(III) complexes are presented in Table 2.10.

Table 2.10: Electronic spectral data of some oxovanadium (IV) complexes.

Compound	$^5B_{1g}{\rightarrow}^5A_{1g}$ (ν_1) (cm^{-1})	$^5B_{1g}{\rightarrow}^5B_{2g}$ (ν_2) (cm^{-1})	$^5B_{1g}{\rightarrow}^5E_g$ (ν_3) (cm^{-1})
$[Cr(H_2O)_6]^{2+}$	9,500	14,000	–
$[Mn(H_2O)_6]^{3+}$	–	–	21,000
CrF_2	11,500	14,700	–
$[MnF_6]^{3-}$	9,000	17,400	19,600

2.4 Correlation diagram: ordering of energy states

So far, we have discussed how free ion terms split on lowering the symmetry of chemical environment in metal complexes using only weak ligand field approximation. The next question obviously arises about the nature and exact ordering of the energy levels when the metal is subject to varying crystal fields starting from weak field limit to the infinitely strong field limit. It may be possible to find ligands capable of producing the crystal fields of all strengths. So, we can have a logical and gradual transition from the weak through medium to strong ligand fields. At the extremely strong fields, the metal ion configuration is subject first to crystal field interaction producing what are called "strong field interactions", which are in turn affected by the inter-electronic repulsion perturbation. The sequence of these interactions lead to the energy levels which must correlate in "number" and "nature" with those obtained from the weak field approximation.

The diagram demonstrating correspondence/conformity between the energy levels derived from the weak and strong field treatments of the metal ion configuration is called a *correlation diagram*. The lines joining both the extremes thus represent the *energy level ordering for the complexes* of medium or intermediate field strength.

This diagram will show how free ion terms are splitted on lowering the symmetry and how the energy of these terms vary with the varying ligand field. In this diagram, *x*-axis (abscissa) represents ligand field strength Δ_o or Δ_t and y-axis (ordinate) represents energy of the terms. On the extreme left of the diagram, we have energies of the free ion terms (spherical symmetry) where Δ_o or Δ_t is zero. On the extreme right of the diagram, we have energies of the states under the strong ligand field environment, that is, separation between e_g and t_{2g} (or *e* and t_2) orbitals will be maximum and inter-electronic repulsion between electrons of e_g and t_{2g}

(or e and t_2) orbitals will be minimum or negligible. Let us construct correlation diagram for d^n configuration in octahedral (O_h) and tetrahedral (T_d) ligand fields.

(i) Correlation diagram for d^2- configuration in O_h field

For d^2 configuration, free ion terms are ^3F, ^3P, ^1D, ^1G and ^1S. We have already discussed earlier in the previous section the splitting of these states in weak field. When ligand field is very strong, the five degenerate d-orbitals of free metal ion splits into t_{2g} and e_g set of orbitals. In this situation, the possible arrangements of two electrons can be written as $t_{2g}^2e_g^0$ (ground state), $t_{2g}^1e_g^1$ (first excited state) and $t_{2g}^0e_g^2$ (second excited state).

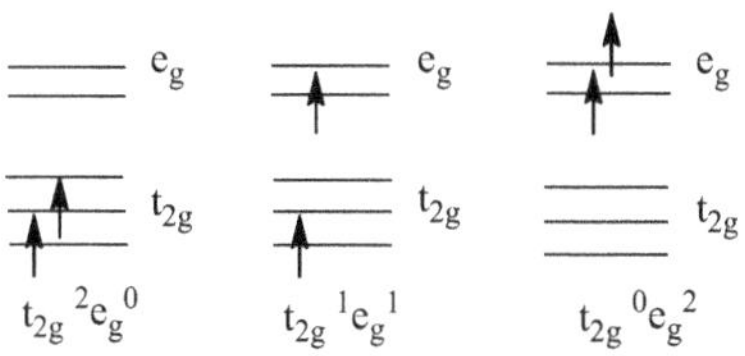

The electrons in different configuration will have spin-orbital interactions giving rise to microstates. We are aware that the d^2-configuration gives rise to a total of 45 microstates. So, if there are no other strong field configurations missing from what were written above, then the sum of the microstates resulting from each of these configurations should be equal to 45. We can use the following formula for calculating the number of microstates for each of these configurations:

$$\text{No. of microstates} = \frac{n!}{r!\,(n-r)!}$$

Where n is twice the number of orbitals and r is the number of electrons and ! is the sign of factorial. Thus, we get the number of microstates for

$$t_{2g}^2e_g^0\text{configuration} = \frac{2\times 3!}{2!\,(2\times 3-2)!} = \frac{6\times 5\times 4\times 3\times 2\times 1}{2\times 1\times 4\times 3\times 2\times 1} = 15$$

$$t_{2g}^1e_g^1\text{configuration} = \left[\frac{2\times 3!}{1!\,(2\times 3-1)!}\right]\left[\frac{2\times 2!}{1!\,(2\times 2-1)!}\right] = \left[\frac{6\times 5\times 4\times 3\times 2\times 1}{1\times 5\times 4\times 3\times 2\times 1}\right]\left[\frac{4\times 3\times 2\times 1}{1\times 3\times 2\times 1}\right]$$

$$= 6\times 4 = 24$$

$$t_{2g}^0e_g^2\text{configuration} = \left[\frac{2\times 2!}{2!\,(2\times 2-2)!}\right] = \left[\frac{4\times 3\times 2\times 1}{2\times 1\times 2\times 1}\right] = 6$$

$$\text{Total number of microstates} = 15 + 24 + 6 = 45$$

This shows that the strong field configurations have been completely written. The next step is to determine the symmetry/nature and number of the states obtained in each strong configuration. For this, we first split each strong field configuration into one-electron wave functions as follows:

$$t_{2g}^{\,2} = t_{2g}^{\,1} \times t_{2g}^{\,1} = T_{2g} \times T_{2g}$$
$$t_{2g}^{\,1}e_g^{\,1} = t_{2g}^{\,1} \times e_g^{\,1} = T_{2g} \times E_g$$
$$e_g^{\,2} = e_g^{\,1} \times e_g^{\,1} = E_g \times E_g$$

In order to obtain the number and symmetry of the states, we first multiply the characters of the operations of the corresponding irreducible representation in the character table of O_h point group. This gives the total character of reducible representation for two electrons. Then applying the reduction formula, the total character of the two electrons is reduced into sum of the irreducible representation of the group.

(i) t_{2g} configuration

Let us first consider the $t_{2g}^{\,2}$ configuration in which both the electrons have characters for each operation of the T_{2g} irreducible representation in the character table of O_h point group. One can consider either only rotation group O instead of full O_h point group or full O_h point group to determine the total character of reducible representation for two electrons in $t_{2g}^{\,2}$ by direct product method.

A theorem on which the direct product method is based is that "the characters of representation of a direct product are equal to the product of characters of the representations based on the individual set of functions", that is,

$$\chi_{\text{direct product}}(R) = \chi_a(R).\chi_b(R).\chi_c(R).\chi_d(R)\ldots\ldots\chi_i(R)$$

So, we can get the direct product $T_{2g} \times T_{2g}$ in the O_h point group and obtain the total character of the reducible representation $\Gamma_{T2g.T2g}$ as:

O_h	E	$8C_3$	$3C_2$	$6C_4$	$6C_2'$	i	$6S_4$	$8S_6$	$3\sigma_h$	$6\sigma_d$
T_{2g}	3	0	−1	−1	1	3	−1	0	−1	1
$T_{2g}.T_{2g}$	9	0	1	1	1	9	1	0	1	1

The above reducible representation ($T_{2g}.T_{2g}$) can be reduced to irreducible representations using the reduction formula given below:

$$N_i = \frac{1}{h}\sum_R \chi(R)\,.\,n\,.\chi_i(R)$$

Where N_i is the number of times the ith irreducible representation occurs in a reducible representation, h is the order of group (number of symmetry operations), $\chi(R)$ is the character of a particular operation in the reducible representation, n is the number of operation of that type and $\chi_i(R)$ is the character of the same operation in the irreducible representation.

Reducible representation for in question

O_h	E	$8C_3$	$3C_2(=C_4^2)$	$6C_4$	$6C_2'$	i	$6S_4$	$8S_6$	$3\sigma_h$	$6\sigma_d$
$T_{2g}.T_{2g}$	9	0	1	1	1	9	1	0	1	1

Character Table of O_h Point Group

O_h	E	$8C_3$	$3C_2$	$6C_4$	$6C_2'$	i	$6S_4$	$8S_6$	$3\sigma_h$	$6\sigma_d$		
A_{1g}	1	1	1	1	1	1	1	1	1	1		$x^2+y^2+z^2$
A_{2g}	1	1	1	−1	−1	1	−1	1	1	−1		
E_g	2	−1	2	0	0	2	0	−1	2	0		$(2z^2-x^2-y^2, x^2-y^2)$
T_{1g}	3	0	−1	1	−1	3	1	0	−1	−1	(R_x, R_y, R_x)	
T_{2g}	3	0	−1	−1	1	3	−1	0	−1	1		(xy, yz, xy)
A_{1u}	1	1	1	1	1	−1	−1	−1	−1	−1		
A_{2u}	1	1	1	−1	−1	−1	1	−1	−1	1		
E_u	2	−1	2	0	0	−2	0	1	−2	0		
T_{1u}	3	0	−1	1	−1	−3	−1	0	1	1	(x,y,z)	
T_{2u}	3	0	−1	−1	1	−3	1	0	1	−1		

Using the reduction formula, we get:

$$NA_{1g} = 1/48\ [(9.1.1) + (0.8.1) + (1.3.1) + (1.6.1) + (1.6.1) + (9.1.1) + (1.6.1) + (0.8.1) + (1.3.1) + (1.6.1)$$

$$= 1/48[(9+0+3+6+6+9+6+0+3+6)]$$

$$= 1/48[48] = 1$$

$$NA_{2g} = 1/48\ [(9.1.1) + (0.8.1) + (1.3.1) + (1.6.-1) + (1.6.-1) + (9.1.1) + (1.6.-1) + (0.8.1) + (1.3.1) + (1.6.-1)$$

$$= 1/48[(9+0+3-6-6+9-6+0+3-6)]$$

$$= 1/48[0] = 0$$

$$NE_g = 1/48\ [(9.1.2) + (0.8.-1) + (1.3.2) + (1.6.0) + (1.6.0) + (9.1.2) + (1.6.0) + (0.8.-1) + (1.3.2) + (1.6.0)$$

$$= 1/48[(18+0+6+0+0+18+0+0+6+0)]$$

$$= 1/48[48] = 1$$

$$NT_{1g} = 1/48\ [(9.1.3) + (0.8.0) + (1.3.-1) + (1.6.1) + (1.6.-1) + (9.1.3) + (1.6.1) + (0.8.0) + (1.3.-1) + (1.6.-1)$$

$$= 1/48[(27+0-3+6-6+27+6+0-3-6)]$$

$$= 1/48[48] = 1$$

$$\begin{aligned} NT_{2g} &= 1/48\ [(9.1.3) + (0.8.0) + (1.3.-1) + (1.6.-1) + (1.6.1) + (9.1.3) + (1.6.-1) \\ &\quad + (0.8.0) + (1.3.-1) + (1.6.1) \\ &= 1/48[(27 + 0 - 3 - 6 + 6 + 27 - 6 + 0 - 3 + 6)] \\ &= 1/48[48] = 1 \end{aligned}$$

$$\begin{aligned} NA_{1u} &= 1/48\ [(9.1.1) + (0.8.1) + (1.3.1) + (1.6.1) + (1.6.1) + (9.1.-1) + (1.6.-1) \\ &\quad + (0.8.-1) + (1.3.-1) + (1.6.-1) \\ &= 1/48[(9 + 0 + 3 + 6 + 6 - 9 - 6 + 0\ - 3 - 6)] \\ &= 1/48[0] = 0 \end{aligned}$$

$$\begin{aligned} NA_{2u} &= 1/48\ [(9.1.1) + (0.8.1) + (1.3.1) + (1.6.-1) + (1.6.-1) + (9.1.-1) + (1.6.1) \\ &\quad + (0.8.-1) + (1.3.-1) + (1.6.1) \\ &= 1/48[(9 + 0 + 3 - 6 - 6 - 9 + 6 + 0\ - 3 + 6)] \\ &= 1/48[0] = 0 \end{aligned}$$

$$\begin{aligned} N_{Eu} &= 1/48\ [(9.1.2) + (0.8.0) + (1.3.2) + (1.6.0) + (1.6.0) + (9.1.-2) + (1.6.0) \\ &\quad + (0.8.1) + (1.3.-2) + (1.6.0) \\ &= 1/48[(18 + 0 + 6 + 0 + 0\ - 18 + 0 + 0 - 6 + 0)] \\ &= 1/48[0] = 0 \end{aligned}$$

$$\begin{aligned} NT_{1u} &= 1/48\ [(9.1.3) + (0.8.0) + (1.3.-1) + (1.6.1) + (1.6.-1) + (9.1.-3) + (1.6.-1) \\ &\quad + (0.8.0) + (1.3.1) + (1.6.1) \\ &= 1/48[(27 + 0 - 3 + 6 - 6 - 27 - 6 + 0\ + 3 + 6)] \\ &= 1/48[0] = 0 \end{aligned}$$

$$\begin{aligned} NT_{2u} &= 1/48\ [(9.1.3) + (0.8.0) + (1.3.-1) + (1.6.-1) + (1.6.1) + (9.1.-3) + (1.6.1) \\ &\quad + (0.8.0) + (1.3.1) + (1.6.-1) \\ &= 1/48[(27 + 0 - 3 - 6 + 6 - 27 + 6 + 0\ + 3 - 6)] \\ &= 1/48[0] = 0 \end{aligned}$$

The above calculations suggest that the reducible representation ($T_{2g} \times T_{2g}$) in question is reduced to:

$$\Gamma_{T2g.T2g} = \Gamma_{T2g}{}^2 = A_{1g} + E_g + T_{1g} + T_{2g}$$

These are the orbital symmetries of the spectral states in $t_{2g}{}^2$ case.

(ii) $t_{2g}^1\ e_g^1$configuration

Here one electron is excited to higher energy level. To get the term arising from this configuration, we take the direct product $T_{2g}\times E_g$ as done above.

O_h	E	8	3C$_2$	6C$_4$	6C$_4$'	i	6S$_4$	8S$_6$	3σ$_h$	6σ$_d$
T_{2g}	3	0	−1	−1	1	3	−1	0	−1	1
E_g	2	−1	2	0	0	2	0	−1	2	0
T_{2g} x E_g	6	0	−2	0	0	6	0	0	−2	0

The above reducible representation ($T_{2g}\times E_g$) can be reduced to irreducible representations using the reduction formula given below:

$$N_i = \frac{1}{h}\sum_R \chi(R).n.\chi_i(R)$$

Reducible representation for in question

O_h	E	8C$_3$	3C$_2$	6C$_4$	6C$_2$'	i	6S$_4$	8S$_6$	3σ$_h$	6σ$_d$
T_{2g}xE_g	6	0	−2	0	0	6	0	0	−2	0

Character Table of O_h Point Group

O_h	E	8C$_3$	3C$_2$	6C$_4$	6C$_2$'	i	6S$_4$	8S$_6$	3σ$_h$	6σ$_d$		
A_{1g}	1	1	1	1	1	1	1	1	1	1		$x^2+y^2+z^2$
A_{2g}	1	1	1	−1	−1	1	−1	1	1	−1		
E_g	2	−1	2	0	0	2	0	−1	2	0		$(2z^2-x^2-y^2, x^2-y^2)$
T_{1g}	3	0	−1	1	−1	3	1	0	−1	−1	(R_x, R_y, R_z)	
T_{2g}	3	0	−1	−1	1	3	−1	0	−1	1		(xy, yz, xy)
A_{1u}	1	1	1	1	1	−1	−1	−1	−1	−1		
A_{2u}	1	1	1	−1	−1	−1	1	−1	−1	1		
E_u	2	−1	2	0	0	−2	0	1	−2	0		
T_{1u}	3	0	−1	1	−1	−3	−1	0	1	1	(x,y,z)	
T_{2u}	3	0	−1	−1	1	−3	1	0	1	−1		

Using the reduction formula, we get:

$$\begin{aligned} NA_{1g} &= 1/48\ [(6.1.1)+(0.8.1)+(-2.3.1)+(0.6.1)+(0.6.1)+(6.1.1)+(0.6.1)+(0.8.1) \\ &\quad +(-2.3.1)+(0.6.1) \\ &= 1/48[(6+0\ -6+0+0+6+0+0\ -6+0)] \\ &= 1/48[0]=0 \end{aligned}$$

$$NA_{2g} = 1/48\ [(6.1.1) + (0.8.1) + (-2.3.1) + (0.6.-1) + (0.6.-1) + (6.1.1) + (0.6.-1) + (0.8.1) + (-2.3.1) + (0.6.-1)$$

$$= 1/48[(6+0-6+0+0+6+0+0-6+0)]$$

$$= 1/48[0] = 0$$

$$NE_g = 1/48\ [(6.1.2) + (0.8.-1) + (-2.3.2) + (0.6.0) + (0.6.0) + (6.1.2) + (0.6.0) + (0.8.-1) + (-2.3.2) + (0.6.0)$$

$$= 1/48[(12+0-12+0+0+12+0+0-12+0)]$$

$$= 1/48[0] = 0$$

$$NT_{1g} = 1/48\ [(6.1.3) + (0.8.0) + (-2.3.-1) + (0.6.\ 1) + (0.6.-1) + (6.1.3) + (0.6.1) + (0.8.0) + (-2.3.-1) + (0.6.-1)$$

$$= 1/48[(18+0+6+0+0+18+0+0+6+0)]$$

$$= 1/48[48] = 1$$

$$NT_{2g} = 1/48\ [(6.1.3) + (0.8.0) + (-2.3.-1) + (0.6.-1) + (0.6.-1) + (6.1.3) + (0.6.-1) + (0.8.0) + (-2.3.-1) + (0.6.1)$$

$$= 1/48[(18+0+6+0+0+18+0+0+6+0)]$$

$$= 1/48[48] = 1$$

$$NA_{1u} = 1/48\ [(6.1.1) + (0.8.1) + (-2.3.1) + (0.6.1) + (0.6.1) + (6.1.-1) + (0.6.-1) + (0.8.-1) + (-2.3.-1) + (0.6.-1)$$

$$= 1/48[(6+0-6+0+0-6+0+0+6+0)]$$

$$= 1/48[0] = 0$$

$$NA_{2u} = 1/48\ [(6.1.1) + (0.8.1) + (-2.3.1) + (0.6.-1) + (0.6.-1) + (6.1.-1) + (0.6.1) + (0.8.-1) + (-2.3.-1) + (0.6.1)$$

$$= 1/48[(6+0-6+0+0-6+0+0+6+0)]$$

$$= 1/48[0] = 0$$

$$N_{Eu} = 1/48\ [(6.1.2) + (0.8.-1) + (-2.3.2) + (0.6.0) + (0.6.0) + (6.1.-2) + (0.6.0) + (0.8.1) + (-2.3.-2) + (0.6.0)$$

$$= 1/48[(12+0-12+0+0-12+0+0+12+0)]$$

$$= 1/48[0] = 0$$

$$NT_{1u} = 1/48\ [(6.1.3) + (0.8.0) + (-2.3.-1) + (0.6.1) + (0.6.-1) + (6.1.-3) + (0.6.-1)$$
$$+ (0.8.0) + (-2.3.1) + (0.6.1)$$
$$= 1/48[(18 + 0 + 6 + 0 + 0 - 18 + 0 + 0 - 6 + 0)]$$
$$= 1/48[0] = 0$$

$$NT_{2u} = 1/48\ [(6.1.3) + (0.8.0) + (-2.3.-1) + (0.6.-1) + (0.6.1) + (6.1.-3) + (0.6.1)$$
$$+ (0.8.0) + (-2.3.1) + (0.6.-1)$$
$$= 1/48[(18 + 0 + 6 + 0 + 0 - 18 + 0 + 0 - 6 + 0)]$$
$$= 1/48[0] = 0$$

The above calculations indicate that the reducible representation ($T_{2g}\times E_g$) in question is reduced to:

$$\Gamma_{T2g.Eg} = T_{1g} + T_{2g}$$

These are the orbital symmetries of the spectral states in $\mathbf{t_{2g}}^{1}\ \mathbf{e_g}^{1}$ case.

(iii) e_g^2 configuration

Here both the electrons are present in the excited energy level e_g. To get the term arising from this configuration, we take the direct product $E_g\times E_g$ as done above.

O_h	E	$8C_3$	$3C_2$	$6C_4$	$6C_2'$	i	$6S_4$	$8S_6$	$3\sigma_h$	$6\sigma_d$
E_g	2	−1	2	0	0	2	0	−1	2	0
E_g	2	−1	2	0	0	2	0	−1	2	0
$E_g \times E_g$	4	1	4	0	0	4	0	1	4	0

The reducible representation ($E_g\times E_g$) can be reduced to irreducible representations using the reduction formula given below:

$$N_i = \frac{1}{h}\sum_R \chi(R)\,.\,n.\chi_i(R)$$

Reducible representation for in question

O_h	E	$8C_3$	$3C_2$	$6C_4$	$6C_2'$	i	$6S_4$	$8S_6$	$3\sigma_h$	$6\sigma_d$
$E_g \times E_g$	4	1	4	0	0	4	0	1	4	0

Character Table of O_h Point Group

O_h	E	$8C_3$	$3C_2$	$6C_4$	$6C_2'$	i	$6S_4$	$8S_6$	$3\sigma_h$	$6\sigma_d$		
A_{1g}	1	1	1	1	1	1	1	1	1	1		$x^2+y^2+z^2$
A_{2g}	1	1	1	−1	−1	1	−1	1	1	−1		
E_g	2	−1	2	0	0	2	0	−1	2	0		$(2z^2-x^2-y^2, x^2-y^2)$
T_{1g}	3	0	−1	1	−1	3	1	0	−1	−1	(R_x, R_y, R_z)	
T_{2g}	3	0	−1	−1	1	3	−1	0	−1	1		(xy, yz, xy)
A_{1u}	1	1	1	1	1	−1	−1	−1	−1	−1		
A_{2u}	1	1	1	−1	−1	−1	1	−1	−1	1		
E_u	2	−1	2	0	0	−2	0	1	−2	0		
T_{1u}	3	0	−1	1	−1	−3	−1	0	1	1	(x,y,z)	
T_{2u}	3	0	−1	−1	1	−3	1	0	1	−1		

Using the reduction formula, we get:

$$NA_{1g} = 1/48\ [(4.1.1) + (1.8.1) + (4.3.1) + (0.6.1) + (0.6.1) + (4.1.1) + (0.6.1) + (1.8.1) + (4.3.1) + (0.6.1)$$

$$= 1/48[(4 + 8 + 12 + 0 + 0 + 4 + 0 + 8 + 12 + 0)]$$

$$= 1/48[48] = 1$$

$$NA_{2g} = 1/48\ [(4.1.1) + (1.8.1) + (4.3.1) + (0.6. - 1) + (0.6. - 1) + (4.1.1) + (0.6. - 1) + (1.8.1) + (4.3.1) + (0.6. - 1)$$

$$= 1/48[(4 + 8 + 12 + 0 + 0 + 4 + 0 + 8 + 12 + 0)]$$

$$= 1/48[48] = 1$$

$$NE_g = 1/48[(4.1.2) + (1.8.-1) + (4.3.2) + (0.6.0) + (0.6.0) + (4.1.2) + (0.6.0) + (1.8.-1) + (4.3.2) + (0.6.0)$$

$$= 1/48[(8-8 + 24 + 0 + 0 + 8 + 0-8 + 24 + 0)]$$

$$= 1/48[48] = 1$$

$$NT_{1g} = 1/48[(4.1.3) + (1.8.0) + (4.3.-1) + (0.6.1) + (0.6.-1) + (4.1.3) + (0.6.1) + (1.8.0) + (4.3.-1) + (0.6.-1)$$

$$= 1/48[(12 + 0-12 + 0 + 0 + 12 + 0 + 0-12 + 0)]$$

$$= 1/48[0] = 0$$

$$NT_{2g} = 1/48[(4.1.3) + (1.8.0) + (4.3.-1) + (0.6.-1) + (0.6.-1) + (4.1.3) + (0.6.-1) + (1.8.0) + (4.3.-1) + (0.6.1)$$

$$= 1/48[(12 + 0 - 12 + 0 + 0 + 12 + 0 + 0 - 12 + 0)]$$

$$= 1/48[0] = 0$$

$$NA_{1u} = 1/48[(4.1.1) + (1.8.1) + (4.3.1) + (0.6.1) + (0.6.1) + (4.1.-1) + (0.6.-1) + (1.8.-1) + (4.3.-1) + (0.6.-1)$$

$$= 1/48[(4 + 8 + 12 + 0 + 0 - 4 + 0 - 8 - 12 + 0)]$$

$$= 1/48[0] = 0$$

$$NA_{2u} = 1/48[(4.1.1) + (1.8.1) + (4.3.1) + (0.6.-1) + (0.6.-1) + (4.1.-1) + (0.6.1) + (1.8.-1) + (4.3.-1) + (0.6.1)$$

$$= 1/48[(4 + 8 + 12 + 0 + 0 - 4 + 0 - 8 - 12 + 0)]$$

$$= 1/48[0] = 0$$

$$N_{Eu} = 1/48[(4.1.2) + (1.8.-1) + (4.3.2) + (0.6.0) + (0.6.0) + (4.1.-2) + (0.6.0) + (1.8.1) + (4.3.-2) + (0.6.0)$$

$$= 1/48[(8 - 8 + 24 + 0 + 0 - 8 + 0 + 8 - 24 + 0)]$$

$$= 1/48[0] = 0$$

$$NT_{1u} = 1/48[(4.1.3) + (1.8.0) + (4.3.-1) + (0.6.1) + (0.6.-1) + (4.1.-3) + (0.6.-1) + (1.8.0) + (4.3.1) + (0.6.1)$$

$$= 1/48[(12 + 0 - 12 + 0 + 0 - 12 + 0 + 0 + 12 + 0)]$$

$$= 1/48[0] = 0$$

$$NT_{2u} = 1/48[(4.1.3) + (1.8.0) + (4.3.-1) + (0.6.-1) + (0.6.1) + (4.1.-3) + (0.6.1) + (1.8.0) + (4.3.1) + (0.6.-1)$$

$$= 1/48[(12 + 0 - 12 + 0 + 0 - 12 + 0 + 0 + 12 + 0)]$$

$$= 1/48[0] = 0$$

The above calculations indicate that the reducible representation ($E_g \times E_g$) in question is reduced to

$$\Gamma_{Eg.Eg} = A_{1g} + A_{2g} + E_g$$

Thus, e_g^2 configuration gives the states as $A_{1g} + A_{2g} + E_g$.

The orbital symmetries of all states so obtained from t_{2g}^2, $t_{2g}^1 e_g^1$ and e_g^2 configurations are summarized in the table below:

Configuration	Orbital Symmetry of States obtained
t_{2g}^{2}	$A_{1g} + E_g + T_{1g} + T_{2g}$
$t_{2g}^{1}e_g^{1}$	$T_{1g} + T_{2g}$
e_g^{2}	$A_{1g} + A_{2g} + E_g$

The orbital symmetries and the multiplicity of the spectral states derived from each configuration are thus assigned. The next task is to assign the spin-multiplicities for each of these states in a configuration so that the total multiplicity of the states matches with that of the configuration. The assignment of the spin-multiplicities can be achieved by two methods: (a) a trial-and-error method and (b) method of descending symmetry.

(a) Trial-and-error method

(i) t_{2g}^{2} configuration

Let us first apply this method for the d^2 case taking one of the strong field configurations, t_{2g}^{2}, thus

$$t_{2g}^{2} = t_{2g}^{1} \times t_{2g}^{1} = T_{2g} \times T_{2g} = {}^{a}A_{1g} + {}^{b}E_g + {}^{c}T_{1g} + {}^{d}T_{2g}$$

Let the spin-multiplicities be a, b, c and d as shown above. As we have earlier calculated, the total number of microstates for t_{2g}^{2}configuration as 15, the sum total of the product of the spin-multiplicity and orbital degeneracy, that is, Σ (spin-multiplicity × orbital degeneracy) of the states must be 15. Then, we have

$$a \times 1 + b \times 2 + c \times 3 + d \times 3 = 15$$

When we are dealing with two-electron system, the spin-multiplicity (a, b, c, and d) of the states can be either 1 or 3 (singlet or triplet). With these (singlet or triplet) restrictions, the possible solutions are shown in the table below:

t_{2g}^{2} case	a	b	c	d	States with spin multiplicities solution	Total degeneracy
(I)	1	1	1	3	${}^{1}A_{1g} + {}^{1}E_g + {}^{1}T_{1g} + {}^{3}T_{2g}$	15
(II)	1	1	3	1	${}^{1}A_{1g} + {}^{1}E_g + {}^{3}T_{1g} + {}^{1}T_{2g}$	15
(III)	3	3	1	1	${}^{3}A_{1g} + {}^{3}E_g + {}^{1}T_{1g} + {}^{1}T_{2g}$	15

We will see that the possibility (II) is correct when we finally construct the correlation diagram.

(ii) $t_{2g}^{1}e_g^{1}$ configuration

Let us now apply the trial-and-error method for the d^2 case taking strong field configurations, $t_{2g}^{1}e_g^{1}$, thus

$$t_{2g}{}^1 e_g{}^1 = t_{2g}{}^1 \times e_g{}^1 = T_{2g} \times E_g = T_{1g} + T_{2g}$$

So, if spin-multiplicity of T_{1g} and T_{2g} is *a* and *b*, respectively, then we can write,

$$^aT_{1g} + {}^bT_{2g} = a \times 3 + b \times 3 = 24$$

The probable value of *a* and *b* should be both 1 and 3, that is, both T_{1g} and T_{2g} states should have singlet and triplet multiplicities for the sum total of the product of the spin-multiplicity and orbital degeneracy to be 24. So, we can write these states with their degeneracy considering as if there are four states,

$$^1T_{1g} + {}^3T_{1g} + {}^1T_{2g} + {}^3T_{2g} = 24$$
$$\text{or } 1 \times 3 + 3 \times 3 + 1 \times 3 + 3 \times 3 = 24$$

(iii) $e_g{}^2$configuration

Proceeding in the same way as above, we can write

$$e_g{}^2 = e_g{}^1 \times e_g{}^1 = E_g \times E_g = A_{1g} + A_{2g} + E_g$$

Since, the total number of microstates calculated for $e_g{}^2$ configuration as 6, we may write

$$^aA_{1g} + {}^bA_{2g} + {}^cE_g = 6$$
$$\text{or} \quad a \times 1 + b \times 1 + c \times 2 = 6 \quad \text{or} \quad a + b + 2c = 6$$

Again a, b and c can have 1 or 3 spin-multiplicity value. The possible solutions are given in the table below:

$e_g{}^2$ case	a	b	c	States with spin multiplicities solution	Total degeneracy
(I)	1	3	1	$^1A_{1g} + {}^3A_{2g} + {}^1E_g$	6
(II)	3	1	1	$^3A_{1g} + {}^1A_{2g} + {}^1E_g$	6

We will see that the possibility (I) is correct when we finally construct the correlation diagram.

It is notable here that the energy order of states derived from an infinitely strong field configuration can be obtained by the following two Hund's rules:

(i) States with highest spin-multiplicity will be of the lowest energy.
(ii) If the states are of equal spin-multiplicity, the states with highest orbital degeneracy will tend to lie at lower position, that is, the order of degeneracy is T < *E* < A. Thus, T will have the lowest energy.

If any problem comes after using these two Hund's rules, then that problem may be resolved by quantum mechanical calculations.

Thus, the Hund's rules help us in fixing the energy order of various strong field and ground state levels. For examples,

$$(t_{2g})^2 \quad : {}^3T_{1g} < {}^1T_{2g} < {}^1E_g < {}^1A_{1g}$$

$$(t_{2g})^1(e_g)^1 \quad : {}^3T_{1g} < {}^3T_{2g} < {}^1T_{1g} < {}^1T_{2g}$$

$$(e_g)^2 \quad : {}^3A_{2g} < {}^1E_g < {}^1A_{1g}$$

Of the ${}^3T_{1g}$ and ${}^3T_{2g}$ of $(t_{2g})^1(e_g)^1$, we have been guided by the higher symmetry of ${}^3T_{1g}$ over ${}^3T_{2g}$.

Before going to construct correlation diagram in which on left-side extreme is the free ion (d^2) terms with zero ligand field and the right-side extreme is the d^2 ion (in complex) in a strong octahedral field, the following points are notable about terms/states in correlation diagrams.

(a) There must be one-to-one correspondence between the terms/states at the two extremes of the abscissa, that is, the free ion states (extreme left) must match with states obtained from t_{2g}^2, $t_{2g}^1e_g^1$ and e_g^2 configurations on extreme right. Thus, the total number of states must be equal and same on both sides. Here, there are 11 states at both extremes.

(b) As the symmetry is lowered from free ion to O_h or to lower symmetry, the spin-multiplicity of states (degenerate or nondegenerate state arising from free ion term) remains the same.

(c) As the strength of ligand field strength changes in going from free ion to strong field, the states of same multiplicity do not cross each other. This is called non-crossing rule.

Following above points into consideration, the correlation diagram for a d^2 ion in an octahedral ligand environment may be qualitatively drawn as given in Fig. 2.51.

On the extreme left side of the diagram, we have shown the terms/states of the free ion. Immediately to the right, we have shown the states into which these free ion states split under the influence of the weak ligand field octahedral environment. Here, we know the *spin multiplicity* of all the *eleven states*. At the extreme right, there are the states in the case of an infinitely strong field interaction with the octahedral environment. Immediately to the left of them, there are the distinct states which exist in the case of a very strong, but not infinitely strong field interaction. In order that each state on the left go over into a state of the same kind on the right without violation of the noncrossing rule, the connecting lines can be drawn only in the manner shown in the Fig. 2.51.

The details of the architecture of the diagram may be described as follows:

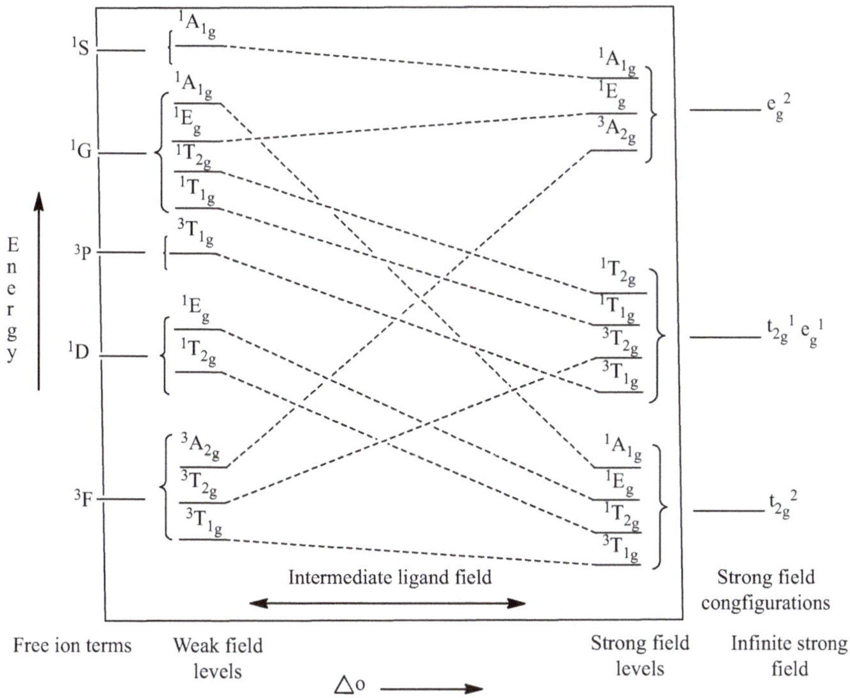

Fig. 2.51: Correlation diagram for a d^2-ion in octahedral ligand field.

(i) There are two $^1A_{1g}$ states on the left-hand side (in the weak field) that arise from ^1G and ^1S states and no $^3A_{1g}$ state. Thus, on the right-hand side (in the strong field), both the A_{1g} states obtained, one from e_g^2 and other from t_{2g}^2, configuration must have same spin multiplicity, that is, this should be $^1A_{1g}$. With this observation, we can very easily find the solution for spin multiplicity a, b and c of states in e_g^2 configuration. Solution (II), which considers $^3A_{1g}$ state and that is not there on the left-hand side, is not correct. Hence, solution (I) is correct. Thus, the states arising from e_g^2configuration with desired spin multiplicity are: $^1A_{1g}$ + $^3A_{2g}$ + ^1E_g [solution (I)]. By now, we could fix the spin multiplicity of e_g^2 and $t_{2g}^1e_g^1$ configurations that are summarized as:

$$e_g^2 \rightarrow {}^1A_{1g} + {}^3A_{2g} + {}^1E_g \text{ and}$$

$$t_{2g}^1e_g^1 \rightarrow {}^1T_{1g} + {}^3T_{1g} + {}^1T_{2g} + {}^3T_{2g}.$$

(ii) Now, the next step is to find the spin multiplicity of the states arising out of t_{2g}^2 configuration. As there are two $^1A_{1g}$ states with spin multiplicity equals to 1 and there is no term of the type $^3A_{1g}$ on the left, hence solution (III) for a, b, c and d can be ignored. Consequently, solution and (I) and (II) may be correct. Now, it is to be decided that out of (I) and (II) which solution is correct.

An observation of the states on the left-hand side shows that there are two $^3T_{1g}$ states: one from ^3P and other from ^3F terms. Following the noncrossing rule, the higher $^3T_{1g}$ state from ^3P term on left-hand side must be connected to higher $^3T_{1g}$ state arising from $t_{2g}{}^1e_g{}^1$ configuration on right-hand side. Then, there will remain a possibility of connecting other weak (left-hand side) $^3T_{1g}$ state with the T_{1g} state from arising from $t_{2g}{}^2$ configuration. Hence, this state must be a triplet and, thus, the possibility solution (II): $^1A_{1g} + {}^1E_g + {}^3T_{1g} + {}^1T_{2g}$ is correct.

(ii) Correlation diagram for d^2- configuration in T_d field

We can use the same procedure in developing the correlation diagram in the present case as used for d^2 configuration in octahedral chemical environment. It is notable here that the ordering of energy states obtained from both weak and strong field configurations will be inverted in d^2 tetrahedral case as compared to that of d^2 octahedral case.

The ordering of electronic configuration as energy point of view in strong field is $e^2 < e^1t_2{}^1 < t_2{}^2$, that is, e^2 configuration is lowest in energy. We know that in tetrahedral ligand field, splitting of d-orbitals is $t_2 < e$. So, the states arising out of e^2, $e^1t_2{}^1$and $t_2{}^2$ will also be inverted in energy as shown below with ordering,

$$e^2 = {}^3A_2 < {}^1E < {}^1A_1$$
$$e^1t_2{}^1 = {}^3T_1 < {}^3T_2 < {}^1T_1 < {}^1T_2 \quad \text{and}$$
$$t_2{}^2 = {}^3T_1 < {}^1T_2 < {}^1E < {}^1A_1$$

The energy states obtained from e^2, $e^1t_2{}^1$ and $t_2{}^2$ configurations are similar to that obtained from these configurations in octahedral field. The ordering of these configurations as well as the terms arising out of these is also inverted in energy in tetrahedral environment.

The states on the left-hand side are also inverted in energy as given below:

$$^3F = {}^3A_2 < {}^3T_2 < {}^3T_1$$
$$^1D = {}^1E < {}^1T_2$$
$$^3P = {}^3T_1$$
$$^1G = {}^1A_1 < {}^1E < {}^1T_2 < {}^1T_1$$
$$^1S = {}^1A_1$$

As the tetrahedral molecules do not have center of symmetry, hence *g* and *u* subscripts are dropped in such cases. The complete correlation diagram for d^2 configuration in tetrahedral field is shown in the Fig. 2.52.

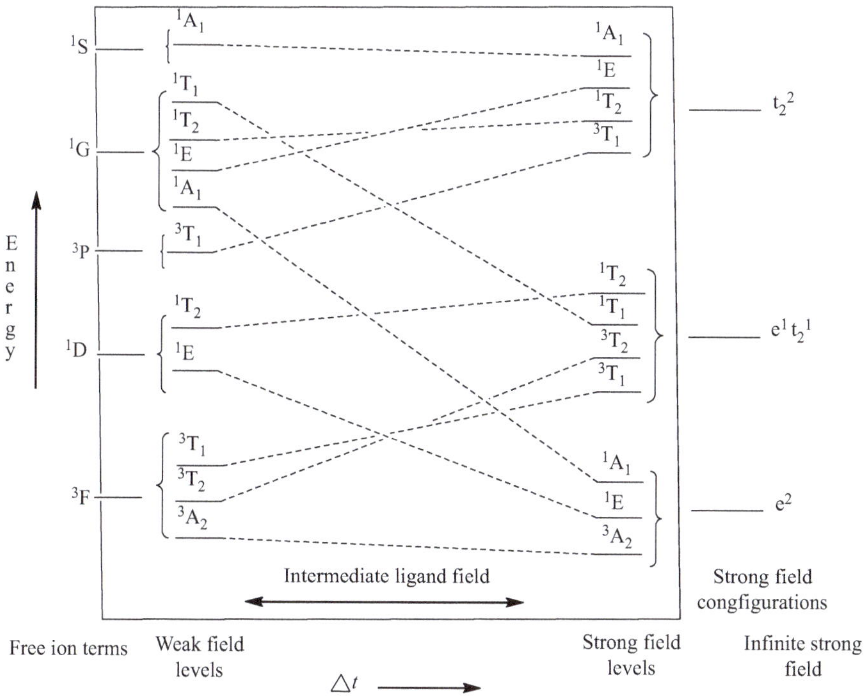

Fig. 2.52: Correlation diagram for a d^2-ion in tetrahedral ligand field.

(b) The method of descending symmetry

The trial-and error method used earlier for assigning spin multiplicities to states arising from strong field configurations by correlation with states from term splitting is practical only for very simple cases. In the d^n configurations, as the value of n increases, the number of microstates and terms also increases, that is, from d^1 to d^5 configurations, and thereafter decreases due to electron pairing. Construction of such correlation diagram is a very tedious job and lot of labor is involved. Therefore, there is a need for a general method which can be used easily for any d^n configuration in any symmetry environment of transition metal complexes. The most general and systematic method which can be used here is called the *method of descending symmetry*. It was first introduced by H. Bethe and further developed by C. K. Jorgenson and C. J. Balhausen. This method depends on the removal of degeneracies of the levels obtained on lowering symmetry until we get the nondegenerate levels or their sums so that the spins of the nondegenerate levels can be clearly assigned.

Let us illustrate this method taking the case of d^2 in an octahedral crystal field. In strong field, the two electrons can occupy as t_{2g}^2, $t_{2g}^1e_g^1$ and e_g^2 configurations. The energy states obtained from their direct products, as we have discussed earlier, are reproduced in the table below:

Configuration/product	States obtained
$t_{2g}^2 = t_{2g}^1 \times t_{2g}^1 = T_{2g} \times T_{2g}$	$A_{1g} + E_g + T_{1g} + T_{2g}$
$t_{2g}^1 e_g^1 = t_{2g}^1 \times e_g^1 = T_{2g} \times E_g$	$T_{1g} + T_{2g}$
$e_g^2 = e_g^1 \times e_g^1 = E_g \times E_g$	$A_{1g} + A_{2g} + E_g$

(i) t_{2g}^2 configuration

This ground state configuration gives rise to $A_{1g} + E_g + T_{1g} + T_{2g}$ states with the total degeneracy equal to 15. We can work out the spin multiplicity by choosing a subgroup of O_h group that removes the orbital degeneracy of t_{2g}, preferably to give different non-degenerate orbitals. The correlation table given below (Table 2.11) reveals that either the C_{2v} or C_{2h} can satisfy this requirement. Let us lower the symmetry of O_h to C_{2v} in the present case so that

$$\begin{aligned} O_h &\rightarrow C_{2v} \\ T_{2g} &\rightarrow A_1 + B_1 + B_2 \end{aligned}$$

For t_{2g} orbitals: $t_{2g} \rightarrow a_1 + b_1 + b_2$ (lower-case letters for orbital symmetry)

Table 2.11: Correlation table of O_h with most of the subgroups.

O_h	O	T_d	D_{4h}	D_{2d}	C_{4v}	C_{2v}	D_{3d}	D	C_{2h}
A_{1g}	A_1	A_1	A_{1g}	A_1	A_1	A_1	A_{1g}	A_1	A_g
A_{2g}	A_2	A_2	B_{1g}	B_1	B_1	A_2	A_{2g}	A_2	B_g
E_g	E	E	$A_{1g}+B_{1g}$	A_1+B_1	A_1+B_1	A_1+A_2	E_g	E	A_g+B_g
T_{1g}	T_1	T_1	$A_{2g}+E_g$	A_2+E	A_2+E	$A_2+B_1+B_2$	$A_{2g}+E_g$	A_2+E_g	A_g+2B_g
T_{2g}	T_2	T_2	$B_{2g}+E_g$	B_2+E	B_2+E	$A_1+B_1+B_2$	$A_{1g}+E_g$	A_1+E	$2A_g+B_g$
A_{1u}	A_1	A_2	A_{1u}	B_1	A_2	A_2	A_{1u}	A_1	A_u
A_{2u}	A_2	A_1	B_{1u}	A_1	B_2	A_1	A_{2u}	A_2	B_u
E_u	E	E	$A_{1u}+B_{1u}$	A_1+B_1	A_2+B_2	A_1+A_2	E_u	E	A_u+B_u
T_{1u}	T_1	T_2	$A_{2u}+E_u$	B_2+E	A_1+E	$A_1+B_1+B_2$	$A_{2u}+E_u$	A_2+E	A_u+2B_u
T_{2u}	T_2	T_1	$B_{2u}+E_u$	A_2+E	B_1+E	$A_2+B_1+B_2$	$A_{1u}+E_u$	A_1+E	$2A_u+B_u$

The direct product $(t_{2g}^2) = (t_{2g}^1)(t_{2g}^1)$ in O_h can now be written as $(a_1 + b_1 + b_2)(a_1 + b_1 + b_2)$ in C_{2v} after lowering the symmetry. Thus, the two electrons can be put in any of the combinations given by the direct products: $(a_1 + b_1 + b_2)(a_1 + b_1 + b_2)$ with the following details:

Product type	**Orbital configuration**	**Spin multiplicity**
$a_1 \times a_1 = A_1$	a_1^2	$^1A_1$
$b_1 \times a_1 = a_1 \times b_1 = B_1$	$a_1^1b_1^1$	$^1B_1$, $^3B_1$
$b_2 \times a_1 = a_1 \times b_2 = B_2$	$a_1^1b_2^1$	$^1B_2$, $^3B_2$
$b_1 \times b_1 = A_1$	b_1^2	$^1A_1$
$b_1 \times b_2 = b_2 \times b_1 = A_2$	$b_1^1b_2^1$	$^1A_2$, $^3A_2$
$b_2 \times b_2 = A_1$	b_2^2	$^1A_1$

The product $a_1 \times a_1 = A_1$, $a_1 \times b_1 = B_1$, and so on can be obtained by multiplying the characters of respective irreducible representation ($A_1.A_1$ or $A_1.B_1$) of C_{2v} character table.

The A_1 state must be only singlet since there are two electrons in one orbital (such as a_1^2, b_1^2 or b_2^2). In other orbital configurations, there are two electrons (such as $a_1^1b_1^1$, $a_1^1b_2^1$ or $b_1^1b_2^1$); hence, they give a singlet and a triplet in each case (such as $^1B_1$ and $^3B_1$, etc.)

The correlation table also shows that the O_h states correlate with C_{2v} states as follows:

$$\begin{array}{ll} O_h & C_{2v} \\ A_{1g} & \rightarrow A_1 \\ E_g & \rightarrow A_1 + A_2 \\ T_1g & \rightarrow A_2 + B_1 + B_2 \\ T_2g & \rightarrow A_1 + B_1 + B_2 \end{array}$$

As all the A_1 states are singlets, the corresponding O_h states must also the singlets ($^1A_{1g}$, ^1E_g and $^1T_{2g}$). Hence, corresponding C_{2v} states will also be singlets as given below:

$$\begin{array}{ll} O_h & C_{2v} \\ ^1A_{1g} & \rightarrow {}^1A_1 \\ ^1E_g & \rightarrow {}^1A_1 + {}^1A_2 \\ ^1T_{2g} & \rightarrow {}^1A_1 + {}^1B_1 + {}^1B_2 \end{array}$$

This is due to the fact that the spin multiplicity of the states is unaffected due to lowering in symmetry from $O_h \rightarrow C_{2V}$. The remaining triplet state (T_{1g}) of O_h must then belong to $^3T_{1g}$.

Thus,

$$^{3}T_{1g}(O_h)\rightarrow{}^{3}A_2+{}^{3}B_1+{}^{3}B_2(C_{2V})$$

Hence, of the states obtained from the $(t_{2g})^2$ configuration, only the T_{1g} is a triplet (i.e., $^{3}T_{1g}$) and the rest are singlet. So, we get

$$(t_{2g})^2=T_{2g}\times T_{2g}={}^{1}A_{1g}+{}^{1}E_g+{}^{3}T_{1g}+{}^{1}T_{2g}(\text{Total degeneracy}=1+2+9+3=15)$$

This result is same as we have obtained earlier by trial-and-error method.

(ii) e_g^2 configuration

This configuration gives rise to A_{1g} + A_{2g} + E_g states with the total degeneracy equal to 6. We can assign the spin multiplicity by choosing a subgroup of O_h group that removes the degeneracy of E_g. The correlation table indicates that we can lower the symmetry of O_h to either C_{4v} or C_{2v}. Let us take C_{2v} subgroup and accordingly,

$$\begin{array}{ll} O_h & C_{2v} \\ E_g \rightarrow & A_1+A_2 \end{array}$$

or $e_g \rightarrow a_1 + a_2$ (lower-case letters for orbital symmetry)

The direct product $(e_g^2) = (e_g^1)(e_g^1) = E_g\times E_g$ in O_h can now be written as $(a_1 + a_2)(a_1 + a_2)$ in C_{2v}, for which one can write the direct products with possible symmetries as follows:

Product type	Orbital configuration	Spin multiplicity
$a_1 \times a_1 = A_1$	a_1^2	$^{1}A_1$
$a_1 \times a_2 = a_2 \times a_1 = A_2$	$a_1^1a_2^1$	$^{1}A_2$, $^{3}A_2$
$a_2 \times a_2 = A_1$	a_2^2	$^{1}A_1$

The A_1 state must be only singlet since there are two electrons in one orbital (such as a_1^2 or a_2^2). In other orbital configurations, there are two electrons (such as $a_1^1a_2^1$); hence, they give a singlet and a triplet (such as $^{1}A_2$ and $^{3}A_2$). The correlation table also indicates that the O_h states correlate with C_{2v} states as follows:

$$\begin{array}{ll} O_h & C_{2v} \\ A_{1g} \rightarrow & A_1 \\ A_{2g} \rightarrow & A_2 \\ E_g \rightarrow & A_1+A_2 \end{array}$$

Since all the A_1 states are singlets, the corresponding O_h states must also the singlets ($^1A_{1g}$ and ^1E_g). The remaining A_{2g} state of O_h must then belong to triplet, that is, $^3A_{2g}$ to have the total degeneracy equal to 6. Hence, the states obtained from $E_g \times E_g$ direct product can be given the spin multiplicity as

$$^1A_{1g} + {}^3A_{2g} + {}^1E_g$$

(iii) $t_{2g}{}^1e_g{}^1$ configuration

In the present case $t_{2g}{}^1e_g{}^1 = T_{2g} \times E_g = T_{1g} + T_{2g}$ in which electrons occupy two different orbitals, the states may be both singlet and triplet types. Therefore,

$$T_{2g} \times E_g = {}^1T_{1g} + {}^3T_{1g} + {}^1T_{2g} + {}^3T_{2g}$$

The result is so straightforward that there is no need of applying the method of descending symmetry for assigning the spin multiplicities for the states of this configuration.

Energy order of states

After the orbital and spin degeneracies of the states arising from a given strong field configuration, the next step is to know the energy ordering of these states, which is possibly independent of the nature of crystal field symmetry. This is in contrast to the dependence of the term splitting into states on the crystal field environment. In order to know the qualitative ordering of states, the following two points are worth considering:

(i) In a strong field environment, the highest spin multiplicity term lies lowest in energy.
(ii) Of the states with same multiplicity, the state which has the highest orbital degeneracy and more symmetry will be of lower energy, that is, T < *E* < A.

Thus, for various configurations, the ordering of states is as follows:

$$(t_{2g})^2: \quad {}^3T_{1g} < {}^1T_{2g} < {}^1E_g < {}^1A_{1g}$$
$$(t_{2g})^1(e_g)^2: \quad {}^3T_{1g} < {}^3T_{2g} < {}^1T_{1g} < {}^1T_{2g}$$
$$(e_g)^2: \quad {}^3A_{2g} < {}^1E_g < {}^1A_{1g}$$

Here, T_{1g} is more symmetric than T_{2g}.

So far, we have constructed correlation diagram for d^2 case in octahedral and tetrahedral crystal fields. In the d^n configurations, as the value of n increases beyond two, the number of microstates and terms increases to a great extent. Therefore, the construction of correlation diagram for d^3, d^4, d^5 . . .and so on. configurations requires a lot of labor. This method can be simplified if one takes care of general relation that exists between configurations, terms and coordination geometry. Bethe has described this general relation by using the concept of the Hole formalism given below:

2.5 Correlation diagram and Hole formalism

An inversion principle known for the Orgel diagram can be applied here for d^n and d^{10-n} configurations. Consequently, their correlation diagrams can be easily adapted from the already-built ones. For instance, a d^1 case must behave just like a d^9 (one positron) except reversal of the sign of the energy of interaction. In other words, *n*-electrons behave like *n*-positrons (10-*n* electrons) according to Hole formalism. The consequence of this formalism is that the energy levels on the right side of the correlation diagram for a d^n configuration will be according to the inversion in the strong field configurations, $t_{2g}{}^x e_g{}^y$ of d^{10-n} configuration and inversion in the splitting of terms on the left side. Thus, for instance, the correlation diagram given in Fig. 2.51 for a d^2-system in an octahedral field can be used for a d^8-system in tetrahedral field. Similarly, the correlation diagram presented in Fig. 2.52 for a d^2-system in a tetrahedral field can be used for a d^8-configuration in an octahedral field. In general, the diagram of

$$\mathrm{d^n(octahedral)} \equiv \mathrm{d^{10-n}(Tetrahedral)}$$
$$\text{and} \quad \mathrm{d^n(Tetrahedral)} \equiv \mathrm{d^{10-n}(Octahedral)}$$

$$\mathrm{d^n(Octahedral)\ is\ inverse\ to\ d^n(Tetrahedral)}$$
$$\text{and} \quad \mathrm{d^{10-n}(Octahedral)\ is\ inverse\ to\ d^{10-n}(Tetrahedral)}$$

Thus, for *18 possible cases, d^1 to d^9 in octahedral and d^1 to d^9 in tetrahedral*, the correlation diagrams for all these cases can be worked out if one knows only the correlation diagrams for d^1, d^2, d^3, d^4 and d^5 cases in the octahedral environment. The correlation diagram for d^5 case is a special one because d^5 (octahedral) $\equiv$ d^5 (Tetrahedral).

2.5.1 Uses of correlation diagrams

The correlation diagram we have just constructed is qualitative and shows the splitting of terms and, hence, gives information about how the energy changes among split terms with change of the ligand field. The obvious question arises in our mind that what is the use of these correlation diagrams. One of the most important uses of correlation diagrams is the interpretation of the spectral properties of the transition metal complexes. That is, we can assign the various electronic transitions for the observed absorption bands in the UV-Vis. spectrum of a complex.

The correlation diagrams are useful for predicting d-d transitions in metal complexes in a qualitative manner. For quantitative prediction, we have to use another kind of diagram, called Tanabe-Sugano diagram correlation diagram. These diagrams were developed by Y. Tanabe and S. Sugano in 1954 [2], which are used by most of the chemists in actual practice for octahedral complexes.

2.6 Tanabe-Sugano correlation diagram

Orgel diagrams presented for weak field complexes and the correlation diagrams discussed above for a smooth transition from the weak field to the strong field limits have only been qualitative. It is necessary to have some quantitative results from these energy level diagrams for the interpretation of spectra. But at the same time, it is difficult to calculate the energies of all the energy levels of a d^n system for a given strength of the crystal field (10Dq) and in presence of inter-electronic repulsions (B and C: Racah Parameters). An attempt at the quantitative aspects of these parameters has been made by two Japanese workers, Y. Tanabe and S. Sugano, in 1954 [2], who devised some useful correlation diagrams called Tanabe-Sugano (T-S) diagrams. Most of the chemists use these diagrams for interpretation of electronic spectra of both weak and strong field systems. Charge transfer and intra-ligand transitions are not predicted from these diagrams.

In T-S diagram, energies (E) of the levels of a d^n system are plotted as the vertical coordinate, in units of the inter-electronic repulsion parameter B, that is, E/B, while crystal field strength is the horizontal coordinate in the units of 10 Dq/B. The restriction on the usefulness of these diagrams lies in the fact that it requires two Racah parameters B and C, to describe the inter-electronic repulsions for a d-electron system. The diagram can only be constructed if the ratio C/B is specified. The diagrams have been compiled for various values of C/B, which are considered most likely for some first transition series ions of the configurations concerned. The value of C/B is stated in each diagram.

For M^{2+} and M^{3+} of the first transition series, the probable C/B values are known to be in the range of 3.2–4.8. However, the diagrams are not very sensitive to the C/B ratio and a value of 4 for this ratio is a good approximation for ions of the first transition series. Moreover, the ratio C/B only affects the inter-scaling between the states of different multiplicities. But for spin-allowed transitions, we are concerned only with terms of the same multiplicity as that of the ground state. For the terms of the same multiplicity within a configuration, the energy difference is a function of the parameter B only, and independent of the ratio C/B for any particular value of B. For instance, for the free ion, the ^4F and ^4P separation is 15B, while the ^4F and ^2G separation is 4B + 3C. For most of the transition metal ions, C is ≈ 4B.

In these diagrams, the zero of energy is always taken as that of the lowest term. Hence, when there is change of ground term, the diagram is discontinuous. The discontinuity always takes the form of an increase in the slope of term energies above a critical value of 10Dq/B.

The Tanabe-Sugano diagrams were drawn up primarily for use with ligand fields of O_h symmetry. However, it is possible to use the diagrams for tetrahedral ligand fields considering the equivalence $d^n = d^{10-n}$. For instance, the diagram for d^2 may be used for tetrahedral d^8 (Ni^{2+}).

It is noteworthy that different electronic configurations require different Tanabe-Sugano diagrams for a given stereochemistry. The diagram for a d^2 (V^{3+}) case is shown in Fig. 2.53. The energy levels on the right-hand side of diagram are identical to those obtained from strong field configurations of d^2 system. Thus, the $^3T_{1g}$ is identified as derived from $t_{2g}{}^2$ configuration. The other three singlets, $^1T_{2g}$, ^1E_g and $^1A_{1g}$, are also from $t_{2g}{}^2$ configuration. The next higher energy strong field configuration $(t_{2g}{}^1)(e_g{}^1)$ gives $^3T_{2g}$, $^3T_{1g}$, $^1T_{2g}$ and $^1T_{1g}$. Similarly, the highest energy levels $^3A_{2g}$, ^1E_g and $^1A_{1g}$ must have been derived from the highest field configuration $(e_g{}^2)$. The diagrams shown in Figs. 2.54 and 2.55 correspond to d^3 and d^8 configurations and have features similar to that described for d^2 configuration.

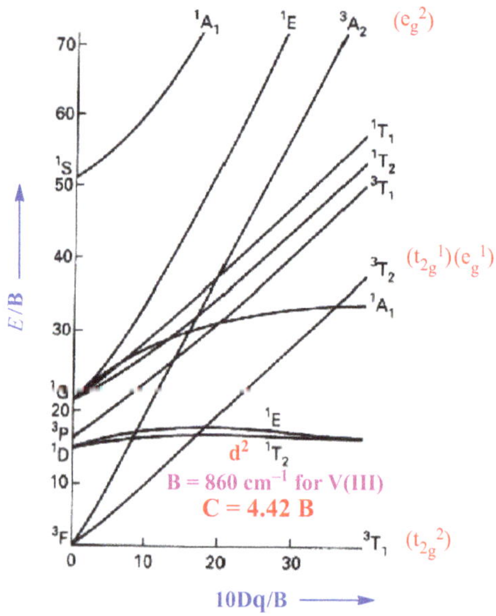

Fig. 2.53: T-S energy level diagram for d^2 configuration.

The Tanabe-Sugano diagrams for d^4, d^5, d^6 and d^7 configurations are shown in Figs. 2.56, 2.57, 2.58 and 2.59, respectively. In all these cases, the diagrams are divided into two parts separated by a vertical line. The left side of the vertical line applies to the high-spin (weak field) complex and resembles the appropriate Orgel diagram for the state of maximum multiplicity. The right side of the diagram applies

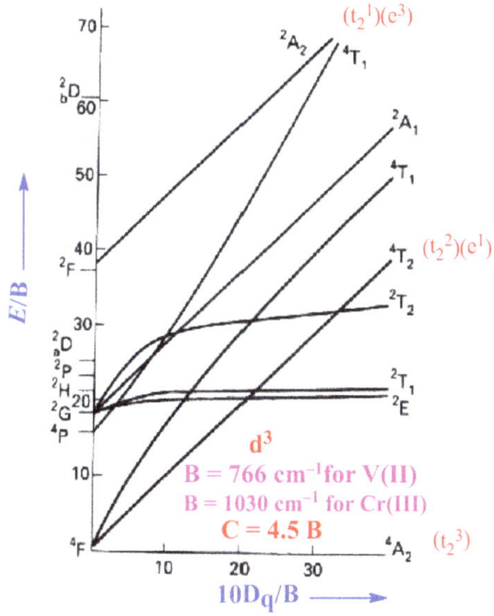

Fig. 2.54: T-S energy level diagram for d^3 configuration.

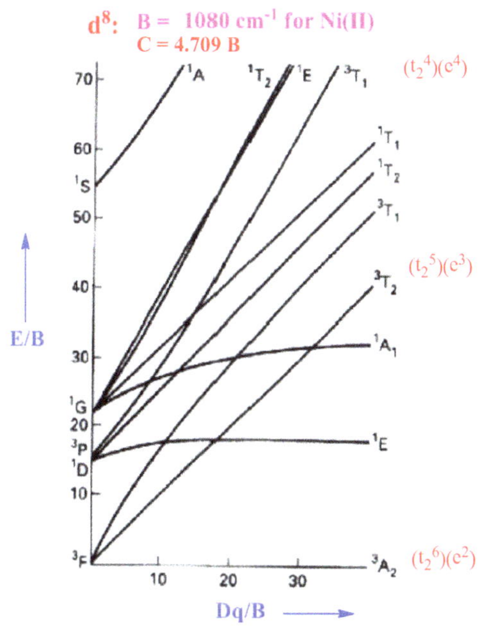

Fig. 2.55: T-S energy level diagram for d^8 configuration.

to the low-spin (strong field) complex where the ground state is different from that of the weak field case. In fact, each of these diagrams can be considered as combination of two separate diagrams representing the extreme end-fields that share the

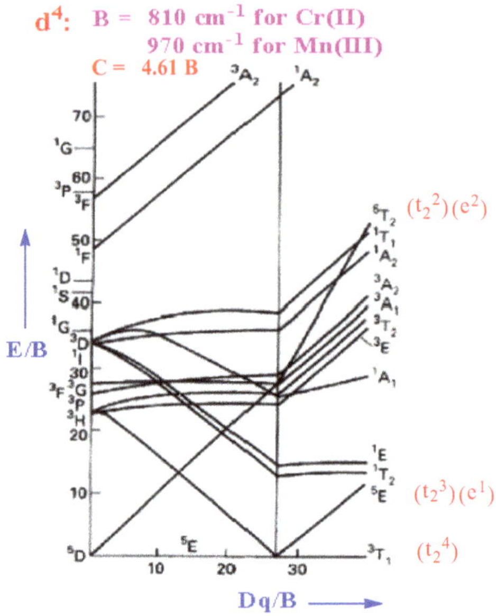

Fig. 2.56: T-S energy level diagram for d^4 configuration.

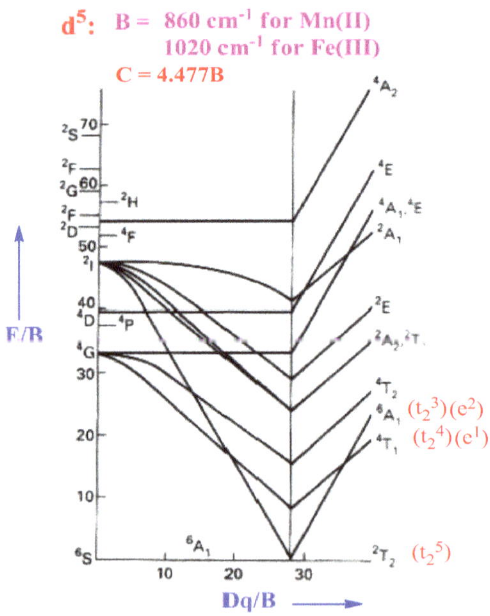

Fig. 2.57: T-S energy level diagram for d^5configuration.

same energy states but in different order. Their energy ordering is same at Dq/B equal to the energy required for spin-pairing.

The labels without the subscript *g* or *u* are given at the end of each line (energy level) so that the same diagram is applicable for the tetrahedral complexes. The

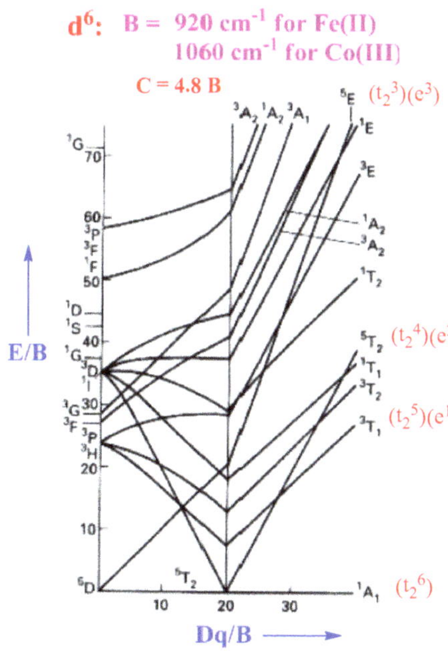

Fig. 2.58: T-S energy level diagram for d^6 configuration.

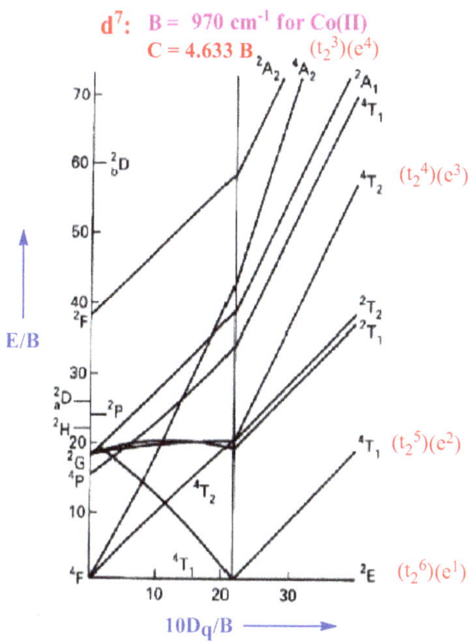

Fig. 2.59: T-S energy level diagram for d^7 configuration.

labels on the right are for O_h symmetry. For example, in the d^4 diagram (Fig. 2.56), the ground state of the free ion is ^{5}D, which gives the ground level/state as ^{5}E in weak octahedral crystal field and $^5T_2$ as the only excited state. But beyond 10 Dq/B ≈ 20.8, the field is strong enough to force spin pairing and $^3T_1$ becomes the ground state with ^{5}E level increasing its energy. In the weak field case of d^4, the spin-allowed transitions are from ^{5}E ground state, whereas in strong field case, the spin-allowed transitions are from $^3T_1$ to the other excited triplet levels. Several of these triplet states originate from ^{3}H, ^{3}P, ^{3}F, ^{3}G and so on terms. The diagrams for d^5, d^6 and d^7 configurations bear similar features.

Orgel and Tanabe-Sugano (T-S) diagrams

(i) Both the diagrams are correlation diagrams. In the Orgel diagrams, the splitting of the free ion terms in weak field is considered, while in T-S diagrams, all terms arising from weak and strong fields are considered. Moreover, the ground state terms of the complex is obtained from the ground state terms of the free ion. Hence, the multiplicity of the ground state in the complex is same as that of the free ion ground state.

(ii) In the Orgel diagrams, the splitting of the higher energy terms is not considered, whereas in T-S this has been considered.

(iii) In Orgel diagrams, the splitting of ground state term is given importance, whereas the splitting of ground state and the higher energy states of the free ion as a function of Dq in weak and strong fields are given due considerations in T-S diagrams.

(iv) In T-S diagram, E/B is plotted against Dq/B (or Δ_o/B). The value of E is obtained as in the Orgel diagram and the value of B for metal ion is obtained from its emission spectrum. Hence, the position of the spectral states are shown as a function of two parameters, Dq and B, and, therefore, the T-S diagrams are valid for all central metal ions of a particular d^n configuration. So, the T-S diagram d^5 configuration may be used for the Mn(II) and Fe(III) both; however, the values of B for these metal ions are different.

(v) In T-S diagram, the ground state term is taken as the horizontal line and energy of this term is taken as zero. So that the energy of the electronic transition from ground to the excited state can be evaluated by the vertical distance from the base line. In Orgel diagram, such evaluation is difficult because split-up terms are not taken as horizontal base line.

(vi) In T-S diagram, some spectral states are shown by curved lines rather than straight lines. This is due to interaction between states of the same spin and orbital multiplicity. For example, this effect can be seen in the T-S diagram of d^2 octahedral case (Fig. 2.60) for the two ^{1}E (spin multiplicity 1 and orbital multiplicity 2), two $^1T_2$ (spin multiplicity 1 and orbital multiplicity 3) and the two $^1A_1$ (spin multiplicity 1 and orbital multiplicity 1).

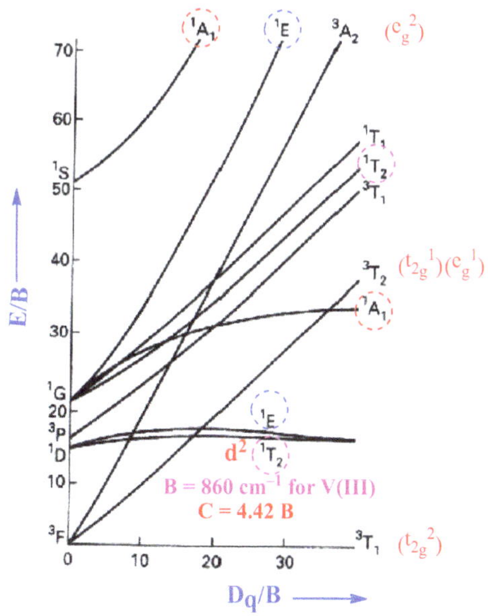

Fig. 2.60: T. S. Energy level diagram for d^2-configuration in O_h crystal field.

2.7 Variation in Racah parameter B: nephelauxetic series

The values of the Racah parameter B, for free metal ions determined from the emission spectra of the free metal ions are not the same as those obtained for their metal complexes from the absorption spectra. The values of Racah parameter obtained for complexes (B′) are always less than B (for the free metal ion). The low value of B′ in a complex as compared to B in the free ion indicates that the inter-electronic repulsion has been lowered in the complex, that is, delocalization of metal d-electrons in the complex. This is possible only if there is expansion of the d-electrons cloud of the metal ion in the complex. The lower value of B′ in complex is an evidence of the interaction (overlap) of metal ion and ligand orbitals, and, consequently, metal electrons are delocalized over ligand orbitals. This provides an experimental evidence for ligand field theory. The expansion of electron cloud is called the *nephelauxetic effect*. It is a Greek word that means cloud expanding.

The extent of lowering in the value of B′ is a measure of the extent of overlap between metal ion and ligand orbitals. Thus, the covalent interaction can be understood in terms of lowered value of B′ in the complex.

The covalency factor or (nephelauxetic parameter) β or β^0 may be defined as

$$\beta = \frac{B'}{B}$$
$$\text{and } \beta^0 = \left\{\frac{(B - B')}{B}\right\} \times 100$$
$$\text{or } \beta^0 = (1 - \beta) \times 100$$

From the above equations, it appears that lower the value of β or higher the value β^0, more will be the metal-orbital overlap. The values of β depend on the kind of metal ion and the ligand.

The ordering of various ligands with decreasing value of β can be made, and thus a series can be obtained called nephelauxetic series. For a particular metal ion in a particular oxidation state, the following series of ligands can be ordered as:

$$F^- > H_2O > \text{urea} > NH_3 > \text{Oxalate}^{2-} > \text{en} > NCS^- > Cl^- > Br^- > S^{2-} \sim I^- > R_2PSe_2$$

This series is apparently parallel with the degree of electronegativity of the donor atoms in the ligands:

$$F > O > N > Cl > Br > S \sim I > Se$$

For metal ion, a similar nephelauxetic series is:

$$M(0) > M(I) \;\; > M(II) > M(III) \ldots\ldots.$$

$$Mn^{2+} \sim Pt^{2+} > Ni^{2+} > Cr^{3+} > Fe^{3+} > Ir^{3+} \sim Rh^{3+} > Co^{3+} > Pt^{4+}$$

Lager β, small overlap   Small β, large overlap

(Ionic)   (Covalent)

Thus, β decreases with increase in oxidation state of the metal ion.

If B = B′, then β = 1 and the metal-ligand bond is 100% ionic. The fact that F^- and H_2O are at the left end of the series, their β values are expected to be close to 1, suggesting that their complexes are more ionic. On the other hand, Br^- and I^- lie in the middle and right end of the series, suggesting that they give more covalent complexes with their β values much less than one (1).

The significant reduction observed in the magnitude of Racah parameter in complexes can be explained in terms of Jorgenson's argument as follows:

(a) Electrons are transferred to a lesser or a greater degree to the metal ion. Consequently, effective nuclear charge (Z^*) on the central metal ion is thus lowered compared to the actual charge (Z). This is called "central field covalency".
(b) Once the metal-ligand bond is established, a delocalization of d-electrons in the direction of the ligands may occur by a π-mechanism leading to "symmetry restricted covalency".

2.7.1 Evaluation of Dq, B′ and β parameters

Two methods are normally used for the evaluation of electronic parameters: Dq, B′ and β. The first one is *experimental method* taking help of Tanabe-Sugano (T-S) diagram and second one is *Graphical method* based on theoretical plots of transition energy ratios corresponding to several Dq/B values for complexes of T_1 and A_2 ground states. One more method known as *Konig's numerical method* is also used for evaluation of electronic parameters of complexes of d^2, d^8, d^3 and d^7 configurations.

(i) Experimental method

The T-S diagram is helpful in determining the value of Dq and B′ from the experimental electronic spectrum of a complex with more than one d-electrons in which electron-electron repulsion is significant. The method may be illustrated by taking some examples for which we will deduce splitting parameter Dq and inter-electronic repulsion parameter B′ for the respective complex involving d^n configuration.

(i) $[Cr(NH_3)_6]^{3+}$ involving d^3 configuration

The T-S diagram for d^3configuration in O_h crystal field is shown in Fig. 2.61.

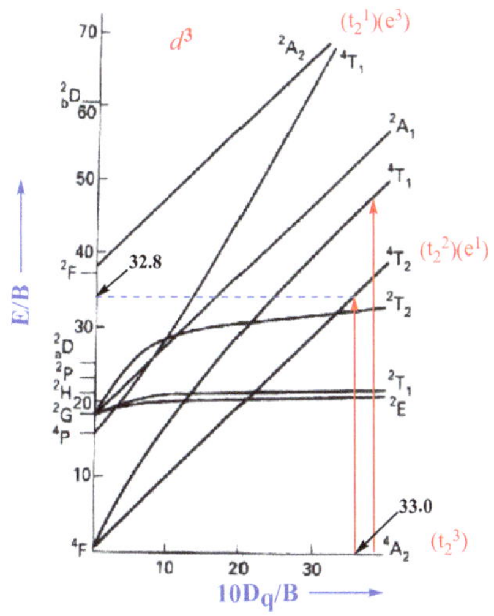

Fig. 2.61: T-S diagram for d^3 configuration in O_h crystal field.

For d^3 ion, we are concerned only to the spin-allowed transitions, which can be *three* in the present case and these are:

$$^4A_{2g} \rightarrow {}^4T_{2g}(F)$$
$$^4A_{2g} \rightarrow {}^4T_{1g}(F)$$
$$^4A_{2g} \rightarrow {}^4T_{1g}(P)$$

However, the experimentally recorded electronic spectrum of $[Cr(NH_3)_6]^{3+}$ ion shows only two low energy ligand field bands at 21,550 and 28, 500 cm^{-1}. These two bands correspond to the two low energy transitions $^4A_{2g} \rightarrow {}^4T_{2g}(F)$ and $^4A_{2g} \rightarrow {}^4T_{1g}$ (F) as shown in the diagram by two vertical lines. The ratio of the energies of these two transitions is 28,500/21,550 = 1.32. This ratio fits at a particular point in abscissa of the T-S diagram, and this point is on the far right side of abscissa. From the T-S diagram we can read the value of 10 Dq/B or Δ_o/B for this point as 33.0., that is,

$$\Delta_o/B = 33.0 \tag{2.4}$$

Now, from the head of the arrow of the lowest energy transition, the corresponding value of energy comes out as 32.8, that is, E/B = 32.8 or E = 32.8B. So, we have

$$E = 32.8\,B = 21,550\,cm^{-1}$$

$$\text{or}\;\; B = 21,550/32.8 = 657 cm^{-1}$$

On substituting the value of B in eq. (2.4), we get the splitting parameter of the complex ion, Δ_o as

$$\Delta_o = 10Dq = 33.0 \times B = 33.0 \times 657$$

$$= 21,681\,cm^{-1}$$

$$\text{or}\;\; Dq = 2168\,cm^{-1}$$

As β = B′/B (B = Racah parameter for free Cr^{3+}ion and B′ = Racah parameter for Cr^{3+} ion in the complex). Hence,

$$\beta = 657/918 = 0.71$$

The values of Racah parameter (B) for some transition metal ions are given in Table 2.12.

Table 2.12: Values of Racah parameter (B) for some transition metal ions and their ions.

Metals	**M^{2+}**	**M^{3+}**
Ti	695	–
V	755	861
Cr	810	918
Mn	860	965
Fe	917	1015
Co	971	1065
Ni	1030	1115

Note: Dash (–) in the table indicates that the value of B is unknown.

$[V(H_2O)_6]^{3+}$ involving d^2 configuration

The electronic spectrum of a d^2-complex may exhibits three spin-allowed bands corresponding to the following transitions:

$$^3T_{1g} \rightarrow {}^3T_{2g}(F): \nu_1$$
$$^3T_{1g} \rightarrow {}^3T_{1g}(P): \nu_2$$
$$^3T_{1g} \rightarrow {}^3A_{2g}(F): \nu_3$$

The $[V(H_2O)_6]^{3+}$ ion in ammonium alum exhibits only two bands at 17,800 (ν_1) and 25,700 (ν_2) cm^{-1}. The third high energy band in the spectrum is usually obscured by more intense charge transfer band occurring in this region of the spectrum. The spectrum of the complex ion and T-S diagram of d^2-configuration are shown in Fig. 2.62.

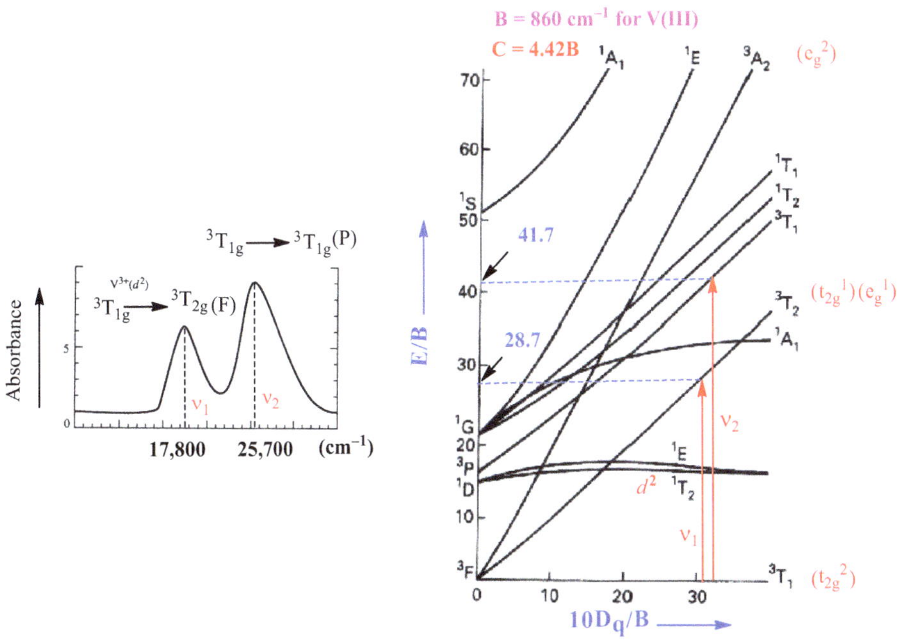

Fig. 2.62: Electronic spectrum of $[V(H_2O)_6]^{3+}$ and T. S. Energy level diagram for d^2-configuration in Oh crystal field.

So, ν_1/ν_2 = 17,800/25,700 = 0.693. This can only be fitted to T-S diagram at 10 Dq/B = ~31 when ν_1/B = 28.7 and ν_2/B = 41.7. So, we have

$$\nu_1/B = 28.7 \quad \text{or} \quad 17,800/B = 28.7$$
$$\text{or} \quad B = 17,800/28.7 = 620 cm^{-1}$$

In the present case, 10 Dq/B = ~31. So, 10 Dq = 31 × B = 31 × 620 = 19,200 cm^{-1}.

As β = B′/B (B = Racah parameter for free V^{3+}ion and B′ = Racah parameter for V^{3+}ion in the complex), hence

$$\beta = 620/861 = 0.72 \quad [\text{Racah parameter for free } V^{3+} \text{ ion} = 861]$$

$$\beta^0 = (1-\beta) \times 100 = (1-0.72) \times 100 = 28\%$$

(ii) Graphical method

This method is most liked by many chemists in calculations of electronic parameters. An inspection of all the energy level diagrams (both Orgel and Tanabe-Sugano diagrams of d^n configurations) reveals that the ground electronic state in most of the complexes is either "A" type or "T" type. Moreover, of all the energy level diagrams, only T-S diagrams are presented as quantitative. These diagrams are constructed with an assumption that C≅4B and the exact C/B ratio used in the construction of each diagram are stated.

Of all the configurations, the one with d^2, d^8, d^3 and d^7 configurations are important in context of the utility of this graphical method. It is possible to calculate the theoretical transition energy ratios amongst various possible spin-allowed transitions from the T-S diagrams. Thus, the transition energy ratios corresponding to several Dq/B values for complexes of T_1 ground states are enlisted in Table 2.13. Similarly, transition energy ratios corresponding to several Dq/B values for complexes of A_2 ground states are enlisted in Table 2.14. The graphs given in Figs. 2.63 and 2.64 are theoretical plots of Dq/B (on x-axis) versus v_2/v_1, v_3/v_1 and v_3/v_2 for T_1 and A_2 ground states, respectively.

Table 2.13: Dq/B versus transition energy ratios for metal complexes of T_1 ground states.

Dq/B′	v_2/v_1	v_3/v_1	v_3/v_2	v_3/B
0.1	2.234	19.279	8.629	15.620
0.2	2.220	9.930	4.473	16.279
0.3	2.207	6.828	3.094	16.971
0.4	2.195	5.288	2.408	17.692
0.5	2.185	4.370	2.000	18.439
0.6	2.175	3.763	1.730	19.209
0.7	2.167	3.333	1.538	20.000
0.8	2.159	3.014	1.396	20.809
0.9	2.151	2.768	1.286	21.633
1.0	2.145	2.572	1.199	22.472
1.1	2.138	2.414	1.129	23.324
1.2	2.133	2.283	1.071	24.187
1.3	2.127	2.173	1.022	25.060
1.4	2.123	2.080	0.980	25.942

Table 2.13 (continued)

Dq/B′	ν_2/ν_1	ν_3/ν_1	ν_3/ν_2	ν_3/B
1.5	2.118	2.000	0.944	26.833
1.6	2.114	1.930	0.913	27.731
1.7	2.110	1.869	0.866	28.636
1.8	2.106	1.816	0.862	29.547
1.9	2.103	1.768	0.841	30.463
2.0	2.099	1.725	0.822	31.385
2.1	2.096	1.687	0.805	32.311
2.2	2.093	1.652	0.789	33.242
2.3	2.091	1.621	0.775	34.176
2.4	2.088	1.592	0.762	35.114
2.5	2.086	1.566	0.751	36.056
2.6	2.083	1.542	0.740	37.000
2.7	2.081	1.519	0.730	37.947
2.8	2.079	1.499	0.721	38.897
2.9	2.077	1.480	0.713	39.85
3.0	2.075	1.462	0.705	40.804
3.1	2.073	1.446	0.697	41.761
3.2	2.072	1.431	0.691	42.720
3.3	2.070	1.416	0.684	43.681
3.4	2.068	1.403	0.678	44.643
3.5	2.067	1.390	0.673	45.607
3.6	2.066	1.378	0.667	46.573
3.7	2.064	1.367	0.662	47.539
3.8	2.063	1.357	0.658	48.508
3.9	2.062	1.347	0.653	49.477
4.0	2.060	1.337	0.649	50.448
4.1	2.059	1.328	0.645	51.420
4.2	2.058	1.320	0.641	52.393
4.3	2.057	1.312	0.638	53.367
4.4	2.056	1.304	0.634	54.342
4.5	2.055	1.297	0.631	55.317
4.6	2.054	1.290	0.628	56.294
4.7	2.053	1.283	0.625	57.271
4.8	2.052	1.277	0.622	58.249
4.9	2.051	1.271	0.619	59.228
5.0	2.050	1.265	0.617	60.208

Calculation of electronic parameters in complexes of T_1 ground states

The octahedral complexes of d^2, d^7 and tetrahedral complexes of d^3, d^8 configurations give T_1 ground state. Let us illustrate the method for calculation of electronic parameters: Dq, B′ and β for octahedral $d^2$and d^7 complexes using the experimental data available on them.

Table 2.14: Dq/B versus transition energy ratios for metal complexes of A_2 ground states.

Dq/B′	v_2/v_1	v_3/v_1	v_3/v_2	v_3/B
0.1	1.799	16.211	9.062	16.210
0.2	1.777	8.723	4.909	17.446
0.3	1.764	6.236	3.535	18.708
0.4	1.750	5.000	2.857	20.000
0.5	1.735	4.265	2.458	21.325
0.6	1.719	3.781	2.199	22.685
0.7	1.702	3.440	2.021	24.083
0.8	1.685	3.190	1.893	25.521
0.9	1.667	3.000	1.800	27.000
1.0	1.648	2.852	1.731	28.521
1.1	1.629	2.735	1.679	30.083
1.2	1.610	2.640	1.640	31.685
1.3	1.590	2.563	1.612	33.325
1.4	1.571	2.500	1.591	35.000
15	1.553	2.447	1.576	36.708
1.6	1.535	2.403	1.566	38.446
1.7	1.517	2.365	1.559	40.211
1.8	1.500	2.333	1.556	42.000
1.9	1.484	2.306	1.554	43.810
2.0	1.468	2.282	1.552	45.639
2.1	1.453	2.261	1.556	47.485
2.2	1.439	2.243	1.559	49.346
2.3	1.425	2.227	1.563	51.220
2.4	1.412	2.213	1.567	53.105
2.5	1.400	2.200	1.571	55.000
2.6	1.388	2.189	1.576	56.904
2.7	1.377	2.178	1.582	58.817
2.8	1.367	2169	1.587	60.736
2.9	1.356	2.161	1.593	62.662
3.0	1.347	2.153	1.599	64.593
3.1	1.338	2.146	1.604	66.530
3.2	1.329	2.140	1.610	68.471
3.3	1.321	2.134	1.616	70.416
3.4	1.313	2.128	1.621	72.365
3.5	1.305	2.123	1.627	74.318
3.6	1.298	2.119	1.632	76.273
3.7	1.291	2.114	1.638	78.232
3.8	1.284	2.110	1.643	80.192
3.9	1.278	2.107	1.648	82.155
4.0	1.272	2.103	1.653	84.121
4.1	1.266	2.100	1.658	86.088
4.2	1.261	2.097	1.663	88.057
4.3	1.255	2.094	1.668	90.028
4.4	1.250	2.091	1.673	92.000

Table 2.14 (continued)

Dq/B′	ν_2/ν_1	ν_3/ν_1	ν_3/ν_2	ν_3/B
4.5	1.245	2.088	1.677	93.974
4.6	1.240	2.086	1.682	95.949
4.7	1.236	2.084	1.686	97.925
4.8	1.231	2.081	1.690	99.902
4.9	1.227	2.079	1.695	101.881
5.0	1.223	2.077	1.699	103.860

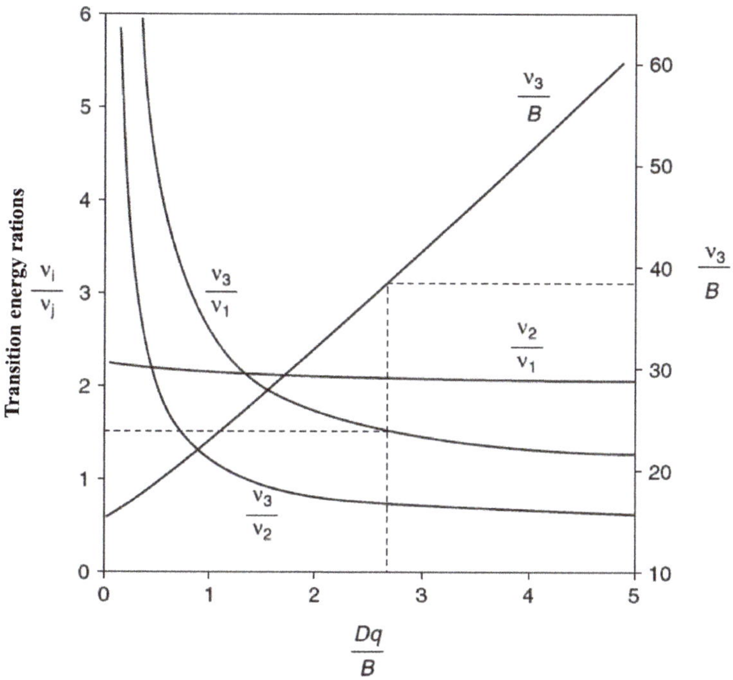

Fig. 2.63: Plots of theoretical transition energy ratios against Dq/B for complexes with T_1 ground states (drawn using data presented by A. B. P. Lever, 1968).

(i) d^2- configuration $[V(H_2O)_6]^{3+}$ O_h complex

For octahedral complexes of d^2 ion, three spin-allowed transitions expected (as shown by three arrows in Fig. 2.65) are: ${}^3T_{1g}(F)\rightarrow{}^3T_{2g}(F)(\nu_1)$, ${}^3T_{1g}(F)\rightarrow{}^3A_{2g}(F)$ (ν_2) ${}^3T_{1g}$ (F)→ ${}^3T_{1g}$ (P)(ν_3). But actually only two absorption bands located at 17, 800 cm^{-1} (ν_1) and 25, 700 cm^{-1} (ν_3) are observed in weak field electronic spectrum of $[V(H_2O)_6]^{3+}$ in alum crystals.

In the strong field configuration, the ${}^3T_{1g}$ (F) → ${}^3A_{2g}$ (F) (ν_2) band corresponds to a two-electron excitation ($t_{2g}{}^2\rightarrow e_g{}^2$) and hence is ought to be forbidden. It is thus

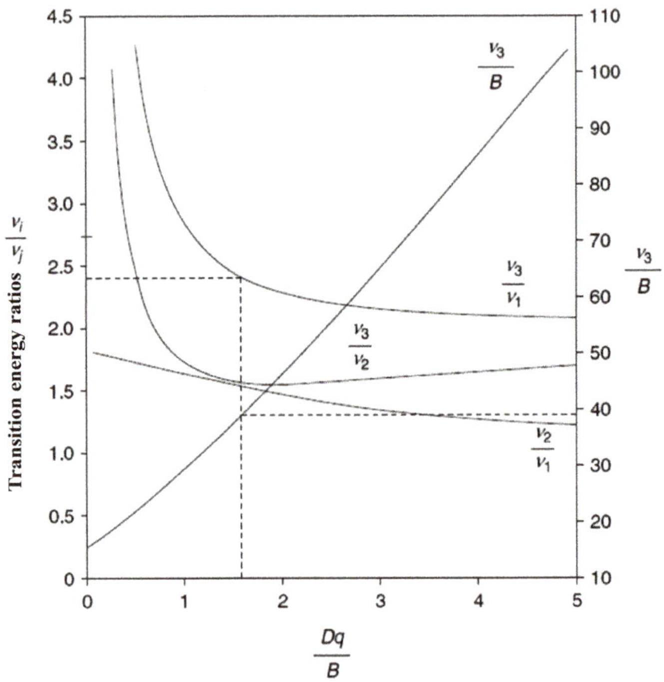

Fig. 2.64: Plots of theoretical transition energy ratios against Dq/B for complexes with A_2 ground states (drawn using data presented by A. B. P. Lever, 1968).

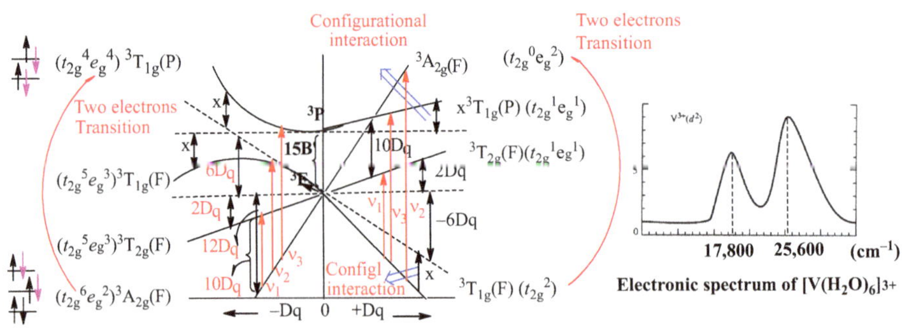

Fig. 2.65: Orgel diagram for d^2 O_h crystal field.

relatively less probable. It is not observed owing to the intense charge transfer band occurring in this region of the spectrum and also probably due to its low intensity.

As per Orgel diagram, the assignments of three expected absorption bands are as follows:

$$^3T_{1g} \rightarrow {}^3T_{2g}(F)(\nu_1) = 8Dq + x$$

$$^3T_{1g} \rightarrow {}^3A_{2g}(F)(\nu_2) = 18Dq + x$$

$$^3T_{1g} \rightarrow {}^3T_{1g}(P)(\nu_3) = 15B' + 6Dq + x$$

Hence,

$\nu_3/\nu_1 = 25,700/17,800 = 1.443$. This gives $Dq/B' = 3.1$ (From Table 2.13).

But from Table 2.13, $Dq/B' = 3.1$ is equivalent to $\nu_3/B' = 41.761$. So that

$$B' = \nu_3/41.761 = 25,\ 700/41.761 = 615.4$$

But $Dq/B' = 3.1$, therefore, $Dq = 3.1 \times B' = 3.1 \times 615.4 = 1907.4$ or $10\ Dq = 19,\ 074$

$$\beta = B'/B = 615.4/861 = 0.71$$

$$\beta^0 = (1 - \beta) \times 100 = (1 - 0.71) \times 100 = 29\%$$

Now from $(\nu_1) = 8Dq + x$, the bending parameter (x) can be calculated. So,

$$17800 = 8 \times 1907.4 + x$$
$$\text{or } x = 17800 - 15259 = 2541\ cm^{-1}$$

Now, we can calculate the ν_2 band using equation

$$\nu_2 = 18Dq + x = 18 \times 1907.4 + 2541$$
$$= 36874\ cm^{-1}$$

As already mentioned, this band is masked by the charge transfer bands. Obviously, this is a case where $^3T_{1g}$ (P) is lower than $^3A_{2g}$ (F) and the second band that appeared in the spectrum must be assigned to be ν_3.

It is notable here that electronic parameters, 10Dq, β and β^0 for d^2 octahedral complex calculated using d^2-Tanabe-Sugano diagram differ slightly from those calculated by graphical method. This may be due to an error in reading the ν_1/B from the diagram at $10Dq/B = 31$.

(ii) d^7- configuration $[Co(H_2O)_6]^{2+}$ O_h complex

The six-coordinate high-spin Co(II) is a commonly encountered $3d^7$ system. The aqueous solutions of cobaltous (Co^{2+}) salts containing $[Co(H_2O)_6]^{2+}$ complex ion are pale pink. A weak, well-resolved absorption band is observed around 8000 cm^{-1} and a multiple absorption band comprising of three overlapping peaks at about 20,000 cm^{-1}.

The right half of the Orgel diagram (Fig. 2.66) is applicable for assignment of these bands. The energy terms arising out of d^7 configuration of free Co^{2+} ion are ^4F, ^4P, ^2H, ^2G, ^2F, ^2D, ^2P. The lowest energy band at 8, 000 cm^{-1} is assigned to the $^4T_{1g}$ (F)$\rightarrow$$^4T_{2g}$(F)($\nu_1$) transition. The multiple bands have three peaks at about 16,000, 19,

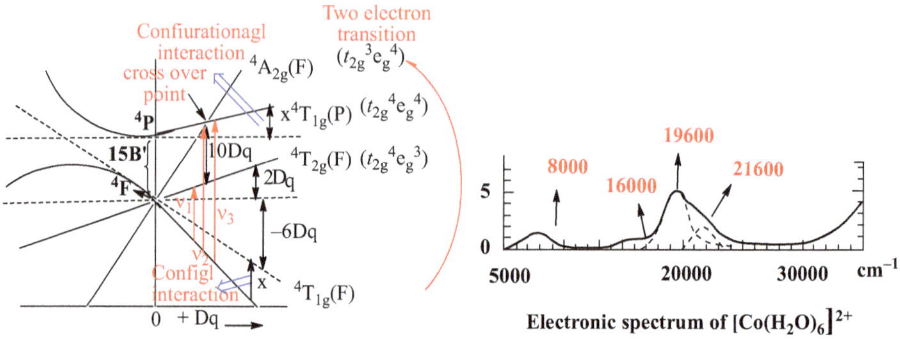

Fig. 2.66: Orgel diagram for d^7 metal ion in O_h crystal field.

400 and 21,600 cm^{-1}. Two of these low energy peaks at 16, 000, 19, 400 cm^{-1} are generally accepted as due to $^4T_{1g}(F)\rightarrow{}^4A_{2g}(F)$ (ν_2) and $^4T_{1g}(F)\rightarrow{}^4T_{1g}(P)$ (ν_3), respectively. Since, these two peaks are close together, it indicates that in this complex, these two transitions are close to the crossover point between the $^4A_{2g}(F)$ and $^4T_{1g}$ (P) on the energy diagram. The fourth band at 21, 600 cm^{-1} is attributed either to spin-orbit coupling effects or to transitions to doublet states, ^2H, ^2G, ^2F, ^2D and ^2P.

As per Orgel diagram, the assignments of three expected absorption bands are as follows:

$$8,000\,cm^{-1}:\quad {}^4T_{1g}\rightarrow{}^4T_{2g}(F)(\nu_1) = 8Dq + x$$
$$16,000\,cm^{-1}:\quad {}^4T_{1g}\rightarrow{}^4A_{2g}(F)(\nu_2) = 18Dq + x$$
$$19,400\,cm^{-1}:\quad {}^4T_{1g}\rightarrow{}^4T_{1g}(P)(\nu_3) = 15B' + 6Dq + 2x$$

Now,

$\nu_3/\nu_1 = 19,400/8,000 = 2.425$. This gives $Dq/B' = 1.1$ (From Table 2.13).

But from Table 2.13, $Dq/B' = 1.1$ is equivalent to $\nu_3/B' = 23.324$. So that

$$B' = \nu_3/23.324 = 19,\ 400/23.324 = 832$$

But $Dq/B' = 1.1$ or $Dq = B' \times 1.1 = 832 \times 1.1 = 915\,cm^{-1}$

Hence, $10Dq = 9,\ 150\,cm^{-1}$

$$\beta = B'/B = 832/971 = 0.85\ (\text{For } Co^{2+}, B = 971)$$

$$\beta^0 = (1-\beta) \times 100 = (1-0.85) \times 100 = 15\%$$

Now from $\nu_1 = 8Dq + x$, the bending parameter (x) *can be calculated. So,*

$$8000 = 8 \times 915 + x$$

$$\text{or}\quad x = 8000 - 7320 = 680\,cm^{-1}$$

Now, we can theoretically calculate the values of ν_2 and ν_3 using the values of Dq, B′ and x calculated above. So, that

$$\nu_2 = 18\text{Dq} + \text{x} = 18 \times 915 + 680 = 17,150\ \text{cm}^{-1}$$

$$\nu_3 = 15\text{B}' + 6\text{Dq} + 2\text{x} = 15 \times 832 + 6 \times 915 + 2 \times 680 = 12480 + 5490 + 1360 = 19,330\ \text{cm}^{-1}$$

Thus, the theoretical calculations of ν_2 and ν_3 prove that $^4A_{2g}(F)$ (ν_2) level is placed lower than $^4T_{1g}(P)$ (ν_3) level unlike that was found in d^2 (V^{3+}) case.

Calculation of electronic parameters in complexes of A_2 ground states

The octahedral complexes of d^3, d^8 and tetrahedral complexes of d^2, d^7 configurations give A_2 ground state. Let us illustrate the method for calculation of electronic parameters: Dq, B′ and β for octahedral d^3 and d^8 complexes using the appropriate experimental data.

(i) d^3- configuration $[Cr(H_2O)_6]^{3+}$ O_h complex

The salts of Cr(III) in water having $[Cr(H_2O)_6]^{3+}$ ion, give a light green solution. The color is the result of absorption in the yellow and blue parts of the electromagnetic radiation. The electronic spectrum of the complex $[Cr(H_2O)_6]^{3+}$ ion shows two bands at 17,400 (ε = 14) and 24,600 (ε = 15) cm^{-1}. There is a weak absorption band at 37,800 cm^{-1} (ε = 4). The low intensity of the band corresponds to a two-electron transition as expected. The left half of the Orgel diagram (Fig. 2.67) may be considered for assigning these absorption bands. A simplified version of this diagram along with full details is shown in Fig. The free ion has ^4F as the ground state term. The other term of the same multiplicity is ^4P.

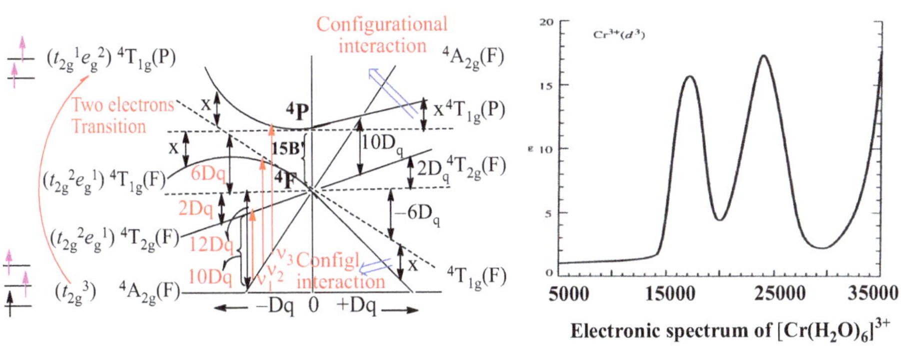

Fig. 2.67: Orgel diagram for d^3 metal ion in O_h crystal field.

As per Orgel diagram, the assignments of three expected absorption bands are as follows:

$$
\begin{aligned}
17,400\mathrm{cm}^{-1}&:\quad {}^4A_{2g}\rightarrow{}^4T_{2g}(F)(\nu_1)=10Dq\\
24,600\mathrm{cm}^{-1}&:\quad {}^4A_{2g}\rightarrow{}^4T_{2g}(F)(\nu_2)=18Dq-x\\
37,800\mathrm{cm}^{-1}&:\quad {}^4A_{2g}\rightarrow{}^4T_{2g}(F)(\nu_3)=12Dq+15B'+x
\end{aligned}
$$

Now,

ν_2/ν_1 = 24, 600/ 17,400 = 1.413. This gives Dq/B′ = 2.4 (From Table 2.14). But from Table 2.14, Dq/B′ = 2.4 is equivalent to ν_3/B′ = 53.105. So that

$$B'=\nu_3/53.105=37,800/53.105=712\ \mathrm{cm}^{-1}$$

$$\text{As}\ \nu_1=10\ Dq,\ \text{hence}\ 10\ Dq=17,400\ \mathrm{cm}^{-1}\ \text{and}\ Dq=1740$$

$$\beta=B'/B=712/918=0.77(\text{For}\ Cr^{3+},B=918)$$

$$\beta^0=(1-\beta)\times100=(1-0.77)\times100=23\%$$

Now from ν_2 = 18Dq – x, the bending parameter (x) can be calculated. So,

$$24,\ 600=18\times1740-x=31,\ 320-x$$

$$\text{or}\quad x=31,\ 320-24,\ 600=6720\ \mathrm{cm}^{-1}$$

Then, the ν_3 band is expected to be at

$$
\begin{aligned}
\nu_3&=12Dq+15B'+x=12\times1740+15\times712+6720\\
&=20,\ 880+10,\ 680+6720\\
&=38,\ 280\ \mathrm{cm}^{-1}
\end{aligned}
$$

This result suggests a good fit of the data.

(ii) d^8- configuration $[Ni(H_2O)_6]^{2+}$ O_h complex

In d^8 Ni(II), there are two holes in the e_g level. Hence, promoting an electron is equivalent to transferring a hole from e_g to t_{2g} level. This is the inverse of the d^2 case. Hence, one can consider the left half of the Orgel diagram (Fig. 2.68) for

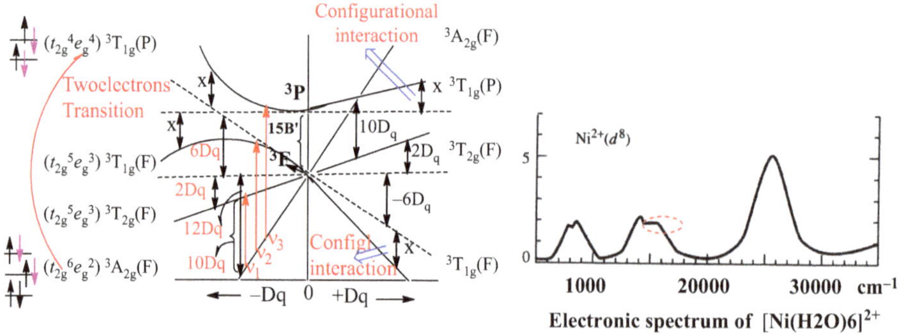

Fig. 2.68: Orgel diagram for d^8 metal ion in O_h crystal field.

interpretation/assignment of absorption bands for Ni(II) complexes. The d^8 Ni^{2+} ion has energy terms, ^3F, ^3P, ^1G, ^1D, ^1S, as a free metal ion with d^2 configuration has. The ground term is ^3F and the other term of the same multiplicity is ^3P.

The three bands in the electronic spectrum of $[Ni(H_2O)_6]^{2+}$ at 8,700, 14,500 are 25,300 cm^{-1} are identified. The weak bands in the region 15,000–18,000 cm^{-1} are attributed to spin-forbidden triplet $^3A_{2g}(F)$→singlet (^1G, ^1D, ^1S) transitions.

As per Orgel diagram, the assignments of three expected absorption bands are as follows:

$$8,700\,cm^{-1}:\quad {}^3A_{2g}\rightarrow{}^3T_{2g}(F)(\nu_1) = 10Dq$$

$$14,500\,cm^{-1}:\quad {}^3A_{2g}\rightarrow{}^3T_{1g}(F)(\nu_2) = 18Dq - x$$

$$25,300\,cm^{-1}:\quad {}^4A_{2g}\rightarrow{}^4T_{1g}(P)(\nu_3) = 12Dq + 15B' + x$$

Now,

ν_3/ν_1 = 25, 300/ 8,700 = 2.9. This gives Dq/B′ = 0.95 (From Table 2.14).

But from Table 2.14, Dq/B′ = 0.95 is equivalent to ν_3/B′ = 27.760. So that

$$B' = \nu_3/27.760 = 25,300/27.760 = 916\ cm^{-1}$$

$$\text{As } \nu_1 = 10\ Dq,\ \text{hence } 10\ Dq = 8,700\ cm^{-1} \text{ and } Dq = 870$$

$$\beta = B'/B = 916/1030 = 0.89 (\text{For } Ni^{2+}, B = 1030)$$

$$\beta^0 = (1-\beta)\times 100 = (1-0.89)\times 100 = 11\%$$

Now from ν_2 = 18Dq – x, the bending parameter (x) can be calculated. So,

$$14,\ 500 = 18\times 870 - x = 15,\ 660 - x$$

$$\text{or}\quad x = 15,\ 660 - 14,\ 500 = 1160\,cm^{-1}$$

Then, the ν_3 band is expected to be at

$$\begin{aligned}\nu_3 &= 12Dq + 15B' + x = 12\times 870 + 15\times 916 + 1160\\ &= 10,\ 440 + 13,\ 740 + 1160\\ &= 25,\ 340\,cm^{-1}\end{aligned}$$

This result justifies a good fit of the data.

(iii) Konig's numerical method

This method is used for calculation of electronic parameters for configurations d^2, d^8, d^3 and d^7. Each of these configurations under O_h symmetry will be taken separately.

(i) d^2-configuration with O_h complexes

Let us consider the d^2 system. From the Orgel diagram given below, it is notable that the energies of ^3F and ^3P terms are as follows:

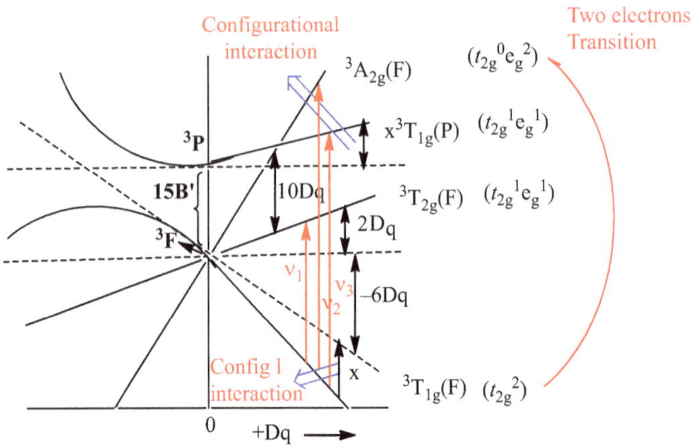

This configuration gives rise the following three spin-allowed transitions, which are labelled as ν_1, ν_2 and ν_3 in the increasing order of energy:

$$^3T_{1g}{\to}^3T_{2g}(F)\ (\nu_1)$$
$$^3T_{1g}{\to}^3T_{1g}(P)\ (\nu_2)$$
$$^3T_{1g}{\to}^3A_{2g}(F)\ (\nu_3)$$

(a) Without interaction between the terms

$$E\left[^3T_{1g}(F)\right] = -6Dq$$
$$E\left[^3T_{1g}(P)\right] = 15B'$$
$$E\left[^3T_{2g}(F)\right] = 2Dq$$
$$E\left[^3A_{2g}(F)\right] = 12Dq$$

(b) With interaction between the T_{1g} terms

$$E\left[^3T_{1g}(F)\right] = -6Dq - x$$
$$E\left[^3T_{1g}(P)\right] = 15B' + x$$
$$E\left[^3T_{2g}(F)\right] = 2Dq$$
$$E\left[^3A_{2g}(F)\right] = 12Dq$$

Here x is the energy of interaction.

The secular determinant is as follows:

$$\begin{vmatrix} -6Dq - E & x \\ x & 15B' - E \end{vmatrix} = 0 \tag{2.5}$$

This determinant on expansion gives:

$$(-6Dq - E)\,(15B' - E) - x^2 = 0 \tag{2.6}$$

Under the strong field limits, the inter-electron repulsions become negligible (i.e., $B' = 0$), so that the eq. (2.6) can be written as

$$\begin{aligned} &(-6Dq - E)\,(-E) - x^2 = 0 \quad (B' = 0) \\ \text{or}\quad &E^2 + 6DqE - x^2 = 0 \end{aligned} \tag{2.7}$$

If $x = 4Dq$, then eq. (2.7) becomes

$$E^2 + 6DqE - 16Dq^2 = 0$$

$$\text{So,}\ E = \frac{-6Dq \pm \sqrt{36\,Dq^2 + 64\,Dq^2}}{2} = \frac{-6Dq \pm 10Dq}{2}$$

This quadratic equation gives $E = -8$ Dq and 2 Dq.

Substituting $x = 4Dq$ in eq. (2.6), we get

$$\begin{aligned} &(-6Dq - E)\,(15B' - E) - 16Dq^2 = 0 \\ \text{or}\quad &-90DqB' + 6DqE - E15B' + E^2 - 16Dq^2 = 0 \end{aligned}$$

$$\text{or}\quad \underbrace{E^2}_{a=1} + E\,\underbrace{(6Dq - 15B')}_{b} - \underbrace{90\,DqB' - 16\,Dq^2}_{c} \qquad \left[\begin{array}{l} ax^2 + bx + c = 0 \\ x = \frac{-b \pm \sqrt{b^2 - 4ac}}{2a} \end{array}\right]$$

$$E = \frac{-b \pm \sqrt{b^2 - 4ac}}{2a}$$

This is a quadratic equation, which gives us two roots corresponding to the energies of two $^3T_{1g}$ levels as follows:

$$E = \frac{-(6Dq - 15B') \pm \sqrt{36\,Dq^2 + 225\,B'^2 - 180\,DqB' - 4(-90DqB' - 16\,Dq^2)}}{2}$$

$$\text{or}\quad E = \frac{15B' - 6Dq \pm \sqrt{36\,Dq^2 + 225\,B'^2 - 180\,DqB' + 360DqB' + 64\,Dq^2)}}{2}$$

$$\text{or}\quad E = \frac{15B' - 6Dq \pm \sqrt{100\,Dq^2 + 225\,B'^2 + 180\,DqB'}}{2}$$

$$\text{or}\quad E = \tfrac{1}{2}\left[15B' - 6Dq \pm \{225B'^2 + 100\,Dq^2 + 180\,DqB'\}^{1/2}\right]$$

Thus,

$$E[^3T_{1g}(F)] = \frac{1}{2}\left[15B' - 6Dq - \{225B'^2 + 100\,Dq^2 + 180\,DqB'\}^{1/2}\right]$$

$$E[^3T_{1g}(P)] = \frac{1}{2}\left[15B' - 6Dq + \{225B'^2 + 100\,Dq^2 + 180\,DqB'\}^{1/2}\right]$$

The $^3T_{1g}$ (F) is lowered, whereas the $^3T_{1g}$ (P) level is raised due to the interaction. This interaction introduces curvature into the two $^3T_{1g}$ levels. The energy of the other unchanged noninteracting levels giving straight lines will be the same as given above, which are again presented here as:

$$E[^3T_{2g}(F)] = 2Dq$$
$$E[^3A_{2g}(F)] = 12Dq$$

The same discussion can be extended to $^3T_{1g}$ levels of d^8 and to the $^4T_{1g}$ levels of d^3 and d^7-configurations.

Using these new energy expressions for $^3T_{1g}$ (F) and $^3T_{1g}$ (P) for a d^2-octahedral complex, we get the transition energies as given below:

$$\begin{aligned}
\nu_1{:}[^3T_{1g}(F)\rightarrow{}^3T_{2g}(F)] &= 2Dq - \tfrac{1}{2}[15B' - 6Dq - \{225B'^2 + 100Dq^2 + 180DqB'\}^{1/2}]\\
&= \tfrac{1}{2}4Dq - \tfrac{1}{2}[15B' - 6Dq - \{225B'^2 + 100Dq^2 + 180DqB'\}^{1/2}]\\
&= \tfrac{1}{2}[10Dq - 15B' + \{225B'^2 + 100Dq^2 + 180DqB'\}^{1/2}]
\end{aligned}$$

$$\begin{aligned}
\nu_2 : [^3T_{1g}(F) \rightarrow 3T_{1g}(P)] &= \frac{1}{2}\left[15B'-6Dq + \{225B'^2 + 100\,Dq^2 + 180\,DqB'\}^{1/2}\right]\\
&\quad -\frac{1}{2}\left[15B'-6Dq-\{225B'^2 + 100\,Dq^2 + 180\,DqB'\}^{1/2}\right]\\
&= \frac{1}{2}\{225B'^2 + 100\,Dq^2 + 180\,DqB'\}^{1/2}\\
&\quad + \frac{1}{2}\{225B'^2 + 100\,Dq^2 + 180\,DqB'\}^{1/2}\\
&= \{225B'^2 + 100\,Dq^2 + 180\,DqB'\}^{1/2}
\end{aligned}$$

$$\begin{aligned}
\nu_3{:}[^3T_{1g}(F)\rightarrow{}^3A_{2g}(F)] &= 12Dq - \tfrac{1}{2}\left[15B' - 6Dq - \{225B'^2 + 100\,Dq^2 + 180\,DqB'\}^{1/2}\right]\\
&= \tfrac{1}{2}\,24\,Dq - \tfrac{1}{2}\left[15B' - 6Dq - \{225B'^2 + 100\,Dq^2 + 180\,DqB'\}^{1/2}\right]\\
&= \tfrac{1}{2}\left[30\,Dq - 15B' + \{225B'^2 + 100\,Dq^2 + 180\,DqB'\}^{1/2}\right]
\end{aligned}$$

On dividing these energy relations by B′ (Racah parameter), we see that the transition energies can be plotted with E/B' versus Dq/B' to get the Tanabe-Sugano diagrams. In terms of Dq and B′, the values of ν_1, ν_2 and ν_3 are given by

$$\nu_1 = \frac{1}{2}(10D_q - 15B') + \frac{1}{2}[(10D_q + 15B')^2 - 120D_qB']^{1/2} \quad [(10D_q + 15B')^2 = 100D_q^2 + 225B'^2 + 300D_qB']$$

$$\nu_2 = [(10D_q + 15B')^2 - 120D_qB']^{1/2}$$

$$\nu_3 = \frac{1}{2}(30D_q - 15B') + \frac{1}{2}[(10D_q + 15B')^2 - 120D_qB']^{1/2}$$

From the above equations, it can be observed that the ν_1 and ν_3 contain ν_2 as their second part of the equations. Often, it is not possible to observe all the three absorption bands as evidenced from the experimental data available. However, depending on the bands observed, suitable modification of the above equations helps to obtain the values of 10Dq and B′.

(a) Using ν_1 and ν_2 bands

In case first and second bands are observed, the corresponding equations can be manipulated to give

$$10Dq = 2\nu_1 - \nu_2 + 15B' \tag{2.8}$$

$$30B' = \left[(\nu_2 - 2\nu_1) \pm (-\nu_1^2 + \nu_2^2 + \nu_1\nu_2)^{1/2}\right]$$

$$\text{or} \quad B' = 1/30\left[(\nu_2 - 2\nu_1) \pm (-\nu_1^2 + \nu_2^2 + \nu_1\nu_2)^{1/2}\right] \tag{2.9}$$

Or, alternatively, we can get

$$10Dq = 1/2\left[(2\nu_1 - \nu_2) + (-\nu_1^2 + \nu_2^2 + \nu_1\nu_2)^{1/2}\right] \tag{2.10}$$

$$15B' = (\nu_2 - 2\nu_1 + 10Dq)$$

$$\text{or} \quad B' = 1/15(\nu_2 - 2\nu_1 + 10Dq) \tag{2.11}$$

Use of either of the above sets of equations leads to the same values of 10 Dq and B′. It is notable here that only the positive sign in eq. (2.9) gives the meaningful results.

(b) Using ν_1 and ν_3 bands

When the first and third d-d transitions are observed in the spectrum, we can use the equations.

$$10Dq = \nu_3 - \nu_1 \tag{2.12}$$

$$B' = (2\nu_1^2 - \nu_1\nu_3)/(12\nu_3 - 27\nu_1) \tag{2.13}$$

(c) Using ν_2 and ν_3 bands

When only ν_2 and ν_3 bands are observed, manipulation of ν_2 and ν_3 leads to

$$10Dq = 1/3(2\nu_3 - \nu_2) + 5B' \tag{2.14}$$

$$B' = (1/510)\left[7(\nu_2 - 2\nu_3) \pm 3\{81\nu_2^2 - 16\nu_3(\nu_3 - \nu_2)\}^{1/2}\right] \tag{2.15}$$

(d) Using ν_1, ν_2 and ν_3 bands

When all the three d-d transitions are observed, the following simplified expressions could be used.

$$10Dq = \nu_3 - \nu_1 \tag{2.16}$$

$$15B' = (\nu_2 + \nu_3 - 3\nu_1) \tag{2.17}$$

As an example the application of the above sets of equation to V^{3+} in Al_2O_3 is given below.

The observed transitions are: $\nu_1 = 17, 400$, $\nu_2 = 25, 200$ and $\nu_3 = 34, 500\ cm^{-1}$

Method (d)

$$10Dq = \nu_3 - \nu_1 = 34,500 - 17,400 = 17,100 cm^{-1}$$

$$15B' = (\nu_2 + \nu_3 - 3\nu_1) = 25,200 + 34,500 - 3 \times 17,400$$

$$= 59,700 - 52,200 = 7500 cm^{-1}$$

$$B' = 7500/15 = 500 cm^{-1}$$

(ii) d^7-configuration with O_h complexes

The energy equations of d^2 and d^7 are identical excepting that the energy states are spin- quartets. The following spin-allowed states are possible:

$$^4T_{1g} \rightarrow {}^4T_{2g}(F) \quad (\nu_1)$$

$$^4T_{1g} \rightarrow {}^4A_{2g}(F) \quad (\nu_2)$$

$$^4T_{1g} \rightarrow {}^4T_{1g}(P) \quad (\nu_3)$$

As long as the configuration remains high-spin, the ground state is $^4T_{1g}$ and is similar to that of d^2. However, the energy ordering of d^7 differs from that of d^2 in respect of ν_2 and ν_3. In fact, $^4T_{1g}$ (P) state has been found experimentally at higher energy than $^4A_{2g}$ (F).

$$\nu_1 = \frac{1}{2}(10Dq - 15B') + \frac{1}{2}\left[(10Dq + 15B')^2 - 120\ DqB'\right]^{1/2}$$

$$\nu_2 = \frac{1}{2}(30Dq - 15B') + \frac{1}{2}\left[(10Dq + 15B')^2 - 120\ DqB'\right]^{1/2}$$

$$\nu_3 = \left[(10Dq + 15B')^2 - 120\ DqB'\right]^{1/2}$$

It is notable here that ν_2 of d^2 had now become ν_3 of d^7. The parameters 10 Dq and B can be calculated by any of the following methods.

(a) Using ν_1 and ν_2 bands

In case only two bands are observed, then we get

$$10Dq = \nu_2 - \nu_1 \tag{2.18}$$

$$B' = (2\nu_1^2 - \nu_1\nu_2)/(12\nu_2 - 27\nu_1) \tag{2.19}$$

(b) Using ν_1 and ν_3 bands

When ν_1 and ν_3 bands are observed, we can use the following relations:

$$10Dq = 2\nu_1 - \nu_3 + 15B' \tag{2.20}$$

$$30B' = [(\nu_3 - 2\nu_1) \pm (-\nu_1^2 + \nu_3^2 + \nu_1\nu_3)^{1/2}]$$
$$\text{or} \quad B' = 1/30[(\nu_3 - 2\nu_1) \pm (-\nu_1^2 + \nu_3^2 + \nu_1\nu_3)^{1/2}] \tag{2.21}$$

(c) Using ν_2 and ν_3 bands

When ν_2 and ν_3 bands are observed, we can use the following relations:

$$10Dq = 1/3(2\nu_2 - \nu_3) + 5B' \tag{2.22}$$

$$B' = (1/510)[7(\nu_3 - 2\nu_2) \pm 3\{81\nu_3^2 - 16\nu_2(\nu_2 - \nu_3)\}^{1/2}] \tag{2.23}$$

(d) Using ν_1, ν_2 and ν_3 bands

When all the three bands are observed, the following relations can be used:

$$10Dq = \nu_2 - \nu_1 \tag{2.24}$$

$$15B' = (\nu_2 + \nu_3 - 3\nu_1) \tag{2.25}$$

Application to Co^{2+} (d^7) in $[Co(H_2O)_6]^{2+}$

The observed transitions are: $\nu_1 = 8,100$, $\nu_2 = 16,000$ and $\nu_3 = 19,400\ cm^{-1}$

Method (d)

$$\mathbf{10Dq} = \nu_2 - \nu_1 = 16,000 - 8,100 = 7,900 cm^{-1}$$

$$\mathbf{15B'} = (\nu_2 + \nu_3 - 3\nu_1) = 16,000 + 19,400 - 3 \times 8,100$$
$$= 35,400 - 24,300 = 11,100 cm^{-1}$$
$$\mathbf{B'} = 11,100/15 = 740 cm^{-1}$$

Similarly other methods (a), (b) and (c) may be used for such calculations.

(iii) d^3 and d^8-configuration with O_h complexes

The energy expressions and order of transitions are identical for both the d^3 and d^8 octahedral systems except for different multiplicity. There are three possible spin-allowed transitions as given below:

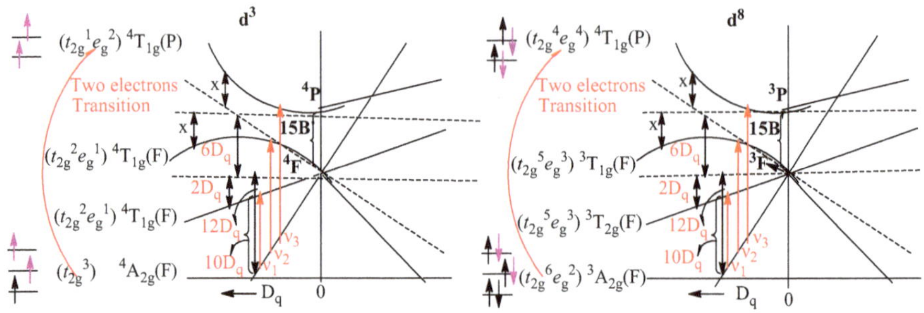

$$^{n}A_{2g} \rightarrow {}^{n}T_{2g}(F)\ (\nu_1)$$
$$^{n}A_{2g} \rightarrow {}^{n}T_{1g}(F)\ (\nu_2)$$
$$^{n}A_{2g} \rightarrow {}^{n}T_{1g}(P)\ (\nu_3)$$

Here $n = 4$ for d^3 and 3 for d^8 configurations. Their transition energies can be similarly obtained as:

$$\nu_1 = 10\text{Dq}$$
$$\nu_2 = 1/2(15\text{B}' + 30\text{Dq}) - 1/2[(15\text{B}' - 10\text{Dq})^2 + 120\text{DqB}']^{1/2}$$
$$\nu_3 = 1/2(15\text{B}' + 30\text{Dq}) + 1/2[(15\text{B}' - 10\text{Dq})^2 + 120\text{DqB}']^{1/2}$$

The above equations can be recombined to get the relevant expressions for the calculation of 10Dq and B′ parameters from the experimental spectrum.

(a) Using ν_1 and ν_2 bands

$$10\text{Dq} = \nu_1 \quad (2.26)$$

$$\text{B}' = (2\nu_1^2 + \nu_2^2 - 3\nu_1\nu_2)/(15\nu_2 - 27\nu_1) \quad (2.27)$$

(b) Using ν_1 and ν_3 bands

$$10\text{Dq} = \nu_1 \quad (2.28)$$

$$\text{B}' = (2\nu_1^2 + \nu_3^2 - 3\nu_1\nu_3)/(15\nu_3 - 27\nu_1) \quad (2.29)$$

(c) Using ν_1, ν_2 and ν_3 bands

$$10\text{Dq} = \nu_1 \quad (2.30)$$

$$15\text{B}' = (\nu_2 + \nu_3 - 3\nu_1) \quad (2.31)$$

Application to V^{3+} (d^3) in $[V(H_2O)_6]^{3+}$

The observed transitions/bands are: ν_1 = 12, 300, ν_2 = 18, 500 and ν_3 = 27, 900 cm^{-1}

Method (c)

$$\mathbf{10Dq} = \nu_1 = 12,300\text{cm}^{-1}$$

$$\mathbf{15B}' = (\nu_2 + \nu_3 - 3\nu_1) = 18,500 + 27,900 - 3 \times 12,300$$

$$= 46,400 - 36,900 = 9,500\text{cm}^{-1}$$

$$\mathbf{B}' = 9,500/15 = 633\text{cm}^{-1}$$

Similarly other methods (a) and (b) may be used for such calculations.

Application to Ni^{2+} (d^8) in $[Ni(H_2O)_6]^{2+}$

The observed transitions/bands are: ν_1 = 8,500, ν_2 = 13,800 and ν_3 = 25,300 cm^{-1}

$$\mathbf{10Dq} = \nu_1 = 8,500\text{cm}^{-1}$$

$$\mathbf{15B}' = (\nu_2 + \nu_3 - 3\nu_1) = 13,800 + 25,300 - 3 \times 8,500$$

$$= 39,\ 100 - 25,\ 500 = 13,\ 600\text{cm}^{-1}$$

$$\mathbf{B}' = 13,\ 600/15 = 906.7\text{cm}^{-1}$$

Similarly other methods (a) and (b) may be used for such calculations.

Exercises

Multiple choice questions/fill in the blanks

1. The molecular orbitals of an organic molecule are in the order:
   (a) $\pi < \sigma < n < \pi^* < \sigma^*$ (b) $\sigma < \pi < n < \sigma^* < \pi^*$
   (c) $\sigma < \pi < n < \pi^* < \sigma^*$ (d) $n < \pi < \sigma < \pi^* < \sigma^*$

2. In centrosymmetric molecules, transitions between electronic states of same symmetry parity are called...............transitions.

3. Which one of the following transitions is spin forbidden?
   (a) singlet to singlet (b) doublet to doublet
   (c) triplet to triplet (d) doublet to quartet

4. Among the following involving charge transfer transitions, which one is colorless
   (a) CdS (b) As_2S_3 (c) Sb_2S_3 (d) ReO_4^-

5. The term symbol of atoms having empty subshells and fully filled subshells is..........

6. In subshells containing one electron in each (np^1nd^1), the number of microstates will be:
(a) 40 (b) 50 (c) 60 (d) 70

7. The ground state term for d^2 for configuration among $^3F_4$, $^3F_3$, $^3F_2$ is. . . . .

8. The additional broadening feature in the electronic spectrum of the complex ion, $[Ti(H_2O)_6]^{3+}$ is due to . . . .

9. The Tanabe-Sugano correlation diagram is useful for predicting d-d transitions in metal complexes in a
(a) qualitative manner (b) quantitative manner
(c) both qualitative and quantitative anner (d) none of these

10. When the Racah parameter for a free metal ion (B) is equal to the Racah parameter of it complex (B′), the metal ligand bond will be
(a) 100% ionic (b) 100% covalent (c) 50% ionic (d) 50% covalent

11. In case of $[Ni(H_2O)_6]^{2+}$, 10 Dq equals to:
(a) ν_3 (b) ν_2 (c) ν_2- ν_1 (d) none of these

Short answer type questions

1. How is Group Theory helpful in predicting allowed and forbidden transitions in organic compounds? Explain with a suitable example.
2. What is vibronic coupling and how does it arise? Explain.
3. Charge transfer spectra are also called redox spectra. Justify this statement.
4. What are Russell–Saunders and j-j coupling? Explain.
5. What is term symbol? Work out the term symbols of H-atom in ground and excited states.
6. How microstates are worked out for electronic configurations involving inequivalent electrons? Explain with examples.
7. Constructing a pigeon hole diagram, work out the term symbols and microstates for d^2 configuration.
8. What are three Hund's rules for the determination of ground state terms for many electron atoms/ions?
9. Explain Hole formulations with suitable examples.
10. Develop the Orgel diagram for d^1 and d^9 electronic configurations in O_h field.
11. Three spin-allowed transitions are expected in octahedral $[V(H_2O)_6]^{3+}$ but only two transition are observed, why? Explain with reasoning.
12. Differentiate between Orgel and Tanabe-Sugano diagrams.

Long answer type questions

1. Based on the concept of Group Theory, discuss the electronic spectrum of a molecule belonging to C_{2v} point group. Also point out whether each of the electronic transition is allowed or not.
2. Present a detailed account of electronic spectrum of benzene molecule belonging to D_{6h} point group using the concept of Group Theory.
3. Describe in detail the charge transfer spectra in simple and coordination compounds.
4. How term symbols and microstates are worked out for atoms having equivalent and nonequivalent electronic configurations?
5. Present a detailed view of pigeon hole method to work out term symbols and microstates for p^2 and d^2 electronic configurations.
6. Present a detailed view for the determination of ground state terms for many electron atoms.
7. Present an explanatory view of octahedral complexes with d^2, d^8 and T_d complexes with d^2 electronic configurations.
8. Highlight the electronic spectra of octahedral complexes with d^3, d^7 and T_d complexes with d^2, $d^{3,}$ d^7, d^8 electronic configurations.
9. Present an account of the effect of Jahn-Teller distortion on electronic spectra of metal complexes.
10. Present a comparative view of correlation diagram for d^2 configuration in octahedral and tetrahedral field.
11. Giving the salient features of Tanabe-Sugano diagrams, discuss their utility in quantitative interpretation of electronic spectra of octahedral and tetrahedral complexes.
12. Present a detailed view of experimental method for evaluation of Dq, B′ and β parameters in metal complexes.
13. Highlight the utility of graphical method in calculation of electronic parameters in complexes of A_2 and T_1 ground states.
14. Describe Konig's numerical method for calculation of electronic parameters for d^2, d^8, d^3 and d^7 electronic configurations under O_h symmetry.

3 Molecular symmetry and group theory to vibrational spectroscopy

3.1 Introduction

A molecule has three types of internal energy. These are electronic, vibrational and rotational energies in the decreasing order of their magnitudes. In the previous chapter, we have dealt with the use of symmetry properties for understanding the electronic states of various kinds of molecules and ions. Rotational energy states have no symmetry properties of importance in ordinary chemical processes, and so it is not considered here. Then, we are left with the molecular vibrations (molecular vibrational energy), to which symmetry arguments may be fruitfully applied.

Each molecule is constantly executing vibrational motions at all temperatures, including even the absolute zero with some energy known as zero point energy ($E = ½\ h\nu$). In such vibrational motions of the molecule, its bond lengths and internal angles change periodically without producing any net translation of the center of mass of the molecule or imparting any net angular momentum (rotator motion) to the molecule. Although a superficial look at a vibrating molecule might suggest that its vibratory motion is random, a close inspection and proper analysis reveals a basic regularity and simplicity. It is the underlying basis for this simplicity that we shall formulate in this chapter. We shall also develop working methods by which all the analysis of molecular motions which symmetry alone allows may be rapidly and reliably performed.

Spectroscopy deals with the use of electromagnetic radiations in the study of structure and shapes of molecules. The infrared and Raman spectra comprise a major part of spectroscopic techniques. Molecular vibrations can interact with infrared radiations and this interaction is the subject matter of infrared or vibrational spectroscopy. The use of visible radiations to induce vibrational-rotational transitions in molecules is the basic idea underlying in Raman spectroscopy. The infrared technique is a direct measure of vibrational frequencies in the infrared region while the Raman technique measures the difference between frequency of incident and scattered light and this difference corresponds to vibrational frequencies. Use of infrared and Raman spectroscopy, which are complimentary to one another, in conjunction, is an extraordinarily efficient tool in the elucidation of molecular structure.

The presence of certain symmetry in molecules greatly reduces the efforts in predicting the molecular properties of molecules. We know that how the reducible representation Γ_{3N} can be obtained and reduced to irreducible representation of the point group to which the molecule belongs. In this chapter, we will make use of these principles to have a good deal of information about IR and Raman spectra of the molecules.

https://doi.org/10.1515/9783110635034-003

In a molecule with N number of atoms, each atom has three degrees of freedom along x, y and z axes. Hence, there should be $3N$ degrees of freedom in the molecule (Fig. 3.1), which involve translational, rotational and vibrational motions.

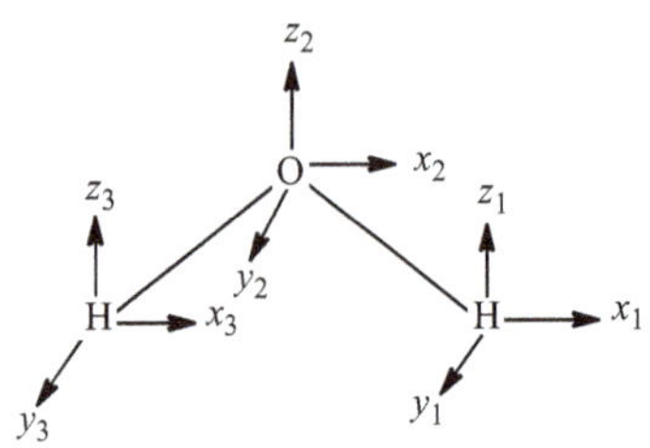

Fig. 3.1: Directions of the nine degrees of freedom in H_2O ($N = 3$) molecule.

All the molecules, irrespective of linear or non-linear, will have three degrees of translational motion along x-, y- and z-axes considering whole molecule as a rigid unit. A linear molecule will have two degrees of rotational motion (Fig. 3.2a) because rotation only about the two axes perpendicular to the bond axis (Molecular axis) constitutes rotations of the system. The rotation about the molecular axis is forbidden. Thus, the remaining 3N-(3)-(2) = 3N-5 degrees of freedom will be associated with the vibrational motion. A non-linear molecule can rotate about three mutually perpendicular axes that pass through the center of gravity of the molecule and thus have three degrees of freedom for its rotational motion (Fig. 3.2b). The remaining 3N-(3)-(3) = 3N-6 degrees of freedom will be associated with the vibrational motion.

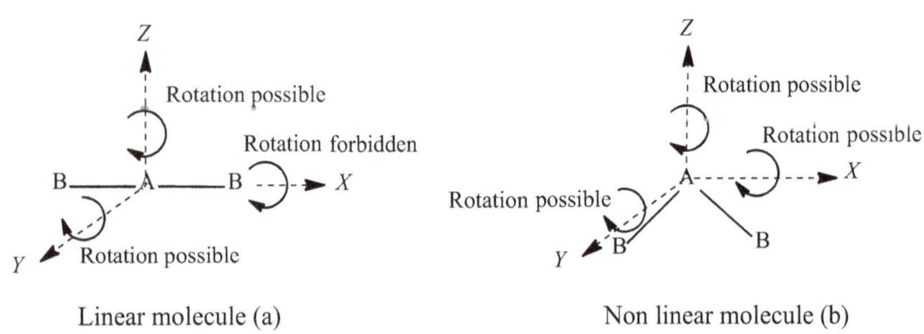

Fig. 3.2: Degrees of rotational motion in linear and non-linear molecules.

These (3N-5 or 3N-6) molecular vibrations are called normal or fundamental modes which are further divided into stretching and bending modes. Of the 3N-5 fundamental modes in linear molecules, (N-1) are stretching modes and (2N-4) are deformation or bending modes. Likewise, of the 3N-6 fundamental modes in non-linear molecules, (N-1) are stretching modes and (2N-5) are deformation/bending modes. Certain modes for linear (CO_2) and non-linear (H_2O) molecules are presented in Fig. 3.3.

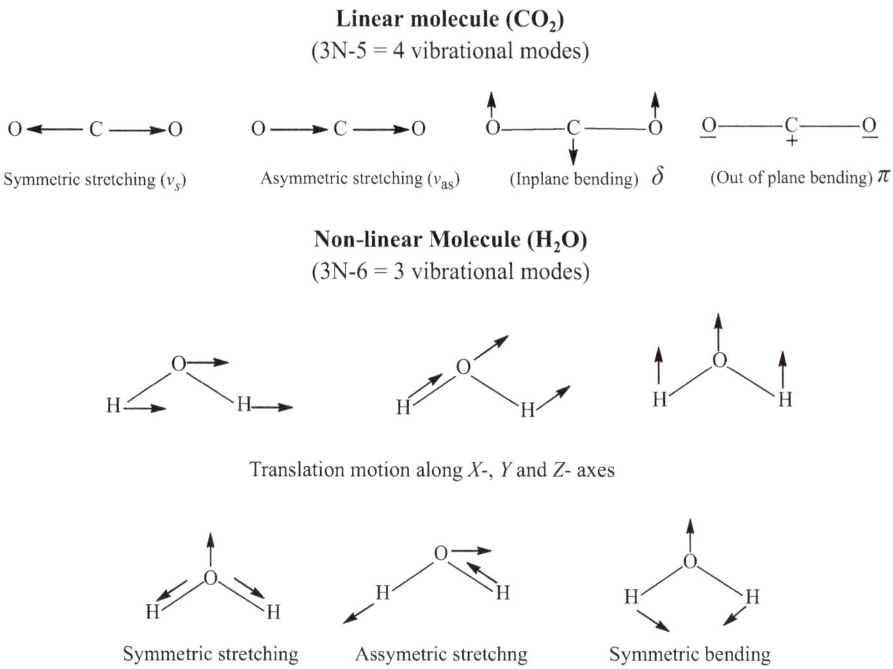

Fig. 3.3: Some modes of vibrations for CO_2 and H_2O molecules.

The complex vibrations of the molecules are due to the superimposition of relatively simple vibrations. Each of these vibrational motions involves displacement of atoms or deformation of angles. These independent displacements/deformations are called normal modes of vibrations of the molecules. These are generally divided in bond stretchings and angle of deformations. Each of these vibrational modes occurs at a characteristic energy/frequency that is dependent on the strength of the restoring force called restoring force constant.

The question now arises how the values of these characteristics frequency are obtained. In fact, there are two complimentary techniques available for obtaining these characteristics of frequencies, and these are:

(i) Infrared (IR) Spectroscopy
(ii) Raman Spectroscopy

Both these techniques when used together provide a lot of information about the molecular bonding/structure. In IR spectroscopy, if a sample is irradiated with infrared light, then those frequencies that match the corresponding frequencies of vibrational modes will be absorbed and can be measured. For a vibrational mode to be active in IR region, there should be change in the dipole moment (μ) of the molecule during vibration. The molecule is not required to have a permanent dipole. This change in dipole moment can be resolved into x-, y-, z-components and these components will transform in the same way as vectors in the x-, y-, z-directions or as p_x, p_y, p_z-orbitals transform. The frequency with which the molecule vibrates depends upon: (i) the force between the combining atoms, (ii) the masses of the atoms and (iii) geometry of the molecule.

For a vibrational mode to be active in Raman region, there must be change in polarizability (α) of the molecule during vibrations. Molecule is irradiated with UV/Visible light. The "α" is less easy to visualize than the change in dipole moment and it is usually associated with the more symmetric vibrational modes.

The polarizability is described by a tensor, that is, by a (3×3) matrix that is given by the terms α_{xx}, α_{xy}, α_{xz}, α_{yx}, α_{yy}, α_{yz}, α_{zx}, α_{zy}, α_{zz}. If a polarized excited source (a gas laser) is used for Raman experiment, then those modes that retain this polarization are totally symmetrical, that is, belong to A type representation. All the rest will be depolarized. These polarization and depolarization can easily be measured experimentally.

The fundamental difference between IR and Raman techniques is that in the former case absorption of photon is studied while in the latter case scattering of photons (elastic and non-elastic) is studied. The detailed discussions of these techniques are not being given here. However, fundamental principles of these are being given in brief.

In IR spectroscopy, quanta of radiation in IR region have energy comparable to that required for vibrational transitions in molecules. This technique is based on two fundamental principles/laws: (i) Hook's law and (ii) Franck–Condon principle.

According to Hook's law, a diatomic system may be assumed to be equivalent to a system of two masses A and B joined by a spring or a spring hooked on a wall. If the spring is stretched and let it go, then it will start vibrating about an equilibrium position. Hook's law is mathematically expressed as:

$$F = -kx$$

where F is the restoring force acting on the system to bring back it to the original position, x is the distance from equilibrium position, k is the force constant that describes the stiffness of the spring. Negative sigh means restoring force. Diatomic system A–B may be treated in a similar manner. The restoring force that acts on two atomic masses is related as

$$\Delta E = \frac{h}{2\pi}\sqrt{\left(\frac{k}{\mu}\right)}$$

where ΔE is the energy difference between ground vibrational quantum state and first excited vibrational quantum state, that is, $Ev_0 \rightarrow Ev_1$; v represents vibrational quantum states, k is force constant and μ is the reduced mass that is given below:

$$\mu = m_A m_B/(m_A + m_B)$$

Here, m_A and m_B are atomic masses of atoms A and B, respectively. As most of the molecules are in v_0 vibrational quantum state at room temperature, most transition will occur from $Ev_0 \rightarrow Ev_1$. The frequency corresponding to this transition energy is called fundamental frequency as shown in the Fig. 3.4. Therefore, first selection rule in IR spectroscopy is that absorption of radiation will occur for transition for which $\Delta v = \pm 1$ only. The second rule says that electric field component of electromagnetic radiations in IR region will interact with electric field component of oscillating electric dipole moment in the molecule as it vibrates. Any change in dipole moment or any change in its direction will give different oscillating field components of dipole moment that can interact with electric field components of electromagnetic radiations and absorption of radiation can occur.

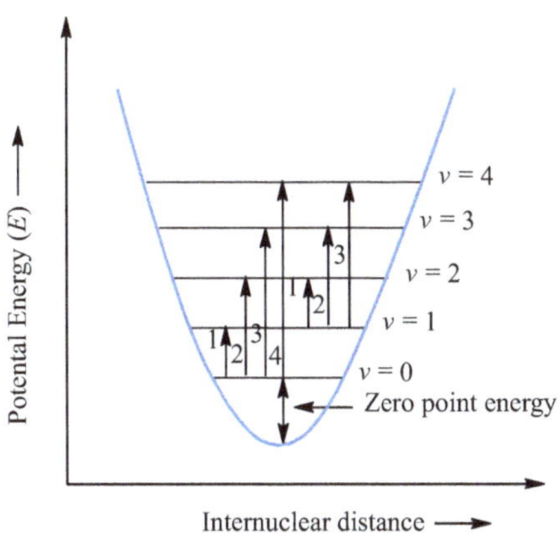

Fig. 3.4: Vibrational states corresponding to normal vibrational modes in harmonic oscillator.

As shown in the Fig. the transition (1) is fundamental frequency corresponding to a normal mode of vibration. Transitions (2) and (3) are non-fundamental frequencies.

In fact, these frequencies are first and second overtone's frequencies. The intensities of overtones are generally less than that of the fundamental frequency.

At molecular level, it is not so simple and this model does not hold good. According to Franck–Condon principle, transition tend to take place between two vibrational states in which nuclear configurations remain same in both the states and these tend to occur when nuclear kinetic energies are very small. This model is generally not observed and deviations from above model are observed.

Raman spectroscopy is complementary to IR spectroscopy and deals with the rotational and vibrational transitions in the molecule. When both the techniques are used together, a lot of information can be obtained for the molecule.

In Raman spectroscopy, a monochromatic beam of light (generally a gas laser beam) strikes the sample surface and scattered photons (elastic and non-elastic) are measured at the right angle to the incident beam of light. Photons when scattered elastically are known as Rayleigh scattering, and the scattered photons have same wavelength as that of absorbed light.

Raman Effect deals with inelastic photons scatterings by the molecules. Thus in Raman Effect, the energies of the incident and scattered photons are different. If the energy of the scattered photon is more (frequency is more) than that of the incident photons, then the lines observed in Raman spectrum are known as **Antistokes lines.** If the energy of the scattered photon is less than that of the incident photons, then the lines observed are called **Stokes lines**. The occurrence of Raman Effect is shown in the following Fig. 3.5.

The selection rules for Raman spectroscopy are different than that of IR spectroscopy. There must be change in polarizability of the molecule during vibrations. Symmetric stretch of CO_2 which is IR inactive is Raman active (appears at 1340 cm^{-1}) because during symmetric stretches there is change in polarizability of the molecule. The induced dipole during vibration is related to the polarizability tensor by the following equations:

$$D_x = \alpha_{xx}E_x + \alpha_{xy}E_y + \alpha_{xz}E_z$$

$$D_y = \alpha_{yx}E_x + \alpha_{yy}E_y + \alpha_{yz}E_z$$

$$D_z = \alpha_{zx}E_x + \alpha_{zy}E_y + \alpha_{zz}E_z$$

where E is the electric field and α is the polarizability that is defined as $D = \alpha E$. Here, D represents the induced dipole moment.

Whether or not a particular vibration will bring about a dipole moment change or polarization in a molecule can be understood easily in the case of simple molecules. But as the complexity of the molecule increases, it becomes more and more difficult to understand the change in properties during molecular vibration. Hence, it may not easy always to find out whether a particular vibration is IR and/or Raman active or not.

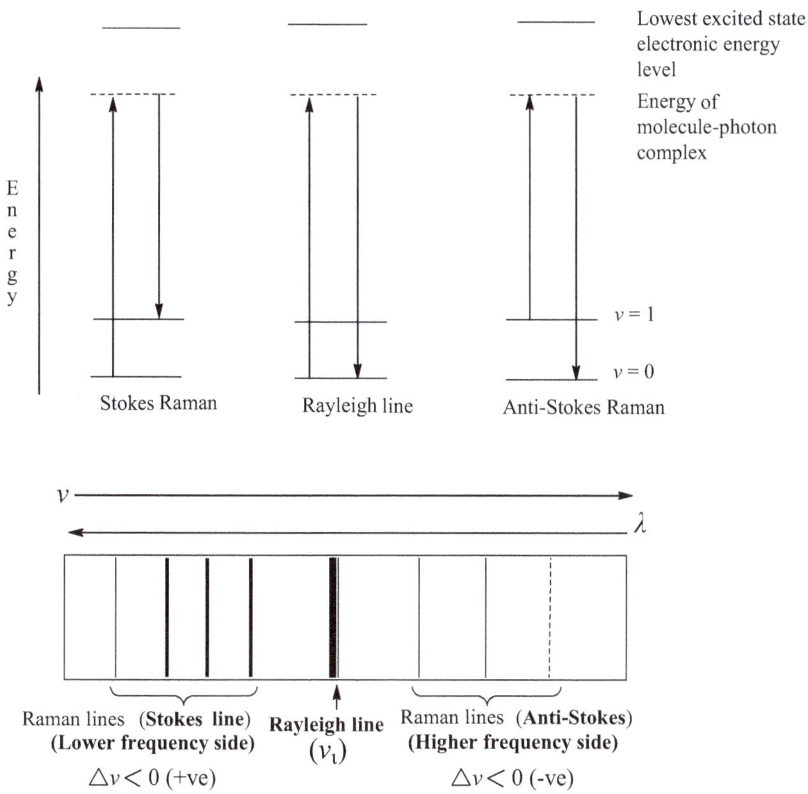

Fig. 3.5: Appearance of Raman spectrum.

Group theory can be successfully applied to predict the Infrared and Raman activity of various vibrations associated with molecules. It also helps us to understand the complementarities of Raman and IR spectra, and thus we can have better insight into structure and shapes of molecules.

3.2 Generation of reducible representation

(i) Reducible representation using 3*N* vectors as basis: calculation of total character of a reducible representation

We are aware that how representation of a point group (for various symmetry operations) is obtained by choosing various types of basis vectors. As already stated, a molecule containing *N* atoms will have 3*N* degrees of freedom. Initially, if the *N* atoms are assumed to be independent, each atom is associated with three degrees of translational freedom. Thus, for *N* atoms, there will be 3*N* degrees of translational freedom. The number of degrees of freedom remains conserved, even when the atoms join together to form the molecule. However, there will be only three degrees of translational freedom for the molecule; others will be rotational and vibrational degrees of freedom. Thus, the three degrees of freedom of each atom in a molecule may be represented by three mutually perpendicular arrows, each of which is a vector in x, y or z direction. These $3N$ vectors may be taken as basis for evaluating the total character associated with the $3N$ degrees of freedom. *If we subtract the characters relating to the translational and rotational degrees of freedom from the above total character, we will get the total character relating to vibrational degrees of freedom only.*

Let us consider the H_2O molecule belonging to C_{2v} point group with symmetry operations E, C_2, $\sigma_{v(xz)}$ and $\sigma_{v(yz)}$. The basis vectors of the atoms representing the degrees of freedom may be denoted as $x_1, y_1, z_1, x_2, y_2, z_2$ and x_3, y_3, z_3 as given in Fig. 3.6.

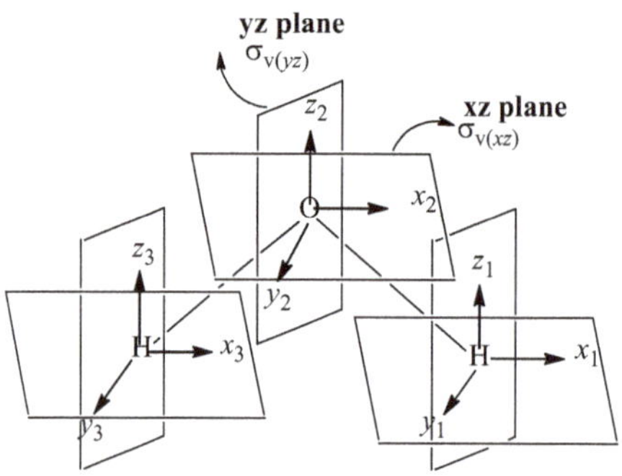

Fig. 3.6: Directions of the nine degrees of freedom.

From the transformations of these vectors during each of the above point group symmetry operations, one can set up nine dimensional matrices from which the corresponding characters can be worked out. The matrix product indicating the transformations will appear as

$$\begin{bmatrix} & & \\ & \text{Transformation matrix} & \\ & & \end{bmatrix} \begin{bmatrix} x_1 \\ y_1 \\ z_1 \\ \vdots \\ z_3 \end{bmatrix} = \begin{bmatrix} x_1' \\ y_1' \\ z_1' \\ \vdots \\ z_3' \end{bmatrix}$$

(i) On applying E operation, all the vectors are retained as such, and hence $x_1' = x_1$, $y_1' = y_1$, $z_1' = z_1$, etc.

$$E\begin{bmatrix}x_1\\y_1\\z_1\\x_2\\y_2\\z_2\\x_3\\y_3\\z_3\end{bmatrix}=\begin{bmatrix}x_1'\\y_1'\\z_1'\\x_2'\\y_2'\\z_2'\\x_3'\\y_3'\\z_3'\end{bmatrix};\;\begin{array}{l}x_1'=x_1+0+0+0+0+0+0+0+0\\y_1'=0+y_1+0+0+0+0+0+0+0\\z_1'=0+0+z_1+0+0+0+0+0+0\\x_2'=0+0+0+x_2+0+0+0+0+0\\y_2'=0+0+0+0+y_2+0+0+0+0\\z_2'=0+0+0+0+0+z_2+0+0+0\\x_3'=0+0+0+0+0+0+x_3+0+0\\y_3'=0+0+0+0+0+0+0+y_3+0\\z_3'=0+0+0+0+0+0+0+0+z_3\end{array};\;\begin{bmatrix}\mathbf{1}&0&0&0&0&0&0&0&0\\0&\mathbf{1}&0&0&0&0&0&0&0\\0&0&\mathbf{1}&0&0&0&0&0&0\\0&0&0&\mathbf{1}&0&0&0&0&0\\0&0&0&0&\mathbf{1}&0&0&0&0\\0&0&0&0&0&\mathbf{1}&0&0&0\\0&0&0&0&0&0&\mathbf{1}&0&0\\0&0&0&0&0&0&0&\mathbf{1}&0\\0&0&0&0&0&0&0&0&\mathbf{1}\end{bmatrix}\begin{bmatrix}x_1\\y_1\\z_1\\x_2\\y_2\\z_2\\x_3\\y_3\\z_3\end{bmatrix}$$

$$\chi=9$$

(ii) The operation C_2 brings about the following changes: $x_1'=-x_3$, $y_1'=-y_3$, $z_1'=z_3$; $x_2'=-x_2$, $y_2'=-y_2$, $z_2'=z_2$; $x_3'=-x_1$, $y_3'=-y_1$, $z_3'=z_1$.

$$C_2\begin{bmatrix}x_1\\y_1\\z_1\\x_2\\y_2\\z_2\\x_3\\y_3\\z_3\end{bmatrix}=\begin{bmatrix}x_1'\\y_1'\\z_1'\\x_2'\\y_2'\\z_2'\\x_3'\\y_3'\\z_3'\end{bmatrix};\;\begin{array}{l}x_1'=0+0+0+0+0+0-x_3+0+0\\y_1'=0+0+0+0+0+0+0-y_3+0\\z_1'=0+0+0+0+0+0+0+0+z_3\\x_2'=0+0+0-x_2+0+0+0+0+0\\y_2'=0+0+0+0-y_2+0+0+0+0\\z_2'=0+0+0+0+0+z_2+0+0+0\\x_3'=-x_1+0+0+0+0+0+0+0+0\\y_3'=0-y_1+0+0+0+0+0+0+0\\z_3'=0+0+z_1+0+0+0+0+0+0\end{array};\;\begin{bmatrix}0&0&0&0&0&0&-1&0&0\\0&0&0&0&0&0&0&-1&0\\0&0&0&0&0&0&0&0&1\\0&0&0&-1&0&0&0&0&0\\0&0&0&0&-1&0&0&0&0\\0&0&0&0&0&1&0&0&0\\-1&0&0&0&0&0&0&0&0\\0&-1&0&0&0&0&0&0&0\\0&0&1&0&0&0&0&0&0\end{bmatrix}\begin{bmatrix}x_1\\y_1\\z_1\\x_2\\y_2\\z_2\\x_3\\y_3\\z_3\end{bmatrix}$$

$$\chi=-1$$

(iii) The operation $\sigma_{v(xz)}$ brings about the following changes: $x_1'=x_3$, $y_1'=-y_3$, $z_1'=z_3$; $x_2'=x_2$, $y_2'=-y_2$, $z_2'=z_2$; $x_3'=x_1$, $y_3'=-y_1$, $z_3'=z_1$.

$$\sigma_{v(xz)}\begin{bmatrix}x_1\\y_1\\z_1\\x_2\\y_2\\z_2\\x_3\\y_3\\z_3\end{bmatrix}=\begin{bmatrix}x_1'\\y_1'\\z_1'\\x_2'\\y_2'\\z_2'\\x_3'\\y_3'\\z_3'\end{bmatrix};\;\begin{array}{l}x_1'=0+0+0+0+0+0+x_3+0+0\\y_1'=0+0+0+0+0+0+0-y_3+0\\z_1'=0+0+0+0+0+0+0+0+z_3\\x_2'=0+0+0+x_2+0+0+0+0+0\\y_2'=0+0+0+0-y_2+0+0+0+0\\z_2'=0+0+0+0+0+z_2+0+0+0\\x_3'=x_1+0+0+0+0+0+0+0+0\\y_3'=0-y_1+0+0+0+0+0+0+0\\z_3'=0+0+z_1+0+0+0+0+0+0\end{array};\;\begin{bmatrix}0&0&0&0&0&0&1&0&0\\0&0&0&0&0&0&0&-1&0\\0&0&0&0&0&0&0&0&1\\0&0&0&1&0&0&0&0&0\\0&0&0&0&-1&0&0&0&0\\0&0&0&0&0&1&0&0&0\\1&0&0&0&0&0&0&0&0\\0&-1&0&0&0&0&0&0&0\\0&0&1&0&0&0&0&0&0\end{bmatrix}\begin{bmatrix}x_1\\y_1\\z_1\\x_2\\y_2\\z_2\\x_3\\y_3\\z_3\end{bmatrix}$$

$$\chi=1$$

(iv) The operation $\sigma_{v(yz)}$ brings about the following changes: $x_1' = -x_1$, $y_1' = y_1$, $z_1' = z_1$; $x_2' = -x_2$, $y_2' = y_2$, $z_2' = z_2$; $x_3' = -x_3$, $y_3' = y_3$, $z_3' = z_3$.

$$\sigma_{v(yz)}\begin{bmatrix} x_1 \\ y_1 \\ z_1 \\ x_2 \\ y_2 \\ z_2 \\ x_3 \\ y_3 \\ z_3 \end{bmatrix} = \begin{bmatrix} x_1' \\ y_1' \\ z_1' \\ x_2' \\ y_2' \\ z_2' \\ x_3' \\ y_3' \\ z_3' \end{bmatrix}; \begin{matrix} x_1' = -x_1 0+0+0+0+0+0+0+0 \\ y_1' = 0+y_1+0+0+0+0+0+0+0 \\ z_1' = 0+0+z_1+0+0+0+0+0+0 \\ x_2' = 0+0+0-x_2+0+0+0+0+0 \\ y_2' = 0+0+0+0+y_2+0+0+0+0 \\ z_2' = 0+0+0+0+0+z_2+0+0+0 \\ x_3' = 0+0+0+0+0+0-x_3+0+0 \\ y_3' = 0+0+0+0+0+0+0+y_3+0 \\ z_3' = 0+0+0+0+0+0+0+0+z_3 \end{matrix}; \begin{bmatrix} -1 & 0 & 0 & 0 & 0 & 0 & 0 & 0 & 0 \\ 0 & 1 & 0 & 0 & 0 & 0 & 0 & 0 & 0 \\ 0 & 0 & 1 & 0 & 0 & 0 & 0 & 0 & 1 \\ 0 & 0 & 0 & -1 & 0 & 0 & 0 & 0 & 0 \\ 0 & 0 & 0 & 0 & 1 & 0 & 0 & 0 & 0 \\ 0 & 0 & 0 & 0 & 0 & 1 & 0 & 0 & 0 \\ 0 & 0 & 0 & 0 & 0 & 0 & -1 & 0 & 0 \\ 0 & 0 & 0 & 0 & 0 & 0 & 0 & 1 & 0 \\ 0 & 0 & 0 & 0 & 0 & 0 & 0 & 0 & 1 \end{bmatrix}\begin{bmatrix} x_1 \\ y_1 \\ z_1 \\ x_2 \\ y_2 \\ z_2 \\ x_3 \\ y_3 \\ z_3 \end{bmatrix}$$

$$\chi = 3$$

Thus, we have obtained 9×9 matrices for each symmetry operation of C_{2v} point group using total nine unit vectors (three on each atom). If we take the character of each matrix corresponding to each operation, then we will get a reducible representation in the tabular form as give below:

C_{2v}	E	C_2	$\sigma_{v(xz)}$	$\sigma_{v(yz)}$
Γ_{3N}	9	−1	1	3

This reducible representation Γ_{3N} can be reduced into smaller irreducible representations to get the total degrees of freedom (vibrational, rotational and translational) of different symmetries. This process/procedure of calculating total character of the reducible representation is very cumbersome and time-consuming. For complicated molecule, this procedure becomes further more difficult to be followed. For instance, if we want to find Γ_{3N} for benzene (C_6H_6), then we have to deal with 36×36 matrices. So, there is a need for simplified procedure for determining Γ_{3N}.

(ii) Simplified procedure for determining Γ_{3N}

The simplified procedure involves in finding character of matrix corresponding to symmetry operation *R* [i.e. χ(R)] without constructing 3N×3N matrices. By matrix multiplication rules, it is well clear that if an atom or its associated vector is shifted in space by/on applying any symmetry operation, then the atom/its associated vectors will not contribute to the character of the matrix or such vector will contribute zero to matrix character $\chi(R)$. Only unshifted atom or unshifted vector contributes $\chi(R) = +1$, and vector whose direction is reversed contributes $\chi(R) = -1$.

Based on the above idea, the overall procedure for determining Γ_{3N} involves the following steps:

(i) Find the number of unshifted atoms during a particular symmetry operation taking *one symmetry operation from each class.*

(ii) Workout the $\chi(R)$ for every unshifted atom for every symmetry operation taking one from each class.

(iii) The contributions of these characters will be the same for the symmetry operation (*R*) in every point group when taken 3N vectors as the basis.

(iv) Γ_{3N} can be written in the tabular form as shown earlier

Character $\chi(R)$ for a symmetry operation *R*, for unshifted atoms

Let us take various symmetry operations and a set of 3N unit vectors along *x*, *y* and *z* axes **on a point** and perform symmetry operation one by one to know $\chi(R)$ for unshifted atom.

(i) **Identity operation, *E*** The effect of identity operation is shown below.

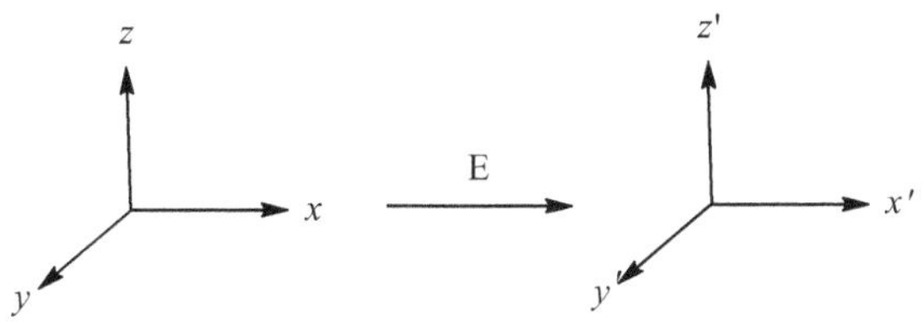

$$E\begin{bmatrix} x \\ y \\ z \end{bmatrix} = \begin{bmatrix} x' \\ y' \\ z' \end{bmatrix} \quad \text{i.e.} \quad \begin{matrix} x' = x \\ y' = y \\ z' = z \end{matrix} \quad \text{or} \quad E = \begin{bmatrix} 1 & 0 & 0 \\ 0 & 1 & 0 \\ 0 & 0 & 1 \end{bmatrix} \quad \chi = +3$$

(ii) **Inversion operation, *i*** The effect of inversion operation is shown below.

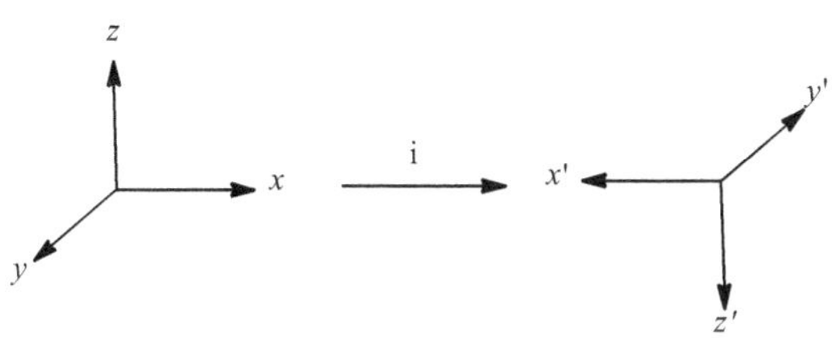

$$i\begin{bmatrix} x \\ y \\ z \end{bmatrix} = \begin{bmatrix} x' \\ y' \\ z' \end{bmatrix} \quad \text{i.e.} \quad \begin{matrix} x' = -x \\ y' = -y \\ z' = -z \end{matrix} \quad \text{or} \quad i = \begin{bmatrix} -1 & 0 & 0 \\ 0 & -1 & 0 \\ 0 & 0 & -1 \end{bmatrix} \quad \chi = -3$$

(iii) **Reflection operation, σ** The effects of reflection operations σ_{xz}, σ_{yz}, σ_{xy} (σ_h) are shown below.

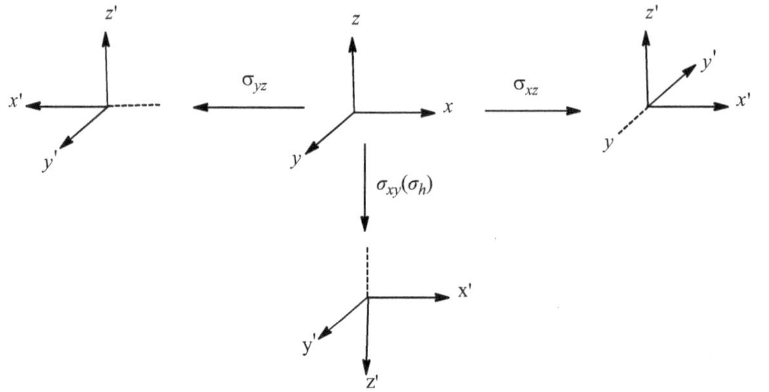

$$\sigma_{xz} = \begin{bmatrix} x \\ y \\ z \end{bmatrix} = \begin{bmatrix} x' \\ y' \\ z' \end{bmatrix} \quad \text{i.e.} \quad \begin{matrix} x' = x \\ y' = -y \\ z' = z \end{matrix} \quad \text{or} \quad \sigma_{xz} = \begin{bmatrix} 1 & 0 & 0 \\ 0 & -1 & 0 \\ 0 & 0 & 1 \end{bmatrix} \quad \chi = 1$$

$$\sigma_{yz} = \begin{bmatrix} x \\ y \\ z \end{bmatrix} = \begin{bmatrix} x' \\ y' \\ z' \end{bmatrix} \quad \text{i.e.} \quad \begin{matrix} x' = -x \\ y' = y \\ z' = z \end{matrix} \quad \text{or} \quad \sigma_{xz} = \begin{bmatrix} -1 & 0 & 0 \\ 0 & 1 & 0 \\ 0 & 0 & 1 \end{bmatrix} \quad \chi = 1$$

$$\sigma_{xy} = \begin{bmatrix} x \\ y \\ z \end{bmatrix} = \begin{bmatrix} x' \\ y' \\ z' \end{bmatrix} \quad \text{i.e.} \quad \begin{matrix} x' = x \\ y' = y \\ z' = -z \end{matrix} \quad \text{or} \quad \sigma_{xz} = \begin{bmatrix} 1 & 0 & 0 \\ 0 & 1 & 0 \\ 0 & 0 & -1 \end{bmatrix} \quad \chi = 1$$

(iv) **C_2 operation along z-axis** In this operation, molecule is rotated through 180°, and its effect is shown below.

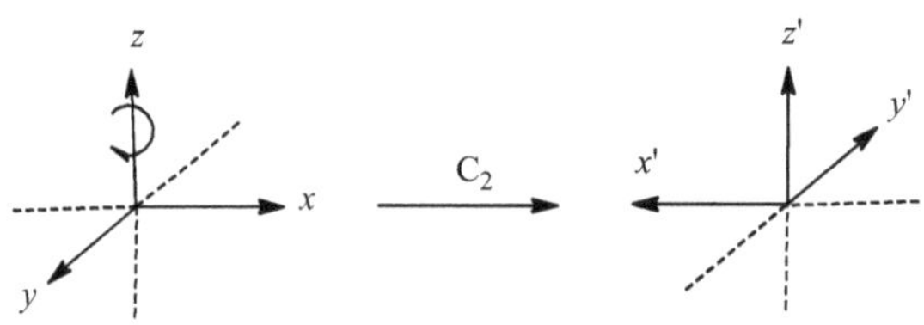

$$C_2 = \begin{bmatrix} x \\ y \\ z \end{bmatrix} = \begin{bmatrix} x' \\ y' \\ z' \end{bmatrix} \quad \text{i.e.} \quad \begin{matrix} x' = -x \\ y' = -y \\ z' = z \end{matrix} \quad \text{or} \quad E = \begin{bmatrix} -1 & 0 & 0 \\ 0 & -1 & 0 \\ 0 & 0 & 1 \end{bmatrix} \quad \chi = -1$$

(v) **C_n operation along z-axis** In this operation, molecule is rotated through certain angle θ, and its effect is shown in the form of C_n matrix below.

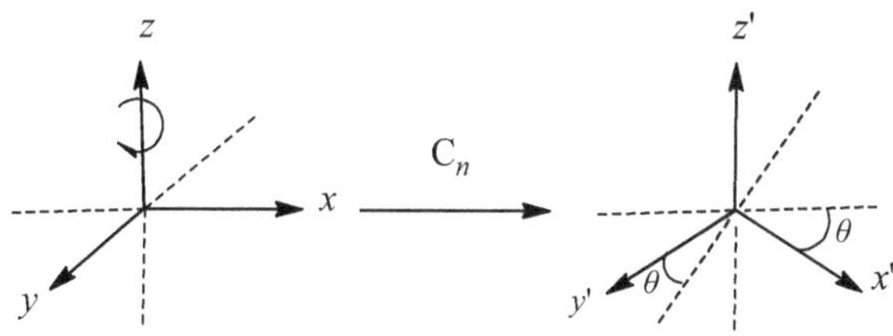

$$\begin{bmatrix} \cos\theta & \sin\theta & 0 \\ -\sin\theta & \cos\theta & 0 \\ 0 & 0 & 1 \end{bmatrix} \chi = 2\cos\theta + 1$$

Same results are obtained for $C_n^{n\text{-}1}$ and same expression is obtained for C_n^m.

(vi) **Improper rotation, S_n** This is equivalent to C_n operation through z-axis followed by reflection through a plane perpendicular to the rotation axis, that is, σ_{xy} (σ_h).

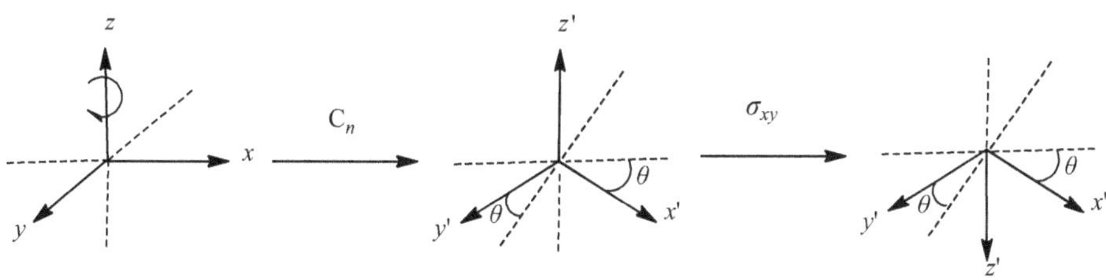

Here, the symmetry operation C_n changes x to x' and y to y' only and σ_{xy} (σ_h) changes z to z', that is, di-reaction is reversed or sign is changed.

The effect of improper operation is thus the multiplication of C_n and σ_{xy} matrices as shown below.

$$\overset{C_n}{\begin{bmatrix} \cos\theta & \sin\theta & 0 \\ -\sin\theta & \cos\theta & 0 \\ 0 & 0 & 1 \end{bmatrix}} \cdot \overset{\sigma_{xy}}{\begin{bmatrix} 1 & 0 & 0 \\ 0 & 1 & 0 \\ 0 & 0 & -1 \end{bmatrix}} = \overset{S_n}{\begin{bmatrix} \cos\theta & \sin\theta & 0 \\ -\sin\theta & \cos\theta & 0 \\ 0 & 0 & -1 \end{bmatrix}} \chi = 2\cos\theta - 1$$

The results so obtained are quite general and can be applied to any point group. For convenience of learners, Table 3.1 summarizes the various symmetry operations and characters of their matrices using 3N vectors as the basis set.

Table 3.1: The character $\chi(R)$ of various symmetry operations using 3N unit vectors as the basis.

Symmetry Operation (*R*)	$\chi(R)$ of matrices for operation *R*
E	+3
i	−3
$\sigma(\sigma_{xz}, \sigma_{yz}, \sigma_{xy})$	+1
C_2	−1
C_3^1 (θ= 120°), C_3^2 (θ= 240°)	0
C_4^1 (θ= 90°), C_4^3(θ= 270°)	+1
C_6^1 (θ= 60°), C_6^5(θ= 300°)	+2
S_3^1 (θ= 120°), S_3^5(θ= 240°)	−2
S_4^1 (θ= 90°), S_4^3(θ= 270°)	−1
S_6^1 (θ= 60°), S_6^5(θ= 300°)	0
In general, proper axis	$2\cos\theta + 1$
In general, improper axis	$2\cos\theta - 1$

This simple method can be explained by taking examples of molecules belonging to different point groups:

(a) C_{2v} point group (H_2O or SO_2)

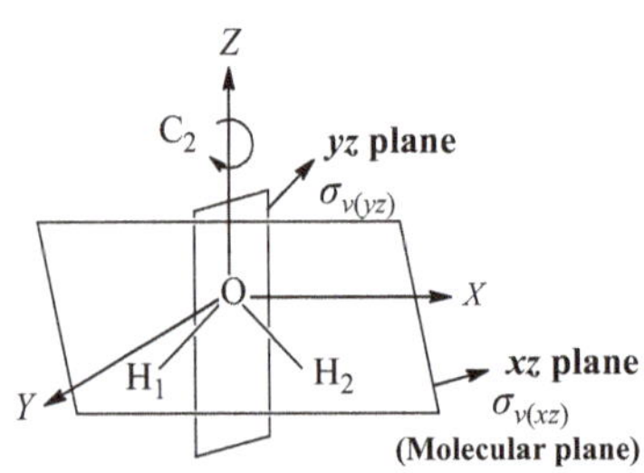

Let us take the H_2O molecule belonging to this point group having symmetry operations, E, C_2, $\sigma_{v(xz)}$ and $\sigma_{v(yz)}$.

We will apply these symmetry operations one by one and see how many atoms are not shifting, details of which are given below:

(i) E does not shift any atom, so unshifted atoms are **3**.
(ii) C_2 interchanges H_1 and H_2 but does not shift O. So unshifted atom is **1**.
(iii) $\sigma_{v(xz)}$ does not shift any of the atoms as all lie in $\sigma_{v(xz)}$ (molecular plane) plane. So, unshifted atoms equal to **3**.
(iv) $\sigma_{v(yz)}$ interchanges H_1 and H_2 but does not shift O. So unshifted atom is **1.**

These results can be tabulated as:

C_{2v}	E	C_2	$\sigma_{v(yz)}$	$\sigma_{v(xz)}$
No. of unshifted atoms (*n*)	3	1	1	3
$\chi(R)$ from Table 3.1	3	−1	1	1
Γ_{3N} [$n\times\chi(R)$]	9	−1	1	3

(b) D_{3h} point group (BF_3, CO_3^{2-}, SO_3, NO_3^-)

Let us consider BF_3 molecule of D_{3h} point group with symmetry operations, *E*, $2C_3$, $3C_2$, σ_h, $2S_3$, and $3\sigma_v$. On applying symmetry operations to BF_3, we observe the following:

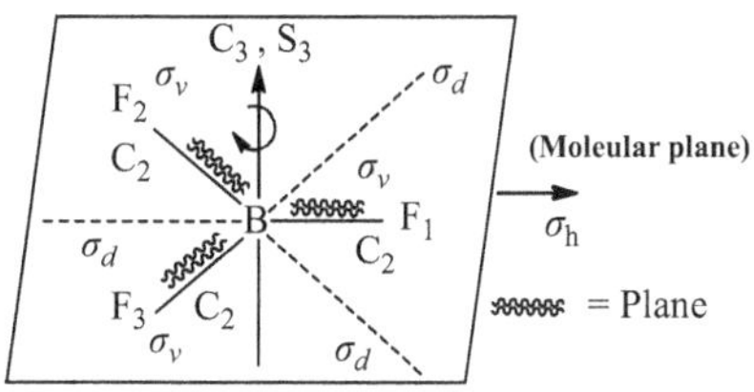

(i) *E* does not shift any atom, so unshifted atoms are **4**.
(ii) C_2 and σ_v interchange 2F's but does not shift B and one F. So unshifted atoms is **2**.
(iii) C_3 and S_3 shift 3F's but only B is unshifted.
(iv) Molecular plane σ_h does not shift any atom, so unshifted atoms are **4**.

D_{3h}	E	$2C_3$	$3C_2$	S_h	$2S_3$	$3\sigma_v$
No. of unshifted atoms (*n*)	4	1	2	4	1	2
$\chi(R)$ from Table 3.1	3	0	−1	1	−2	1
Γ_{3N} [$n\times\chi(R)$]	12	0	−2	4	−2	2

The reducible representation (Γ_{3N}) can be reduced to irreducible representations applying the reduction formula given below using character table of D_{3h} point group as follows:

$$N_i = \frac{1}{h}\sum_R \chi(R)\,.\,n\,.\,\chi_i(R)$$

where N_i is the number of times the *i*th irreducible representation occurs in a reducible representation, *h* is the order of group (number of symmetry operations), $\chi(R)$ is the character of a particular operation in the reducible representation, *n* is the number of operation of that type and $\chi_i(R)$ is the character of the same operation in the irreducible representation.

Reducible representation in question

D_{3h}	E	$2C_3$	$3C_2$	σ_h	$2S_3$	$3\sigma_v$
T_{3N}	12	0	−2	4	−2	2

Character table of D_{3h} point group

D_{3h}	E	$2C_3$	$3C_2$	σ_h	$2S_3$	$3\sigma_v$		
A_1'	1	1	1	1	1	1		x^2+y^2, z^2
A_2'	1	1	−1	1	1	−1	R_z	
E'	2	−1	0	2	−1	0	(x,y)	(x^2-y^2, xy)
A_1''	1	1	1	−1	−1	−1		
A_2''	1	1	−1	−1	−1	1	z	
E''	2	−1	0	−2	1	0	R_x, R_y	(xz, yz)

$$
\begin{aligned}
NA_1' &= 1/12\,[(12.1.1) + (0.2.1) + (-2.3.1) + (4.1.1) + (-2.2.1) + (2.3.1)\\
&= 1/12\,[12 + 0 - 6 + 4 - 4 + 6]\\
&= 1/12\,[12] = 1\\
NA_2' &= 1/12\,[(12.1.1) + (0.2.1) + (-2.3.-1) + (4.1.1) + (-2.2.1) + (2.3.-1)\\
&= 1/12\,[12 + 0 + 6 + 4 - 4 - 6]\\
&= 1/12\,[12] = 1\\
E' &= 1/12\,[(12.1.2) + (0.2.-1) + (-2.3.0) + (4.1.2) + (-2.2.-1) + (2.3.0)\\
&= 1/12\,[24 + 0 + 0 + 8 + 4 + 0]\\
&= 1/12\,[36] = 3\\
NA_1'' &= 1/12\,[(12.1.1) + (0.2.1) + (-2.3.1) + (4.1.-1) + (-2.2.-1) + (2.3.-1)\\
&= 1/12\,[12 + 0 - 6 - 4 + 4 - 6]\\
&= 1/12\,[0] = 0\\
NA_2'' &= 1/12\,[(12.1.1) + (0.2.1) + (-2.3.-1) + (4.1.-1) + (-2.2.-1) + (2.3.1)\\
&= 1/12\,[12 + 0 + 6 - 4 + 4 + 6]\\
&= 1/12\,[24] = 2\\
E'' &= 1/12\,[(12.1.2) + (0.2.-1) + (-2.3.0) + (4.1.-2) + (-2.2.1) + (2.3.0)\\
&= 1/12\,[24 + 0 + 0 - 8 - 4 + 0]\\
&= 1/12\,[12] = 1
\end{aligned}
$$

The above calculations show that the reducible representation Γ_{3N} is reduced to:

$$\Gamma_{3N} = A_1' + A_2' + 3E' + 2A_2'' + E''$$

$$\Gamma_{transl} = E' + A_2''; \quad \Gamma_{rot} = A_2' + E'' \quad \text{(from inspection of character table)}$$

So,

$$\begin{aligned}\Gamma_{vib} &= \Gamma_{3N} - (\Gamma_{transl} + \Gamma_{rot}) \\ &= A_1' + A_2' + 3E' + 2A_2'' + E'' - E' - A_2'' - A_2' - E'' \\ &= A_1' + A_2'' + 2E'\end{aligned}$$

(c) C_{3v} point group (NH_3, $CHCl_3$ or $POCl_3$)

Let us take the $POCl_3$ molecule belonging to C_{3v} point group having symmetry operations, E, $2C_3$ and $3\sigma_v$. On applying symmetry operations, we observe as follows:

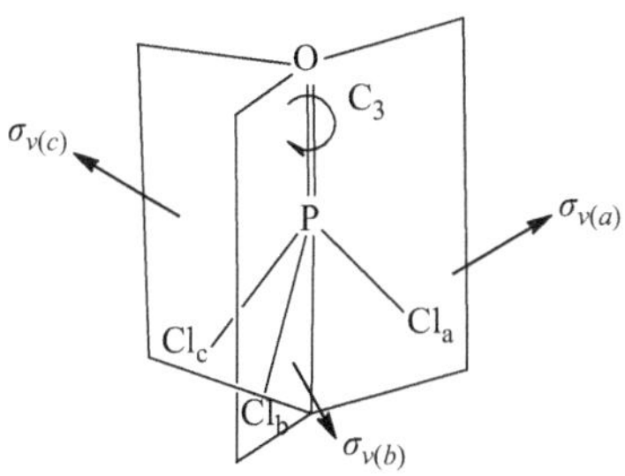

(i) Symmetry operation E does not shift any of the atoms. So, number of unshifted atoms are 5.
(ii) C_3^1 along P = O bond shifts all the three Cl's but does not shift atoms P and O. So unshifted atoms are 2.
(iii) σ_v containing P-O and one of the Cl's shifts only two Cl's. So unshifted atoms are 3.

We can tabulate these results as:

C_{3v}	E	$2C_3$	$3\sigma_v$
No. of unshifted atoms (n)	5	2	3
$\chi(R)$ from Table 3.1	3	0	1
Γ_{3N} [$n \times \chi(R)$]	15	0	3

(d) D_{4h} point group [XeF_4, PtX_4^{2-} (X = halogen), $AuCl_4^-$]

$PtCl_4^{2-}$ may be taken as a representative example belonging to D_{4h} point group with symmetry operations, E, $2C_4$, C_2, $2C_2'$, $2C_2''$, i, $2S_4$, σ_h, $2\sigma_v$ and $2\sigma_d$. On applying symmetry operations, we observe as follows:

(i) As usual the number of unshifted atoms against E/σ_h operation is **5**.

(ii) C_4 operation shifts all Cl's and only Pt atom is unshifted. So, unshifted atom is **1**.

(iii) The C_2/S_4 coinciding with C_4 has the same effect as C_4, and so the number of unshifted atom is **1**.

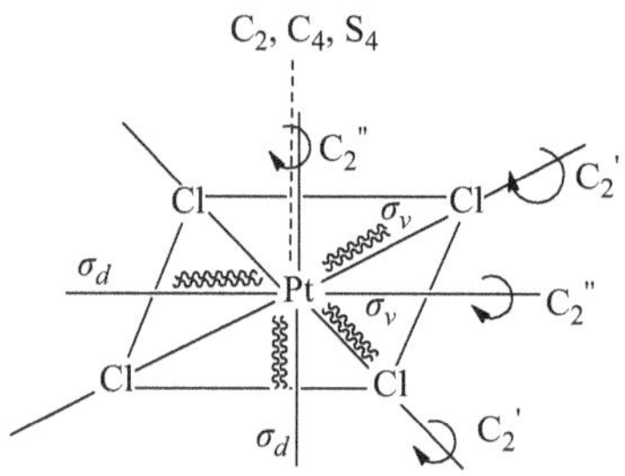

Symmetry operations in $PtCl4^{2-}$

(iv) The C_2' or σ_v along diagonal does not shift Pt and 2Cl's lying along diagonal, and so unshifted atoms are **3**.

(v) The C_2'' or σ_d bisecting the opposite sides shifts all Cl's except Pt. So, the unshifted atom is **1**.

(vi) In inversion operation only the central metal Pt is unshifted.

We can write the results as:

D_{4h}	E	$2C_4$	C_2	$2C_2'$	$2C_2''$	i	$2S_4$	σ_h	$2\sigma_v$	$2\sigma_d$
No. of unshifted atoms (n)	5	1	1	3	1	1	1	5	3	1
$\chi(R)$ from Table 3.1	3	1	−1	−1	−1	−3	−1	1	1	1
Γ_{3N} [$n\times\chi(R)$]	15	1	−1	−3	−1	−3	−1	5	3	1

(e) T_d point group [CH_4, CCl_4, $NiCl_4^{2-}$, $Ni(CO)_4$, $CoCl_4^{2-}$, $Zn(CN)_4^{2-}$]

Taking CH_4 as an example of T_d point symmetry with symmetry operations, E, $8C_3$, $3C_2$, $6S_4$ and $6\sigma_d$, we observed the following when symmetry operation is applied one by one.

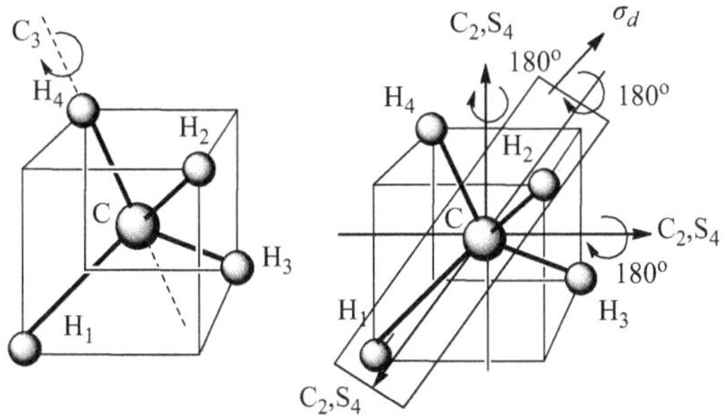

Symmetry operations in CH_4

(i) As usual the number of unshifted atoms against E operation is **5**.
(ii) As C_3 lies along each C–H bond and center of the triangle formed by the other 3H's, it shifts three base H's. So, one C and one H are not shifted.
(iii) Each of the Cartesian axes, x, y and z bisecting two HCH bond angles, is the C_2 axis, so, it shifts all the 4H's except central C atom. So, the number of unshifted atom is **1**.
(iv) The S_4 coinciding with C_2 has the same effect as C_2, and so the number of unshifted atom is **1**.
(v) As σ_d contains central C and 2H's, it shifts rest 2H's only. So, the number of unshifted atoms is **3**.

The summary of the results are tabulated below.

T_d	E	$8C_3$	$3C_2$	$6S_4$	$6\sigma_d$
No. of unshifted atoms (n)	5	2	1	1	3
$\chi(R)$ from Table 3.1	3	0	−1	−1	1
Γ_{3N} [$n \times \chi(R)$]	15	0	−1	−1	3

(f) Square pyramidal AB_5 molecule of C_{4v} point group

Let us take $WOCl_4$ as an example of $\mathbf{C_{4v}}$ point symmetry with symmetry operations, E, $2C_4$, C_2, $2\sigma_v$ and $2\sigma_d$. We observed the following when symmetry operation is applied one by one.

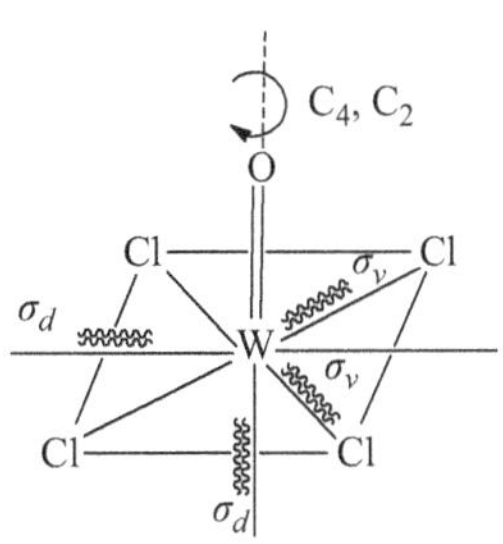

Symmetry operations in $WOCl_4$ molecule

(i) As usual the number of unshifted atoms in E operation is **6.**
(ii) C_2/C_4 operation shifts all Cl's and only W and O atoms are unshifted. So, unshifted atom is **2.**
(iii) The σ_v along diagonal reflects/shifts 2Cl's but does not shift W, 2Cl's and O atoms lying on the plane, and so unshifted atoms are **4.**
(iv) The σ_d bisecting the opposite sides shifts/reflects all Cl's except W and O atoms lying on the plane. So, the unshifted atoms are **2.**

The summary of the results are tabulated below.

C_{4v}	E	$2C_4$	C_2	$6\sigma_v$	$2\sigma_d$
No. of unshifted atoms (n)	6	2	2	4	2
$\chi(R)$ from Table3.1	3	1	−1	1	1
Γ_{3N} [$n \times \chi(R)$]	18	2	−2	4	2

The reducible representation (Γ_{3N}) can be reduced to irreducible representations applying the reduction formula given below using character table of C_{4v} point group as follows:

Reducible representation in question

C_{4v}	E	$2C_4$	C_2	$2\sigma_v$	$2\sigma_d$
T_{3N}	18	2	−2	4	2

Character table of C_{4v} point group

C_{4v}	E	$2C_4$	C_2	$2\sigma_v$	$2\sigma_d$		
A_1	1	1	1	1	1	z	x^2+y^2, z^2
A_2	1	1	1	−1	−1	R_z	
B_1	1	−1	1	1	−1		x^2-y^2
B_2	1	−1	1	−1	1		xy
E	2	0	−2	0	0	$(x,y)(R_x,R_y)$	xz, yz

$$\begin{aligned}
NA_1 &= 1/8\,[(18.1.1) + (2.2.1) + (-2.1.1) + (4.2.1) + (2.2.1)\\
&= 1/8\,[18 + 4 - 2 + 8 + 4]\\
&= 1/8\,[32] = 4\\
NA_2 &= 1/8\,[(18.1.1) + (2.2.1) + (-2.1.1) + (4.2.-1) + (2.2.-1)\\
&= 1/8\,[18 + 4 - 2 - 8 - 4]\\
&= 1/8\,[8] = 1\\
NB_1 &= 1/8\,[(18.1.1) + (2.2.-1) + (-2.1.1) + (4.2.1) + (2.2.-1)\\
&= 1/8\,[18 - 4 - 2 + 8 - 4]\\
&= 1/8\,[16] = 2\\
NB_2 &= 1/8\,[(18.1.1) + (2.2.-1) + (-2.1.1) + (4.2.-1) + (2.2.1)\\
&= 1/8\,[18 - 4 - 2 - 8 + 4]\\
&= 1/8\,[8] = 1\\
NE &= 1/8\,[(18.1.2) + (2.2.0) + (-2.1.-2) + (4.2.0) + (2.2.0)\\
&= 1/8\,[36 + 0 + 4 + 0 + 0]\\
&= 1/8\,[40] = 5
\end{aligned}$$

The reducible representation (Γ_{3N}) is thus reduced to: $4A_1 + A_2 + 2B_1 + B_2 + 5E$

$$\Gamma_{transl} = A_1 + E; \Gamma_{rot} = A_2 + E$$

So,

$$\begin{aligned}
\Gamma_{vib} &= \Gamma_{3N} - (\Gamma_{transl} + \Gamma_{rot})\\
&= 4A_1 + A_2 + 2B_1 + B_2 + 5E - A_1 - E - A_2 - E\\
&= 3A_1 + 2B_1 + B_2 + 3E\ (3N - 6 = 18 - 6 = 12)
\end{aligned}$$

(iii) Reducible representations using bond vectors

We have just used base vectors *x*, *y*, *z* of unit length along *x*, *y*, *z* axes for generation of reducible representation of several molecules of different point group. Similarly, we can use bond vectors as the basis for generating reducible representation of a point group. These types of representations are important in dealing with vibrational spectra of molecules.

The following rules are taken into consideration when bond vectors are taken as basis for generating reducible representations.

(i) The bond vectors that are shifted by the symmetry operations contribute nothing to the character of reducible representation.

(ii) The unshifted bond vectors contribute + 1 to the character of reducible representation.

(iii) The bond vectors that are unshifted but only its direction is reversed during symmetry operation contribute –1.

Let us apply these rules to some molecules of different point groups.

(a) H_2O molecule of C_{2v} point group

There are two O-H bond vectors. The effects of symmetry operations [E, C_2, $\sigma_{v(xz)}$ and $\sigma_{v(yz)}$] of $\mathbf{C_{2v}}$ point group on these vectors are tabulated below:

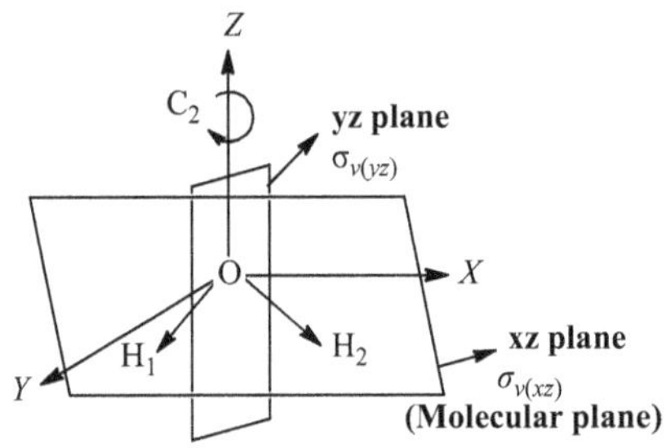

C_{2v}	E	C_2	$\sigma_{v(yz)}$	$\sigma_{v(xz)}$
No. of unshifted bond vectors (n)	2	0	0	2
$\chi(R)$	+ 1	0	0	+ 1
$\Gamma_{\text{bond vector}}$ [$n \times \chi(R)$]	2	0	0	2

(b) CH_4 molecule of T_d point group

There are four C-H bond vectors. The effects of symmetry operations (E, $8C_3$, $3C_2$, $6S_4$ and $6\sigma_d$) of T_d point group on these vectors are tabulated below:

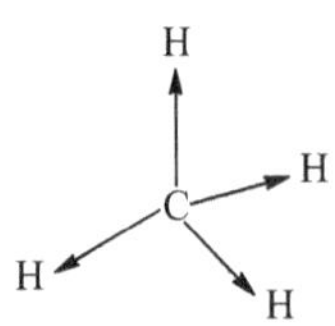

T_d	E	$8C_3$	$3C_2$	$6S_4$	$6\sigma_d$
No. of unshifted vectors (n)	4	1	0	0	2
$\chi(R)$	+1	+1	0	0	+1
$\Gamma_{\text{bond vector}}$ [$n \times \chi(R)$]	4	1	0	0	2

(c) XeF_4 molecule of D_{4h} point group

There are four Xe-F bond vectors. The effect of symmetry operations (E, $2C_4$, C_2, $2C_2'$, $2C_2''$, i, $2S_4$, σ_h, $2\sigma_v$, $2\sigma_d$) of D_{4h} point group on these four vectors are given below:

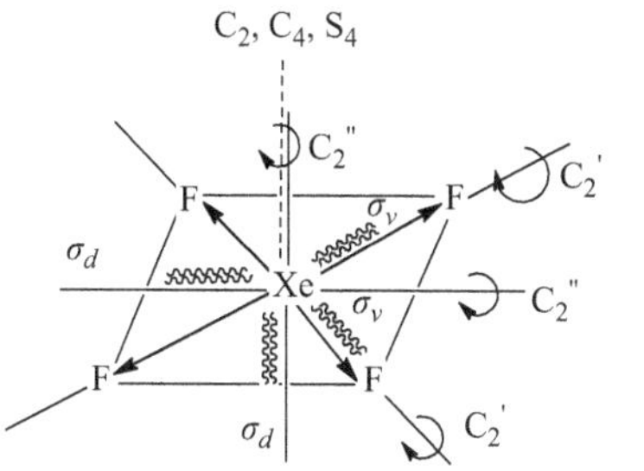

D_{4h}	E	$2C_4$	C_2	$2C_2'$	$2C_2''$	i	$2S_4$	σ_h	$2\sigma_v$	$2\sigma_d$
No. of unshifted vectors (n)	4	0	0	2	0	0	0	4	2	0
$\chi(R)$	1	0	0	1	0	0	0	1	1	0
$\Gamma_{\text{bond vector}}$ $[n\times\chi(R)]$	4	0	0	2	0	0	0	4	2	0

(d) PF_5 molecule of D_{3h} point group

There are five P-F bond vectors. The effects of symmetry operations (E, $2C_3$, $3C_2$, $2S_3$ and $3\sigma_v$) of D_{4h} point group on these four vectors are given below:

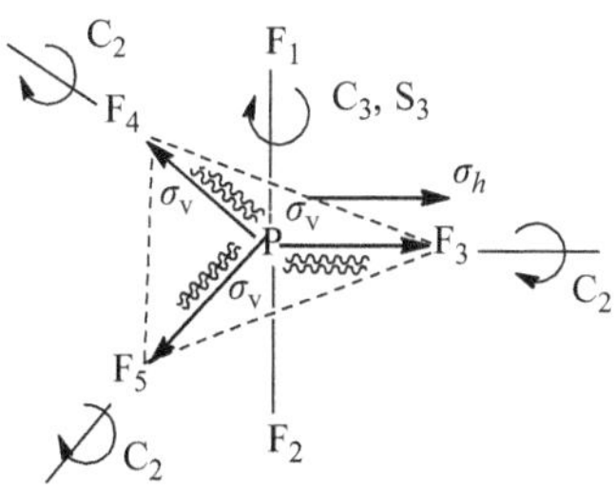

(i) It is notable that C_3 operation along two axial F's passing through central atom P shifts all three equatorial P-F bond vectors. So, only two vectors are not shifted.

(ii) The C_2 operation interchanges (shifts) two equatorial P-F and two axial P-F bond vectors. So, only one vector is not shifted.

(iii) σ_h interchanges (shifts) two axial P-F bond vectors while three equatorial P-F bond vectors are unshifted.

(iv) S_3 (C_3 followed by σ_h) shifts all the P-F vectors, so the number of unshifted bond vector is zero.

(v) σ_v passes through P and three F atoms F_1 F_2 F_3, F_1 F_2 F_4 or F_1 F_2F_5. So, it shifts two equatorial P-F bond vectors but two axial and one equatorial P-F bond vectors are unshifted.

These results are tabulated below:

D_{3h}	E	$2C_3$	$3C_2$	σ_h	$2S_3$	$3\sigma_v$
No. of unshifted vectors (n)	5	2	1	3	0	3
$\chi(R)$	+ 1	+ 1	+ 1	+ 1	0	+ 1
$\Gamma_{bond\ vector}$ $[n\times\chi(R)]$	5	2	1	3	0	3

(iv) Reducible representations generated by set of three p-orbitals

Three p-orbitals form a function space of three functions. These functions have radial and angular part. As radial part is totally symmetric and does not change by any symmetry operation, hence only angular part of these functions needs to be considered. Omitting constant terms, these orbitals can be written as $p_x = x$, $p_y = y$ and $p_z = z$. From symmetry point of view, these behave exactly as x, y, z unit vectors. Thus, Γ_p derived using p-orbitals as the basis will be very similar to Γ_{3N} as discussed before in this chapter. Also characters will be identical to that given in Table 3.1.

Let us explain how reducible representation is generated using p-orbitals as the basis taking examples of some molecules belonging to different point groups.

(a) H_2O molecule of C_{2v} point group

Here, we take three p-orbitals of oxygen. Only p_x orbital is shown in the Fig. These three p-orbitals under different symmetry operations (E, C_2, σ_{yz}, σ_{xz}) will be affected as:

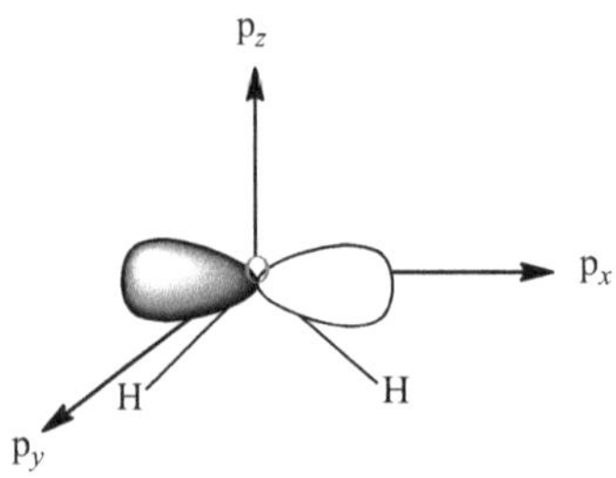

(a) Symmetry operation E does not shift any p-orbitals, so all three p-orbitals are not shifted. Net contribution is (+ 1) + (+ 1) + (+ 1) = + 3
(b) The C_2 operation reverses the directions of p_x and p_y orbitals but p_z orbital remains unchanged as it lies on z-axis. Net contribution is (−1) + (−1) + 1 = −1.
(c) The reflection operation σ_{xz} does not shift p_x and p_z orbitals as these orbitals lie in xz plane but direction of p_y-orbital is reversed. So, net contribution is

(+1) + (−1) + (+1) = +1. On the other hand, σ_{yz} reverse the direction of p_x orbital but does not shift the directions of p_y and p_z orbitals. Again the net contribution is (−1) + (+1) + (+1) = +1

These results are tabulated below:

C_{2v}	E	C_2	$\sigma_{v(yz)}$	$\sigma_{v(xz)}$
Net unshifted p–orbitalsby the symmetry operation	3	−1	1	1
Γ_p	3	−1	0	+1

Or

C_{2v}	E	C_2	$\sigma_{v(yz)}$	$\sigma_{v(xz)}$
Unshifted atom, only O atom is considered that has three p-orbitals (n)	1	1	1	1
$\chi(R)$ from Table 3.1	3	−1	+1	+1
Γ_p $[n\times\chi(R)]$	3	−1	1	1

(b) NH_3 molecule of C_{3v} point group

The three p-orbitals on N-atom are shown in the diagram by vectors along the x, y, z axes. Arrow head indicates the positive lobe of the orbital. The three p-orbitals on N-atom will behave under the symmetry operations as 3N unit vectors. The results of symmetry operations (E, $2C_3$, $3\sigma_v$) will not shift N-atom. Results are tabulated below.

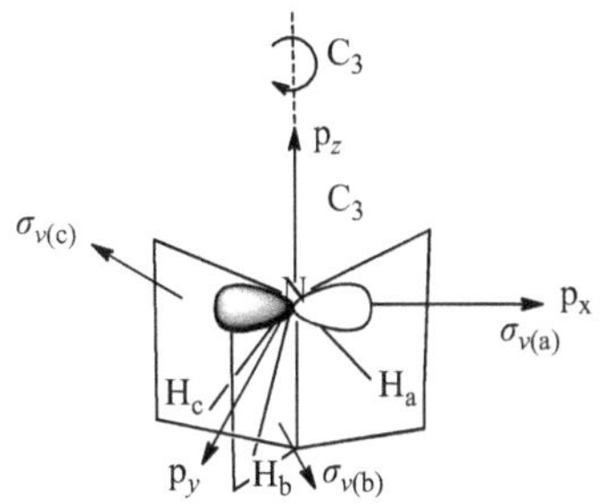

C_{3v}	E	$2C_3$	$3\sigma_v$
Unshifted atom, only N atom is considered that has three p-orbitals (*n*)	1	1	1
$[n\times\chi(R)]$ from Table 3.1	3	0	1
Γ_p $[n\times\chi(R)]$	3	0	1

(c) SF_6 molecule of O_h point group

The three p-orbitals on S-atom are shown in the diagram by vectors along the x, y, and z axes. Arrow head indicates the positive lobe of the orbital. The symmetry operations (E, $8C_3$, $3C_2$, $6C_4$, $6C_2'$, i, $6S_4$, $8S_6$, $3_{\sigma h}$, $6_{\sigma d}$) will not shift S-atom. Results are tabulated below.

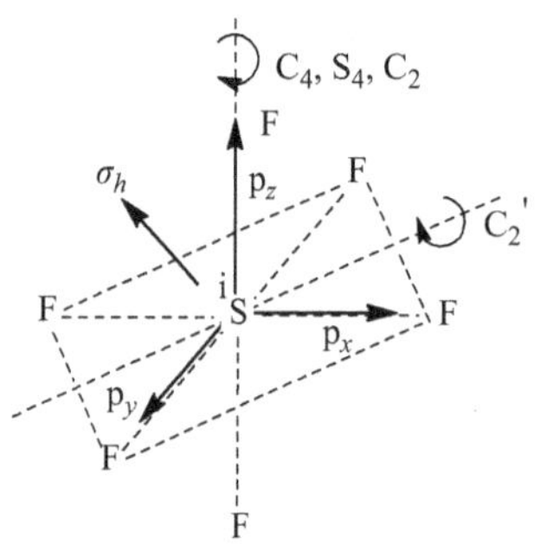

O_h	E	$8C_3$	$3C_2$	$6C_4$	$6C_2'$	i	$6S_4$	$8S_6$	$3\sigma_h$	$6\sigma_d$
Net unshifted p-orbitals	3	0	−2 + 1 = −1	1	−2 + 1 = −1	−3	−1	0	2−1 = 1	1
Γ_p	3	0	1	1	−1	−3	−1	0	1	1

Note: (–) sign means that direction of p-orbital is reversed.

(v) Reducible representations generated by set of five d-orbitals

Omitting all constants, the angular part of the five d-orbitals is:

$$d_1 = (x^2 - y^2)/2;\ d_2 = xy;\ d_3 = xz;\ d_4 = yz;\ d_5 = z^2$$

Similar to three p-orbitals, d-orbitals function space form the bases for representation of the point group. The character per unshifted atom for five d-orbitals are obtained by taking character of the matrices generated by d-orbitals and these characters are:

$$\chi(E) = 5$$
$$\chi(i) = 5$$
$$\chi(\sigma) = 1$$
$$\chi(C_n^{\ 1}) = 4\cos^2(360°/n) + 2\cos(360°/n) - 1$$
$$\chi(S_n^{\ 1}) = 4\cos^2(360°/n) - 2\cos(360°/n) - 1$$

Thus, five d-orbitals can also be used as the basis function for representation of any point group. Using set of five d-orbitals, the following results are obtained for some molecules of C_{3v} and O_h point groups.

(a) $POCl_3$ molecule of C_{3v} point group

$$^{15}P = 1s^2 2s^2 2p^6 3s^2 3p^3 3d^0$$

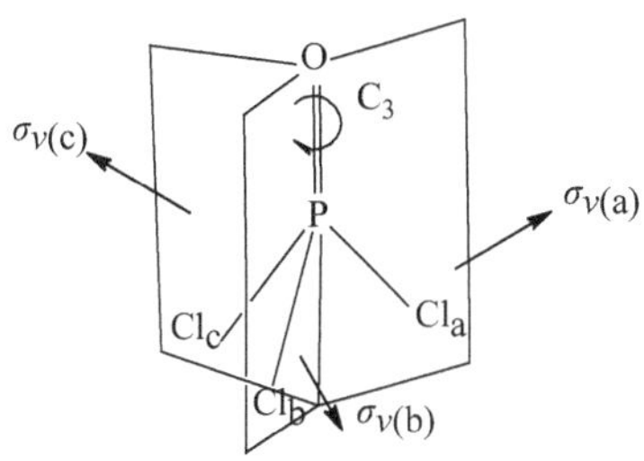

C_{3v}	E	$2C_3$	$3\sigma_v$
Unshifted atom, only P atom is considered that has five d-orbitals (*n*)	1	1	1
$\chi(R)$ from given results above	5	−1*	1
Γ_d $[n \times \chi(R)]$	5	−1	1

$$^*\chi(C_n^{\ 1}) = 4\cos^2(360°/n) + 2\cos(360°/n) - 1$$

For C_3, angle $= 360°/3 = 120°$

$$\therefore \chi(C_3^{\ 1}) = 4\cos^2(120°) + 2\cos(120°) - 1 \qquad [\cos 120° = -1/2]$$
$$= 4(-1/2)^2 + 2(-1/2) - 1$$
$$= 4.1/4 - 1 - 1 = 1 - 1 - 1 = -1$$

(b) SF_6 molecule of O_h point group

$$^{16}S = 1s^2 2s^2 2p^6 3s^2 3p^4 3d^0$$

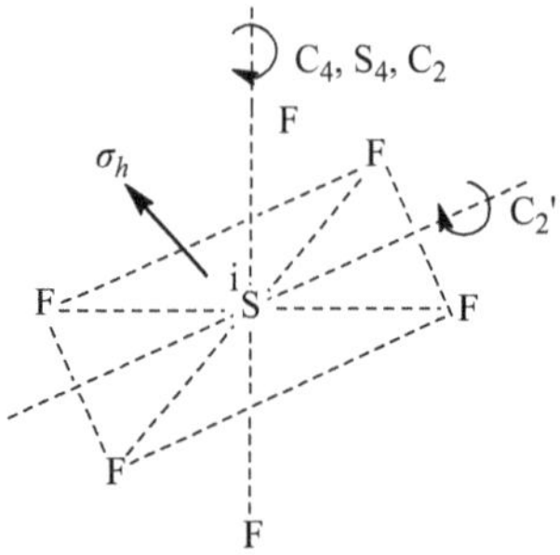

O_h	E	$8C_3$	$3C_2$	$6C_4$	$6C_2'$	i	$6S_4$	$8S_6$	$3\sigma_h$	$6\sigma_d$
Unshifted S atom (*n*)	1	1	1	1	1	1	1	1	1	1
$\chi(R)$ from given results above	5	−1	1^*	$-1^@$	1	5	-1^{**}	$-1^\#$	1	1
Γ_d $[n\times\chi(R)]$	5	−1	1	−1	1	5	−1	−1	1	1

$$^*\chi(C_n^{\ 1}) = 4\cos^2(360°/n) + 2\cos(360°/n) - 1$$

For C_2, angle = 360°/2 = 180°

$$\therefore \chi(C_2^{\ 1}) = 4\cos^2(180°) + 2\cos(180°) - 1 \qquad [\cos 180° = -1]$$

$$= 4(-1)^2 + 2(-1) - 1$$

$$= 4.1 - 2 - 1 = 4 - 3 = 1$$

$$^@\chi(C_4^{\ 1}) = 4\cos^2(90°) + 2\cos(90°) - 1 \qquad [\cos 90° = 0]$$

$$= 4(0)^2 + 2(0) - 1$$

$$= 0 + 0 - 1 = -1$$

$$\chi(S_n^{\ 1}) = 4\cos^2(360°/n) - 2\cos(360°/n) - 1 \qquad [\cos 90° = 0]$$

$$^{**}\chi(S_4^{\ 1}) = 4\cos^2(90°) - 2\cos(90°) - 1$$

$$= 4(0)^2 - 2(0) - 1 = -1$$

$$^\#\chi(S_6^{\ 1}) = 4\cos^2(60°) - 2\cos(60°) - 1 \qquad [\cos 60° = 1/2]$$

$$= 4(1/2)^2 - 2(1/2) - 1$$

$$= 1 - 1 - 1 = -1$$

(vi) Reducible representations using Internal coordinates as the basis: classification of Vibrational modes

A normal or fundamental mode consists of the movement of many or even all atoms of the molecule and is difficult to visualize. Because of this, we chemists try to simplify the description of normal mode by focusing the attention on just few atoms that seem to dominate the normal modes. Chemists have learnt to understand geometry and conformation of molecules in terms of localized internal coordinates (bond lengths, bond angles, etc). Therefore, it will be useful to discuss vibrational spectra in terms of these internal coordinates that are localized and are counterpart of delocalized normal modes (that are difficult to visualize).

The internal coordinates can be of the following type:

(i) Stretching – a change in bond length
(ii) Bending – a change in bond angle or angle deformation
(iii) Rocking – a change in angle between a group of atoms and rest of the molecule
(iv) Twisting – a change in angle between the plane of two groups of atoms
(v) Out of plane deformation– the movement of the central atom in and out of plane of the rest of the atoms
(vi) Wagging – a change in the angle between the plane of a group of atoms and a plane through the rest of atoms

In rocking, twisting or wagging, the involved coordinates, the bond angles and bond lengths, within the groups do not change. Rocking may be distinguished from wagging by the fact that the atoms in the group stay in the same plane. Some of these modes of vibrations are given in the following Fig. 3.7.

The most important internal coordinates are: (i) change in bond lengths (stretching) and (ii) angle deformation/bond angle (bending). Only these two will be used as the basis for representations to get Γ_{str} and Γ_{bend}. The symmetry species of each of these can be obtained by reducing these to irreducible representations. These results will be correlated with those that were obtained using Γ_{3N} method, that is $\Gamma_{vib} = \Gamma_{3N} - \Gamma_T - \Gamma_R$.

This method of generating reducible representations can be explained by taking examples of molecules belonging to different point groups:

(a) H_2O molecule of C_{2v} point group

In the following Fig., two O-H bonds/stretches are shown by arrows that pointing from O to H while the bond angle HOH is shown by curved double-headed arrow.

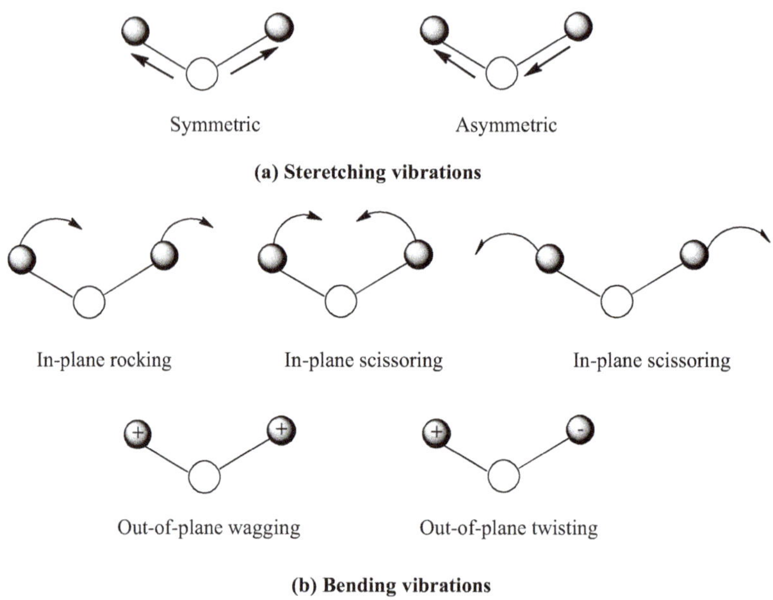

Fig. 3.7: Some normal modes of vibrations. (+) indicates out of plane motion and (–) indicates back of plane motion.

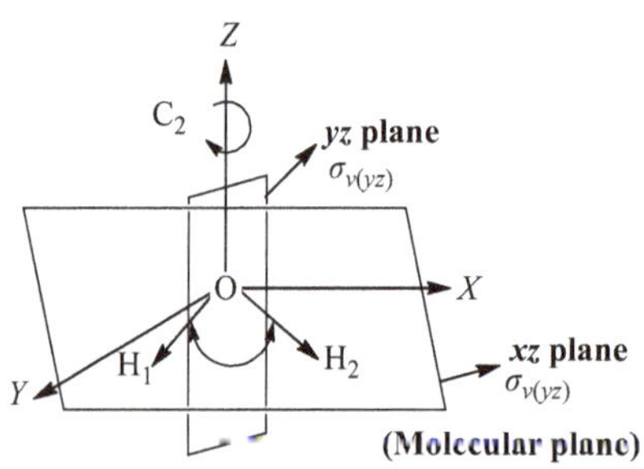

First we find the number of unshifted bond vectors and then follow the rules given below to find χ which contributes towards Γ_{str}.

(i) A vector which is unshifted by the symmetry operations of the point group will contribute + 1, that is, $\chi(R) = +1$.

(ii) When a vector is shifted to another position by applying a symmetry operation of the point group, then the contribution will be taken as zero, that is, $\chi(R) = 0$.

(iii) A vector which is unshifted but only its direction is reversed by the symmetry operation of the point group, its contribution will be $\chi(R) = -1$.

Based on the above rules, we will have the following results for Γ_{str} and Γ_{ben} in case of H_2O molecule as given below:

Representation Γ_{str}

The effects of the symmetry operations on OH-stretches are:

(i) Symmetry operation E does not shift O-H stretches. So the number of unshifted stretches is 2.

(ii) C_2 and $\sigma_{v(yz)}$ interchanges both O-H vectors, so the number of unshifted stretches is 0.

(iii) $\sigma_{v(xz)}$ does not shift O-H stretches. So the number of unshifted stretches is 2.

C_{2v}	E	C_2	$\sigma_{v(yz)}$	$\sigma_{v(xz)}$
No. of unshifted O-H stretch (*n*)	2	0	0	2
$\chi(R)$	+1	+1	+1	+1
Γ_{str} $[n \times \chi(R)]$	2	0	0	2

Representation Γ_{bend}

The effects of the symmetry operations on double-headed arrow are:

(i) All the symmetry operations of C_{2v} point group do not shift double-headed arrow. So, the number of unshifted double-headed arrow in each case is 1.

C_{2v}	E	C_2	$\sigma_{v(yz)}$	$\sigma_{v(xz)}$
No. of unshifted double-headed arrow	1	1	1	1
$\chi(R)$	+1	+1	+1	+1
Γ_{bend} $[n \times \chi(R)]$	1	1	1	1

The symmetry species of these two reducible representations (Γ_{str} and Γ_{bend}) can be obtained by reducing these reducible representations to irreducible representations as follows:

The reducible representation (Γ_{str}) can be reduced to irreducible representations applying the reduction formula given below using character table of C_{4v} point group as follows:

$$N_i = \frac{1}{h}\sum_R \chi(R).n.\chi_i(R)$$

Reducible representation in question

C_{2v}	E	C_2	$\sigma_{v(xz)}$	$\sigma_{v(yz)}$
Γstr	2	0	0	2

Character table of C_{2v} point group

C_{2v}	E	C_2	$\sigma_{v(xz)}$	$\sigma_{v(yz)}$		
A_1	1	1	1	1	z	x^2, y^2, z^2
A_2	1	1	−1	−1	R_z	xy
B_1	1	−1	1	−1	x, R_y	xz
B_2	1	−1	−1	1	y, R_x	yz

$$\begin{aligned} NA_1 &= 1/4\,[(2.1.1) + (0.1.1) + (0.1.1) + (2.1.1) \\ &= 1/4\,[(2+0+0+2] \\ &= 1/4\,[4] = 1 \\ NA_2 &= 1/4\,[(2.1.1) + (0.1.1) + (0.1.-1) + (2.1.-1) \\ &= 1/4\,[(2+0+0-2] \\ &= 1/4\,[0] = 0 \\ NB_1 &= 1/4\,[(2.1.1) + (0.1.-1) + (0.1.1) + (2.1.-1) \\ &= 1/4\,[(2+0+0-2] \\ &= 1/4\,[0] = 0 \\ NB_2 &= 1/4\,[(2.1.1) + (0.1.-1) + (0.1.-1) + (2.1.1) \\ &= 1/4\,[(2+0+0+2] \\ &= 1/4\,[4] = 1 \end{aligned}$$

The above calculations suggest that the reducible representation ($T\sigma$) in question is reduced to

$$\Gamma_{str} = A_1 + B_2$$

Similarly, the reducible representation (Γ_{bend}) can be reduced to irreducible representations applying the reduction formula and using character table of C_{2v} point group as follows:

Reducible representation in question

C_{2v}	E	C_2	$\sigma_{v(xz)}$	$\sigma_{v(yz)}$
Γ_{str}	1	1	1	1

Character table of C_{2v} point group

C_{2v}	E	C_2	$\sigma_{v(xz)}$	$\sigma_{v(yz)}$		
A_1	1	1	1	1	z	x^2, y^2, z^2
A_2	1	1	−1	−1	R_z	xy
B_1	1	−1	1	−1	x, R_y	xz
B_2	1	−1	−1	1	y, R_x	yz

$$
\begin{aligned}
NA_1 &= 1/4\,[(1.1.1) + (1.1.1) + (1.1.1) + (1.1.1)\\
&= 1/4\,[(1 + 1 + 1 + 1]\\
&= 1/4\,[4] = 1\\
NA_2 &= 1/4\,[(1.1.1) + (1.1.1) + (1.1. - 1) + (1.1. - 1)\\
&= 1/4\,[(1 + 1 - 1 - 1]\\
&= 1/4\;[0] = 0\\
NB_1 &= 1/4\,[(1.1.1) + (1.1. - 1) + (1.1.1) + (1.1. - 1)\\
&= 1/4\,[(1 - 1 + 1 - 1]\\
&= 1/4\;[0] = 0\\
NB_2 &= 1/4\,[(1.1.1) + (1.1. - 1) + (1.1. - 1) + (1.1.1)\\
&= 1/4\,[(1 - 1 - 1 + 1]\\
&= 1/4\;[0] = 0
\end{aligned}
$$

Thus, the reducible representation Γ_{bend} is reduced to A_1.

Hence,

$$
\begin{aligned}
\Gamma_{Internalcoordinate} &= \Gamma_{vib} = \Gamma_{str} + \Gamma_{bend} = A_1 + B_2 + A_1\\
&= 2A_1 + B_2
\end{aligned}
$$

(b) NH_3, $POCl_3$, $CHCl_3$ molecule of C_{3v} point group

$CHCl_3$

Let us consider first the case of $CHCl_3$ with symmetry operations E, $2C_3$, $3\sigma_v$. The bond stretches and the angle deformation in this molecule are shown by straight arrows and double-headed arrows.

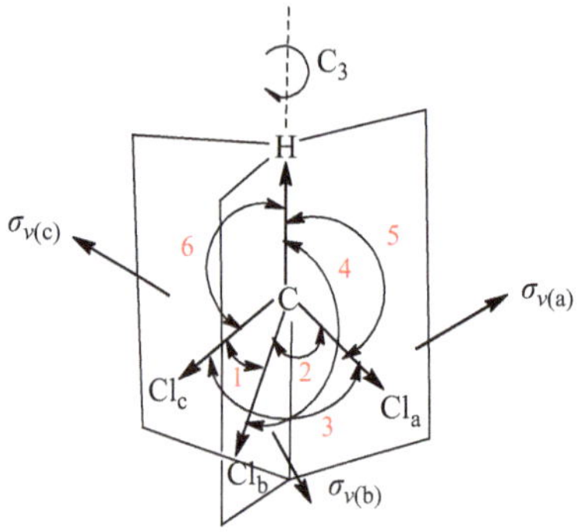

There are three C-Cl and one C-H stretches and six angle deformations as shown in the Fig. given above. We can use these stretches and bending vectors as the basis representation to develop $\Gamma_{C\text{-}Cl}$, $\Gamma_{C\text{-}H}$ and Γ_{bend} as follows:

(i) $\Gamma_{C\text{-}Cl}$

C_{3v}	E	$2C_3$	$3\sigma_v$
No. of unshifted C-Cl stretch (n)	3	0	1
$\chi(R)$	+1	+1	+1
$\Gamma_{C\text{-}Cl}$ $[n\times\chi(R)]$	3	0	1

(ii) $\Gamma_{C\text{-}H}$

C_{3v}	E	$2C_3$	$3\sigma_v$
No. of unshifted C-H stretch (n)	1	1	1
$\chi(R)$	+1	+1	+1
$\Gamma_{C\text{-}H}$ $[n\times\chi(R)]$	1	1	1

(iii) Γ_{bend}

C_3 axis shifts all the angles, so $n = 0$. For σ_v, angle Cl_aCH (axial) is not changed as it is contained in σ_v. Angle Cl_aCCl_b is half bisected and only transforms into another half. So, it is also not changed. Thus, two angles are not changed.

C_{3v}	E	$2C_3$	$3\sigma_v$
No. of unshifted double-headed arrow vectors (n)	6	0	2
$\chi(R)$	+1	+1	+1
Γ_{bend} [$n\times\chi(R)$]	6	0	2

$POCl_3$

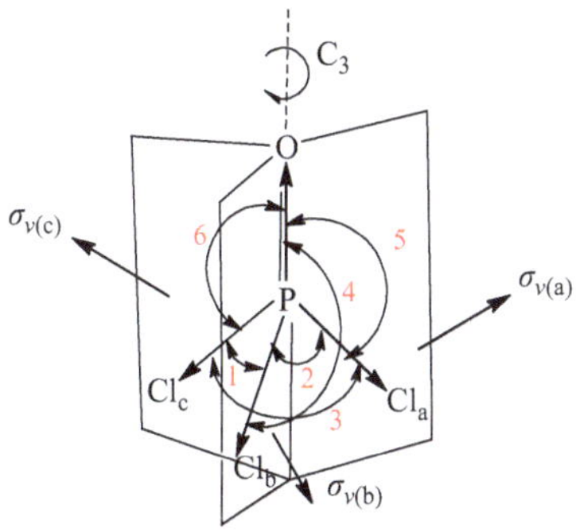

The case of $POCl_3$ is similar to $CHCl_3$. So, one will get same results for $\Gamma_{P\text{-}Cl}$, $\Gamma_{P\text{-}O}$ and Γ_{bend} as in case of $CHCl_3$.

NH_3

The bond stretches and the angle deformation in NH_3 are shown by straight arrows and double-headed arrows, respectively. There are three N-H stretches and three angle deformations as shown in the Fig. given below. We can use these stretches and bending vectors as the basis representation to develop $\Gamma_{N\text{-}H,}$ and Γ_{bend} as follows:

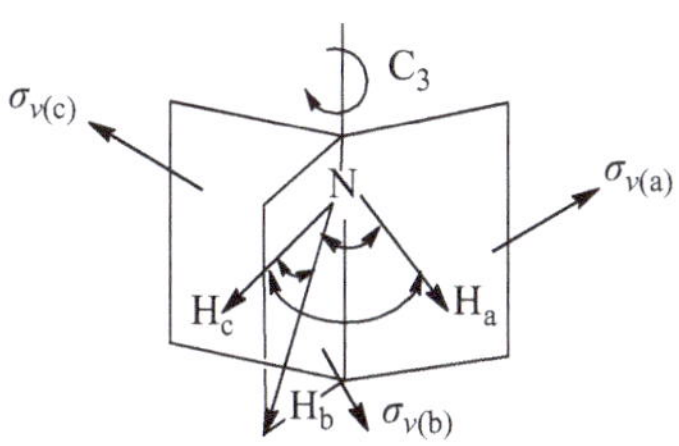

(i) Γ_{str}

C_{3v}	E	$2C_3$	$3\sigma_v$
No. of unshifted N-H stretch (n)	3	0	1
$\chi(R)$	+1	+1	+1
Γ_{str} [$n\times\chi(R)$]	3	0	1

(iv) Γ_{bend}

C_3 axis shifts all the angles, so $n = 0$. For σ_v, angle H_aNH_b is half bisected and transforms into another half. So, it is also not changed. Thus, only one angle is not changed.

C_{3v}	E	$2C_3$	$3\sigma_v$
No. of unshifted double-headed arrow vectors (n)	3	0	1
$\chi(R)$	+1	+1	+1
Γ_{bend} $[n\times\chi(R)]$	3	0	1

The symmetry species of these two reducible representations (Γ_{str} and Γ_{bend}) can be obtained by reducing these reducible representations to irreducible representations as follows:

The reducible representation (Γ_{str}) can be reduced to irreducible representations applying the reduction formula given below using character table of C_{2v} point group as follows:

$$N_i = \frac{1}{h}\sum_R \chi(R).n.\chi_i(R)$$

Reducible represntation in question

C_{3v}	E	$2C_3$	$3\sigma_v$
Γ_{3N}	3	0	1

Character table of C_{3v} Point Group

C_{3v}	E	$2C_3$	3σ		
A_1	1	1	1	z	x^2+y^2, z^2
A_2	1	1	−1	R_z	
E	2	−1	0	(x,y) (R_x, R_y)	$(x^2-y^2, xy)(xz, yz)$

$$
\begin{aligned}
NA_1 &= 1/6\,[(3.1.1) + (0.2.1) + (1.3.1)] \\
&= 1/6[(3 + 0 + 3] \\
&= 1/6[6] = 1 \\
NA_2 &= 1/6\,[(3.1.1) + (0.2.1) + (1.3.-1)] \\
&= 1/6[(3 + 0 - 3] \\
&= 1/6[0] = 0
\end{aligned}
$$

$$\begin{aligned} NE &= 1/6\,[(3.1.2) + (0.2.-1) + (1.3.0)] \\ &= 1/6[(6+0+0] \\ &= 1/6[6] = 1 \end{aligned}$$

The above calculations suggest that the reducible representation (Γ_{str}) in question is reduced to

$$\Gamma_{str} = A_1 + E$$

Similarly, the reducible representation (Γ_{bend}) can be reduced to irreducible representations by applying the reduction formula and using character table of C_{3v} point group. As the reducible representation (Γ_{bend}) is equal to the reducible representation (Γ_{str}), (Γ_{bend}) is reduced to

$$\Gamma_{bend} = A_1 + E$$

Hence,

$$\begin{aligned} \Gamma_{Internalcoordinate} &= \Gamma_{vib} = \Gamma_{str} + \Gamma_{bend} = A_1 + E + A_1 + E \\ &= 2A_1 + 2E \end{aligned}$$

(c) PCl_5 molecule of D_{3h} point group

The bond stretches and the angle deformation in PCl_5 are shown by straight arrows and double-headed arrows in the following Fig. There are 5 P-Cl bond vectors (two axial and three equatorial), **three** equatorial ClPCl bond angles at 120° and six axial ClPCl bond angles at 90°. Using these stretches and bending vectors as the basis, we can generate Γ_{str} [$\Gamma_{P\text{-}Cl\ (axial)}$ and $\Gamma_{P\text{-}Cl\ (equatorial)}$] and Γ_{bend} [$\Gamma_{bend\ (axial)}$ and $\Gamma_{bend\ (equatorial)}$]

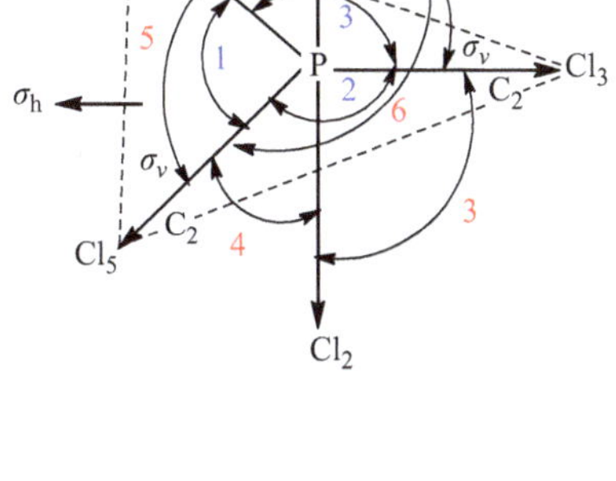

$\Gamma_{P\text{-}Cl}$ (axial):

D_{3h}	E	$2C_3$	$3C_2$	σ_h	$2S_3$	$3\sigma_v$
No. of unshifted axial P-Cl stretch (*n*)	2	2	0	0	0	2
$\chi(R)$	+ 1	+ 1	+ 1	+ 1	+ 1	+ 1
$\Gamma_{\text{P-Cl (axial)}}$ $[n\times\chi(R)]$	2	2	0	0	0	2

$\Gamma_{P\text{-}Cl}$ (equatorial):

D_{3h}	E	$2C_3$	$3C_2$	σ_h	$2S_3$	$3\sigma_v$
No. of unshifted equatorial P-Cl stretch (*n*)	3	0	1	3	0	1
$\chi(R)$	+ 1	+ 1	+ 1	+ 1	+ 1	+ 1
$\Gamma_{\text{P-Cl (equatorial)}}$ $[n\times\chi(R)]$	3	0	1	3	0	1

Γ_{bend} (axial):

D_{3h}	E	$2C_3$	$3C_2$	σ_h	$2S_3$	$3\sigma_v$
No. of unshifted axial double-headed arrow vectors (*n*)	6	0	0	0	0	2
$\chi(R)$	+ 1	+ 1	+ 1	+ 1	+ 1	+ 1
$\Gamma_{\text{bend (axial)}}$ $[n\times\chi(R)]$	6	0	0	0	0	2

Γ_{bend} (equatorial):

D_{3h}	E	$2C_3$	$3C_2$	σ_h	$2S_3$	$3\sigma_v$
No. of unshifted equatorial double-headed arrow vectors (*n*)	3	0	$1^{\#}$	3	0	$1^{\#}$
$\chi(R)$	+ 1	+ 1	+ 1	+ 1	+ 1	+ 1
$\Gamma_{\text{bend (equatorial)}}$ $[n\times\chi(R)]$	3	0	1	3	0	1

Half angle goes into other half.

The symmetry species of these two reducible representations can be obtained by reducing these reducible representations to irreducible representations.

Hence,

$$\Gamma_{\text{Internalcoordinate}} = \Gamma_{\text{vib}} = \Gamma_{\text{str}} + \Gamma_{\text{bend}}$$

$$\text{But } \Gamma_{\text{bend}} = \Gamma_{\text{bend (axial)}} + \Gamma_{\text{bend (equatorial)}}$$

So,

$$\Gamma_{\text{vib}} = \Gamma_{\text{str}} + \Gamma_{\text{bend (axial)}} + \Gamma_{\text{bend (equatorial)}}$$

(d) $PtCl_4^{2-}$/ XeF_4 molecule of D_{4h} point group

Let us consider the case of $PtCl_4^{2-}$ with symmetry operations E, $2C_4$, C_2, $2C_2'$, $2C_2''$, i, $2S_4$, σ_h, $2\sigma_v$, $2\sigma_d$). The bond stretches and the angle deformation in this molecule are shown by straight arrows and double-headed arrows.

Internal coordinates are: four Pt-Cl bond vectors r_1, r_2, r_3 and r_4 and four ClPtCl bond angles β_1, β_2, β_3 and β_4 shown as double-headed arrows (in plane deformation or ipd) in the Fig. Let us take these internal coordinates as bases for representations $\Gamma_{\text{P-Cl}}$ and Γ_{bend}.

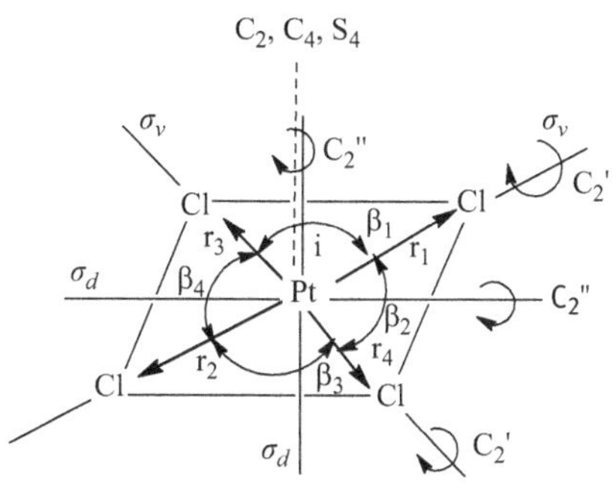

$\Gamma_{\text{P-Cl}}$:

D_{4h}	E	$2C_4$	C_2	$2C_2'$	$2C_2''$	i	$2S_4$	σ_h	$2\sigma_v$	$2\sigma_d$
No. of unshifted Pt-Cl stretches (n)	4	0	0	2	0	0	0	4	2	0
$\chi(R)$	1	1	1	1	1	1	1	1	1	1
$\Gamma_{\text{P-Cl}}$ $[n\times\chi(R)]$	4	0	0	2	0	0	0	4	2	0

The reducible representation ($\Gamma_{\text{P-Cl}}$) can be reduced to irreducible representations applying the reduction formula given below using character table of D_{4h} point group as follows:

$$N_i = \frac{1}{h}\sum_R \chi(R).n.\chi_i(R)$$

Reducible representation in question

D_{4h}	E	$2C_4$	C_2	$2C_2'$	$2C_2''$	i	$2S_4$	σ_h	$2\sigma_v$	$2\sigma_d$
Γ_σ	4	0	0	2	0	0	0	4	2	0

Character table of D_{4h} point group

D_{4h}	E	$2C_4$	C_2	$2C_2'$	$2C_2''$	i	$2S_4$	σ_h	$2\sigma_v$	$2\sigma_d$		
A_{1g}	1	1	1	1	1	1	1	1	1	1		$(x^2+y^2+z^2)$
A_{2g}	1	1	1	−1	−1	1	1	1	−1	−1	R_z	
B_{1g}	1	−1	1	1	−1	1	−1	1	1	−1		(x^2-y^2)
B_{2g}	1	−1	1	−1	1	1	−1	1	−1	1		xy
E_g	2	0	−2	0	0	2	0	−2	0	0	(R_x,R_y)	
A_{1u}	1	1	1	1	1	−1	−1	−1	−1	−1		(xz,yz)
A_{2u}	1	1	1	−1	−1	−1	−1	−1	1	1	z	
B_{1u}	1	−1	1	1	−1	−1	1	−1	−1	1		
B_{2u}	1	−1	1	−1	1	−1	1	−1	1	−1		
E_u	2	0	−2	0	0	−2	0	2	0	0	(x,y)	

$$
\begin{aligned}
NA_{1g} &= 1/16\,[(4.1.1) + (0.2.1)) + (0.1.1) + (2.2.1) + (0.2.1) + (0.1.1) + (0.2.1) \\
&\quad + (4.1.1) + (2.2.1) + (0.2.1)] \\
&= 1/16\,[(4+0+0+4+0+0+0+4+4+0\,] \\
&= 1/16\,[16] = 1
\end{aligned}
$$

$$
\begin{aligned}
NA_{2g} &= 1/16[(4.1.1) + (0.2.1)) + (0.1.1) + (2.2.-1) + (0.2.-1) + (0.1.1) + (0.2.1) \\
&\quad + (4.1.1) + (2.2.-1) + (0.2.-1)] \\
&= 1/16\,[(4+0+0-4+0+0+0+4-4+0] \\
&= 1/16[0] = 0
\end{aligned}
$$

$$
\begin{aligned}
NB_{1g} &= 1/16[(4.1.1) + (0.2.-1)) + (0.1.1) + (2.2.1) + (0.2.-1) + (0.1.1) + (0.2.-1) \\
&\quad + (4.1.1) + (2.2.1) + (0.2.-1)] \\
&= 1/16\,[(4+0+0+4+0+0+0+4+4+0\,] \\
&= 1/16\,[16] = 1
\end{aligned}
$$

$$\begin{aligned}
NB_{2g} &= 1/16[(4.1.1) + (0.2.-1)) + (0.1.1) + (2.2.-1) + (0.2.1) + (0.1.1) + (0.2.-1)\\
&\quad + (4.1.1) + (2.2.-1) + (0.2.1)]\\
&= 1/16[(4 + 0 + 0 - 4 + 0 + 0 + 0 + 4 - 4 + 0\,]\\
&= 1/16[0] = 0
\end{aligned}$$

$$\begin{aligned}
NE_{g} &= 1/16[(4.1.2) + (0.2.0)) + (0.1.-2) + (2.2.0) + (0.2.0) + (0.1.2) + (0.2.0)\\
&\quad + (4.1.-2) + (2.2.0) + (0.2.0)]\\
&= 1/16[(8 + 0 + 0 + 0 + 0 + 0 + 0 - 8 + 0 + 0\,]\\
&= 1/16[0] = 0
\end{aligned}$$

$$\begin{aligned}
NA_{1u} &= 1/16[(4.1.1) + (0.2.1)) + (0.1.1) + (2.2.1) + (0.2.1) + (0.1.-1) + (0.2.-1)\\
&\quad + (4.1.-1) + (2.2.-1) + (0.2.-1)]\\
&= 1/16[(4 + 0 + 0 + 4 + 0 + 0 + 0 - 4 - 4 + 0\,]\\
&= 1/16[0] = 0
\end{aligned}$$

$$\begin{aligned}
NA_{2u} &= 1/16[(4.1.1) + (0.2.1)) + (0.1.1) + (2.2.-1) + (0.2.-1) + (0.1.-1) + (0.2.-1)\\
&\quad + (4.1.-1) + (2.2.1) + (0.2.1)]\\
&= 1/16[(4 + 0 + 0 - 4 + 0 + 0 + 0 - 4 + 4 + 0]\\
&= 1/16[0] = 0
\end{aligned}$$

$$\begin{aligned}
NB_{1u} &= 1/16[(4.1.1) + (0.2.-1)) + (0.1.1) + (2.2.1) + (0.2.-1) + (0.1.-1) + (0.2.1)\\
&\quad + (4.1.-1) + (2.2.-1) + (0.2.1)]\\
&= 1/16[(4 + 0 + 0 + 4 + 0 + 0 + 0 - 4 - 4 + 0\,]\\
&= 1/16[0] = 1
\end{aligned}$$

$$\begin{aligned}
NB_{2u} &= 1/16[(4.1.1) + (0.2.-1)) + (0.1.1) + (2.2.-1) + (0.2.1) + (0.1.-1) + (0.2.1)\\
&\quad + (4.1.-1) + (2.2.1) + (0.2.-1)]\\
&= 1/16[(4 + 0 + 0 - 4 + 0 + 0 + 0 - 4 + 4 + 0\,]\\
&= 1/16[0] = 0
\end{aligned}$$

$$\begin{aligned}
NE_{u} &= 1/16[(4.1.2) + (0.2.0)) + (0.1.-2) + (2.2.0) + (0.2.0) + (0.1.-2) + (0.2.0)\\
&\quad + (4.1.2) + (2.2.0) + (0.2.0)]\\
&= 1/16[(8 + 0 + 0 + 0 + 0 + 0 + 0 + 8 + 0 + 0\,]\\
&= 1/16[16] = 1
\end{aligned}$$

The above calculations show the reducible representation is reduced to

$$\Gamma_{P-Cl} = A_{1g} + B_{1g} + E_u.$$

Γ_{ipd}:

D_{4h}	E	$2C_4$	C_2	$2C_2'$	$2C_2''$	i	$2S_4$	σ_h	$2\sigma_v$	$2\sigma_d$
No. of unshifted double-headed arrow vectors (*n*)	4	0	0	0	2*	0	0	4	0	2*
$\chi(R)$	1	1	1	1	1	1	1	1	1	1
Γ_{ipd} $[n \times \chi(R)]$	4	0	0	0	2	0	0	4	0	2

* Half angle goes into other half, so overall angle is not changed.

The reducible representation (Γ_{ipd}) can be reduced to irreducible representations applying the reduction formula and using character table of D_{4h} point group as follows:

Reducible representation in question

D_{4h}	E	$2C_4$	C_2	$2C_2'$	$2C_2''$	i	$2S_4$	σ_h	$2\sigma_v$	$2\sigma_d$
T_σ	4	0	0	0	2	0	0	4	0	2

Character table of D_{4h} point group

D_{4h}	E	$2C_4$	C_2	$2C_2'$	$2C_2''$	i	$2S_4$	σ_h	$2\sigma_v$	$2\sigma_d$		
A_{1g}	1	1	1	1	1	1	1	1	1	1		$(x^2+y^2+z^2)$
A_{2g}	1	1	1	–1	–1	1	1	1	–1	–1	R_z	
B_{1g}	1	–1	1	1	–1	1	–1	1	1	–1		(x^2-y^2)
B_{2g}	1	–1	1	–1	1	1	–1	1	–1	1		xy
E_g	2	0	–2	0	0	2	0	–2	0	0	(R_x, R_y)	
A_{1u}	1	1	1	1	1	–1	–1	–1	–1	–1		(xz, yz)
A_{2u}	1	1	1	–1	–1	–1	–1	–1	1	1	z	
B_{1u}	1	–1	1	1	–1	–1	1	–1	–1	1		
B_{2u}	1	–1	1	–1	1	–1	1	–1	1	–1		
E_u	2	0	–2	0	0	–2	0	2	0	0	(x, y)	

$$\begin{aligned} NA_{1g} &= 1/16\,[(4.1.1) + (0.2.1)) + (0.1.1) + (0.2.1) + (2.2.1) + (0.1.1) + (0.2.1) \\ &\quad + (4.1.1) + (0.2.1) + (2.2.1)] \\ &= 1/16[(4+0+0+0+4+0+0+4+0+4] \\ &= 1/16[16] = 1 \end{aligned}$$

$$NA_{2g} = 1/16\,[(4.1.1) + (0.2.1)) + (0.1.1) + (0.2.-1) + (2.2.-1) + (0.1.1) + (0.2.1) + (4.1.1) + (0.2.-1) + (2.2.-1)]$$

$$= 1/16[(4 + 0 + 0 + 0 - 4 + 0 + 0 + 4 + 0 - 4]$$

$$= 1/16[0] = 0$$

$$NB_{1g} = 1/16\,[(4.1.1) + (0.2.-1)) + (0.1.1) + (0.2.1) + (2.2.-1) + (0.1.1) + (0.2.-1) + (4.1.1) + (0.2.1) + (2.2.-1)]$$

$$= 1/16[(4 + 0 + 0 + 0 - 4 + 0 + 0 + 4 + 0 - 4]$$

$$= 1/16[0] = 0$$

$$NB_{2g} = 1/16\,[(4.1.1) + (0.2.-1)) + (0.1.1) + (0.2.-1) + (2.2.1) + (0.1.1) + (0.2.-1) + (4.1.1) + (0.2.-1) + (2.2.1)]$$

$$= 1/16[(4 + 0 + 0 + 0 + 4 + 0 + 0 + 4 + 0 + 4]$$

$$= 1/16[16] = 1$$

$$NE_{g} = 1/16\,[(4.1.2) + (0.2.0)) + (0.1.-2) + (0.2.0) + (2.2.0) + (0.1.2) + (0.2.0) + (4.1.-2) + (0.2.0) + (2.2.0)]$$

$$= 1/16[(8 + 0 + 0 + 0 + 0 + 0 + 0 - 8 + 0 + 0]$$

$$= 1/16[0] = 0$$

$$NA_{1u} = 1/16\,[(4.1.1) + (0.2.1)) + (0.1.1) + (0.2.1) + (2.2.1) + (0.1.-1) + (0.2.-1) + (4.1.-1) + (0.2.-1) + (2.2.-1)]$$

$$= 1/16[(4 + 0 + 0 + 0 + 4 + 0 + 0 - 4 + 0 - 4]$$

$$= 1/16[0] = 0$$

$$NA_{2u} = 1/16\,[(4.1.1) + (0.2.1)) + (0.1.1) + (0.2.-1) + (2.2.-1) + (0.1.-1) + (0.2.-1) + (4.1.-1) + (0.2.1) + (2.2.1)]$$

$$= 1/16[(4 + 0 + 0 + 0 - 4 + 0 + 0 - 4 + 0 + 4]$$

$$= 1/16[0] = 0$$

$$NB_{1u} = 1/16\,[(4.1.1) + (0.2.-1)) + (0.1.1) + (0.2.1) + (2.2.-1) + (0.1.-1) + (0.2.1) + (4.1.-1) + (0.2.-1) + (2.2.1)]$$

$$= 1/16[(4 + 0 + 0 + 0 - 4 + 0 + 0 - 4 + 0 + 4]$$

$$= 1/16[0] = 0$$

$$NB_{2u} = 1/16\,[(4.1.1) + (0.2.-1)) + (0.1.\,1) + (0.2.-1) + (2.2.1) + (0.1.-1) + (0.2.1) + (4.1.-1) + (0.2.1) + (2.2.-1)]$$
$$= 1/16[(4 + 0 + 0 + 0 + 4 + 0 + 0 - 4 + 0 - 4]$$
$$= 1/16[0] = 0$$

$$NE_u = 1/16\,[(4.1.2) + (0.2.0)) + (0.1.-2) + (0.2.0) + (2.2.0) + (0.1.-2) + (0.2.0) + (4.1.\,2) + (0.2.0) + (2.2.0)]$$
$$= 1/16[(8 + 0 + 0 + 0 + 0 + 0 + 0 + 8 + 0 + 0]$$
$$= 1/16[16] = 1$$

The above calculations suggest the reducible representation is reduced to:

$$\Gamma_{ipd} = A_{1g} + B_{2g} + E_u$$

Here, A_{1g} mode is redundant (not needed) because all the four bond angles cannot be increased or decreased simultaneously (in plane deformation) and so it can be dropped.

Hence,

$$\Gamma_{ipd} = B_{2g} + E_u$$

Now

$$\Gamma_{P-Cl} + \Gamma_{ipd} = A_{1g} + B_{1g} + E_u + B_{2g} + E_u$$
$$= A_{1g} + B_{1g} + B_{2g} + 2E_u$$

The Γ_{vib} obtained using 3N vectors as basis are: A_{1g} + B_{1g} + B_{2g} + $2E_u$ + A_{2u} + B_{2u}. Thus, we are in short of A_{2u} + B_{2u}, and we have to account for these two irreducible representation.

So far, we have considered only in plane deformation of ClPCl bond angles. Let us consider out of plane deformation (opd) of ClPCl bond angles, wherein all Pt-Cl bond vectors may move out of plane, and thus out of plane ClPCl bond angles deformation may also occur. Here, we represent opd by new type of vectors that lift up these Pt-Cl bond vectors (that is by vertical arrows).

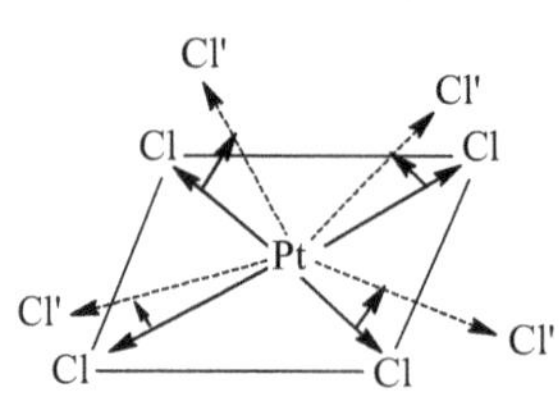

Taking these vertical vectors as the basis, we can find Γ_{opd} as

Γ_{opd}:

D_{4h}	E	$2C_4$	C_2	$2C_2'$	$2C_2''$	i	$2S_4$	σ_h	$2\sigma_v$	$2\sigma_d$
No. of unshifted out of plane vertical vectors (*n*)	4	0	0	−2*	0	0	0	−4*	2	0
$\chi(R)$	1	1	1	1	1	1	1	1	1	1
Γ_{opd} $[n\times\chi(R)]$	4	0	0	−2	0	0	0	−4	2	0

*Symmetry operations C_2' and σ_h reverse the direction of vectors.

Again, this reducible representation Γ_{opd} can be reduced in the similar manner as described above as follows:

Reducible representation in question

D_{4h}	E	$2C_4$	C_2	$2C_2'$	$2C_2''$	i	$2S_4$	σ_h	$2\sigma_v$	$2\sigma_d$
Γ_σ	4	0	0	−2	0	0	0	−4	2	0

Character table of D_{4h} point group

D_{4h}	E	$2C_4$	C_2	$2C_2'$	$2C_2''$	i	$2S_4$	σ_h	$2\sigma_v$	$2\sigma_d$		
A_{1g}	1	1	1	1	1	1	1	1	1	1		$(x^2+y^2+z^2)$
A_{2g}	1	1	1	−1	−1	1	1	1	−1	−1	R_z	
B_{1g}	1	−1	1	1	−1	1	−1	1	1	−1		(x^2-y^2)
B_{2g}	1	−1	1	−1	1	1	−1	1	−1	1		xy
E_g	2	0	−2	0	0	2	0	−2	0	0	(R_x, R_y)	
A_{1u}	1	1	1	1	1	−1	−1	−1	−1	−1		(xz, yz)
A_{2u}	1	1	1	−1	−1	−1	−1	−1	1	1	z	
B_{1u}	1	−1	1	1	−1	−1	1	−1	−1	1		
B_{2u}	1	−1	1	−1	1	−1	1	−1	1	−1		
E_u	2	0	−2	0	0	−2	0	2	0	0	(x, y)	

$$
\begin{aligned}
NA_{1g} &= 1/16\,[(4.1.1)+(0.2.1))+(0.1.1)+(-2.2.1)+(0.2.1)+(0.1.1)+(0.2.1)\\
&\quad +(-4.1.1)+(2.2.1)+(0.2.1)]\\
&= 1/16[(4+0+0-4+0+0+0-4+4+0]\\
&= 1/16[0]=0
\end{aligned}
$$

$$
\begin{aligned}
NA_{2g} &= 1/16\,[(4.1.1)+(0.2.1))+(0.1.1)+(-2.2.-1)+(0.2.-1)+(0.1.1)+(0.2.1)\\
&\quad +(-4.1.1)+(2.2.-1)+(0.2.-1)]\\
&= 1/16[(4+0+0+4+0+0+0-4-4+0]\\
&= 1/16[0]=0
\end{aligned}
$$

$$\begin{aligned}NB_{1g} &= 1/16\,[(4.1.1)+(0.2.-1))+(0.1.1)+(-2.2.1)+(0.2.-1)+(0.1.1)+(0.2.-1)\\&\quad+(-4.1.1)+(2.2.1)+(0.2.-1)]\\&=1/16[(4+0+0-4+0+0+0-4+4+0]\\&=1/16[0]=0\end{aligned}$$

$$\begin{aligned}NB_{2g} &= 1/16\,[(4.1.1)+(0.2.-1))+(0.1.1)+(-2.2.-1)+(0.2.1)+(0.1.1)+(0.2.-1)\\&\quad+(-4.1.1)+(2.2.-1)+(0.2.1)]\\&=1/16[(4+0+0+4+0+0+0-4-4+0]\\&=1/16[0]=0\end{aligned}$$

$$\begin{aligned}NE_{g} &= 1/16\,[(4.1.2)+(0.2.0))+(0.1.-2)+(-2.2.0)+(0.2.0)+(0.1.2)+(0.2.0)\\&\quad+(-4.1.-2)+(2.2.0)+(0.2.0)]\\&=1/16[(8+0+0-4+0+0+0+8+0+0]\\&=1/16[16]=1\end{aligned}$$

$$\begin{aligned}NA_{1u} &= 1/16\,[(4.1.1)+(0.2.1))+(0.1.1)+(-2.2.1)+(0.2.1)+(0.1.-1)+(0.2.-1)\\&\quad+(-4.1.-1)+(2.2.-1)+(0.2.-1)]\\&=1/16[(4+0+0-4+0+0+0+4-4+0]\\&=1/16[0]=0\end{aligned}$$

$$\begin{aligned}NA_{2u} &= 1/16\,[(4.1.1)+(0.2.1))+(0.1.1)+(-2.2.-1)+(0.2.-1)+(0.1.-1)+(0.2.-1)\\&\quad+(-4.1.-1)+(2.2.1)+(0.2.1)]\\&=1/16[(4+0+0+4+0+0+0+4+4+0]\\&=1/16[16]=1\end{aligned}$$

$$\begin{aligned}NB_{1u} &= 1/16\,[(4.1.1)+(0.2.-1))+(0.1.1)+(-2.2.1)+(0.2.-1)+(0.1.-1)+(0.2.1)\\&\quad+(-4.1.-1)+(2.2.-1)+(0.2.1)]\\&=1/16[(4+0+0-4+0+0+0+4-4+0]\\&=1/16[0]=0\end{aligned}$$

$$\begin{aligned}NB_{2u} &= 1/16\,[(4.1.1)+(0.2.-1))+(0.1.1)+(-2.2.-1)+(0.2.1)+(0.1.-1)+(0.2.1)\\&\quad+(-4.1.-1)+(2.2.1)+(0.2.-1)]\\&=1/16[(4+0+0+4+0+0+0+4+4+0]\\&=1/16[16]=1\end{aligned}$$

$$NE_u = 1/16\,[(4.1.2) + (0.2.0)) + (0.1. - 2) + (-2.2.0) + (0.2.0) + (0.1. - 2) + (0.2.0) + (-4.1.2) + (2.2.0) + (0.2.0)]$$
$$= 1/16[(8 + 0 + 0 + 0 + 0 + 0 + 0 - 8 + 0 + 0]$$
$$= 1/16[0] = 0$$

Thus, Γ_{opd} is reduced to

$$E_g + A_{2u} + B_{2u}$$

In the Γ_{vib} obtained using 3N vectors as basis, there is no E_g reducible representation. So, E_g is redundant. Thus, $\Gamma_{opd} = A_{2u} + B_{2u}$.

By noting the difference between Γ_{3N} and $\Gamma_{internal\ coordinate}$, one can find the redundant irreducible representation.

Now $\Gamma_{P-Cl} + \Gamma_{ipd} + \Gamma_{opd} = A_{1g} + B_{1g} + B_{2g} + 2E_u + A_{2u} + B_{2u}$, that is, equal to Γ_{vib}.

(e) BF_3 molecule of D_{3h} point group

The internal coordinates (bond stretches and the angle deformation) in this molecule are shown by straight arrows and double-headed arrows in the following Figure. Using these internal coordinates, let us generate reducible representations.

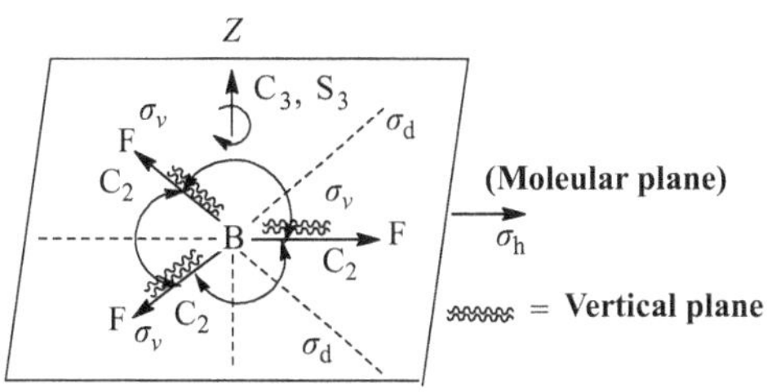

Γ B-F

D_{3h}	E	$2C_3$	$3C_2$	σ_h	$2S_3$	$3\sigma_v$
No. of unshifted B-F stretches (*n*)	3	0	1	3	0	1
$\chi(R)$	1	1	1	1	1	1
$\Gamma_{B\text{-}F}$ $[n\times\chi(R)]$	3	0	1	3	0	1

The reducible representation ($\Gamma_{B\text{-}F}$) can be reduced to irreducible representations applying the reduction formula and using character table of D_{3h} point group as follows:

$$N_i = \frac{1}{h}\sum_R \chi(R).n.\chi_i(R)$$

Reducible Representation in question

D_{3h}	E	$2C_3$	$3C_2$	σ_h	$2S_3$	$3\sigma_v$
$\Gamma_{B\text{-}F}$	3	0	1	3	0	1

Character table of D_{3h} point group

D_{3h}	E	$2C_3$	$3C_2$	σ_h	$2S_3$	$3\sigma_v$		
A_1'	1	1	1	1	1	1		$x^2+y^2+z^2$
A_2'	1	1	−1	1	1	−1	R_z	
E'	2	−1	0	2	−1	0	(x, y)	(x^2+y^2+xy)
A_1''	1	1	1	−1	−1	−1		
A_2''	1	1	−1	−1	−1	1	z	
E''	2	−1	0	−2	1	0	R_x, R_y	(xz, yz)

$$NA_1' = 1/12\,[(3.1.1) + (0.2.1) + (1.3.1) + (3.1.1) + (0.2.1) + (1.3.1)$$
$$= 1/12\,[3 + 0 + 3 + 3 + 0 + 3]$$
$$= 1/12\,[12] = 1$$

$$NA_2' = 1/12\,\,[(3.1.1) + (0.2.1) + (1.3. - 1) + (3.1.1) + (0.2.1) + (1.3. - 1)$$
$$= 1/12\,[3 + 0 - 3 + 3 + 0 + -3]$$
$$= 1/12[0] = 0$$

$$NE' = 1/12\,[(3.1.2) + (0.2. - 1) + (1.3.0) + (3.1.2) + (0.2. - 1) + (1.3.0)$$
$$= 1/12[6 + 0 + 0 + 6 + 0 + 0] = 1$$

$$NA_1'' = 1/12\,[(3.1.1) + (0.2.1) + (1.3.1) + (3.1. - 1) + (0.2. - 1) + (1.3. - 1)$$
$$= 1/12\,[3 + 0 + 3 - 3 + 0 - 3]$$
$$= 1/12[0] = 0$$

$$NA_2'' = 1/12\,[(3.1.1) + (0.2.1) + (1.3. - 1) + (3.1. - 1) + (0.2. - 1) + (1.3.\ 1)$$
$$= 1/12\,[3 + 0 - 3 - 3 + 0 + 3]$$
$$= 1/12[0] = 0$$

$$NE'' = 1/12\,[(3.1.2) + (0.2. - 1) + (1.3.0) + (3.1. - 2) + (0.2.1) + (1.3.0)$$
$$= 1/12\,[6 + 0 + 0 - 6 + 0 + 0]$$
$$= 1/12[0] = 0$$

Thus, $\Gamma_{B\text{-}F}$ is reduced to

$$A_1' + E'$$

Γ_{ipd}: (ipd = In plane deformation)

D_{3h}	E	$2C_3$	$3C_2$	σ_h	$2S_3$	$3\sigma_v$
No. of unshifted double-headed arrow vectors (*n*)	3	0	1*	3	0	1*
$\chi(R)$	1	1	1	1	1	1
Γ_{ipd} $[n \times \chi(R)]$	3	0	1	3	0	1

*Half angle goes into other half, so overall angle is not changed.

The reducible representation (Γ_{ipd}) being equal to $\Gamma_{B\text{-}F}$ on reduction will give

$$\Gamma_{ipd} = A_1' + E'$$

The mode A_1' in plane deformation is redundant as all three angles cannot be increased or decreased simultaneously. In this situation, A_1' in plane deformation is neglected.

Thus, total $\Gamma_{\text{internal coordinate}} = \Gamma_{B-F} + \Gamma_{ipd} = A_1' + E' + E' = A_1' + 2E'$

If we compare $\Gamma_{\text{internal coordinate}}$ (A_1' + $2E'$) with Γ_{vib} (A_1' + A_2''+ $2E'$) obtained through T_{3N} method, it is one mode less. This missing mode (A_2'') is opd of the three B-F bonds shown below. Let us find the missing mode through finding Γ_{opd} for generating reducible representation.

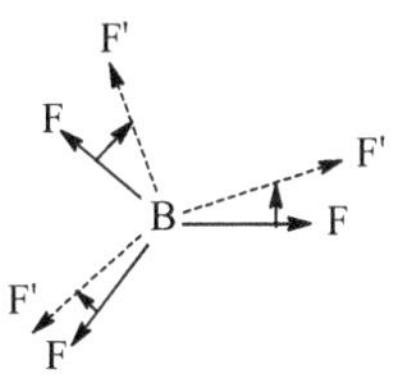

Out-of-plane deformation

Γ_{opd}:

D_{3h}	E	$2C_3$	$3C_2$	σ_h	$2S_3$	$3\sigma_v$
No. of unshifted out of plane vertical vectors (*n*)	3	0	−1	−3	0	1
$\chi(R)$	1	1	1	1	1	1
Γ_{opd} $[n \times \chi(R)]$	3	0	−1	−3	0	1

The reducible representation (Γ_{opd}) can be reduced to irreducible representations applying the reduction formula and using character table of D_{3h} point group as follows:

$$N_i = \frac{1}{h}\sum_R \chi(R).n.\chi_i(R)$$

Reducible Representation in question

D_{3h}	E	$2C_3$	$3C_2$	σ_h	$2S_3$	$3\sigma_v$
$\Gamma_{B\text{-}F}$	3	0	−1	−3	0	1

Character table of D_{3h} point group

D_{3h}	E	$2C_3$	$3C_2$	σ_h	$2S_3$	$3\sigma_v$		
A_1'	1	1	1	1	1	1		x^2+y^2, z^2
A_2'	1	1	−1	1	1	−1	R_z	
E'	2	−1	0	2	−1	0	(x, y)	(x^2-y^2, xy)
A_1''	1	1	1	−1	−1	−1		
A_2''	1	1	−1	−1	−1	1	z	
E''	2	−1	0	−2	1	0	R_x, R_y	(xz, yz)

$$\begin{aligned} NA_1' &= 1/12\,[(3.1.1)+(0.2.1)+(-1.3.1)+(-3.1.1)+(0.2.1)+(1.3.1)\\ &= 1/12\,[3+0-3-3+0+3]\\ &= 1/12\,[0]=0 \end{aligned}$$

$$\begin{aligned} NA_2' &= 1/12\,[(3.1.1)+(0.2.1)+(-1.3.-1)+(-3.1.1)+(0.2.1)+(1.3.-1)\\ &= 1/12\,[3+0+3-3+0-3]\\ &= 1/12\,[0]=0 \end{aligned}$$

$$\begin{aligned} NE' &= 1/12\,[(3.1.2)+(0.2.-1)+(-1.3.0)+(-3.1.2)+(0.2.-1)+(1.3.0)\\ &= 1/12\,[6+0+0-6+0+0]\\ &= 1/12\,[0]=0 \end{aligned}$$

$$\begin{aligned} NA_1'' &= 1/12\,[(3.1.1)+(0.2.1)+(-1.3.1)+(-3.1.-1)+(0.2.-1)+(1.3.-1)\\ &= 1/12\,[3+0-3+3+0-3]\\ &= 1/12\,[0]=0 \end{aligned}$$

$$\begin{aligned} NA_2'' &= 1/12\,[(3.1.1)+(0.2.-1)+(-1.3.-1)+(-3.1.-1)+(0.2.1)+(1.3.1)\\ &= 1/12\,[3+0+3+3+0+3]\\ &= 1/12\,[12]=1 \end{aligned}$$

$$NE'' = 1/12\,[(3.1.2) + (0.2.-1) + (-1.3.0) + (-3.1.-2) + (0.2.1) + (1.3.0)$$
$$= 1/12\,[6 + 0 + 0 + 6 + 0 + 0]$$
$$= 1/12\,[12] = 1$$

Thus, Γ_{opd} is reduced to $A_2'' + E''$. So, total $\Gamma_{\text{internal coordinate}} = \Gamma_{B\text{-}F} + \Gamma_{ipd} + \Gamma_{opd} = A_1' + 2E' + A_2'' + E''$. On comparing this result with that of Γ_{vib} $(A_1' + A_2'' + 2E')$ obtained by T_{3N} method, it is well clear that E″ mode is extra and it is redundant and comes out from opd. Therefore,

$$\Gamma_{\text{internal coordinate}} = \Gamma_{B-F} + \Gamma_{ipd} + \Gamma_{opd} = A_1' + 2E' + A_2'' \; (3N - 6 = 4 \times 3 - 6 = 6)$$

(f) $WOCl_4$ molecule of C_{4v} point group

The bond stretches and the angle deformation in this molecule are shown by straight arrows and double-headed arrows.

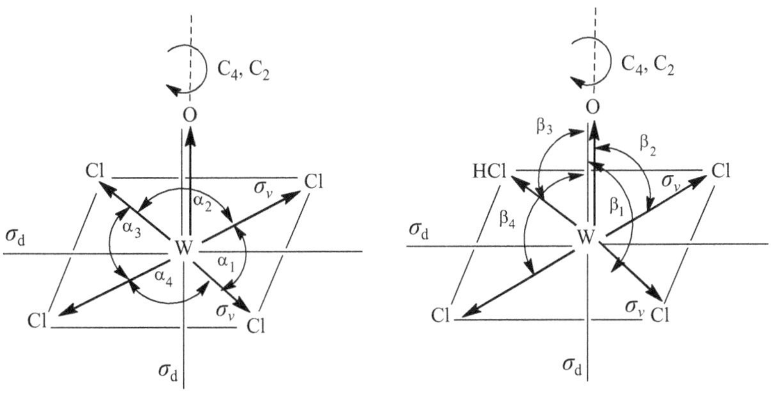

There is five bond vectors (4W–Cl, 1W –O), four in-plane deformation (ipd) of angles $\alpha_1, \alpha_2, \alpha_3, \alpha_4$ and four (opd) of angles $\beta_1, \beta_2, \beta_3, \beta_4$. These are shown in the structure of $WOCl_4$ molecule. Thus, there are 13 internal coordinates which can be used as bases to generate reducible representations.

C_{4v}	E	$2C_4$	C_2	$2\sigma_v$	$2\sigma_d$
No. of unshifted W-O bond vector (n)	1	1	1	1	1
$\chi(R)]$	1	1	1	1	1
$\Gamma_{W\text{-}O}$ $[n \times \chi(R)]$	1	1	1	1	1

C_{4v}	E	$2C_4$	C_2	$2\sigma_v$	$2\sigma_d$
No. of unshifted W-Cl bond vector (n)	4	0	0	2	0
$\chi(R)$]	1	1	1	1	1
$\Gamma_{W\text{-}Cl}$ [$n\times\chi(R)$]	4	0	0	2	0

The results of $\Gamma_{W\text{-}O}$ and $\Gamma_{W\text{-}C}$ may be combined to get $\Gamma_{stretch}$. Therefore,

C_{4v}	E	$2C_4$	C_2	$2\sigma_v$	$2\sigma_d$
$\Gamma_{stretch}$	5	1	1	3	1

This reducible representation can be reduced to get the reducible representation as follows:

Reducible representation in question

C_{4v}	E	$2C_4$	C_2	$2\sigma_v$	$2\sigma_d$
$\Gamma_{stretch}$	5	1	1	3	1

Character table of C_{4v} point group

C_{4v}	E	$2C_4$	C_2	$2\sigma_v$	$2\sigma_d$		
A_1	1	1	1	1	1	z	x^2+y^2, z^2
A_2	1	1	1	−1	−1	R_z	
B_1	1	−1	1	1	−1		x^2-y^2
B_2	1	1	1	1	1		xy
E	2	0	−2	0	0	$(x, y)(R_x, R_y)$	xz, yz

$$NA_1 = 1/8\,[(5.1.1) + (1.2.1) + (1.1.1) + (3.2.1) + (1.2.1)$$
$$= 1/8\,[5 + 2 + 1 + 6 + 2]$$
$$= 1/8\,[16] = 2$$
$$NA_2 = 1/8\,[(5.1.1) + (1.2.1) + (1.1.1) + (3.2. - 1) + (1.2. - 1)$$
$$= 1/8\,[5 + 2 + 1 - 6 - 2]$$
$$= 1/8\,[0] = 0$$

$$NB_1 = 1/8\,[(5.1.1) + (1.2.-1) + (1.1.1) + (3.2.1) + (1.2.-1)$$
$$= 1/8\,[5 - 2 + 1 + 6 - 2]$$
$$= 1/8\,[8] = 1$$
$$NB_2 = 1/8\,[(5.1.1) + (1.2.-1) + (1.1.1) + (3.2.-1) + (1.2.1)$$
$$= 1/8\,[5 - 2 + 1 - 6 + 2]$$
$$= 1/8\,[0] = 0$$
$$NE = 1/8\,[(5.1.2) + (1.2.0) + (1.1.-2) + (3.2.0) + (1.2.0)$$
$$= 1/8\,[10 + 0 - 2 + 0 + 0]$$
$$= 1/8\,[8] = 1$$

Thus, the reducible representation $\Gamma_{stretch}$ is reduced to:

$$\Gamma_{stretch} = 2A_1 + B_1 + E$$

Γ_{ipd}: (ipd = in plane deformation)

C_{4v}	E	$2C_4$	C_2	$2\sigma_v$	$2\sigma_d$
No. of unshifted double-headed arrow vectors (*n*)	4	0	0	0	2*
χ(R)]	1	1	1	1	1
Γ_{ipd} [n×χ(R)]	4	0	0	0	2

*Half angle goes into other half angle.

Γ_{ipd} is reduced as follows:

Reducible representation in question

C_{4v}	E	$2C_4$	C_2	$2\sigma_v$	$2\sigma_d$
Γ_{ipd}	4	0	0	0	2

Character table of point group

C_{4v}	E	$2C_4$	C_2	$2\sigma_v$	$2\sigma_d$		
A_1	1	1	1	1	1	z	x^2+y^2, z^2
A_2	1	1	1	−1	−1	R_z	
B_1	1	−1	1	1	−1		x^2-y^2
B_2	1	−1	1	−1	1		xy
E	2	0	−2	0	0	$(x,y)(R_x,R_y)$	xz, yz

$$NA_1 = 1/8\,[(4.1.1) + (0.2.1) + (0.1.1) + (0.2.1) + (2.2.1)$$
$$= 1/8\,[4 + 0 + 0 + 0 + 4]$$
$$= 1/8\,[8] = 1$$
$$NA_2 = 1/8\,[(4.1.1) + (0.2.1) + (0.1.1) + (0.2.-1) + (2.2.-1)$$
$$= 1/8\,[4 + 0 + 0 + 0 - 4]$$
$$= 1/8\,[0] = 0$$
$$NB_1 = 1/8\,[(4.1.1) + (0.2.-1) + (0.1.1) + (0.2.1) + (2.2.-1)$$
$$= 1/8\,[4 + 0 + 0 + 0 - 4]$$
$$= 1/8\,[0] = 0$$
$$NB_2 = 1/8\,[(4.1.1) + (0.2.-1) + (0.1.1) + (0.2.-1) + (2.2.1)$$
$$= 1/8\,[4 + 0 + 0 + 0 + 4]$$
$$= 1/8\,[8] = 1$$
$$NE = 1/8\,[(4.1.2) + (0.2.0) + (0.1.-2) + (0.2.0) + (2.2.0)$$
$$= 1/8\,[8 + 0 + 0 + 0 + 0]$$
$$= 1/8\,[8] = 1$$

Thus, reducible representation Γ_{ipd} is reduced to:

$$\Gamma_{ipd} = A_1 + B_2 + E$$

A_1 is redundant as all in-plane bond angles cannot be increased or decreased simultaneously. So,

$$\Gamma_{ipd} = B_2 + E$$

Γ_{opd}: ($_{opd}$ = out of plane deformation)

C_{4v}	E	$2C_4$	C_2	$2\sigma_v$	$2\sigma_d$
No. of unshifted double arrow headed vectors (*n*)	4	0	0	2	0
$\chi(R)$]	1	1	1	1	1
Γ_{opd} $[n \times \chi(R)]$	4	0	0	2	0

Similar to above Γ_{opd} on reduction gives:

$$\Gamma_{opd} = A_1 + B_1 + E$$

Now, total irreducible representation $= \Gamma_{stretch} + \Gamma_{ipd} + \Gamma_{opd}$

$$= (2A_1 + B_1 + E) + (B_2 + E) + (A_1 + B_1 + E\}$$
$$= 3A_1 + 2B_1 + B_2 + 3E$$

This is equivalent to Γ_{vib} obtained from $\Gamma_{3N} = 3N - 6 = 3 \times 6 - 6 = 12$

(g) AB_6 molecule of O_h point group

The internal coordinates (bond stretches and the angle deformation) in this molecule are shown by straight arrows and double-headed arrows in the following Fig.

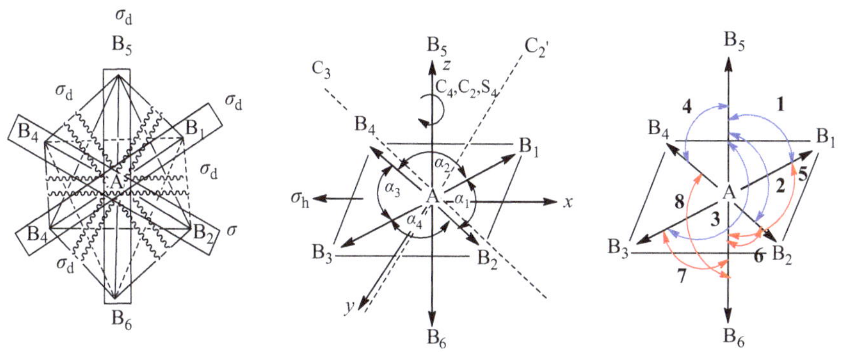

There are six A-B bond vectors, four in-plane deformation (ipd) of angles α_1, α_2, α_3 and α_4 and eight opd angles 1, 2, 3, 4, 5, 6, 7 and 8. These are shown in the structure of AB_6 molecule. Thus, there are 6 + 4 + 8 = 18 internal coordinates while the required modes are 3N-6 = 3×7–6 = 15, that is, three modes are redundant. These can be used as bases to generate reducible representations.

$\Gamma_{A\text{-}B}$

O_h	E	$8C_3$	$3C_2$	$6C_4$	$6C_2'$	i	$6S_4$	$8S_6$	$3\sigma_h$	$6\sigma_d$
No. of unshifted A-B stretches (n)	6	0	2	2	0	0	0	0	4	2
$\chi(R)$	1	1	1	1	1	1	1	1	1	1
$\Gamma_{A\text{-}B}$ [$n \times \chi(R)$]	6	0	2	2	0	0	0	0	4	2

This reducible representation ($\Gamma_{A\text{-}B}$) can be reduced to irreducible representation applying reduction formula and using character table of O_h point group.

Reducible representation in question

O_h	E	$8C_3$	$3C_2(=C_4^2)$	$6C_4$	$6C_2'$	i	$6S_4$	$8S_6$	$3\sigma_h$	$6\sigma_d$
Γ_{A-B}	6	0	2	2	0	0	0	0	4	2

Character Table of O_h Point Group

O_h	E	$8C_3$	$3C_2$	$6C_4$	$6C_2'$	i	$6S_4$	$8S_6$	$3\sigma_h$	$6\sigma_d$		
A_{1g}	1	1	1	1	1	1	1	1	1	1		$x^2+y^2+z^2$
A_{2g}	1	1	1	−1	−1	1	−1	1	1	−1		
E_g	2	−1	2	0	0	2	0	−1	2	0		$(2z^2-x^2-y^2, x^2-y^2)$
T_{1g}	3	0	−1	1	−1	3	1	0	−1	−1	(R_x, R_y, R_z)	
T_{2g}	3	0	−1	−1	1	3	−1	0	−1	1		(xz, yz, xy)
A_{1u}	1	1	1	1	1	−1	−1	−1	−1	−1		
A_{2u}	1	1	1	−1	−1	−1	1	−1	−1	1		
E_u	2	−1	2	0	0	−2	0	1	−2	0		
T_{1u}	3	0	−1	1	−1	−3	−1	0	1	1	(x, y, z)	
T_{2u}	3	0	−1	−1	1	−3	1	0	1	−1		

$$
\begin{aligned}
NA_{1g} &= 1/48\ [(6.1.1)+(0.8.1)+(2.3.1)+(2.6.1)+(0.6.1)+(0.1.1)+(0.6.1)\\
&\quad +(0.8.1)+(4.3.1)+(2.6.1)\\
&= 1/48\ [(6+0+6+12+0+0+0+0+12+12)]\\
&= 1/48\ [48] = 1
\end{aligned}
$$

$$
\begin{aligned}
NA_{2g} &= 1/48\ [(6.1.1)+(0.8.1)+(2.3.1)+(2.6.-1)+(0.6.-1)+(0.1.1)+(0.6.-1)\\
&\quad +(0.8.1)+(4.3.1)+(2.6.-1)\\
&= 1/48\ [(6+0+6-12+0+0+0+0+12-12)]\\
&= 1/48[0] = 0
\end{aligned}
$$

$$
\begin{aligned}
NE_g &= 1/48\ [(6.1.2)+(0.8.-1)+(2.3.2)+(2.6.0)+(0.6.0)+(0.1.2)+(0.6.0)\\
&\quad +(0.8.-1)+(4.3.2)+(2.6.0)\\
&= 1/48\ [(12+0+12+0+0+0+0+0+24+0)]\\
&= 1/48\ [48] = 1
\end{aligned}
$$

$$
\begin{aligned}
NT_{1g} &= 1/48\ [(6.1.3)+(0.8.0)+(2.3.-1)+(2.6.1)+(0.6.-1)+(0.1.3)+(0.6.1)\\
&\quad +(0.8.0)+(4.3.-1)+(2.6.-1)\\
&= 1/48\ [(18+0-6+12+0+0+0+0+0-12-12)]\\
&= 1/48\ [0] = 0
\end{aligned}
$$

$$\begin{aligned} NT_{2g} &= 1/48\ [(6.1.3) + (0.8.0) + (2.3.-1) + (2.6.-1) + (0.6.1) + (0.1.3) + (0.6.-1) \\ &\quad + (0.8.0) + (4.3.-1) + (2.6.1) \\ &= 1/48\ [(18 + 0 - 6 - 12 + 0 + 0 + 0 + 0 - 12 + 12)] \\ &= 1/48\ [0] = 0 \end{aligned}$$

$$\begin{aligned} NA_{1u} &= 1/48\ [(6.1.1) + (0.8.1) + (2.3.1) + (2.6.1) + (0.6.1) + (0.1.-1) + (0.6.-1) \\ &\quad + (0.8.-1) + (4.3.-1) + (2.6.-1) \\ &= 1/48\ [(6 + 0 + 6 + 12 + 0 + 0 + 0 + 0 - 12 - 12)] \\ &= 1/48\ [0] = 0 \end{aligned}$$

$$\begin{aligned} NA_{2u} &= 1/48\ [(6.1.1) + (0.8.1) + (2.3.1) + (2.6.-1) + (0.6.-1) + (0.1.-1) + (0.6.1) \\ &\quad + (0.8.-1) + (4.3.-1) + (2.6.1) \\ &= 1/48\ [(6 + 0 + 6 - 12 + 0 + 0 + 0 + 0 - 12 + 12)] \\ &= 1/48\ [0] = 0 \end{aligned}$$

$$\begin{aligned} N_{Eu} &= 1/48\ [(6.1.2) + (0.8.-1) + (2.3.2) + (2.6.0) + (0.6.0) + (0.1.-2) + (0.6.0) \\ &\quad + (0.8.1) + (4.3.-2) + (2.6.0) \\ &= 1/48\ [(12 + 0 + 12 + 0 + 0 + 0 + 0 + 0 - 24 + 0)] \\ &= 1/48\ [0] = 0 \end{aligned}$$

$$\begin{aligned} NT_{1u} &= 1/48\ [(6.1.3) + (0.8.0) + (2.3.-1) + (2.6.1) + (0.6.1) + (0.1.-3) + (0.6.-1) \\ &\quad + (0.8.0) + (4.3.1) + (2.6.1) \\ &= 1/48\ [(18 + 0 - 6 + 12 + 0 + 0 + 0 + 0 + 0 + 12 + 12)] \\ &= 1/48\ [48] = 1 \end{aligned}$$

$$\begin{aligned} NT_{2u} &= 1/48\ [(6.1.3) + (0.8.0) + (2.3.-1) + (2.6.-1) + (0.6.1) + (0.1.-3) + (0.6.1) \\ &\quad + (0.8.0) + (4.3.1) + (2.6.-1) \\ &= 1/48\ [(18 + 0 - 6 - 12 + 0 + 0 + 0 + 0 + 0 + 12 - 12)] \\ &= 1/48\ [0] = 0 \end{aligned}$$

Thus, $\Gamma_{A\text{-}B}$ is reduced to $A_{1g} + E_g + T_{1u}$.

$\Gamma_{\mathbf{ipd\ +\ opd}}$: (ipd = in plane deformation; opd = out of plane deformation)

O_h	E	$8C_3$	$3C_2$	$6C_4$	$6C_2'$	i	$6S_4$	$8S_6$	$3\sigma_h$	$6\sigma_d$
No. of unshifted double arrow headed vectors (*n*)	12	0	0	0	2	0	0	0	4	2
$\chi(R)$	1	1	1	1	1	1	1	1	1	1
$\Gamma_{ipd+opd}$ $[n\times\chi(R)]$	12	0	0	0	2	0	0	0	4	2

Reducible representation in question

O_h	E	$8C_3$	$3C_2(=C_4^2)$	$6C_4$	$6C_2'$	i	$6S_4$	$8S_6$	$3\sigma_h$	$6\sigma_d$
$\Gamma_{ipd+opd}$	12	0	0	0	2	0	0	0	4	2

Character table of O_h point group

O_h	E	$8C_3$	$3C_2$	$6C_4$	$6C_2'$	i	$6S_4$	$8S_6$	$3\sigma_h$	$6\sigma_d$		
A_{1g}	1	1	1	1	1	1	1	1	1	1		$x^2+y^2+z^2$
A_{2g}	1	1	1	−1	−1	1	−1	1	1	−1		
E_g	2	−1	2	0	0	2	0	−1	2	0		$(2z^2-x^2-y^2, x^2-y^2)$
T_{1g}	3	0	−1	1	−1	3	1	0	−1	−1	(R_x, R_y, R_z)	
T_{2g}	3	0	−1	−1	1	3	−1	0	−1	1		(xz, yz, xy)
A_{1u}	1	1	1	1	1	−1	−1	−1	−1	−1		
A_{2u}	1	1	1	−1	−1	−1	1	−1	−1	1		
E_u	2	−1	2	0	0	−2	0	1	−2	0		
T_{1u}	3	0	−1	1	−1	−3	−1	0	1	1	(x, y, z)	
T_{2u}	3	0	−1	−1	1	−3	1	0	1	−1		

$$
\begin{aligned}
NA_{1g} &= 1/48\,[(12.1.1)+(0.8.1)+(0.3.1)+(0.6.1)+(2.6.1)+(0.1.1)+(0.6.1)\\
&\quad +(0.8.1)+(4.3.1)+(2.6.1)\\
&= 1/48\,[(12+0+0+0+12+0+0+0+12+12)]\\
&- 1/48\,[48]-1
\end{aligned}
$$

$$
\begin{aligned}
NA_{2g} &= 1/48\,[(12.1.1)+(0.8.1)+(0.3.1)+(0.6.-1)+(2.6.-1)+(0.1.1)+(0.6.-1)\\
&\quad +(0.8.1)+(4.3.1)+(2.6.-1)\\
&= 1/48\,[(12+0+0+0-12+0+0+0+12-12)]\\
&= 1/48\,[0]=0
\end{aligned}
$$

$$
\begin{aligned}
NE_{g} &= 1/48\,[(12.1.2)+(0.8.-1)+(0.3.2)+(0.6.0)+(2.6.0)+(0.1.2)+(0.6.0)\\
&\quad +(0.8.-1)+(4.3.2)+(2.6.0)\\
&= 1/48\,[(24+0+0+0+0+0+0+0+24+0)]\\
&= 1/48\,[48]=1
\end{aligned}
$$

$$NT_{1g} = 1/48\,[(12.1.3) + (0.8.0) + (0.3.-1) + (0.6.1) + (2.6.-1) + (0.1.3) + (0.6.1)$$
$$+ (0.8.0) + (4.3.-1) + (2.6.-1)$$
$$= 1/48\,[(36 + 0 + 0 + 0 - 12 + 0 + 0 + 0 - 12 - 12)]$$
$$= 1/48\,[0] = 0$$

$$NT_{2g} = 1/48\,[(12.1.3) + (0.8.0) + (0.3.-1) + (0.6.1) + (2.6.1) + (0.1.3) + (0.6.-1)$$
$$+ (0.8.0) + (4.3.-1) + (2.6.1)$$
$$= 1/48\,[(36 + 0 + 0 + 0 + 12 + 0 + 0 + 0 - 12 + 12)]$$
$$= 1/48\,[48] = 1$$

$$NA_{1u} = 1/48\,[(12.1.1) + (0.8.1) + (0.3.1) + (0.6.1) + (2.6.1) + (0.1.-1) + (0.6.-1)$$
$$+ (0.8.-1) + (4.3.-1) + (2.6.-1)$$
$$= 1/48\,[(12 + 0 + 0 + 0 + 12 + 0 + 0 + 0 - 12 - 12)]$$
$$= 1/48\,[0] = 0$$

$$NA_{2u} = 1/48\,[(12.1.1) + (0.8.1) + (0.3.1) + (0.6.-1) + (2.6.-1) + (0.1.-1) + (0.6.1)$$
$$+ (0.8.-1) + (4.3.-1) + (2.6.1)$$
$$= 1/48\,[(12 + 0 + 0 + 0 - 12 + 0 + 0 + 0 - 12 + 12)]$$
$$= 1/48\,[0] = 0$$

$$NE_u = 1/48\,[(12.1.2) + (0.8.-1) + (0.3.2) + (0.6.0) + (2.6.0) + (0.1.-2) + (0.6.0)$$
$$+ (0.8.1) + (4.3.-2) + (2.6.0)$$
$$= 1/48\,[(24 + 0 + 0 + 0 + 0 + 0 + 0 + 0 - 24 + 0)]$$
$$= 1/48\,[0] = 0$$

$$NT_{1u} = 1/48\,[(12.1.3) + (0.8.0) + (0.3.-1) + (0.6.1) + (2.6.-1) + (0.1.-3) + (0.6.-1)$$
$$+ (0.8.0) + (4.3.1) + (2.6.1)$$
$$= 1/48\,[(36 + 0 + 0 + 0 - 12 + 0 + 0 + 0 + 12 + 12)]$$
$$= 1/48\,[48] = 1$$

$$NT_{2u} = 1/48\,[(12.1.3) + (0.8.0) + (0.3.-1) + (0.6.-1) + (2.6.1) + (0.1.-3) + (0.6.1)$$
$$+ (0.8.0) + (4.3.1) + (2.6.-1)$$
$$= 1/48\,[(36 + 0 + 0 + 0 + 12 + 0 + 0 + 0 + 12 - 12)]$$
$$= 1/48\,[48] = 1$$

The above calculations show that $\Gamma_{ipd\ +\ opd}$ is reduced to $A_{1g} + E_g + T_{2g} + T_{1u} + T_{2u}$
Hence,

$$\Gamma_{vib} = \Gamma_{internal\ coordinate} = \Gamma_{A-B} + \Gamma_{ipd} + \Gamma_{opd}$$
$$= A_{1g} + E_g + T_{1u} + A_{1g} + E_g + T_{2g} + T_{1u} + T_{2u}$$
$$= 2A_{1g} + 2E_g + T_{2g} + 2T_{1u} + T_{2u}\ (2 + 4 + 3 + 6 + 3 = 18)$$

Now, the three excess modes (redundant modes) can be identified. Since A-B bond vectors are totally independent, hence, the redundancy must be in the bond angle representation. By comparing the representation of internal coordinates with genuine internal modes (Γ_{vib}), we can identify that A_{1g} and E_g in $\Gamma_{ipd\ +\ opd}$ are redundant modes. So, after removing these redundant modes from $\Gamma_{ipd\ +\ opd}$ representation, we get

$$\Gamma_{ipd+opd} = T_{2g} + T_{1u} + T_{2u}$$

So, total internal coordinate vibrational representation becomes,

$$\Gamma_{vib} = A_{1g} + E_g + T_{2g} + 2T_{1u} + T_{2u}$$
$$= 1 + 2 + 3 + 6 + 3 = 15\ (3N - 6)$$

3.3 Symmetry selection rules for IR and Raman spectroscopy: identification of IR and Raman active vibrations

(a) Identification of IR active vibrations

After knowing the symmetry of different molecular vibrations/normal or fundamental modes, the next job before us is to work out which of these vibrations are infrared and which are Raman active. We shall consider the infrared activity first.

The time-dependent perturbation theory provides an equation to predict the spectral transition probabilities. The most general form of the equation is

$$P \propto \int_{-\infty}^{+\infty} \Psi_i p^{\wedge} \Psi_f .\ d\tau \qquad \text{(i)}$$

where ψ_i and ψ_f are the wave functions of the initial and final states involved in the transition, P is the probability of transition and $p^{\wedge}$ is the operator depending on the type of spectral transition. The operator $p^{\wedge}$ is called the transition moment operator and whole integral is the transition moment integral. A vibrational mode will be infrared active if the vibration brings about a change in the dipole moment of the molecule. Hence in the infrared spectroscopy, $p^{\wedge}$ will be the dipole moment operator of the molecule during the vibration. So,

$$p^{\wedge} = \mu = \mu^0 + (\delta\mu/\delta q)_{q=0} \cdot q \qquad \text{(ii)}$$

where μ^o is the equilibrium dipole moment of the molecule, μ is the dipole moment at any instant, q is a general coordinate indicating the displacement during vibration and $(\delta\mu/\delta q)_{q=0}$ is the rate of change of dipole moment with vibration at equilibrium. During the vibration, the coordinate q varies and hence μ varies. The dipole moment can be resolved into three components μ_x, μ_y and μ_z, and each component may be represented with a similar equation. For example,

$$\mu_x = \mu_x{}^0 + (\delta\mu/\delta x)_{x=0} \cdot x \qquad \text{(iii)}$$

The probability P also will have three components, P_x, P_y, P_z. Thus,

$$P_x \propto \int_{-\infty}^{+\infty} \Psi_i \, \mu_x \, \Psi_f \, d\tau \qquad \text{(iv)}$$

Since the total probability P is the sum of P_x, P_y and P_z components, and if at least one of the three components of P_x, P_y or P_z is non zero, then there will be a probability of transition. In no case transition will occur, if $P_x = P_y = P_z = 0$.

Putting the value of μ_x from equation (iii) to equation (iv), we have

$$P_x \propto \int_{-\infty}^{+\infty} \Psi_i . \mu_x . \Psi_f . d\tau = \propto \int_{-\infty}^{+\infty} \Psi_i . [\mu_x{}^0 + (\delta\mu/\delta x)_{x=0} . x] \Psi_f . d\tau =$$

$$\propto \mu_x{}^0 \int_{-\infty}^{+\infty} \Psi_i . \Psi_f . d\tau (I) + (\delta\mu/\delta x)_{x=0} \int_{-\infty}^{+\infty} \Psi_i . x \Psi_f . d\tau (II) \qquad \text{(v)}$$

Since $\mu_x{}^o$ and $\delta\mu/\delta x_{x=0}$ are constants, the integrand in the first term **(I)** in the equation (v) is zero because a set of $\psi'_n{}^s$ will be always orthonormal. Then, equation (v) takes the form

$$P_x \propto \int_{-\infty}^{+\infty} \Psi_i . \, x \Psi_f . \, d\tau \qquad \text{(vi)}$$

If the integrand in equation (vi) is zero, $P_x = 0$, then no transition will occur and we call this a vanishing matrix element. If the matrix element does not vanish, $P_x \neq 0$, then a transition will be IR active (that is allowed).

In the Infrared spectroscopy, ψ_i and ψ_f are the wave functions of the initial and final vibrational energy levels, between which transition occurs. It is an easy matter to identify whether the matrix elements representing the probabilities of transition, P_x, P_y or P_z evaluate to zero or not.

Consider the integral of the type $\int_{-\alpha}^{+\alpha} x^n \, dx$.

It can be seen that this integral evaluates to zero if n is odd and to a non-zero value n is even. In general, the value of an integral will be non-zero if the integrand is totally symmetric. Thus, even without evaluating the integral, we can predict whether the final value will be zero or not by ascertaining whether the integrand is totally symmetric or not.

In the integrand, $P_x \infty \int_{-\infty}^{+\infty} \psi_i . x\, \psi_f . dt$, if we have the symmetries of ψ_i, ψ_f and x, we can easily ascertain whether the integrand is totally symmetric or not.

We have already determined the symmetry species of the normal modes of vibration in the case of H_2O molecule of C_{2v} point symmetry. The actual symmetry of a vibrational level can be obtained from these. For example, the symmetries of the various vibrational levels of the vibrational mode transforming as B_2 can be obtained as $(B_2)^v$ where v is the vibrational quantum number. Similarly, we can set the symmetries of any of the vibrational levels of any vibrational mode. If $v = 0$, the vibrational level will always be totally symmetric since any number raised to zero is unity.

For example, in the C_{2v} point group, the characters of B_2 are 1, −1, −1, 1. For the ground vibrational level, $v = 0$ and hence $(B_2)^o = (1)^o, (-1)^o, (-1)^o, (1)^o = 1, +1, +1, 1$. These characters are those of A_1 representation (vide Character table).

Character table of C_{2v} point symmetry

C_{2v}	E	C_2	$\sigma_{v(yz)}$	$\sigma_{v(xz)}$		
A_1	1	1	1	1	z	x^2, y^2, z^2
A_2	1	1	−1	−1	R_z	xy
B_1	1	−1	1	−1	x, R_y	xz
B_2	1	−1	−1	1	y, R_x	yz

It follows that, all the ground vibrational levels will be always totally symmetric. Hence, ψ_i is totally symmetric. In case of water molecule, ψ_i in all three vibrational modes ($2A_1 + B_2$) belong to A_1 representation. Because the characters of A_1 are 1,1,1,1 and $(A_1)^o = (1)^o, (1)^o, (1)^o, (1)^o = 1,1,1,1$ (A_1 characters).

The ψ_f will have A_1 symmetry for fundamental transition in symmetric stretching and symmetric bending, since $(A_1)^1 = A_1$ as $v = 1$. In the anti-symmetric stretching (B_2), the symmetry of four fundamental transition is $(B_2)^1 = B_2$. In the C_{2v} point group, x transforms as B_1 (vide character table of C_{2v} point group).

Hence, for the vibrational mode having A_1 symmetry,

$$P_x \propto \int_{-\infty}^{+\infty} \Psi_i . x \Psi_f . d\tau \text{ (x transforms as } B_1)$$

$$\propto A_1 B_1 A_1 = B_1 \text{ (not symmetric, vide Character Table)}$$

Since this direct product is not totally symmetric, $P_x = 0$.

Similarly,

$$P_y \propto \int_{-\infty}^{+\infty} \Psi_i . \, y \Psi_f . \, d\tau \text{ (y transforms as } B_2 \text{ vide Character Table)}$$
$$\propto A_1 \, B_2 \, A_1 = B_2 \text{ (not symmetric, vide Character Table)}$$

Therefore, $P_y = 0$.

Finally,

$$P_z \propto \int_{-\infty}^{+\infty} \Psi_i . \, z \Psi_f . \, d\tau \text{ (z transforms as } A_1)$$
$$\propto A_1 \, A_1 \, A_1 = A_1$$

As this integrand is totally symmetric, $P_z \neq 0$. Since $P = P_x + P_y + P_z \neq 0$, the corresponding vibration (A_1 symmetry) is infrared active.

For the B_2 vibration,

$$P_x \propto \int_{-\infty}^{+\infty} \Psi_i . \, x \Psi_f . \, d\tau \text{ (x transforms as } B_1)$$
$$\propto A_1 \, B_1 \, B_2 = A_2 \text{ (asymmetric, vide Character Table)}$$
$$P_y \propto \int_{-\infty}^{+\infty} \Psi_i . \, y \Psi_f . \, d\tau \text{ (y transforms as } B_2)$$
$$\propto A_1 \, B_2 \, B_2 = A_1 \text{ (symmetric, vide Character Table)}$$

Thus, the direct product of the species in the integrand for P_x is not totally symmetric and hence $P_x = 0$, whereas it is totally symmetric in P_y. So, P_y is not zero. The vibration B_2 (asymmetric stretching) is IR active.

In the water molecule, all the three vibrations are IR active. Since in these cases, ψ_i is totally symmetric, total symmetry of P_x, P_y or P_z will be decided by the symmetry of x, y or z and that of ψ_f.

Accordingly, only if ψ_f and x belong to the same symmetry species, their product, P_x will be totally symmetric, So also, P_y will be totally symmetric if ψ_f and y are of the same symmetry. This is true for P_z also.

Hence, infrared activity of a fundamental vibration may be stated as follows:

"A fundamental mode will be infrared active (i.e., will give rise to an absorption band) if the normal mode which is excited belongs to the same representations as any one or several of the Cartesian coordinates, and will be inactive if it doesn't."

Thus, the identification of a vibrational mode as IR active reduces simply to see whether at least one of the Cartesian coordinate x, y or z belongs to the same representation as that of vibrational mode.

(b) Identification of Raman active vibrations

Vibrations which bring about a change in polarizability are Raman active. When a molecule is placed in an applied electric field, the molecule undergoes a polarization and an induced dipole moment μ_i is set up on the molecule. This dipole moment is related to the field strength E of the applied field by the equation $\mu_i = \alpha E$, where α is the polarizability of the molecule. This dipole moment can be resolved into three mutually perpendicular components, μ_{ix}, μ_{iy} and μ_{iz}. The applied electric field can also be resolved into three components, E_x, E_y, E_z. However, the induced dipole moment μ_{ix} is set up not only by E_x, but also by E_y and E_z. Thus,

$$\mu_{ix} = \alpha_{xx}E_x + \alpha_{xy}E_y + \alpha_{xz}E_z$$

where α_{xx} is the component of the polarizability in the x-direction brought about by E_x, α_{xy} is the component of the polarizability in the x-direction brought about by E_y and so on.

Similarly,

$$\mu_{iy} = \alpha_{yx}E_x + \alpha_{yy}E_y + \alpha_{yz}E_z \quad \text{and} \quad \mu_{iz} = \alpha_{zx}E_x + \alpha_{zy}E_y + \alpha_{zz}E_z$$

Since the polarizability in one direction is induced by an electric field of other directions also, polarizability is referred as a tensor.

Some of these components of polarizability are equal, that is $\alpha_{xy} = \alpha_{yx}$, $\alpha_{xz} = \alpha_{zx}$ and $\alpha_{yz} = \alpha_{zy}$. Also α_{xx}, α_{yy}, and α_{zz} are usually denoted by α_x^2, α_y^2, α_z^2. Thus, there are six components for polarizability and these may be denoted by α_x^2, α_y^2, α_z^2, α_{xy}, α_{yz}, α_{xz}.

Accordingly, the probability of transition (*P*) will have six components, each of which will be of the type,

$$P_{ab} = \int \Psi_i .\, ab\,\Psi_f .\, d\tau$$

where ψ_i and ψ_f are the wave functions of the initial and final vibrational energy levels and ab is the binary product of the type x^2, y^2, z^2, xy, xz or yz. If any of these components is non-zero, the vibration is Raman active.

For example, if we consider the symmetric stretching or symmetric bending in water molecule,

$$\begin{aligned} P_x^2 &\propto \int \Psi_i\ x^2\,\Psi_f\ d\tau \\ &\propto A_1 .\, A_1 .\, A_1 = A_1\ \text{(symmetric)} \end{aligned}$$

The ψ_i for symmetric stretching or symmetric bending belongs to A_1 symmetry and the ψ_f will have A_1 symmetry for fundamental transition in symmetric stretching or symmetric bending, since $(A_1)^1 = A_1$ as $v = 1$. Since, x^2 also belongs to A_1 (see Character table of C_{2v} point group), the integrand is totally symmetric and hence $P_x^2 \neq 0$. Hence, the symmetric stretching and symmetric bending vibrations are Raman active in water molecule.

Similarly, for asymmetric stretching B_2,

$$\begin{aligned} P_{yz} &\propto \int \Psi_i .\, yz\,\Psi_f .\, d\tau \\ &\propto A_1 .\, B_2 .\, B_2 = A_1\ \text{(symmetric)} \end{aligned}$$

So, the integrand is totally symmetric and hence $P_{yz} \neq 0$. Hence, the asymmetric stretching is also Raman active in water molecule.

Thus if ψ_f and at least any one of the binary products of the Cartesian coordinates belong to the same species, the particular vibration will be Raman active. The rule for Raman activity can be stated as:

"A fundamental transition will be Raman active if the normal mode which is excited belongs to the same representation as one or more of the binary products of the Cartesian coordinates".

3.4 Complementary nature of IR and Raman spectra

The great utility of Raman spectrum lies in the fact that it obeys different selection rules from the IR spectrum. The selection rules for the Raman spectrum depend on the molecular polarizability (α_{ij}) in contrast to the selection rules for the IR, which depend on the dipole moment (μ_i). Mathematically, these statements may be written as:

$$\left(\frac{\partial \alpha ij}{\partial \varphi}\right) \neq 0 \quad \text{Raman selection rule}$$

$$\left(\frac{\partial \mu i}{\partial \varphi}\right) \neq 0 \quad \text{IR selection rule}$$

That is, for a vibration to be Raman active, the derivative of one of the components of polarizability tensor (α_{ij}) with respect to normal coordinate φ (inter nuclear distance, r) should be non-zero and for IR active vibration, the derivative of one of the components of dipole moment (μ_i) with respect to normal coordinate φ should be non-zero. It is notable here that polarizability has six components denoted by α_{xx}, α_{yy}, α_{zz}, α_{xy}, α_{yz} and α_{xz} and dipole moment has three components along *x*, *y* and *z*-axes, μ_x, μ_y and μ_z.

Polarizability is a tensor quantity in contrast to the vector nature of the dipole moment. A tensor is a mathematical entity, that is, the general equivalent in any *n*-dimensional coordinate system of a vector in two- or three-dimensional coordinates. It is represented by a square matrix. These are used to describe that how all the component of a quantity behave under certain transformations, just as a vector can describe a translation from one point to another point in a plane or in a space.

Let us consider the diatomic and polyatomic molecules for a comparison of their IR and Raman activities.

(i) Homonuclear Diatomic Molecules

All homonuclear diatomic molecules, such as, H_2, O_2, N_2, Cl_2, Br_2, etc., are non-polar, that is, they do not possess any dipole moment. Even when they execute vibrations, they will not be able to generate any dipole moment (μ_i), hence they are not active in the IR spectrum. But these molecules change their polarizability during the vibration, and hence they are Raman active. For example, H_2 gas gives a shift of 4156 cm^{-1} and O_2 a shift of 1555 cm^{-1} in their Raman spectra, whereas their IR spectra cannot be obtained.

(ii) Heteronuclear Diatomic Molecules

As the heteronuclear diatomic molecules possess permanent dipole moment and also change their polarizability during the vibrational motion, they are active both in the IR and Raman. For example, HCl gas absorbs at 2886 cm^{-1} in its IR spectrum and also causes a shift of 2886 cm^{-1} in the Raman spectrum.

(iii) Polyatomic Molecules

Polyatomic molecules containing more than two atoms are very complex to consider. However, their IR and Raman activity can be tackled with the use of "group theory," and from the wealth of information so obtained, an important principle/Rule known as the Mutual Exclusion Principle/Rule can be enunciated.

3.5 The mutual exclusion principle/rule

This rule states that, "if a molecule has a center of symmetry (center of inversion, i), then its IR and Raman spectra will be mutually exclusive."

In terms of group theory, groups of such molecules (with the element, *i*) have two sets of irreducible representations. First, the representations, which are symmetric with respect to inversion operation, are called *g* representation and second, those representations which are antisymmetric with respect to *i* are called *u* representations.

It can be observed that the Cartesian coordinates (*x*, *y*, *z*) transform as *u* species, because for example, on the coordinate *x* (or translational vectors T_x, T_y or T_z), if inversion operation is carried out, then *x* becomes $-x$. Same is true for *y* and *z*. Hence, all representations generated by *x*, *y* and *z* must belong to a *u* symmetry species (or *u* representation). However, the binary product of the coordinates, that is, xy, yz or xz, does not change sign under the inversion operation. Therefore, all such binary products, which represent the components of the polarizability tensor, belong to *g* representations (or *g* symmetry species).

From the above discussion of *u* and *g* symmetry species of Cartesian coordinates and binary products, a very useful and interesting rule (or a kind of selection rule) can be stated as:

For centrosymmetric molecules, the vibrational modes which belong to *g* symmetry can be Raman active and those belong to *u* symmetry can be IR active. This is called mutual exclusion rule. In other words, this rule can also be stated as:

Centrosymmetric molecules will never have vibrations which are both Raman and IR active. It means if a vibration is active in IR (of *u* symmetry), then it will not be Raman active (of *g* symmetry) and vice versa. If some vibrations are active in both, Raman and IR, it clearly indicates that molecule does not possess the center of inversion (*i*).

The reverse statement of this rule is also true, that is, if there is no common band in IR and Raman, it implies the presence of center of symmetry in the molecule. Bur here, some care is needed because Raman active vibration may be too weak to be observed. Therefore, some uncertainty may be there to arrive at conclusion for the presence of center of symmetry. But if some vibrations are common in both, Raman and IR, it is certain that center of symmetry is absent in the molecule. Hence, this rule is useful to obtain valuable structural information.

Some of the following examples illustrate the mutual exclusion rule.

(i) In CO_2, the symmetric stretching is inactive in IR but it is Raman active (1340 cm^{-1}). Therefore, from this rule, we immediately decide that the molecule has a center of symmetry.

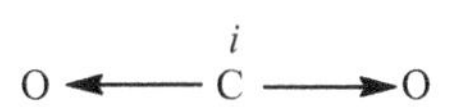

Symmetric stretching (ν_s)
(1340 cm^{-1})

(ii) In case of N_2O, some vibrations are found common in both Raman and IR. Hence, the molecule does not contain a center of symmetry. It actually has N—N—O structure.

(iii) Some molecules are given along with their point groups in the following Table to identify that which symmetry species (irreducible representations of a point group to which the molecule belong) are IR and Raman active.

Molecule	**Point group**	**Symmetry species**
CO_2	$D_{\infty h}$	A_{1g}, A_{1u}, E_{1u} IR active: A_{1u}, E_{1u} Raman active: A_{1g}
C_2H_2	$D_{\infty h}$	$2A_{1g}$, E_{1g}, E_{1u}, A_{1u} IR active: A_{1u}, E_{1u} Raman active: A_{1g}, E_{1g}
trans-N_2F_2	C_{2h}	$3A_g$, A_u, $2B_u$ IR active: A_u, B_u Raman active: A_g

Thus, irreducible representation which has *g* script will be Raman active and those with *u* script are IR active.

The selection rules for vibrational spectroscopy may be summarized as:

For IR spectra:

(i)

$$\left(\frac{\partial \mu i}{\partial \varphi}\right) \neq 0$$

(ii)

$$\Delta v = \pm 1$$

For Raman spectra:

(i)

$$\left(\frac{\partial \alpha ij}{\partial \varphi}\right) \neq 0$$

(ii)

$$\Delta v = \pm 1$$

The polarizability α decreases in the symmetric stretching, whereas it increases in compression of bond. In general, the degree of allowedness of the transition in Raman spectra depends on the magnitude of $(\frac{\partial \alpha}{\partial \varphi})$ at $\varphi = 0$ (equilibrium position). Consequently, symmetric vibration gives intense Raman lines while asymmetric vibrations give rise to weak lines which are sometimes not observed in the Raman spectra.

3.6 Polarization of Raman lines

Totally symmetric normal modes in Raman spectrum can be defined by measuring the depolarization ratio (ρ). The depolarization ratio is defined as

$$\rho = I_{\perp}/I_{\parallel}$$

where $I_{\perp}$ is the intensity of scattered light polarized in the plane perpendicular to the incident light and $I_{\parallel}$ is the intensity of scattered light polarized in the same plane as the incident light. This is when the incident light is plane polarized light.

In case, when the incident light used is unpolarized in Raman experiment, the light scattered out right angle to the incident light will be at least partially polarized when molecule has symmetry less than T_d. The extent of polarization depends on how the polarization varies with vibration. Usually, a plane-polarized light (laser) is used in Raman experiment to measure the extent of polarization. This measurement of polarization gives information about the symmetry of molecular vibrations.

Generally, it can be stated that a symmetric vibration gives rise to a polarized or partially polarized Raman line while a non-symmetric vibration gives a depolarized

Raman line. Theoretically, if the degree of depolarization (ρ) has value less than or equal to 6/7 (or 3/4), then the vibration concerned is symmetric and the Raman line is called polarized, whereas if $\rho > 6/7$, the line is called depolarized and the vibration will be non-symmetric.

One can draw inference from the measurement of depolarization ratio (ρ) that higher the symmetry of the molecule, the smaller will be the degree of depolarization of the Raman line for a particular kind of vibration. The maximum value of ρ is 3/4 for polarized incident light and 6/7 for unpolarized light. If $\rho = 0$, it clearly shows that molecule under investigation is isotropic (having direction independent properties) and the vibration of the molecule is totally symmetric.

For a sample in solution, theory predicts that a depolarized band in Raman spectrum will show $\rho = 3/4$, whereas polarized band will give the values of ρ as $0 \le \rho \le 3/4$. Polarized bands can only occur from vibrations which are totally symmetric (A, A_1, etc.). Hence, the bands which are polarized can soon be assigned to the totally symmetric representations and depolarized bands can be assigned to modes of any other symmetry. Also, the totally symmetric vibrations give the most intense Raman bands. Thus, the polarization of Raman bands/lines tells us the symmetry of vibrations.

3.7 Prediction of IR and Raman active modes in some molecules of different point Groups

Two group theoretical methods, which are based on the selection of basis functions for reducible representation, will be used for prediction of IR and Raman active modes.

(a) Cartesian coordinate method or 3N vector method
(b) Internal coordinate method

(a) Cartesian coordinate method or 3N vector method

In previous section, we have seen that the total reducible representation taking the 3N (or 3×3 = 9) vectors on the three atoms of H_2O molecule (belonging to C_{2v} symmetry) as the basis is as follows:

C_{2v}	E	C_2	$\sigma_{v(yz)}$	$\sigma_{v(xz)}$
Γ_{3N}	9	−1	1	3

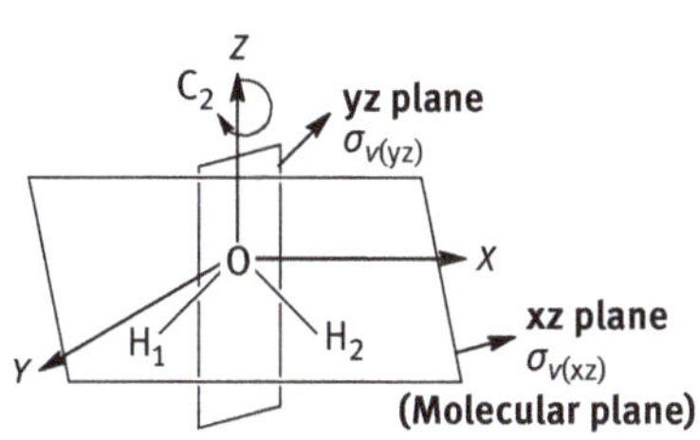

An examination of the characters indicates that only those atoms which retain position during the symmetry operation contribute to the total character. In such atoms, each vector which retains direction contributes a character 1 and each vector which reverses direction contributes a character −1 to the total representation.

Thus, the total character of the reducible representation of the molecule in a particular symmetry operation can be obtained as follows:

Total character $\chi(R)$ = No. of atoms unshifted (n) in a symmetry operation R × {No. of vectors retaining direction (a) – the number of vectors reversing direction in an atom (b)}

$$\text{or } \chi(\text{R}) = \text{n}_\text{R}(\text{a} - \text{b}) \tag{i}$$

For example in H_2O, during E operation all the three atoms retain position, nE = 3, all the vectors are also retaining direction, so, $a = 3$, $b = 0$. Hence,

$$\chi(\text{E}) = 3 \times (3 - 0) = 9$$

On applying C_2 operation, only oxygen atom retains position (because O-atom lies on the z-axis). Likewise, the vectors in the z-direction retain direction while those in the x- and y- directions reverse direction. Hence, $nC_2 = 1$, a = 1 and b = 2. Therefore,

$$\chi(\text{C}_2) = 1 \times (1 - 2) = -1$$

As $\sigma_{v(yz)}$ passes through z-direction containing O-atom, during $\sigma_{v(yz)}$ operation only the oxygen atom retains position. Again during $\sigma_{v(yz)}$, the vectors in y–, z-directions retain the direction, while that in the x-direction reverses direction. Hence,

$$\text{n}\sigma_{\text{v(yz)}} = 1, \ \text{a} = 2 \ \text{ and } \ \text{b} = 1; \ \chi(\sigma_{\text{v(yz)}}) = 1 \times (2 - 1) = 1$$

As $\sigma_{v(xz)}$ passes through all the atoms and hence $n\sigma_{v(xz)} = 3$. During this operation, the vectors in the x- and z-directions retain directions and that in the y-direction is reversed. Hence,

$$\text{n} = 3, \ \text{a} = 2 \ \text{ and } \ \text{b} = 1; \ \chi(\sigma_{\text{v(xz)}}) = 3 \times (2 - 1) = 3$$

Thus, taking the nine vectors on the three atoms of H_2O as the basis, the following reducible representation, T is obtained.

C_{2v}	E	C_2	$\sigma_{v(yz)}$	$\sigma_{v(xz)}$
T_{3N}	9	−1	1	3

In case of molecules having the order of rotation axis > 2, it is not easy to identify whether a particular vector reverses direction or not. For instance, in NH_3 molecule of C_{3v} point symmetry [E, C_3, $3\sigma_v$], a C_3 rotation about the z-axis will rotate the vectors in the x- and y-directions to 120°, not 180°. In such cases, we make use of a more general method for the determination of the total representation

In this method, all symmetry operations of a point group are classified **either as proper rotation** or as **improper rotation**. The identity operation (E) and all types of proper rotations are classified as proper rotations. While the C_n operations are proper rotation through certain angles, the identity operation is a proper rotation through 0°. As the inversion operation, $i = S_2$, it is an improper rotation through 180°. Also $\sigma = \sigma C_1 = S_1$, and hence a reflection operation is an improper rotation through 0°. The S_n operations are improper rotations.

Similar to cases of all symmetry operations, only those points (atoms/vectors) that retain positions during an operation, will contribute to the character, and that the contribution of one point to the total character is $(1 + 2\cos\beta)$ for proper rotation and $(-1 + 2\cos\beta)$ (β = angle of rotation), for improper rotation. This is well clear from the characters of matrix from the representation of C_n and S_n operations.

$$C_n(z) = \begin{bmatrix} \cos\beta & \sin\beta & 0 \\ -\sin\beta & \cos\beta & 0 \\ 0 & 0 & 1 \end{bmatrix} \qquad S_n = \begin{bmatrix} \cos\beta & \sin\beta & 0 \\ -\sin\beta & \cos\beta & 0 \\ 0 & 0 & -1 \end{bmatrix}$$

$$\text{Character} = (1 + 2\cos 2\beta) \qquad \text{Character} = (-1 + 2\cos 2\beta)$$

That is, each point contributes $(\pm 1 + 2\cos\beta)$ to the total character provided the point retains position during proper/improper operation. Based on equation (i) given below, the total character corresponding to an operation R will be,

$$\text{Total character} = n_R(a - b)$$

$$\chi(R) = n_R(\pm 1 + 2\cos\beta)$$

where n_R is the number of atoms retaining position during the operation R, β is the angle of proper or improper rotation and + and – sign will be applicable for proper or improper rotation, respectively.

(b) Internal coordinate method

The most important internal coordinates are: (i) change in bond lengths (stretching) and (ii) angle deformation/ bond angle (bending). Only these two will be used as the basis for representations to get Γ_{str} and Γ_{bend}. The symmetry species of each of these can be obtained by reducing these to irreducible representations. These results will be correlated with those that were obtained using Γ_{3N} method, that is $\Gamma_{vib} = \Gamma_{3N} - \Gamma_T - \Gamma_R$.

These two methods are illustrated here taking examples of molecules belonging to different point groups.

(i) H_2O/SO_2 molecule of C_{2v} point group

Let us consider the H_2O molecule that belongs to $\mathbf{C_{2v}}$ point group having symmetry operations, E, C_2, $\sigma_{v(xz)}$ and $\sigma_{v(yz)}$.

Cartesian coordinate method or 3N vector method:

The symmetry operations in this case are two proper rotations through 0°(E) and 180°(C_2) and two improper rotations through 0° [$\sigma_{v(xz)}$ and $\sigma_{v(yz)}$].

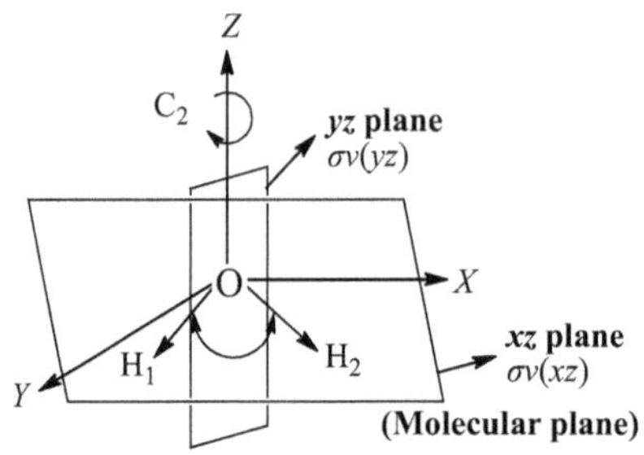

The following Table is constructed to calculate the total characters. The vertical dashed line separates the proper and improper rotations. The corresponding cosβ and also ±1 + 2cosβ values are given in the table. The proper rotations will have (+1 + cosβ) and the improper rotation will have (−1 + cosβ) as the character per unit. The number of atoms retaining position, n_R in each of the operations is 3, 1, 1 and 3.

Table: Determination of total character for NH_3 molecule

C_{2v}	E	C_2	$\sigma_{v(yz)}$	$\sigma_{v(yz)}$
β	0	180	0	0
cosβ	1	−1	1	1
±1+ 2cosβ	3	−1	1	1
n_R	3	1	1	3
Γ_{3N}	**9**	**−1**	**1**	**3**

The above reducible representation (Γ_{3N}) can be reduced to irreducible representations applying reduction formula and using character table of C_{2v} point group as follows:

$$N_i = \frac{1}{h}\sum_R \chi(R)\,.\,n\,.\,\chi_i(R)$$

where N_i is the number of times the ith irreducible representation occurs in a reducible representation, h is the order of group (number of symmetry operations), $\chi(R)$ is the character of a particular operation in the reducible representation, n is the number of operation of that type and $\chi_i(R)$ is the character of the same operation in the irreducible representation.

Reducible representation in question

C_{2v}	E	C_2	$\sigma_{v(yz)}$	$\sigma_{v(xz)}$
Γ_{3N}	9	−1	1	3

Character table of C_{2v} point symmetry

C_{2v}	E	C_3	$\sigma_{v(yz)}$	$\sigma_{v(yz)}$		
A_1	1	1	1	1	z	x^2+y^2, z^2
A_2	1	1	−1	−1	R_z	xy
B_1	1	−1	1	−1	x, R_y	xz
B_2	1	−1	−1	1	y, R_x	yz

$$\begin{aligned}
NA_1 &= 1/4\,[(9.1.1)+(-1.1.1)+(1.1.1)+(3.1.1)\\
&= 1/4\,[9-1+1+3]\\
&= 1/4\,[12]=3\\
NA_2 &= 1/4\,[(9.1.1)+(-1.1.1)+(1.1.-1)+(3.1.-1)\\
&= 1/4\,[9-1-1-3]\\
&= 1/4\,[4]=1\\
NB_1 &= 1/4\,[(9.1.1)+(-1.1.-1)+(1.1.1)+(3.1.-1)\\
&= 1/4\,[9+1+1-3]\\
&= 1/4\,[8]=2\\
NB_2 &= 1/4\,[(9.1.1)+(-1.1.-1)+(1.1.-1)+(3.1.1)\\
&= 1/4\,[9+1-1+3]\\
&= 1/4\,[12]=3
\end{aligned}$$

The above calculations suggest that the reducible representation (Γ_{3N}) in question reduces to:

$$\Gamma_{3N} = 3A_1 + A_2 + 2B_1 + 3B_2$$

Thus, there are nine modes (3 + 1 + 2 + 3 = 9) of degrees of freedom with different symmetries, A_1, A_2, B_1 and B_2. These include the vibrational, rotational and translational degrees of freedom.

If we subtract the rotational and translational degrees of freedom from these, the remaining one will be the vibrational degrees of freedom. The C_{2v} character table indicates that the translational modes have the same symmetry as the coordinates x, y and z, that is, B_1, B_2, A_1, respectively. So also the rotational modes have the same symmetry as the rotation axes, R_x, R_y and R_z, that is, B_2, B_1 and A_2, respectively. Subtracting these translational and rotational degrees of freedom, we get the vibrational degrees of freedom only of appropriate symmetry.

$$\begin{array}{rl} T &= 3A_1 + A_2 + 2B_1 + 3B_2 \\ T_{tr} &= A_1 \qquad\quad + B_1 + B_2 \\ T_{rot} &= \qquad\quad A_2 + B_1 + B_2 \\ \hline T_{vib} &= 2A_1 + B_2 \end{array}$$

Thus, out of three vibrational modes, two belongs to A_1 and one belongs to B_2 irreducible representation.

Internal coordinate method

We have already worked out in the previous section that

$$\Gamma_{stretch} = A_1 + B_2 \text{ and}$$
$$\Gamma_{bend} = A_1.$$

$$\Gamma_{\text{Internal coordinate}} = \Gamma_{vib} = \Gamma_{stretch} + \Gamma_{bend} = (A_1 + B_2) + A_1$$
$$= 2A_1 + B_2$$

This result of Γ_{vib} obtained through internal coordinate method is the same to that we have obtained through Cartesian coordinate method.

Detection/Identification of IR and Raman active vibrations

Now, the next step is to detect that out of $2A_1$ and B_2 vibrations which is IR active and which is Raman active. We shall consider the infrared activity first.

"A fundamental mode will be infrared active (i.e., will give rise to an absorption band) if the normal mode which is excited belongs to the same irreducible representation as any one or several of the Cartesian coordinates, and will be inactive if it doesn't."

Thus, the identification of a vibrational mode as infrared active reduces simply to see whether at least one of *x*, *y* or *z* belongs to the same representation as that of vibrational mode.

From the character table of C_{2v} point group, it is notable that *z*-coordinate belongs to A_1 and *y*-coordinate to B_2. Therefore, all these vibrational modes are IR active.

The rule for Raman activity can be stated as:

"A fundamental transition will be Raman active if the normal mode which is excited belongs to the same representation as one or more of the binary products of the Cartesian coordinates".

From the character table of C_{2v} point symmetry, the A_1 vibrations in the H_2O molecule are Raman active since x^2, y^2, z^2 also transforms as A_1. The B_2 vibration is also Raman active since yz transforms as B_2. Thus, all the vibrations in H_2O are Raman active also.

Similarly in SO_2, all the three vibrational modes are infrared and Raman active.

The following general observations are notable in this regard. These observations are based upon large number of experimental data on large number of molecules and from chemical intuitions.

(i) Stretching frequencies are higher than the bending frequencies involving similar atoms.
(ii) Asymmetric stretching frequencies are higher than symmetric stretching frequencies.
(iii) Symmetric frequencies are always Raman polarized.
(iv) Vibrational frequencies are generally written in the decreasing order within each symmetry species and working further according to the order of symmetry species given in the character table.

Based on the above guidelines, the assignment of actual vibrational spectral data in H_2O and SO_2 is given below.

3657 cm^{-1} (pol.) | 1595 cm^{-1} (pol.) | 3756 cm^{-1} (depol.)

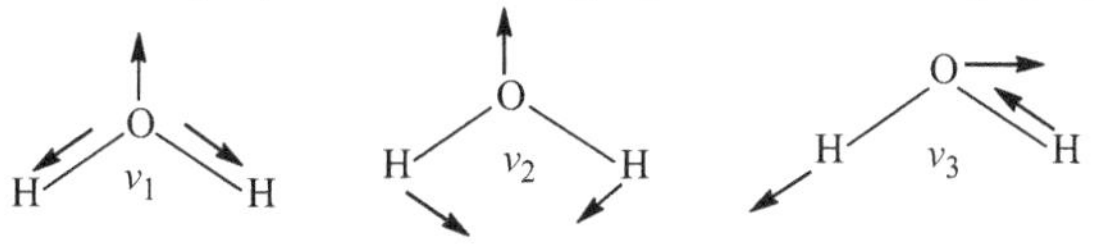

Symm. stretch A_1 | Symm. bending A_1 | Asymm. stretch B_2

1150 cm^{-1}(pol.) (vapor) | 518 cm^{-1}(pol.) (vapor) | 1362 cm^{-1}(pol.) (vapor)
1146 cm^{-1}(pol.) (liquid) | 524 cm^{-1}(pol.) (liquid) | 1336 cm^{-1}(depol.) (liquid)

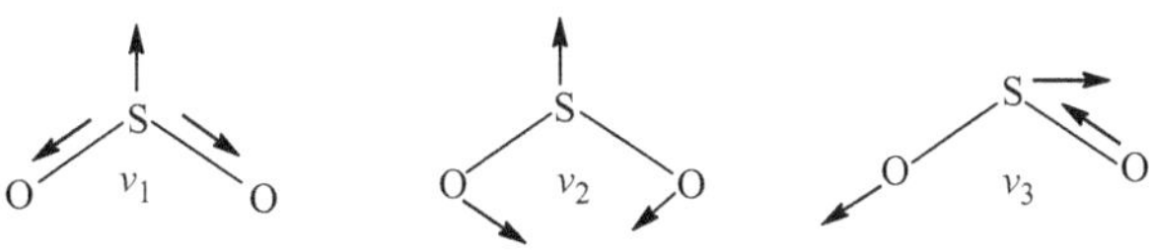

Symm. stretch A_1 | Symm. bending A_1 | Asymm. Stretch B_2

Vibrational spectra of H_2O molecule	
Infrared (vapor) (cm^{-1})	Raman (liquid) (cm^{-1})
3657 (v_1)	3657 (pol.) (v_1)
1595 (v_2)	1595 (pol.) (v_2)
3756 (v_3)	3756 (depol.) (v_3)
Vibrational spectra of SO_2 molecule	
Infrared (vapor) (cm^{-1})	Raman (liquid) (cm^{-1})
1150 (v_1)	1146 (pol.) (v_1)
518 (v_2)	524 (pol.) (v_2)
1362 (v_3)	1336 (depol.) (v_3)

(ii) NH_3/PH_3/ SO_3^{2-} pyramidal molecule of C_{3v} point group

Let us consider NH_3 molecule which has pyramidal geometry and belongs to C_{3v} point group having symmetry operations E, $2C_3$, $3\sigma_v$.

Cartesian coordinate method or Γ_{3N} method

The operations in this case are two proper rotations through 0°(E) and 120° (C_3) and three improper rotations through 0°($3\sigma_v$). The following table provides the calculations of the total characters.

Table: Determination of total character for NH_3 molecule

C_{3v}	E	$2C_3$	$3\sigma_v$
β	0	120	0
cosβ	1	$-\frac{1}{2}$	1
±1+2cosβ	3	0	1
n_R	4	1	2
Γ_{3N}	**12**	**0**	**2**

The vertical dashed line separates the proper and improper rotations. The corresponding angle β, cosβ and also (±1 + 2cosβ) values are tabulated for each operation. The proper rotations have (+1 + 2cosβ) and the improper rotations have (−1 + 2cosβ) values. The number of points retaining position (n_R) for each operation is also given in the table.

The above reducible representation (Γ_{3N}) can be reduced to irreducible representations applying reduction formula and using character table of C_{3v} point group as follows:

$$N_i = \frac{1}{h}\sum_R \chi(R).n.\chi_i(R)$$

Total representatin of NH_3 molecule

C_{3v}	E	$2C_3$	$3\sigma_v$
Γ_{3N}	**12**	**0**	**3**

Character table of C_{3v} point group

C_{3v}	E	$2C_3$	$3\sigma_v$		
A_1	1	1	1	z	x^2+y^2, z^2
A_2	1	1	−1	R_z	
E	2	−1	0	$(x,y)\ R_x,R_y)$	$(x^2-y^2, xy)\ (xz,yz)$

$$NA_1 = 1/6\,[(12.1.1) + 0 + (2.3.1)]$$
$$= 1/6\,[12 + 0 + 6]$$
$$= 1/6\,[18] = 3$$
$$NA_2 = 1/6\,[(12.1.1) + 0 + (2.3.-1)]$$
$$= 1/6\,[12 - 6]$$
$$= 1/6\,[6] = 1$$
$$NE = 1/6\,[(12.1.2) + 0 + 0]$$
$$= 1/6\,[24 + 0 + 0]$$
$$= 1/6\,[24] = 4$$

Thus, the above reducible representation is reduced to:

$$\Gamma_{3N} = 3A_1 + A_2 + 4E$$

Thus, there are 12 (3 + 1 + 4×2 = 12) of degrees of freedom with different symmetries, A_1, A_2 and *E*. These include the vibrational, rotational and translational degrees of freedom.

The C_{3v} character table indicates that the translational modes have the same symmetry as the coordinates *x*, *y* and *z*, that is, A_1 and E. So also the rotational modes have the same symmetry as the rotation axes, R_x, R_y and R_z, that is, A_2 and *E*. Subtracting these translational and rotational degrees of freedom, we get the vibrational degrees of freedom only of appropriate symmetry.

$$
\begin{aligned}
T_{3N} &= 3A_1 + A_2 + 4E \\
T_{tr} &= A_1 \qquad + E \\
T_{rot} &= \qquad A_2 + E \\
\hline
T_{vib} &= 2A_1 + \qquad 2E
\end{aligned}
$$

Thus, out of six vibrational modes [$2(A_1) + 4(E) = 6$], two belongs to A_1 and four belongs to *E* irreducible representation.

Internal coordinate method

We have already worked out in previous section that

$$\Gamma_{stretch} = A_1 + E \text{ and}$$
$$\Gamma_{bend} = A_1 + E.$$

$$
\begin{aligned}
\Gamma_{\text{Internal coordinate}} = \Gamma_{vib} = \Gamma_{stretch} + \Gamma_{bend} &= (A_1 + E) + (A_1 + E) \\
&= 2A_1 + 2E
\end{aligned}
$$

This result of Γ_{vib} obtained through internal coordinate method is the same to that we have obtained through Cartesian coordinate method.

Identification of IR and Raman active vibrations

IR Active Vibrations

From the character table of C_{3V} point group, it is notable that *z*-coordinate belongs to A_1, x-and *y*-coordinate to *E* irreducible representation.

Recalling the rule for IR activity given below, the vibrations, A_1 and E are IR active.

"A fundamental mode will be infrared active (i.e., will give rise to an absorption band) if the normal mode which is excited belongs to the same representation as any one or several of the Cartesian coordinates, and will be inactive if it doesn't."

As shown in the Fig. below, out of six vibrational modes, two vibrations belong to A_1 and four belong to *E* representations. However, four IR bands (3534, 3464, 1765 and 1139 cm^{-1}) should be observed in IR because ν_{3a}/ν_{3b} and ν_{4a}/ν_{4b} are degenerate.

Raman Active Vibrations

Again, the 4th area of character table indicates that $x^2 + y^2$ and z^2 transform as A_1 and $(x^2 - y^2, xy)$ (xz, yz) transform as *E*. Therefore, vibrations A_1 and E are Raman active also, following the rule for Raman activity as given below:

"A fundamental transition will be Raman active if the normal mode which is excited belongs to the same irreducible representation as one or more of the binary products of the Cartesian coordinates."

The assignments of actual vibrational spectral data in NH_3 are given below.

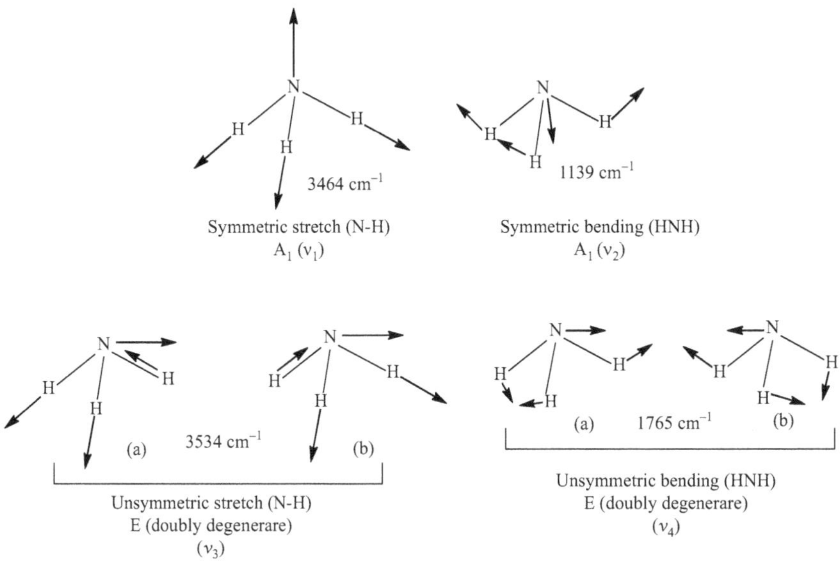

Vibrational spectra of NH_3 molecule	
Infrared (cm^{-1})	**Raman (cm^{-1})**
3464 (ν_1) A_1	3464 (ν_1) A_1 (pol.)
1139 (ν_2) A_1	1139 (ν_2) A_1 (pol.)
3534 (ν_3) E	3534 (ν_3) E (depol.)
1765 (ν_4) E	1765 (ν_4) E (depol.)

(iii) Tetrahedral phosphoryl trichloride ($POCl_3$) molecule of C_{3v} point group

This is a pentavalent tetrahedral molecule which belongs to C_{3v} point group .The C_3 axis passes through the center of the Cl_3 trigonal base. The possible symmetry operations in this molecule are: E, $2C_3$, $3\sigma_v$.

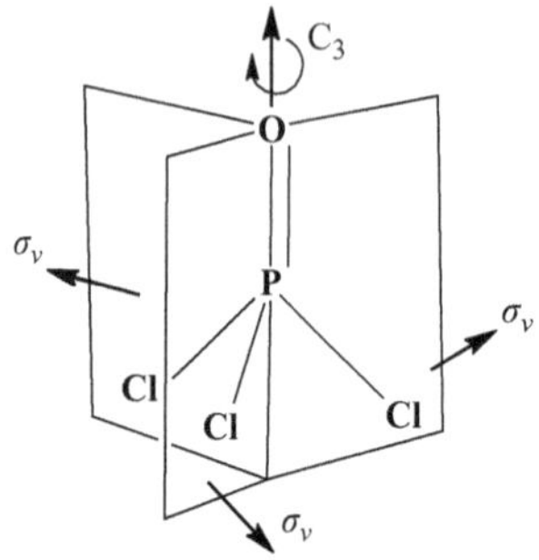

Cartesian coordinate method or 3N vectors method

Out of four symmetry operations, E and C_3 are proper rotations through 0° and 120°, respectively, while σ_v is improper rotations through 0°, respectively. The following table provides the calculations of the total characters.

Table: Determination of total character for $POCl_3$ molecule

C_{3v}	E	$2C_3$	$3\sigma_v$
β	0	120	0
cosβ	1	$-\frac{1}{2}$	1
±1+2cosβ	3	0	1
n_R	5	2	3
Γ_{3N}	**15**	**0**	**3**

The vertical dashed line separates the proper and improper rotations. The corresponding angle β, cosβ and also (±1 + 2cosβ) values are tabulated for each operation. The proper rotations have (+1 + 2cosβ) and the improper rotations have (−1 + 2cosβ) values. The number of atoms retaining position (n_R) for each operation is also given in the table.

The above reducible representation (Γ_{3N}) can be reduced to irreducible representations using the reduction formula given as follows:

$$N_i = \frac{1}{h}\sum_R \chi(R).n.\chi_i(R)$$

Total representatin of $POCl_3$ molecule

C_{3v}	E	$2C_3$	$3\sigma_v$
Γ_{3N}	**15**	**0**	**3**

Character table of C_{3v} Point Group

C_{3v}	E	$2C_3$	$3\sigma_v$		
A_1	1	1	1	z	x^2+y^2, z^2
A_2	1	1	−1	R_z	
E	2	−1	0	(x,y) R_x, R_y)	(x^2-y^2, xy) (xz,yz)

$$NA_1 = 1/6\,[(15.1.1) + 0 + (3.3.\,1)]$$
$$= 1/6\,[15 + 0 + 9]$$
$$= 1/6\,[24] = 4$$
$$NA_2 = 1/6\,[(15.1.1) + 0 + (3.3. - 1)]$$
$$= 1/6\,[15 + 0 - 9]$$
$$= 1/6\,[6] = 1$$
$$NE = 1/6\,[(15.1.2) + 0 + 0]$$
$$= 1/6\,[15 + 0 - 9]$$
$$= 1/6\,[30] = 5$$

Thus, the reducible representation reduces to:

$$\Gamma_{3N} = 4A_1 + A_2 + 5E\ \ (3N = 15)$$

Thus, there are 15 modes (4 + 1 + 10 = 15) of degrees of freedom with different symmetries, A_1, A_2, *E*. These include the vibrational, rotational and translational degrees of freedom.

The character table shows that the translational modes have the same symmetry as the coordinates *x*, *y* and *z*, that is, A_1 and *E*. Similarly, the rotational modes have the same symmetry as the rotation axes, R_x, R_y and R_z, that is, A_2 and *E*. Subtracting these translational and rotational degrees of freedom, we get the vibrational degrees of freedom only of appropriate symmetry.

$$\begin{aligned} \Gamma_{3N} &= 4A_1 + A_2 + 5E \\ \Gamma_{tr} &= \ \ A_1 \qquad\quad\ \ E \\ \Gamma_{rot} &= \qquad\quad A_2 + E \\ \hline \Gamma_{vib} &= 3A_1 + 3E\ (3N - 6 = 9) \end{aligned}$$

Internal coordinate method

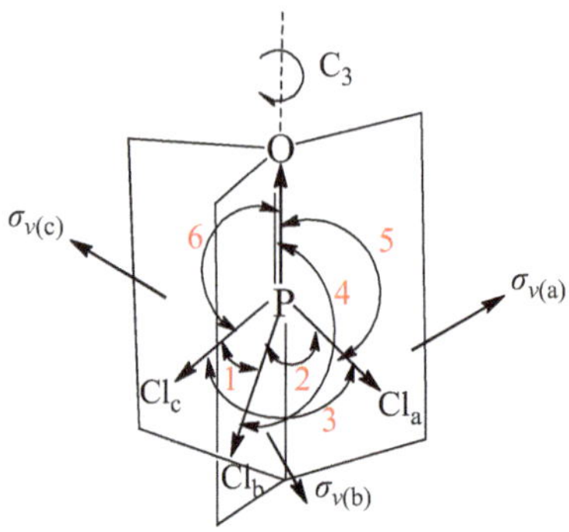

There are three P - Cl and one P - O stretches and six angle deformations as shown in the Fig. given above. We can use these stretches and bending vectors as the basis representation to develop $\Gamma_{P\text{-}Cl}$, $\Gamma_{P\text{-}O}$ and Γ_{bend} as follows:

$\Gamma_{P\text{-}Cl}$:

C_{3v}	E	$2C_3$	$3\sigma_v$
No. of unshifted P-Cl stretch (*n*)	3	0	1
$\chi(R)$	+1	+1	+1
$\Gamma_{P\text{-}Cl}$ [$n \times \chi(R)$]	3	0	1

$\Gamma_{P\text{-}O}$:

C_{3v}	E	$2C_3$	$3\sigma_v$
No. of unshifted P-O stretch (*n*)	1	1	1
$\chi(R)$	+1	+1	+1
$\Gamma_{P\text{-}O}$[$n \times \chi(R)$]	1	1	1

Γ_{bend}

C_3 axis shifts all the angles, so $n = 0$. For σ_v, angle Cl_aPO (axial) is not changed as it is contained in σ_v. Angle Cl_aPCl_b is half bisected and only transforms into another half. So, it is also not changed. Thus, two angles are not changed.

C_{3v}	E	$2C_3$	$3\sigma_v$
No. of unshifted double arrow headed vectors (*n*)	6	0	2
$\chi(R)$	+1	+1	+1
Γ_{bend} $[n\times\chi(R)]$	6	0	2

The reducible representation $\Gamma_{P\text{-}Cl}$, $\Gamma_{P\text{-}O}$, and Γ_{bend} can be reduced to irreducible representations applying the reduction formula and using character table of C_{3v} point group as follows:

$$N_i = \frac{1}{h}\sum_R \chi(R)\,.\,n\,.\,\chi_i(R)$$

$\Gamma_{P\text{-}Cl}$

Reducible representation in question

C_{3v}	E	$2C_3$	$3\sigma_v$
Γ_{P-Cl}	**3**	**0**	**1**

Character table of C_{3v} Point Group

C_{3v}	E	$2C_3$	$3\sigma_v$		
A_1	1	1	1	z	x^2+y^2, z^2
A_2	1	1	−1	R_z	
E	2	−1	0	(x,y) $R_x, R_y)$	(x^2-y^2, xy) (xz,yz)

$$NA_1 = 1/6\,[(3.1.1) + 0 + (1.3.1)]$$
$$= 1/6\,[3 + 0 + 3]$$
$$= 1/6\,[6] = 1$$
$$NA_2 = 1/6\,[(3.1.1) + 0 + (1.3.-1)]$$
$$= 1/6\,[3 + 0 - 3]$$
$$= 1/6\,[0] = 0$$
$$NE = 1/6\,[(3.1.2) + 0 + 0]$$
$$= 1/6\,[6 + 0 + 0]$$
$$= 1/6\,[6] = 1$$

Thus, $\Gamma_{P\text{-}Cl}$ is reduced to:

$$\Gamma_{P-Cl} = A_1 + E$$

$\Gamma_{P\text{-}O}$:

Reducible representation in question

C_{3v}	E	$2C_3$	$3\sigma_v$
Γ_{P-Cl}	**1**	**1**	**1**

Character table of C_{3v} Point Group

C_{3v}	E	$2C_3$	$3\sigma_v$		
A_1	1	1	1	z	x^2+y^2, z^2
A_2	1	1	−1	R_z	
E	2	−1	0	$(x,y)\ R_x, R_y)$	$(x^2-y^2, xy)\ (xz,yz)$

$$\begin{aligned} NA_1 &= 1/6\,[(1.1.1)+(1.2.1)+(1.3.1)] \\ &= 1/6\,[1+2+3] \\ &= 1/6\,[6] = 1 \\ NA_2 &= 1/6\,[(1.1.1)+(1.2.1)+(1.3.-1)] \\ &= 1/6\,[1+2-3] \\ &= 1/6\,[0] = 0 \\ NE &= 1/6\,[(1.1.2)+(1.2.-1)+0] \\ &= 1/6\,[2-2+0] \\ &= 1/6\,[0] = 0 \end{aligned}$$

Thus, $\Gamma_{P\text{-}O}$ is reduced to

$$\Gamma_{P-O} = A_1$$

Thus,

$$\Gamma_{str} = \Gamma_{P-Cl} + \Gamma_{P-O} = A_1 + E + A_1 = 2A_1 + E$$

Γ_{bend}:

Reducible representation in question

C_{3v}	E	$2C_3$	$3\sigma_v$
Γ_{bend}	**6**	**0**	**2**

Character table of C_{3v} Point Group

C_{3v}	E	$2C_3$	$3\sigma_v$		
A_1	1	1	1	z	x^2+y^2, z^2
A_2	1	1	−1	R_z	
E	2	−1	0	$(x,y)\ R_x, R_y)$	$(x^2-y^2, xy)\ (xz,yz)$

$$\begin{aligned} NA_1 &= 1/6\,[(6.1.1) + 0 + (2.3.\,1)] \\ &= 1/6\,[6 + 0 + 6] \\ &= 1/6\,[12] = 2 \\ NA_2 &= 1/6\,[(6.1.1) + 0 + (2.3.-1)] \\ &= 1/6\,[0] \\ &= 1/6\,[0] = 0 \\ NE &= 1/6\,[(6.1.2) + 0 + 0] \\ &= 1/6\,[12 + 0 + 0] \\ &= 1/6\,[12] = 2 \end{aligned}$$

The above calculations show that Γ_{bend} is reduced to:

$$\Gamma_{bend} = 2A_1 + 2E$$

Now $\Gamma_{internal\ coordinate} = \Gamma_{vib} = \Gamma_{P-C} + \Gamma_{P-o} + \Gamma_{bend} = (A_1 + E) + (A_1) + (2A_1 + 2E) = 4A_1 + 3E$

Looking over the result of Γ_{vib} (= $3A_1$ + 3E) obtained through Cartesian coordinate method, it appears that we get one extra A_1 mode using internal coordinate method. One totally symmetric A_1 mode that arises from deformation of all six angles simultaneously is impossible. All the six angles cannot be increased or decreased simultaneously, that is, all the six angles are not independent. Thus, this totally symmetric deformation A_1 mode is redundant mode, and so can be neglected. So, the effective deformation modes,

$$\Gamma_{bend} = (2A_1 + 2E) - (A_1) = A_1 + 2E.$$

So,

$$\Gamma_{vib} = \Gamma_{P-C} + \Gamma_{P-o} + \Gamma_{bend} - A_1 = (4A_1 + 3E) - A_1 = 3A_1 + 3E$$

Identification of IR and Raman active vibrations

IR active vibrations

From the character table of C_{3v} point group, it is notable that *x*, *y*, *z*-coordinates belong to A_1 and *E* irreducible representations. Hence, following the symmetry selection rule for IR, $3A_1$ and 3E vibrational modes will be IR active. Thus, we can expect six frequencies in the IR spectra

Raman active vibrations

An observation of the character table (C_{3v} point group) indicates that the binary products of the Cartesian coordinates belong to A_1 and E irreducible representations. Hence, following symmetry selection rule for Raman, $3A_1$ and 3E vibrational modes are Raman active. Thus, we can expect six frequencies in the Raman spectrum.

Vibrational spectral data of $POCl_3$ molecule is given below.

Vibrational spectra of $POCl_3$ molecule

Infrared (cm^{-1}) (liquid)	Raman (cm^{-1}) (liquid)
1292	1290 A_1 (pol.)
580	581 E (depol.)
487	486 A_1 (pol.)
340	337 E (depol.)
260	267 A_1 (pol.)
Not observed	193 E (depol.)

Based on experimental observations, it is acceptable that
(a) Stretches have higher frequencies than the bending frequencies.
(b) Double bonds have higher frequencies compared to the bending frequencies.
(c) Frequency generally decreases with the increasing atomic mass
(d) Two Pt-Cl stretches should be closed.

Based on these facts, one can assign the observed frequencies to the corresponding modes. The first three frequencies at 1292/1290, 580/581 and 487/486 cm^{-1} are assigned to stretching frequencies. Out of these, the highest frequency at 1292/1290 cm^{-1} is assigned to $P = O$ double bond stretch and the rest two frequencies at 580/581 and 487/486 $cm^{-1,}$ which are closed by are due to P–Cl stretches. Out of the three deformation frequencies, the band at 267 cm^{-1} must be A_1 type as it is Raman polarized, and the other two deformation frequencies are of E type. These modes cannot be distinguished with the available spectral data. For the numbering of the frequencies, we take the symmetric modes and arrange them in descending order, number then and then consider other modes of lower symmetry. Accordingly,

$\nu_1(A_1)$ P = O stretch; 1292/1290 cm^{-1}
$\nu_2(A_1)$ P–Cl stretch; 487/486 cm^{-1}
$\nu_3(A_1)$ symmetric deformation; 260/267 cm^{-1}
ν_4 (E) degenerate anti-symmetric P–Cl stretches; 580/581 cm^{-1}
ν_5 (E) deformation; 337/340 cm^{-1}
ν_6 (E) deformation; 193 cm^{-1}, not observed in IR

(iv) AB_3 Type of Molecules (BF_3, CO_3^{2-}, SO_3, NO_3^-, etc.) of D_3h point symmetry
AB_3 type of molecules or ions is triangular planer. The three *B* atoms are at the corners of the triangular plane at an angle of 120°, and their *σ*-orbitals, σ_1, σ_2 and σ_3 overlap the central atom A along the *x*, *y* and *z* axes, respectively. The molecular point symmetry of the molecule is D_{3h} (D_{nh}: A principal axis C_n, nC_2's ⊥ to C_n, $n\sigma_v$, improper axis S_n and a horizontal plane, *σh*).The total number of 12 symmetry operations in triangular planer AB_3 molecules of D_{3h} point group are: **E, $2C_3$, $3C_2$, $2S_3$, σ_h, $3\sigma_v$** .

Let us consider BF_3 molecule.

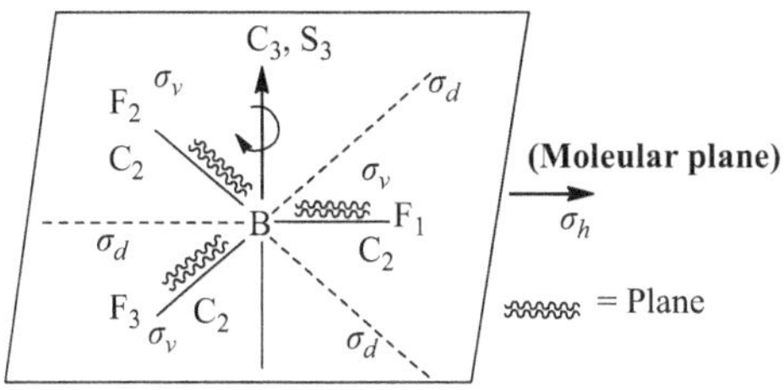

Cartesian coordinate method or 3N vectors method
The symmetry operations in this case are three proper rotations through 0°(E), 120° (C_3) and 180°(C_2) and three improper rotations through 120°(S_3), 0°(σ_h) and 0°(σ_d). The following table provides the calculations of the total characters.

The vertical dashed line separates the proper and improper rotations. The corresponding angle*β*, cos*β* and also (±1 + 2cos*β*) values are tabulated for each operation. The proper rotations have (+1 + 2cos*β*) and the improper rotations have (−1 + 2cos*β*) values. The number of atoms retaining position (n_R) for each operation is also given in the table.

Table: Determination of total character for BF_3 molecule

D_{3h}	E	$2C_3$	$3C_2$	σ_h	$2S_3$	$3\sigma_v$
β	0	120	180	0	120	0
cosβ	1	$-\frac{1}{2}$	−1	1	$-\frac{1}{2}$	1
±1+2cosβ	3	0	−1	1	−2	1
n_R	4	1	2	4	1	2
Γ_{3N}	**12**	**0**	**−2**	**4**	**−2**	**2**

The above reducible representation (Γ_{3N}) can be reduced to irreducible representations using the reduction formula and character table D_{3h} point group as follows:

Reducible representation in question

D_{3h}	E	$2C_3$	$3C_2$	σ_h	$2S_3$	$3\sigma_v$
Γ_{3N}	**12**	**0**	**−2**	**4**	**−2**	**2**

Character table of D_{3h} point group

D_{3h}	E	$2C_3$	$3C_2$	σ_h	$2S_3$	$3\sigma_v$		
A_1'	1	1	1	1	1	1		x^2+y^2, z^2
A_2'	1	1	−1	1	1	−1	R_z	
E'	2	−1	0	2	−1	0	(x,y)	(x^2-y^2, xy)
A_1''	1	1	1	−1	−1	−1		
A_2''	1	1	−1	−1	−1	1	z	
E''	2	−1	0	−2	1	0	R_x, R_y	(xz, yz)

$$\begin{aligned} NA_1' &= 1/12\,[(12.1.1)+0+(-2.3.1)+(4.1.1)\,(-2.2.1)+(2.3.1)] \\ &= 1/12\,[(12+0-6+4-4+6] \\ &= 1/12\,[(12]=1 \end{aligned}$$

$$\begin{aligned} NA_2' &= 1/12\,[(12.1.1)+0+(-2.3.-1)+(4.1.1)\,(-2.2.1)+(2.3.-1)] \\ &= 1/12\,[(12+0+6+4-4-6] \\ &= 1/12\,[(12]=1 \end{aligned}$$

$$\begin{aligned} NE' &= 1/12\,[(12.1.2)+0+0+(4.1.2)\,(-2.2.-1)+0] \\ &= 1/12\,[(24+8+4 \\ &= 1/12\,[(36]=3 \end{aligned}$$

$$\begin{aligned} NA_1'' &= 1/12\,[(12.1.1)+0+(-2.3.1)+(4.1.-1)\,(-2.2.-1)+(2.3.-1)] \\ &= 1/12\,[(12+0+-6-4+4-6] \\ &= 1/12\,[(0]=0 \end{aligned}$$

$$\begin{aligned} NA_2'' &= 1/12\,[(12.1.1)+0+(-2.3.-1)+(4.1.-1)\,(-2.2.-1)+(2.3.1)] \\ &= 1/12\,[(12+0+6-4+4+6] \\ &= 1/12\,[(24]=2 \end{aligned}$$

$$\begin{aligned} NE'' &= 1/12\,[(12.1.2)+0+0+(4.1.-2)\,(-2.2.1)+0] \\ &= 1/12\,[(24-8-4] \\ &= 1/12\,[(12]=1 \end{aligned}$$

Thus, the above reducible representation is reduced to:

$$T = A_1' + A_2' + 3E' + 2A_2'' + E''$$

Thus, there are 12 modes (1 + 1 + 6 + 2 + 2 = 12) of degrees of freedom with different symmetry. These include the vibrational, rotational and translational degrees of freedom.

The D_{3h} character table indicates that the translational modes have the same symmetry as the coordinates *x*, *y* and *z*, that is, E′ and A_2''. So also the rotational modes have the same symmetry as the rotation axes, R_x, R_y and R_z, which is, E″, A_2'. Subtracting these translational and rotational degrees of freedom, we get the vibrational degrees of freedom of appropriate symmetry.

$$\begin{array}{lllllll}
\Gamma = & A_1' & + A_2' & + 3E' & + 2A_2'' & + E'' \\
\Gamma_{tr} = & & & E' & + A_2'' & \\
\Gamma_{rot} = & & A_2' & & & + E'' \\
\hline
\Gamma_{vib} = & A_1' & & + 2E' & + A_2'' &
\end{array}$$

Internal coordinate method

In the previous section, using internal coordinate method, we have seen that

$$\Gamma_{\text{internal coordinate}} = \Gamma_{vib} = \Gamma_{B-F} + \Gamma_{ipd} + \Gamma_{opd} = A_1' + 2E' + A_2'' \ (3N - 6 = 4 \times 3 - 6 = 6)$$

This result of Γ_{vib} obtained is the same to that we have obtained through Cartesian coordinate method.

Identification of IR and Raman active vibrations

IR active vibrations

From the character table of D_{3h} point group, it is notable that *x*, *y* coordinates belong to *E′* and *z*-coordinate belongs to A_2'' representation. Hence, 2E′ and A_2'' vibrational modes will be IR active. As 2E′ fundamental vibrations are two sets of doubly degenerate vibrations, and A_2'' vibrational mode is non-degenerate, only three frequencies will be observed in the IR spectra.

Raman active vibrations

An observation of the character table indicates that the binary products of the Cartesian coordinates belong to the irreducible representations *E′* and A_1' . Hence, two vibrational modes (2E′ and A_1') will be Raman active. As 2E′ fundamental vibrations are two sets of doubly degenerate vibrations, and A_2' vibrational mode is non-degenerate, only three frequencies will be observed in the IR spectra.

The vibrational spectral data of BF_3 molecule are given below.

Vibrational spectra of BF_3 molecule

Infrared (cm^{-1})	Raman (cm^{-1})
1454	1453 (depol.)
–	888 (pol.)
692	–
479	481 (depol.)

Let us assign the symmetry species to these frequencies and find which internal coordinates are responsible for these modes of vibration. Based on the guidelines given earlier that stretches have higher frequencies than angle deformation frequencies. Thus, B-F stretches are higher than FBF angle of deformation.

Let us first see the activities of these modes of vibration which are as follows:

B-F stretches: A_1' (Raman pol.) + E' (Raman depol. and IR)
In-plane FBF angle deformation: E' (Raman depol. and IR)
Out of plane FBF angle deformation: A_2'' (IR)

Vibrational frequencies at 1454/1453 and 888 cm^{-1} are higher, and these must correspond to B-F stretches. Rest of the frequencies at 692 and 479/481 cm^{-1} is due to angle of deformations. As one of the B-F stretch is Raman polarized, it should be A_1' mode. Thus, frequency at 888 cm^{-1} A_1' is type. Another stretching frequency at 1454/1453 must have E' symmetry (as $\Gamma_{B\text{-}F} = A_1' + E'$). The frequency at 692 cm^{-1} appears only in IR and it should be A_2'' due to out of plane FBF angle deformation. Frequency at 479/481 cm^{-1} (which is both IR and Raman active) is thus due to in plane deformation and it has E' symmetry. The frequencies can now be numbered and listed as follows:

$\nu_1(A_1')$ B-F stretch; 888 cm^{-1}
$\nu_2(A_2'')$ opd; 692 cm^{-1}
$\nu_3(E')$ B-F stretch; 1454/1453 cm^{-1}
ν_4 (E') In-plane deformation; 479/481 cm^{-1}

From the analysis of the vibrational frequency, it is concluded that the Raman spectrum provides with a frequency (A_1' is type), which cannot be obtained from the infrared spectrum. This is why Raman spectrum is said to be complementary to infrared spectrum. In fact, the Raman spectrum provides information which the infrared spectrum fails to give. The actual complementarity of the Raman and infrared spectrum depends on the shape and hence the point group of the molecules. Thus, this complementarity can be used to ascertain the point group of the unknown molecules making use of the predicted and experimentally observed spectra of the molecules. The shape of the molecule can be easily deduced once the point group of the molecule is known.

(v) AB_5 type of Molecules

Two common shapes of AB_5 type molecules are trigonal bipyramidal and square pyramidal. We will consider these cases one by one.

(a) AB_5 type molecules of trigonal bipyramidal geometry

Such molecules belong to D_{3h} point symmetry (contains C_3 as the principal axis and three number of C_2 axes ⊥r to C_3 axis, there will be one plane ⊥r to the C_3 axis (σ_h) and three planes all containing the C_3 axis, that is, σ_v planes). The possible symmetry elements/operations in such molecules are: E, $2C_3$, $3C_2$, $2S_3$, σ_h, $3\sigma_v$. PCl_5 molecule may be taken as a suitable molecule of trigonal bipyramidal geometry (TBP).

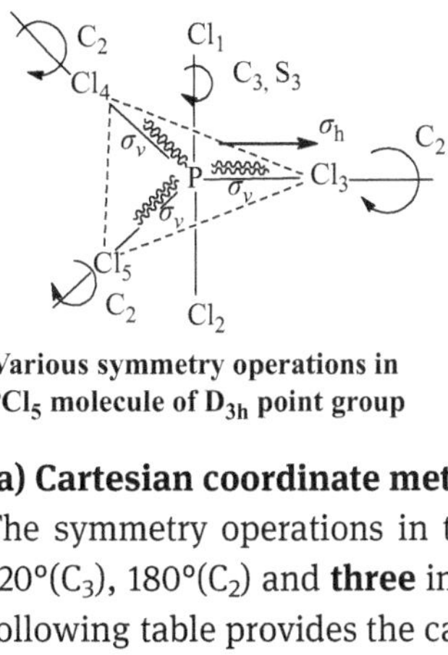

Various symmetry operations in PCl_5 molecule of D_{3h} point group

(a) Cartesian coordinate method or 3N vectors method

The symmetry operations in these case are **three** proper rotations through 0°(E), 120°(C_3), 180°(C_2) and **three** improper rotations through 120°(S_3) and 0°(σ_h, σ_v). The following table provides the calculations of the total characters.

Table: Determination of total character for PCl_5 TBP molecule

D_{3h}	E	$2C_3$	$3C_2$	$2S_3$	σ_h	$3\sigma_v$
β	0°	120°	180°	120°	0°	0°
cosβ	1	$-\frac{1}{2}$	−1	$-\frac{1}{2}$	1	1
±1+ 2cosβ	3	0	−1	−2	1	1
n_R	6	3	2	1	4	4
Γ_{3N}	**18**	**0**	**−2**	**−2**	**4**	**4**

The vertical dashed line separates the proper and improper rotations. The corresponding angle β, cosβ and also (±1 + 2cosβ) values are tabulated for each operation. The proper rotations have (+1 + 2cosβ) and the improper rotations have (−1 + 2cosβ)

values. The number of atoms retaining position (n_R) for each operation is also given in the table.

The above reducible representation (Γ_{3N}) can be reduced to irreducible representations using the reduction formula and character table of D_{3h} point group given as follows:

$$N_i = \frac{1}{h}\sum_R \chi(R).n.\chi_i(R)$$

Reducible representation in question

D_{3h}	E	$2C_3$	$3C_2$	σ_h	$2S_3$	$3\sigma_v$
Γ_{3N}	**18**	**0**	**−2**	**4**	**−2**	**4**

Character table of D_{3h} point group

D_{3h}	E	$2C_3$	$3C_2$	σ_h	$2S_3$	$3\sigma_v$		
A_1'	1	1	1	1	1	1		x^2+y^2, z^2
A_2'	1	1	−1	1	1	−1	R_z	
E'	2	−1	0	2	−1	0	(x,y)	(x^2-y^2, xy)
A_1''	1	1	1	−1	−1	−1		
A_2''	1	1	−1	−1	−1	1	z	
E''	2	−1	0	−2	1	0	R_x, R_y	(xz, yz)

$$\begin{aligned} NA_1' &= 1/12\,[(18.1.1) + 0 + (-2.3.1) + (4.1.1) + (-2.2.1) + (4.3.1) \\ &= 1/12\,[18 - 6 + 4 - 4 + 12] \\ &= 1/12\,[24] = 2 \end{aligned}$$

$$\begin{aligned} NA_2' &= 1/12\,[(18.1.1) + 0 + (-2.3.-1) + (4.1.1) + (-2.2.1) + (4.3.-1) \\ &= 1/12\,[18 + 6 + 4 - 4 - 12] \\ &= 1/12\,[12] = 1 \end{aligned}$$

$$\begin{aligned} NE' &= 1/12\,[(18.1.2) + 0 + 0 + (4.1.2) + (-2.2.-1) + 0 \\ &= 1/12\,[36 + 0 + 0 + 8 + 4] \\ &= 1/12\,[48] = 4 \end{aligned}$$

$$\begin{aligned} NA_1'' &= 1/12\,[(18.1.1) + 0 + (-2.3.1) + (4.1.-1) + (-2.2.-1) + (4.3.-1) \\ &= 1/12\,[18 - 6 - 4 + 4 - 12] \\ &= 1/12\,[0] = 0 \end{aligned}$$

$$
\begin{aligned}
NA_2'' &= 1/12\,[(18.1.1) + 0 + (-2.3.-1) + (4.1.-1) + (-2.2.-1) + (4.3.-1)) \\
&= 1/12\,[18 + 6 - 4 + 4 + 12] \\
&= 1/12\,[36] = 3
\end{aligned}
$$

$$
\begin{aligned}
NE'' &= 1/12\,[(18.1.2) + 0 + 0 + (4.1.-2) + (-2.2.1) + 0 \\
&= 1/12\,[36 + 0 + 0 - 8 - 4] \\
&= 1/12\,[24] = 2
\end{aligned}
$$

Thus, the total representation reduces to:

$$\Gamma_{3N} = 2A_1' + A_2' + 4E' + 3A_2'' + 2E''$$

Thus, there are 18 modes (2 + 1 + 8 + 3 + 4 = 18) of degrees of freedom with different symmetry. These include the vibrational, rotational and translational degrees of freedom.

The D_{3h} character table indicates that the translational modes have the same symmetry as the coordinates *x*, *y* and *z*, that is, E′ and A_2''. So also the rotational modes have the same symmetry as the rotation axes, R_x, R_y and R_z, which is, E″, A_2'. Subtracting these translational and rotational degrees of freedom, we get the vibrational degrees of freedom of appropriate symmetry.

$$
\begin{aligned}
\Gamma &= 2A_1' + A_2' + 4E' + 3A_2'' + 2E'' \\
\Gamma_{tr} &= \qquad\qquad\quad E' + A_2'' \\
\Gamma_{rot} &= \qquad A_2' \qquad\qquad\qquad + E'' \\
\hline
\Gamma_{vib} &= 2A_1' + 3E' + 2A_2'' + E''
\end{aligned}
$$

Internal coordinate method

The bond stretches and the angle deformation in PCl_5 are shown by straight arrows and double-headed arrows in the following Fig.

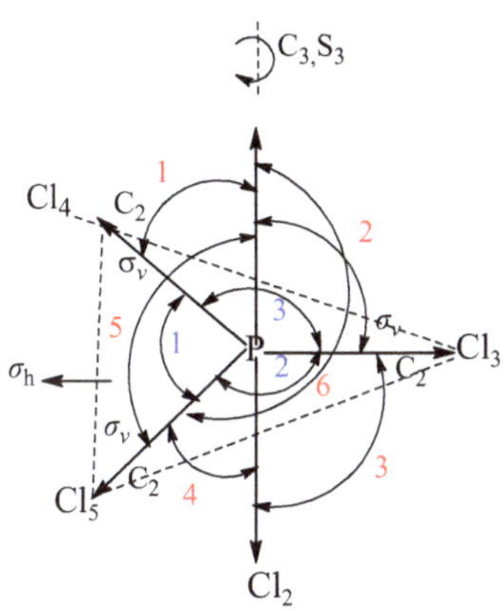

There are five P-Cl bond vectors (three equatorial and two axial), three equatorial ClPCl bond angles at 120° and six axial ClPCl bond angles at 90°. Using these stretches and bending vectors as the basis, we have already worked out the reducible representations for Γ_{str} [$\Gamma_{P\text{-}Cl\ (axial)}$ and $\Gamma_{P\text{-}Cl\ (equatorial)}$] and Γ_{bend} [$\Gamma_{bend\ (axial)}$ and $\Gamma_{bend\ (equatorial)}$].

$\Gamma_{P\text{-}Cl\ (axial)}$:

Reducible representation in question

D_{3h}	E	$2C_3$	$3C_2$	σ_h	$2S_3$	$3\sigma_v$
Γ_{3N}	2	2	0	0	0	2

This reducible representation can be reduced using reduction formula and character table of D_{3h} point group as follows:

$$N_i = \frac{1}{h}\sum_R \chi(R).n.\chi_i(R)$$

D_3h	E	$2C_3$	$3C_2$	σ_h	$2S_3$	$3\sigma_v$		
A_1'	1	1	1	1	1	1		x^2+y^2, z^2
A_2'	1	1	−1	1	1	−1	R_z	
E'	2	−1	0	2	−1	0	(x, y)	(x^2-y^2, xy)
A_1''	1	1	1	−1	−1	1		
A_2''	1	1	−1	−1	−1	1	z	
E''	2	−1	0	−2	1	0	R_x, R_y	(xz, yz)

$$\begin{aligned} NA_1' &= 1/12\,[(2.1.1) + (2.2.1) + 0 + 0 + 0 + (2.3.1)) \\ &= 1/12\,[2+4+0+0+0+6] \\ &= 1/12\,[12] = 1 \end{aligned}$$

$$\begin{aligned} NA_2' &= 1/12\,[(2.1.1) + (2.2.1) + 0 + 0 + 0 + (2.3.-1)) \\ &= 1/12\,[2+4+0+0+0-6] \\ &= 1/12\,[0] = 0 \end{aligned}$$

$$\begin{aligned} NE' &= 1/12\,[(2.1.2) + (2.2.-1) + 0 + 0 + 0 + (2.3.0)) \\ &= 1/12\,[4-4+0+0+0+0] \\ &= 1/12\,[0] = 0 \end{aligned}$$

$$NA_1'' = 1/12\,[(2.1.1) + (2.2.1) + 0 + 0 + 0 + (2.3.-1)$$
$$= 1/12\,[2 + 4 + 0 + 0 + 0 - 6]$$
$$= 1/12\,[0] = 0$$

$$NA_2'' = 1/12\,[(2.1.1) + (2.2.1) + 0 + 0 + 0 + (2.3.1)$$
$$= 1/12\,[2 + 4 + 0 + 0 + 0 + 6]$$
$$= 1/12\,[12] = 1$$

$$NE'' = 1/12\,[(2.1.2) + (2.2.-1) + 0 + 0 + 0 + (2.3.0)$$
$$= 1/12\,[4 - 4 + 0 + 0 + 0 + 0]$$
$$= 1/12\,[0] = 0$$

Thus, $\Gamma_{\text{P-Cl (axial)}}$ is reduced to: $A_1' + A_2''$

$\Gamma_{\text{P-Cl (equatorial)}}$:

Reducible representation in question

D_3h	E	$2C_3$	$3C_2$	σ_h	$2S_3$	$3\sigma_v$
Γ_{3N}	**3**	**0**	**1**	**3**	**0**	**1**

Again, this reducible representation can be reduced using reduction formula and character table of D_{3h} point group as follows:

$$N_i = \frac{1}{h}\sum_R \chi(R)\,.\,n\,.\,\chi_i(R)$$

D_3h	E	$2C_3$	$3C_2$	σ_h	$2S_3$	$3\sigma_v$		
A_1'	1	1	1	1	1	1		x^2+y^2, z^2
A_2'	1	1	−1	1	1	−1	R_z	
E'	2	−1	0	2	−1	0	(x,y)	(x^2-y^2, xy)
A_1''	1	1	1	−1	−1	−1		
A_2''	1	1	1	−1	−1	1	z	
E''	2	−1	0	−2	1	0	R_x, R_y	(xz, yz)

$$NA_1' = 1/12\,[(3.1.1) + 0 + (1.3.1) + (3.1.1) + 0 + (1.3.1)$$
$$= 1/12\,[3 + 0 + 3 + 3 + 0 + 3]$$
$$= 1/12\,[12] = 1$$

$$NA_2{}' = 1/12\,[(3.1.1) + 0 + (1.3. - 1) + (3.1.1) + 0 + (1.3. - 1)$$
$$= 1/12\,[3 + 0 - 3 + 3 + 0 - 3]$$
$$= 1/12\,[0] = 0$$

$$NE' = 1/12\,[(3.1.2) + 0 + 0 + (3.1.2) + 0 + 0$$
$$= 1/12\,[6 + 0 + 0 + 6 + 0 + 0]$$
$$= 1/12\,[12] = 1$$

$$NA_1{}'' = 1/12\,[(3.1.1) + 0 + (1.3.1) + (3.1. - 1) + 0 + (1.3 - .1)$$
$$= 1/12\,[3 + 0 + 3 - 3 + 0 - 3]$$
$$= 1/12\,[0] = 0$$

$$NA_2{}'' = 1/12\,[(3.1.1) + 0 + (1.3. - 1) + (3.1. - 1) + 0 + (1.3.1)$$
$$= 1/12\,[3 + 0 - 3 - 3 + 0 + 3]$$
$$= 1/12\,[0] = 0$$

$$NE'' = 1/12\,[(3.1.2) + 0 + 0 + (3.1. - 2) + 0 + 0$$
$$= 1/12\,[6 + 0 + 0 - 6 + 0 + 0]$$
$$= 1/12\,[0] = 0$$

Thus, $\Gamma_{P-Cl(equatorial)}$ is reduced to: $A_1{}' + E'$

Hence, total

$\Gamma_{stretch} = \Gamma_{P-Cl(axial)} + \Gamma_{P-Cl(equatorial)} = (A_1{}' + A_2{}'') + (A_1{}' + E') = 2A_1{}' + E' + A_2{}''$

$\mathbf{\Gamma_{bend\ (axial)}}$:

Reducible representation in question

D_3h	E	$2C_3$	$3C_2$	σ_h	$2S_3$	$3\sigma_v$
Γ_{3N}	6	0	0	0	0	2

Again, this reducible representation can be reduced using reduction formula and character table of as follows:

$$N_i = \frac{1}{h}\sum_R \chi(R) \,.\, n \,.\, \chi_i(R)$$

D_3h	E	$2C_3$	$3C_2$	σ_h	$2S_3$	$3\sigma_v$		
A_1'	1	1	1	1	1	1		x^2+y^2, z^2
A_2'	1	1	−1	1	1	−1	R_z	
E'	2	−1	0	2	1	0	(x, y)	(x^2-y^2, xy)
A_1''	1	1	1	−1	−1	1		
A_2''	1	1	−1	−1	−1	1	z	
E''	2	−1	0	−2	1	0	R_x, R_y	$(xz\ yz)$

$$\begin{aligned} NA_1' &= 1/12\,[(6.1.1)+0+0+0+0+(2.3.1) \\ &= 1/12\,[6+0+0+0+0+6] \\ &= 1/12\,[12]=1 \end{aligned}$$

$$\begin{aligned} NA_2' &= 1/12\,[(6.1.1)+0+0+0+0+(2.3.-1) \\ &= 1/12\,[6+0+0+0+0-6] \\ &= 1/12\,[0]=0 \end{aligned}$$

$$\begin{aligned} NE' &= 1/12\,[(6.1.2)+0+0+0+0+0 \\ &= 1/12\,[12+0+0+0+0+6] \\ &= 1/12\,[12]=1 \end{aligned}$$

$$\begin{aligned} NA_1'' &= 1/12\,[(6.1.1)+0+0+0+0+(2.3.-1) \\ &= 1/12\,[6+0+0+0+0-6] \\ &= 1/12\,[0]=0 \end{aligned}$$

$$\begin{aligned} NA_2'' &= 1/12\,[(6.1.1)+0+0+0+0+(2.3.1) \\ &= 1/12\,[6+0+0+0+0+6] \\ &= 1/12\,[12]=1 \end{aligned}$$

$$\begin{aligned} NE'' &= 1/12\,[(6.1.2)+0+0+0+0+0 \\ &= 1/12\,[12+0+0+0+0+6] \\ &= 1/12\,[12]=1 \end{aligned}$$

Thus, $\Gamma_{bend(axial)}$ is reduced to: $A_1' + E' + A_2'' + E''$

$\boldsymbol{\Gamma_{bend\ (equatorial)}}$:

Reducible representation in question

D_3h	E	$2C_3$	$3C_2$	σ_h	$2S_3$	$3\sigma_v$
Γ_{3N}	**3**	**0**	**1**	**3**	**0**	**1**

Again, this reducible representation can be reduced using reduction formula and character table of D_{3h} point group as follows:

D_3h	E	$2C_3$	$3C_2$	σ_h	$2S_3$	$3\sigma_v$		
A_1'	1	1	1	1	1	1		x^2+y^2, z^2
A_2'	1	1	−1	1	1	−1	R_z	
E'	2	−1	0	2	−1	0	(x, y)	(x^2-y^2, xy)
A_1''	1	1	1	−1	−1	−1		
A_2''	1	1	−1	−1	−1	1	z	
E''	2	−1	0	−2	1	0	R_x, R_y	(xz yz)

$$\begin{aligned} NA_1' &= 1/12\,[(3.1.1)+0+(1.3.1)+(3.1.1)+0+(1.3.1) \\ &= 1/12\,[3+0+3+3+0+3] \\ &= 1/12\,[12]=1 \end{aligned}$$

$$\begin{aligned} NA_2' &= 1/12\,[(3.1.1)+0+(1.3.-1)+(3.1.1)+0+(1.3.-1) \\ &= 1/12\,[3+0+3-3+0-3] \\ &= 1/12\,[0]=0 \end{aligned}$$

$$\begin{aligned} NE' &= 1/12\,[(3.1.2)+0+0+(3.1.2)+0+(1.3.0) \\ &= 1/12\,[6+0+0+6+0+0] \\ &= 1/12\,[12]=1 \end{aligned}$$

$$\begin{aligned} NA_1'' &= 1/12\,[(3.1.1)+0+(1.3.1)+(3.1.-1)+0+(1.3.-1) \\ &= 1/12\,[3+0+3-3+0-3] \\ &= 1/12\,[0]=0 \end{aligned}$$

$$\begin{aligned} NA_2'' &= 1/12\,[(3.1.1)+0+(1.3.-1)+(3.1.-1)+0+(1.3.1) \\ &= 1/12\,[3+0-3-3+0+3] \\ &= 1/12\,[0]=0 \end{aligned}$$

$$\begin{aligned} NE'' &= 1/12\,[(3.1.2)+0+0+(3.1.-2)+0+(1.3.0) \\ &= 1/12\,[6+0+0-6+0+0] \\ &= 1/12\,[0]=0 \end{aligned}$$

Thus, $\Gamma_{bend(equatorial)}$ is reduced to: $A_1' + E'$

Hence, total

$$\Gamma_{bend} = \Gamma_{bend(axial)} + \Gamma_{bend(equatorial)} = (A_1' + E' + A_2'' + E'') + (A_1' + E')$$
$$= 2A_1' + 2E' + A_2'' + E''$$

Now total modes of vibrations,

$$\Gamma_{vib} = \Gamma_{stretch} + \Gamma_{bend} = (2A_1' + E' + A_2'') + (2A_1' + 2E' + A_2'' + E'')$$
$$= 4A_1' + 3E' + 2A_2'' + E''$$

On comparing the results of Γ_{vib} obtained using Cartesian coordinate or 3N vectors method worked out above ($\Gamma_{vib} = 2A_1' + 3E' + 2A_2'' + E''$), we see that Γ_{vib} in the present method have extra $2A_1'$. These are two redundant modes, one from deformation of equatorial bond angles and other from deformation of axial bond angles. On removing $2A_1'$ from Γ_{vib} obtained using internal coordinates as the bases, we get

$$\Gamma_{vib} = 2A_1' + 3E' + 2A_2'' + E''$$

Identification of IR and Raman active vibrations

IR active vibrations

From the character table of D_{3h} point group, it is clear that *x*, *y*-coordinate belongs to E' and *z*-coordinate belongs to A_2'' irreducible representation. Hence, $3E'$ and $2A_2''$ vibrational modes will be IR active. As $3E'$ fundamental vibrations are three sets of two-fold degenerate vibrations, and $2A_2''$ vibrational mode is non-degenerate, one can expect only five (3 + 2 = 5) frequencies in the IR spectra.

Raman active vibrations

An observation of the character table of D_{3h} point group indicates that the binary products of the Cartesian coordinates belong to representations A_1', E', and E''. Hence $2A_1'$, $3E'$ and E'' vibrational modes are Raman active. As $3E'$ fundamental vibrations are three sets of two-fold degenerate vibrations, E'' is one set of two-fold degenerate vibrations and $2A_1'$ vibrational mode is non-degenerate, one can expect only six (3 + 1 + 2 = 6) frequencies in the Raman spectra.

The vibrational spectral data of PCl_5 molecule is given below.

Infrared (cm^{-1})	Raman (cm^{-1})
581	579 (depol.)
444	–
–	393 (pol.)
–	385 (pol.)
299	–
277	279 (depol.)
–	261(depol.)
100	98 (depol.)

In order to assign the spectral data, we will use general information available for IR/Raman spectra.

The $\Gamma_{vib} = \Gamma_{str} + \Gamma_{bend}$ contains $2A_1'$ which are always Raman polarized and these are at 393 and 385 cm^{-1}. These are the P-Cl stretches only as Γ_{bend} does not contain any A_1' ($2A_1'$ are redundant modes). Based on the results given above, one axial P-Cl stretch is of A_1' type and another equatorial P-Cl stretch is of A_1' type. As axial P-Cl bonds are weaker and longer, so of the two A_1' modes, the mode with higher frequency will belong to P-Cl equatorial stretch (stronger and shorter).

Using character table of D_{3h} point group (given below), we find the activity of these bonds as:

D_3h	E	$2C_3$	$3C_2$	σ_h	$2S_3$	$3\sigma_v$		
A_1'	1	1	1	1	1	1		$x^2 + y^2, z^2$
A_2'	1	1	−1	1	1	−1	R_z	
E'	2	−1	0	2	−1	0	(x, y)	$(x^2 - y^2, xy)$
A_1''	1	1	1	−1	−1	−1		
A_2''	1	1	−1	−1	−1	1	z	
E''	2	−1	0	−2	1	0	R_x, R_y	$(xz\ yz)$

A_1' (Raman pol.), A_2'' (IR), A_1' (Raman pol.), E′ (Raman depol. and IR), E″ (Raman depol.). The complete spectral frequencies are assigned as:

(i) P-Cl (axial) stretch: A_1' (Raman pol.);
(ii) P-Cl (axial) stretch: A_2'' (IR);
(iii) P-Cl (equatorial) stretch: A_1' (Raman pol.);
(iv) P-Cl (equatorial) stretch: E′ (Raman depol., IR);
(v) Cl (equatorial) P-Cl (equatorial) deformation: E′ (Raman depol., IR);

(vi) Cl (axial) P-Cl (equatorial) deformation: E′ (Raman depol., IR);
(vii) Cl (axial) P-Cl (equatorial) deformation: A_2'' (IR);
(viii) Cl (axial) P-Cl (equatorial) deformation: E″ (Raman depol.)

Now, the numbering of frequencies may be done as follows:

ν_1 (A_1') P-Cl (equatorial) stretch: 393 cm^{-1} (Raman pol.)
ν_2 (A_1') P-Cl (axial) stretch: 385 cm^{-1} (Raman pol.)
ν_3 (A_2'') P-Cl (axial) stretch: 444 cm^{-1} (IR)
ν_4 (A_2'') Cl (axial) P-Cl (equatorial) deformation: 299 cm^{-1} (IR)
ν_5 (E′) P-Cl (equatorial) stretch: 581/579 cm^{-1} (IR/ Raman depol.)
ν_6 (E′) Cl (axial) P-Cl (equatorial) deformation: 277/279 cm^{-1} (IR/ Raman depol.)
ν_7 (E′) Cl (equatorial) P-Cl (equatorial) deformation: 100/98 cm^{-1} (IR/ Raman depol.)
ν_8 (E″) Cl (axial) P-Cl (equatorial) deformation: 261 cm^{-1} (Raman depol.)

(b) AB_5 type molecules of square pyramidal geometry

Such molecules belong to C_{4v} point symmetry (contains C_4 as the principal axis and four vertical planes). The possible symmetry elements/operations in such molecules are: E, $2C_4$, C_2, $2\sigma_v$, $2\sigma_d$.

Cartesian coordinate method or 3N vectors method

Let us consider the square pyramidal $WOCl_4$ molecule belonging to C_{4v} point symmetry.

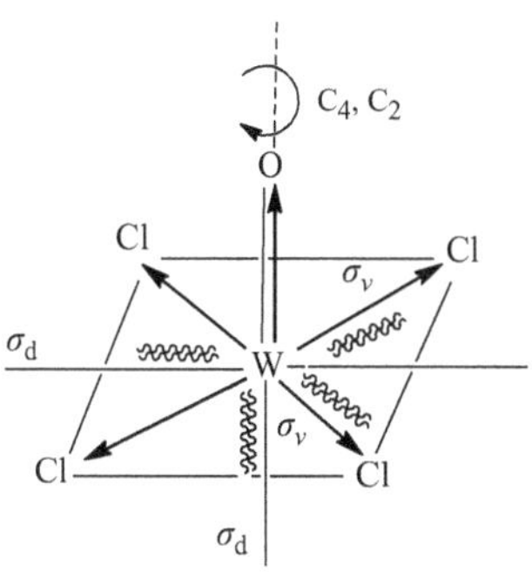

Square pyramidal complex

The symmetry operations in this case are **three** proper rotations through 0°(E), 90° (C_4), 180°(C_2) and **two** improper rotations through 0°(σ_v, σ_d). The following table provides the calculations of the total characters.

Table: Determination of total character for $WOCl_4$ square pyramidal molecule

C_{4v}	E	$2C_4$	C_2	$2\sigma_v$	$2\sigma_d$
β	0	90	180	0	0
cosβ	1	0	−1	1	1
±1+2cosβ	3	1	−1	1	1
n_R	6	2	2	4	2
Γ_{3N}	**18**	**2**	**−2**	**4**	**2**

The vertical dashed line separates the proper and improper rotations. The corresponding angle β, cosβ and also (± 1 + 2cosβ) values are tabulated for each operation. The proper rotations have (+1 + 2cosβ) and the improper rotations have (−1 + 2cosβ) values. The number of atoms retaining position (n_R) for each operation is also given in the table.

The above reducible representation (Γ_{3N}) can be reduced to irreducible representations using the reduction formula and character table of C_{4v} point group as follows:

$$N_i = \frac{1}{h}\sum_R \chi(R)\,.\,n\,.\,\chi_i(R)$$

Table: Determination of total character for $WOCl_4$ square pyramidal molecule

C_{4v}	E	$2C_4$	C_2	$2\sigma_v$	$2\sigma_d$
Γ_{3N}	18	2	−2	4	2

Character table of C_{4v} point group

C_{4v}	E	$2C_4$	C_2	$2\sigma_v$	$2\sigma_d$		
A_1	1	1	1	1	1	z	x^2+y^2, z^2
A_2	1	1	1	−1	−1	R_z	
B_1	1	−1	1	1	−1		x^2-y^2
B_2	1	−1	1	−1	1		xy
E	2	0	−2	0	0	$(x,y)(R_x,R_y)$	xz, yz

$$\begin{aligned} NA_1 &= 1/8\,[(18.1.1)+(2.2.1)+(-2.1.1)+(4.2.1)+(2.2.1)] \\ &= 1/8\,[18+4-2+8+4] \\ &= 1/8\,[32]=4 \end{aligned}$$

$$\begin{aligned} NA_2 &= 1/8\,[(18.1.1) + (2.2.1) + (-2.1.1) + (4.2.-1) + (2.2.-1)\\ &= 1/8\,[18 + 4 - 2 - 8 - 4]\\ &= 1/8\,[8] = 1 \end{aligned}$$

$$\begin{aligned} NB_1 &= 1/8\,[(18.1.1) + (2.2.-1) + (-2.1.1) + (4.2.1) + (2.2.-1)\\ &= 1/8\,[18 - 4 - 2 + 8 - 4]\\ &= 1/8\,[16] = 2 \end{aligned}$$

$$\begin{aligned} NB_2 &= 1/8\,[(18.1.1) + (2.2.-1) + (-2.1.1) + (4.2.-1) + (2.2.1)\\ &= 1/8\,[18 - 4 - 2 - 8 + 4]\\ &= 1/8\,[8] = 1 \end{aligned}$$

$$\begin{aligned} NE &= 1/8\,[(18.1.2) + (2.2.0) + (-2.1.-2) + (4.2.0) + (2.2.0)\\ &= 1/8\,[36 + 0 + 4 + 0 + 0]\\ &= 1/8\,[40] = 5 \end{aligned}$$

Thus, the total representation reduces to:

$$\Gamma_{3N} = 4A_1 + A_2 + 2B_1 + B_2 + 5E\ (3N)$$

Thus, there are 18 modes (4 + 1 + 2 + 1 + 10 = 18) of degrees of freedom with different symmetry. These include the vibrational, rotational and translational degrees of freedom.

The C_{4v} character table indicates that the translational modes have the same symmetry as the coordinates *x*, *y* and *z*, that is, *E* and A_1. So also the rotational modes have the same symmetry as the rotation axes, R_x, R_y and R_z which is *E* and A_2. Subtracting these translational and rotational degrees of freedom, we get the vibrational degrees of freedom only of appropriate symmetry.

$$\begin{aligned} \Gamma &= 4A_1 + A_2 + 2B_1 + B_2 + 5E\\ \Gamma_{tr} &= A_1 \qquad\qquad\qquad\quad + E\\ \Gamma_{rot} &= \qquad\ A_2 \qquad\qquad\quad + E\\ \hline \Gamma_{vib} &= 3A_1 + 2B_1 + B_2 + 3E\ (3N-6) \end{aligned}$$

Internal coordinate method

In the previous section, using internal coordinate method, we have worked out that

$$\begin{aligned} \Gamma_{stretch} &= 2A_1 + B_1 + E\\ \Gamma_{ipd} &= B_2 + E\\ \Gamma_{opd} &= A_1 + B_1 + E \end{aligned}$$

Now,

$$\begin{aligned}\text{total irreducible representation} &= \Gamma_{\text{stretch}} + \Gamma_{\text{ipd}} + \Gamma_{\text{opd}} \\ &= (2A_1 + B_1 + E) + (B_2 + E) + (A_1 + B_1 + E\} \\ &= 3A_1 + 2B_1 + B_2 + 3E\end{aligned}$$

This is equivalent to Γ_{vib} obtained from Cartesian coordinate or 3N vector method.

Identification of IR and Raman active vibrations

IR active vibrations

From the character table of C_{4v} point group, it is clear that *x*, *y*-coordinates belong to *E* and *z*-coordinate belongs to A_1 irreducible representation. Hence, 3E and $3A_1$ vibrational modes will be IR active. As 3E fundamental vibrations are three sets of two-fold degenerate vibrations, and $3A_1$ vibrational mode is non-degenerate, one can expect only six (3 + 3 = 6) frequencies in the IR spectra

Raman active vibrations

An observation of the character table (C_{4v} point group) indicates that the binary products of the Cartesian coordinates belong to representations A_1, B_1, B_2 and E. Hence $3A_1$, $2B_1$, B_2 and 3E vibrational modes are Raman active.

As 3E fundamental vibrations are three sets of two-fold degenerate vibrations, $2B_1$ is two set of non-degenerate vibrations, B_2 is a set of non-degenerate vibration and $3A_1$ vibrational mode is three set of non-degenerate, one can expect only nine (3 + 2 + 1 + 3 = 9) frequencies in the Raman spectra.

The vibrational spectral data of $WOCl_4$ molecule is given below.

Infrared (cm^{-1})	Raman (cm^{-1})
1055	1057 (pol.)
733	732 (pol.)
698	698 (depol.)
–	631 (depol.)
–	328 (depol.)
298	301
–	291
248	248 (pol.)
236	234

In order to analyze the spectral data, we will use general information available for IR/Raman spectra. Double bonds have higher stretching frequency compared to single-bond stretches. Stretching modes generally have higher frequency than the bending mode. Symmetric modes are always Raman polarized.

In the present case, there are four stretches having symmetry $2A_1$, B_1, E. The first four frequencies at 1055/1057, 733/732, 698 and 631 cm^{-1} must belong to these symmetry species. Hence, 1055/1057 cm^{-1} is of A_1 symmetry as it is Raman polarized. Moreover, it is the highest frequency, so it is $\nu_{W=O}$. Also, other $\nu_{W\text{-}Cl}$ stretches from 733 to 631 cm^{-1} are nearly in the same region. As 733 cm^{-1} is Raman polarized, it is A_1 type and is due to symmetric $\nu_{W\text{-}Cl}$ stretch. The 698 cm^{-1} is active moth in IR and Raman, so it is of E type. The remaining stretch at 631 cm^{-1}, which is only Raman active, must be of B_1 symmetry.

Frequencies from 328 to 234 cm^{-1} are that of in-plane and out of plane angle deformations. The frequency at 248 cm^{-1} is Raman polarized, so it is of A_1 type due to opd. The frequency at 298/301 cm^{-1} is both IR and Raman active, so it should be of E type. As in-plane deformations are of higher frequency than that of opd, the frequency at 298/301 cm^{-1} is due to in-plane deformation. Frequency at 328 cm^{-1} is only Raman active, and it is close to 298/301 cm^{-1}, and so it is also from in-plane deformation and belongs to B_2 symmetry.

The remaining frequencies at 291 cm^{-1} (Raman active) and 234/236 cm^{-1} (Raman and IR active) are to be assigned. In fact, these frequencies arise from opd. The former (Raman active) has B_1 (Raman) and the latter (Raman and IR active) *E* symmetry.

(vi) AB_4 type of Molecules

Such molecules can be tetrahedral, square planar, or see-saw shaped. We shall take each category of molecule one by one for prediction of IR and Raman active bands.

(a) Square Planar AB_4 Molecule

This geometry is quite common in inorganic compounds, particularly in metal complexes. Some common examples are: XeF_4, PdX_4^{2-} (X = halogen), $AuCl_4^-$, PtX_4^{2-} (X = halogen), and some Cu(II), Rh(I), Ir(I) complexes.

These molecules belong to D_{4h} point symmetry (contains C_4 as the principal axis and four number of C_2 axes $\perp$r to C_4 axis, there will be one plane $\perp$r to the C_4 axis (σ_h) and four planes all containing the C_4 axis, that is, σ_v planes).

The possible symmetry elements/operations in such molecules are: E, $2C_4$, C_2, $2C_2'$, $2C_2''$, i, $2S_4$, σ_h, $2\sigma_v$, $2\sigma_d$.

Let us take $PtCl_4^{2-}$ as a specific example of this category.

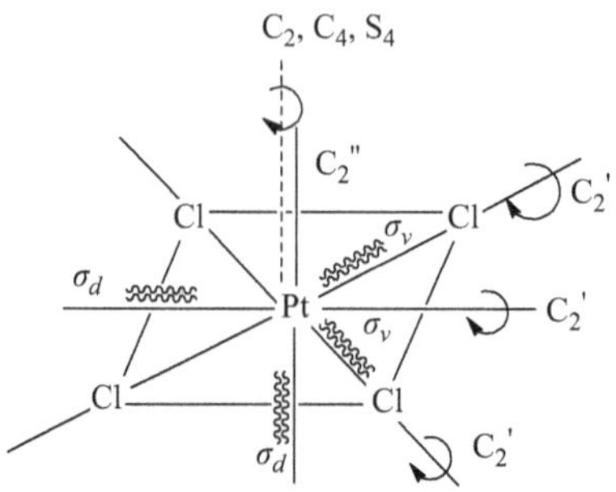

(a) Cartesian coordinate method or 3N vectors method

The symmetry operations in this case are **five** proper rotations through 0°(E), 90° (C_4), 180°(C_2, C_2', C_2'') and **five** improper rotations through 180°(i), 90°(S_4) and 0° (σ_h, σ_v, σ_d).The following table provides the calculations of the total characters.

D_{4h}	E	$2C_4$	C_2	$2C_2'$	$2C_2''$	i	$2S_4$	σ_h	$2\sigma_v$	$2\sigma_d$
β	0	90	180	180	180	180	90	0	0	0
Cosβ	1	0	−1	−1	−1	−1	0	1	1	1
±1 + 2Cos β	3	1	−1	−1	−1	−3	−1	1	1	1
n_R	5	1	1	3	1	1	1	5	3	1
Γ_{3N}	15	1	−1	−3	−1	−3	−1	5	3	1

The vertical dashed line separates the proper and improper rotations. The corresponding angle β, cosβ and also (±1 + 2cosβ) values are tabulated for each operation. The proper rotations have (+1 + 2cosβ) and the improper rotations have (−1 + 2cosβ) values. The number of atoms retaining position (n_R) for each operation is also given in the table.

The above reducible representation (Γ_{3N}) can be reduced to irreducible representations applying the reduction formula and using character table of D_{4h} point group as follows:

$$N_i = \frac{1}{h}\sum_R \chi(R).n.\chi_i(R)$$

Reducible representation in question

D_{4h}	E	$2C_4$	C_2	$2C_2'$	$2C_2''$	i	$2S_4$	σ_h	$2\sigma_v$	$2\sigma_d$
Γ_{3N}	15	1	−1	−3	−1	−3	−1	5	3	1

Character table of D_{4h} point group

D_{4h}	E	$2C_4$	C_2	$2C_2'$	$2C_2''$	i	$2S_4$	σ_h	$2\sigma_v$	$2\sigma_d$		
A_1g	1	1	1	1	1	1	1	1	1	1		$(x^2 + y^2 + z^2)$
A_2g	1	1	1	−1	−1	1	1	1	−1	−1	R_z	
B_1g	1	−1	1	1	−1	1	−1	1	1	−1		$(x^2 - y^2)$
B_2g	1	−1	1	−1	1	1	−1	1	−1	1		xy
Eg	2	0	−2	0	0	2	0	−2	0	0	(R_x, R_y)	
A_1u	1	1	1	1	1	−1	−1	−1	−1	−1		(xz, yz)
A_2u	1	1	1	−1	−1	−1	−1	−1	1	1	z	
B_1u	1	−1	1	1	−1	−1	1	−1	−1	1		
B_2u	1	−1	1	−1	1	−1	1	−1	1	−1		
Eu	2	0	−2	0	0	−2	0	2	0	0	(x,y)	

$$\begin{aligned}
NA_{1g} &= 1/16\,[(15.1.1) + (1.2.1)) + (-1.1.1) + (-3.2.1) + (-1.2.1) + (-3.1.1) + (-1.2.1) \\
&\quad + (5.1.1) + (3.2.1) + (1.2.1) \\
&= 1/16\,[(15 + 2 - 1 - 6 - 2 - 3 - 2 + 5 + 6 + 2] \\
&= 1/16\,[17 - 14 + 13] = 1/16[16] = 1
\end{aligned}$$

$$\begin{aligned}
NA_{2g} &= 1/16\,[(15.1.1) + (1.2.1)) + (-1.1.1) + (-3.2.-1) + (-1.2.-1) + (-3.1.1) + (-1.2.1) \\
&\quad + (5.1.1) + (3.2.-1) + (1.2.-1)] \\
&= 1/16\,[(15 + 2 - 1 + 6 + 2 - 3 - 2 + 5 - 6 - 2] \\
&= 1/16\,[30 - 14] = 1/16[16] = 1
\end{aligned}$$

$$\begin{aligned}
NB_{1g} &= 1/16\,[(15.1.1) + (1.2.-1)) + (-1.1.1) + (-3.2.1) + (-1.2.-1) + (-3.1.1) \\
&\quad + (-1.2.-1) + (5.1.1) + (3.2.1) + (1.2.-1)] \\
&= 1/16\,[(15 - 2 - 1 - 6 + 2 - 3 + 2 + 5 + 6 - 2] \\
&= 1/16\,[30 - 14] = 1/16[16] = 1
\end{aligned}$$

$$\begin{aligned}
NB_{2g} &= 1/16\,[(15.1.1) + (1.2.-1)) + (-1.1.1) + (-3.2.-1) + (-1.2.1) + (-3.1.1) \\
&\quad + (-1.2.-1) + (5.1.1) + (3.2.-1) + (1.2.1)] \\
&= 1/16\,[(15 - 2 - 1 + 6 - 2 - 3 + 2 + 5 - 6 + 2] \\
&= 1/16\,[30 - 14] = 1/16[16] = 1
\end{aligned}$$

$$\begin{aligned}
NE_g &= 1/16\,[(15.1.2)+(0))+(-1.1.-2)+(0)+(0)+(-3.1.2)+(0)\\
&\quad +(5.1.-2)+(0)+(0)]\\
&= 1/16\,[(30+2-6-10]\\
&= 1/16\,[32-16]=1/16[16]=1
\end{aligned}$$

$$\begin{aligned}
NA_{1u} &= 1/16\,[(15.1.1)+(1.2.1))+(-1.1.1)+(-3.2.1)+(-1.2.1)+(-3.1.-1)+(-1.2.-1)\\
&\quad +(5.1.-1)+(3.2.-1)+(1.2.-1)]\\
&= 1/16\,[(15+2-1-6-2+3+2-5-6-2]\\
&= 1/16\,[22-22]=1/16[0]=0
\end{aligned}$$

$$\begin{aligned}
NA_{2u} &= 1/16\,[(15.1.1)+(1.2.1))+(-1.1.1)+(-3.2.-1)+(-1.2.-1)+(-3.1.-1)\\
&\quad +(-1.2.-1)+(5.1.-1)+(3.2.1)+(1.2.1)]\\
&= 1/16\,[(15+2-1+6+2+3+2-5+6+2]\\
&= 1/16\,[38-6]=1/16[32]=2
\end{aligned}$$

$$\begin{aligned}
NB_{1u} &= 1/16\,[(15.1.1)+(1.2.-1))+(-1.1.1)+(-3.2.1)+(-1.2.-1)+(-3.1.-1)+(-1.2.1)\\
&\quad +(5.1.-1)+(3.2.-1)+(1.2.1)]\\
&= 1/16\,[(15-2-1-6+2+3-2-5-6+2]\\
&= 1/16\,[22-22]=1/16[0]=0
\end{aligned}$$

$$\begin{aligned}
NB_{2u} &= 1/16\,[(15.1.1)+(1.2.-1))+(-1.1.1)+(-3.2.-1)+(-1.2.1)+(-3.1.-1)+(-1.2.1)\\
&\quad +(5.1.-1)+(3.2.1)+(1.2.-1)]\\
&= 1/16\,[(15-2-1+6-2+3-2-5+6-2]\\
&= 1/16\,[30-14]=1/16[16]=1
\end{aligned}$$

$$\begin{aligned}
NE_u &- 1/16\,[(15.1.2)+(0))+(-1.1.-2)+(0)+(0)+(-3.1.-2)+(0)\\
&\quad +(5.1.2)+(0)+(0)]\\
&= 1/16\,[(30+2+6+10]\\
&= 1/16\,[48]=3
\end{aligned}$$

Thus, the reducible representation Γ_{3N} reduces to:

$$\Gamma_{3N}=A_{1g}+A_{2g}+B_{1g}+B_{2g}+E_g+2A_{2u}+B_{2u}+3E_u$$

The D_{4h} character table indicates that the translational modes have the same symmetry as the coordinates *x*, *y* and *z*, that is, E_u and A_{2u}. So also the rotational modes have the same symmetry as the rotation axes, R_x, R_y and R_z, that is, E_g and A_{2g}.

Subtracting these translational and rotational degrees of freedom, we get the vibrational degrees of freedom only of appropriate symmetry.

$$\begin{aligned}
\Gamma &= A_{1g} + A_{2g} + B_{1g} + B_{2g} + E_g + 2A_{2u} + B_{2u} + 3E_u \\
\Gamma_{tr} &= A_{2u} + E_u \\
\Gamma_{rot} &= A_{2g} + E_g \\
\hline
\Gamma_{vib} &= A_{1g} + B_{1g} + B_{2g} + A_{2u} + B_{2u} + 2E_u
\end{aligned}$$

Thus, out of nine vibrational modes [$1(A_{1g}) + 1(B_{1g}) + 1(B_{2g}) + 1(A_{2u}) + 1(B_{2u}) + 4(2E_u) = 9$], one belongs to A_{1g}, one belongs to B_{1g}, one belongs to B_{2g}, one belongs to A_{2u}, one belongs to B_{2u}, and four belong to E_u irreducible representation.

(b) Internal Coordinate Method

In the previous section, using internal coordinate method, we have worked out that

$$\begin{aligned}
\Gamma_{P-Cl} &= A_{1g} + B_{1g} + E_u \\
\Gamma_{ipd} &= B_{2g} + E_u \\
\Gamma_{opd} &= A_{2u} + B_{2u}
\end{aligned}$$

Now,

$$\begin{aligned}
\text{total irreducible representation} &= \Gamma_{P-Cl} + \Gamma_{ipd} + \Gamma_{opd} \\
&= (A_{1g} + B_{1g} + E_u) + (B_{2g} + E_u) + (A_{2u} + B_{2u}) \\
&= A_{1g} + B_{1g} + B_{2g} + A_{2u} + B_{2u} + 2E_u
\end{aligned}$$

This is equivalent to Γ_{vib} obtained from Cartesian coordinate or 3N vector method.

Identification of IR and Raman active vibrations

IR active vibrations

From the character table of D_{4h} point group, it is notable that x, y coordinates belong to E_u and z-coordinate belongs to A_{2u} irreducible representation. Hence, E_u and A_{2u} vibrational modes will be IR active. As $2E_u$ fundamental vibrations are two sets of two-fold degenerate vibrations, and A_{2u} vibrational mode is non-degenerate, one can expect only three frequencies in the IR spectra.

Raman active vibrations

From the character table of D_{4h} point group, it is notable that binary products $(x^2+ y^2)$, z^2 belong to A_{1g}, (x^2-y^2) belongs to B_{1g} and xy belongs to B_{2g} irreducible representation. Hence, A_{1g}, B_{1g} and B_{2g} vibrational modes will be IR active. As each of the three vibrational modes is non-degenerate vibration, Raman spectra of square planar molecules are expected to show only three frequencies. It is notable here that B_{2u} vibrational mode is inactive in both IR and Raman.

Since square planar AB_4 molecules of D_{4h} point group are centrosymmetric, all *g* modes are Raman active and all *u* modes are IR active. However, B_{2u} are inactive in both IR and Raman. The seven normal modes of vibration of AB_4 square planar compound can be classified into in-plane and out of plane, and these are shown below:

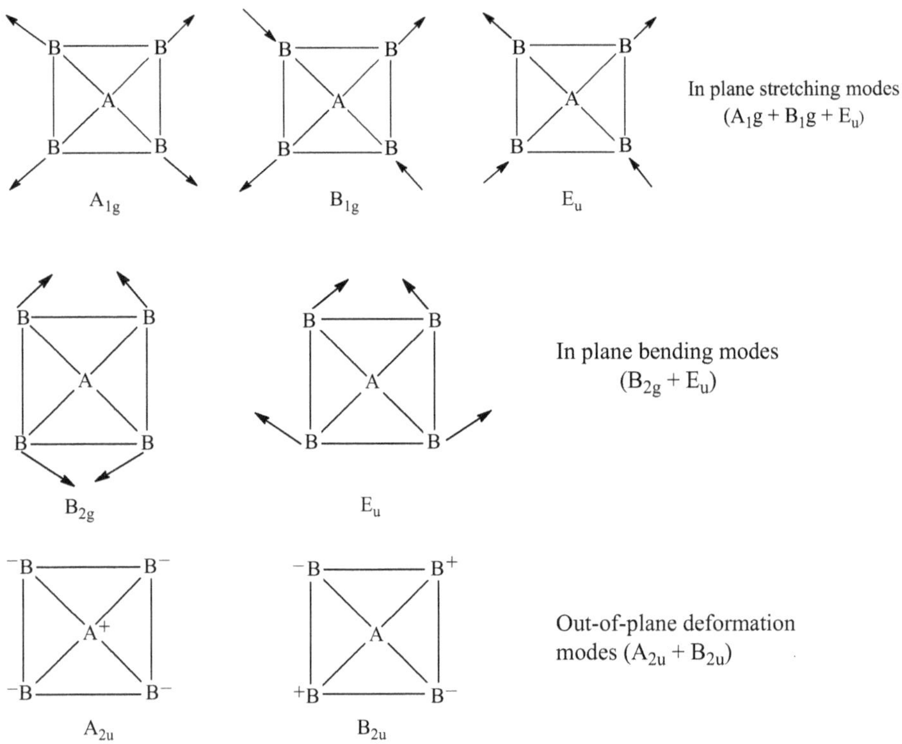

The vibrational spectral data of $PtCl_4^{2-}$ is given below.

Infrared (cm^{-1})	Raman (cm^{-1})
–	332 (pol.)
320	–
–	314 (depol.)
183	–
–	170 (depol.)
93	–

The above vibrational frequency data suggest that there is no common vibrational mode which is active both in IR and Raman. Therefore, the mutual exclusion rule holds well in the present case.

As the stretching modes have higher frequencies than deformation frequencies, the first three frequencies at 332, 320 and 314 cm^{-1}correspond to three Pt-Cl stretches having symmetries A_{1g}, B_{1g} and E_u. The frequency at 332 cm^{-1} is Raman polarized, so it is of A_{1g} type. The frequency 320 cm^{-1} is IR active only, so it is of E_u symmetry. The remaining mode at 314 cm^{-1} (Raman depolarized) corresponds to B_{1g} symmetry.

The rest three vibrational modes at 183, 170 and 93 cm^{-1} having symmetries B_{2g}, E_u, A_{2u}, B_{2u} correspond to angle of deformations. The frequency at 170 cm^{-1} is Raman depolarized, so it is of B_{2g} type due to in-plane deformation. Another frequency which is closed to this (170 cm^{-1}) is 183 cm^{-1}.This should also be due in plane deformation. The symmetry of this band is E_u(IR). The remaining frequency at 93 cm^{-1} is due to opd. The symmetry of this frequency is A_{2u} as B_{2u} is inactive both in IR and Raman spectrum.

(b) Tetrahedral AB_4 Molecules: CH_4, BH_4^-, NH_4^+, SiH_4, ClO_4^-, SO_4^{2-}, MnO_4^-, etc.

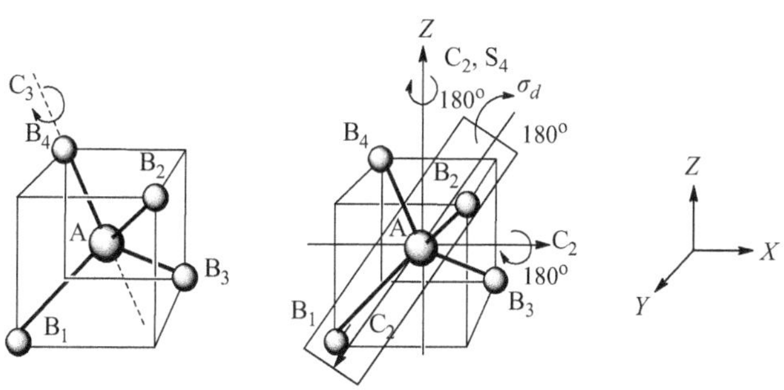

The total number of 24 symmetry operations in tetrahedral molecules of T_d point group is: 1E, $8C_3$, $3C_2$, $6S_4$, $6\sigma_d$.

Cartesian coordinate method or 3N vectors method

The symmetry operations in this case are three proper rotations through 0°(E), 120°(C_3) and 180°(C_2), and two improper rotations through 90°(S_4) and 0°(σ_d). The following table provides the calculations of the total characters.

Table: Determination of total character for AB_4 tetrahedral molecule

T_d	E	$8C_3$	$3C_2$	$6S_4$	$6\sigma_d$
β	0	120	180	90	0
cos β	1	$-\frac{1}{2}$	−1	0	1
±1 + 2cosβ	3	0	−1	−1	1
n_R	5	2	1	1	3
Γ_{3N}	**15**	**0**	**−1**	**−1**	**3**

The vertical dashed line separates the proper and improper rotations. The corresponding angleβ, cosβ and also (±1 + 2cosβ) values are tabulated for each operation. The proper rotations have (+1 + 2cosβ) and the improper rotations have (−1 + 2cosβ) values. The number of atoms retaining position (n_R) for each operation is also given in the table.

The above reducible representation (Γ_{3N}) can be reduced to irreducible representations using the reduction formula and character table of Td point group as follows:

$$N_i = \frac{1}{h}\sum_R \chi(R) \,.\, n \,.\, \chi_i(R)$$

Reducible representation in question

T_d	E	$8C_3$	$3C_2$	$6S_4$	$6\sigma_d$
Γ_{3N}	15	0	−1	−1	3

Character table of Td point group

T_d	E	$8C_3$	$3C_2$	$6S_4$	$6\sigma_d$		
A_1	1	1	1	1	1		$x^2 + y^2 + z^2$
A_2	1	1	1	−1	−1		
E	2	−1	2	0	0		$(2z^2-x^2-y^2, x^2-y^2)$
T_1	3	0	−1	1	−1	(R_x, R_y, R_z)	
T_2	3	0	−1	−1	1	(x, y, z)	(xy, yz, xz)

$$
\begin{aligned}
NA_1 &= 1/24\,[(15.1.1) + 0 + (-1.3.1) + (-1.6.1) + (3.6.1)] \\
&= 1/24\,[(15 + 0 - 3 - 6 + 18] \\
&= 1/24\,[24] = 1 \\
NA_2 &= 1/24\,[(15.1.1) + 0 + (-1.3.1) + (-1.6.-1) + (3.6.-1)] \\
&= 1/24\,[(15 + 0 - 3 + 6 - 18] \\
&= 1/24\,[0] = 0 \\
NE &= 1/24\,[(15.1.2) + 0 + (-1.3.2) + (0) + (0)] \\
&= 1/24\,[(30 + 0 - 6 + 0 + 0] \\
&= 1/24\,[24] = 1 \\
NT_1 &= 1/24\,[(15.1.3) + (0) + (-1.3.-1) + (-1.6.1) + (3.6.-1)] \\
&= 1/24\,[(45 + 0 + 3 - 6 - 18] \\
&= 1/24\,[24] = 1 \\
NT_2 &= 1/24\,[(15.1.3) + (0) + (-1.3.-1) + (-1.6.-1) + (3.6.1)] \\
&= 1/24\,[(45 + 0 + 3 + 6 + 18] \\
&= 1/24\,[72] = 3
\end{aligned}
$$

Thus, the above reducible representation is reduced to:

$$\Gamma_{\mathbf{3N}} = A_1 + E + T_1 + 3T_2$$

Thus, there are 15 modes (1 + 2 + 3 + 9 = 15) of degrees of freedom with different symmetries, A_1, E, T_1 and T_2. These include the vibrational, rotational and translational degrees of freedom.

The T_d character table indicates that the translational modes have the same symmetry as the coordinates x, y and z, that is, T_2. So also the rotational modes have the same symmetry as the rotation axes, R_x, R_y and R_z, that is, T_1. Subtracting these translational and rotational degrees of freedom, we get the vibrational degrees of freedom only of appropriate symmetry.

$$
\begin{aligned}
&\Gamma_{3N} = A_1 + E + T_1 + 3T_2 \\
&\Gamma_{tr} \qquad\qquad\qquad\quad T_2 \\
&\Gamma_{rot} = \qquad\quad T_1 \\
\hline
&T_{vib} = A_1 + E + 2T_2
\end{aligned}
$$

Thus, out of nine vibrational modes [$1(A_1) + 2(E) + 6(T_2) = 9$], one belongs to A_1, two belong to E and six belong to T_2 irreducible representation.

Internal coordinate method

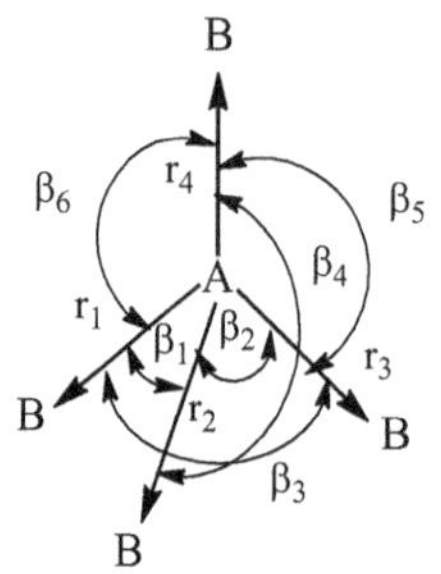

There are four A–B stretches and six angle deformations as shown in the Fig. given above. We can use these stretches and bending vectors as bases for representation to develop $\Gamma_{A\text{-}B}$ Γ_{bend} as follows:

$\Gamma_{A\text{-}B}$:

T_d	E	$8C_3$	$3C_2$	$6S_4$	$6\sigma_d$
No. of unshifted A-B stretch (n)	4	1	0	0	2
$\chi(R)$	+ 1	+ 1	+ 1	+ 1	+ 1
$\Gamma_{A\text{-}B}$ $[n\times\chi(R)]$	4	1	0	0	2

The reducible representation $\Gamma_{A\text{-}B}$ can be reduced to irreducible representation applying standard reduction formula and using character table of T_d point group.

Reducible Representation in question

T_d	E	$8C_3$	$3C_2$	$6S_4$	$6\sigma_d$
$\Gamma_{A\text{-}B}$	4	1	0	0	2

Character table of Td point group

T_d	E	$8C_3$	$3C_2$	$6S_4$	$6\sigma_d$		
A_1	1	1	1	1	1		$x^2+y^2+z^2$
A_2	1	1	1	−1	−1		
E	2	−1	2	0	0		$(2z^2-x^2-y^2, x^2-y^2)$
T_1	3	0	−1	1	−1	(Rx, Ry, Rz)	
T_2	3	0	−1	−1	1	(x, y, z)	(xy, yz, xz)

$$\begin{aligned} NA_1 &= 1/24\,[(4.1.1) + (1.8.1) + (0) + (0) + (2.6.1)] \\ &= 1/24\,[(4 + 8 + 0 + 0 + 12] \\ &= 1/24\,[24] = 1 \end{aligned}$$

$$\begin{aligned} NA_2 &= 1/24\,[(4.1.1) + (1.8.1) + (0) + (0) + (2.6.-1)] \\ &= 1/24\,[(4 + 8 + 0 + 0 - 12] \\ &= 1/24\,[0] = 0 \end{aligned}$$

$$\begin{aligned} NE &= 1/24\,[(4.1.2) + (1.8.-1) + (0) + (0) + (0)] \\ &= 1/24\,[(8 - 8 + 0 + 0 + 0] \\ &= 1/24\,[0] = 0 \end{aligned}$$

$$\begin{aligned} NT_1 &= 1/24\,[(4.1.3) + (0) + (0) + (0) + (2.6.-1)] \\ &= 1/24\,[(12 + 0 + 0 + 0 - 12] \\ &= 1/24\,[0] = 0 \end{aligned}$$

$$\begin{aligned} NT_2 &= 1/24\,[(4.1.3) + (0) + (0) + (0) + (2.6.1)] \\ &= 1/24\,[(12 + 0 + 0 + 0 + 12] \\ &= 1/24\,[24] = 1 \end{aligned}$$

Thus, the above reducible representation is reduced to:

$$\Gamma_{A-B} = A_1 + T_2$$

Γ_{bend}:

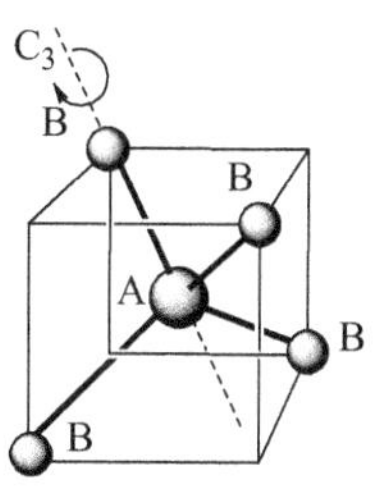

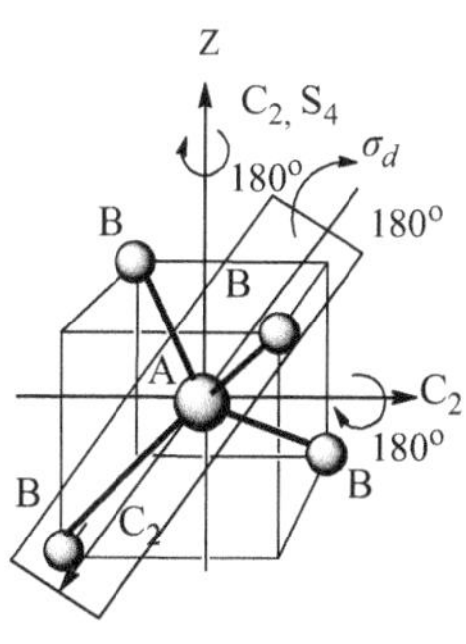

The symmetry operation E does not shift all the six angles, so $n = 6$. C_3 axis shifts all the angles, so $n = 0$. Cartesian axes, x, y and z bisecting **two** BAB bond angles, is the C_2 axis. So symmetry operation C_2 transforms half BAB bond angle into another half. So, it is also not changed. Thus, two angles are not changed. Symmetry operation S_4 shifts all the six angles, so $n = 0$. For σ_d, two BAB angles are not changed as these are contained in σ_d. So, $n = 2$.

T_d	E	$8C_3$	$3C_2$	$6S_4$	$6\sigma_d$
No. of unshifted double headed arrow vectors (*n*)	6	0	2	0	2
$\chi(R)$	+1	+1	+1	+1	+1
Γ_{bend} $[n\times\chi(R)]$	6	0	2	0	2

Again, this reducible representation Γ_{bend} can be reduced to irreducible representation applying standard reduction formula and using character table of T_d point group

$$N_i = \frac{1}{h}\sum_R \chi(R).n.\chi_i(R)$$

Reducible representation in question

T_d	E	$8C_3$	$3C_2$	$6S_4$	$6\sigma_d$
Γ_{A-B}	6	0	2	0	2

Character table of Td point group

T_d	E	$8C_3$	$3C_2$	$6S_4$	$6\sigma_d$		
A_1	1	1	1	1	1		$x^2+y^2+z^2$
A_2	1	1	1	−1	−1		
E	2	−1	2	0	0		$(2z^2-x^2-y^2, x^2-y^2)$
T_1	3	0	−1	1	−1	(R_x, R_y, R_z)	
T_2	3	0	−1	−1	1	(x,y,z)	(xy,yz,xz)

$$\begin{aligned} NA_1 &= 1/24\,[(6.1.1) + (0) + (2.3.1) + (0) + (2.6.1)] \\ &= 1/24\,[(6+0+6+0+12] \\ &= 1/24\,[24] = 1 \end{aligned}$$

$$\begin{aligned} NA_2 &= 1/24\,[(6.1.1) + (0) + (2.3.1) + (0) + (2.6.-1)] \\ &= 1/24\,[(6+0+6+0-12] \\ &= 1/24\,[0] = 0 \end{aligned}$$

$$\begin{aligned} NE &= 1/24\,[(6.1.2) + (0) + (2.3.2) + (0) + (0)] \\ &= 1/24\,[(12+0+12+0+0] \\ &= 1/24\,[24] = 1 \end{aligned}$$

$$
\begin{aligned}
NT_1 &= 1/24\,[(6.1.3) + (0) + (2.3.-1) + (0) + (2.6.-1)] \\
&= 1/24\,[(18 + 0 - 6 + 0 - 12] \\
&= 1/24\,[0] = 0
\end{aligned}
$$

$$
\begin{aligned}
NT_2 &= 1/24\,[(6.1.3) + (0) + (2.3.-1) + (0) + (2.6.1)] \\
&= 1/24\,[(18 + 0 - 6 + 0 + 12] \\
&= 1/24\,[24] = 1
\end{aligned}
$$

Thus, the above reducible representation is reduced to:

$$\Gamma_{bend} = A_1 + E + T_2$$

Now, total irreducible representation

$$
\begin{aligned}
\Gamma_{vib} &= \Gamma_{A-B} + \Gamma_{bend} \\
&= (A_1 + T_2) + (A_1 + E + T_2) \\
&= 2A_1 + E + 2T_2
\end{aligned}
$$

This is not the same result as we have obtained through Cartesian coordinate method but having one excess mode A_1.

Now, we have to decide which redundant mode is. This can be obtained on the basis that how many internal coordinates change independently. In the present case, it is possible for all the A-B bonds to change independently, but it is not possible for all the six bond angles β to change independently. If any five bond angles is arbitrarily changed then the change in sixth bond angle is automatically fixed. Thus, we conclude that A_1 species in Γ_{bend} is redundant because for the totally symmetric vibration A_1, all the six bond angles β should be changed in the same way at the same time which is impossible. Thus, this surplus A_1 irreducible representation can be neglected. So, the only deformation modes

$$\Gamma_{bend} = (A_1 + E + T_2) - A_1 = E + T_2$$

Therefore,

$$
\begin{aligned}
\Gamma_{vib} = \Gamma_{A-B} + \Gamma_{bend} &= (A_1 + T_2) + (E + T_2) \\
&= A_1 + E + 2T_2
\end{aligned}
$$

Identification of IR active vibrations

IR active vibrations

From the character table of T_d point group, it is notable that *x*, *y*, *z*-coordinates belong to T_2 irreducible representation. Hence, T_2 vibrational modes will be IR active. As $2T_2$ fundamental vibrations are two sets of three-fold degenerate vibrations, we can expect only two frequencies in the IR spectra.

Raman active vibrations

From the character table of T_d point symmetry, it is notable that, the A_1, E and T_2 vibrations are Raman active since (x^2, y^2, z^2) transforms as A_1, ($2z^2$-x^2-y^2, x^2-y^2) transforms as E, and (xy, xz, yz) transforms as T_2 irreducible representation. As A_1 is singly degenerate vibration and E is a set of two-fold degenerate vibrations while $2T_2$ are two sets of three old degenerate vibrations, Raman spectra of tetrahedral molecules are expected to give only four frequencies.

IR and Raman active vibrational modes may be summarized as:

Mode	Infrared	Raman	Nature of vibration
A_1	Inactive	Active	Pure stretching
E	Inactive	Active	Pure bending
T_2	Active	Active	Mixed

The vibrational modes in tetrahedral AB_4 molecule are shown below:

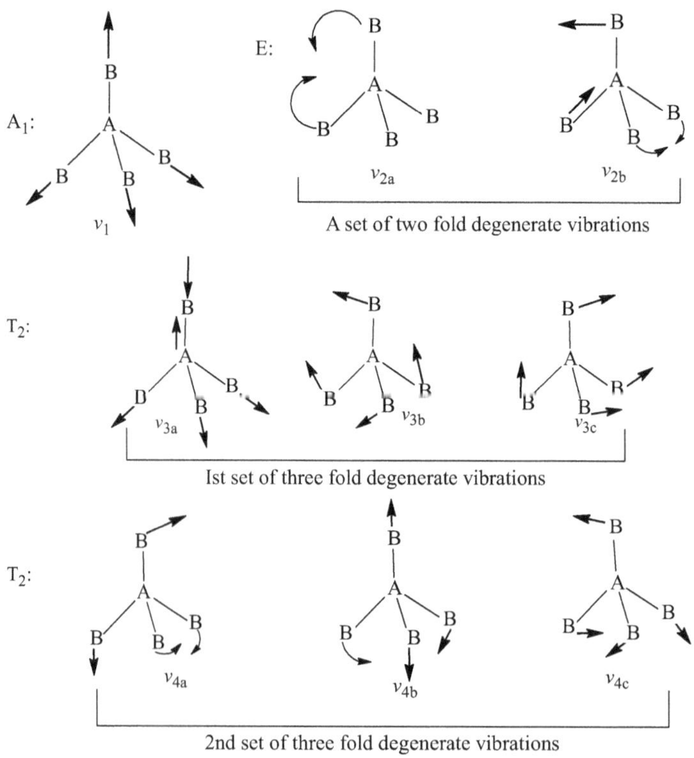

Thus, for the molecules belonging to T_d point group vibrational activity in IR and Raman spectra may be summarized as:

Point group	IR active	Raman active	Polarized	No. of coincidences
T_d	02	04	01	02
	$(2T_2)$	$(A_1 + E + 2T_2)$	A_1	$2T_2$
	ν_3, ν_4	$\nu_1, \nu_2, \nu_3, \nu_4$	ν_1	ν_3, ν_4

(c) See-Saw Shaped AB_4 Molecules

Molecules such as SF_4, XeO_2F_2, SeF_4, etc. have see-saw shaped structure, and belong to C_{2v} point group. The possible symmetry operations in these molecules are: E, C_2, $\sigma_v(xz)$, $\sigma_v(yz)$.

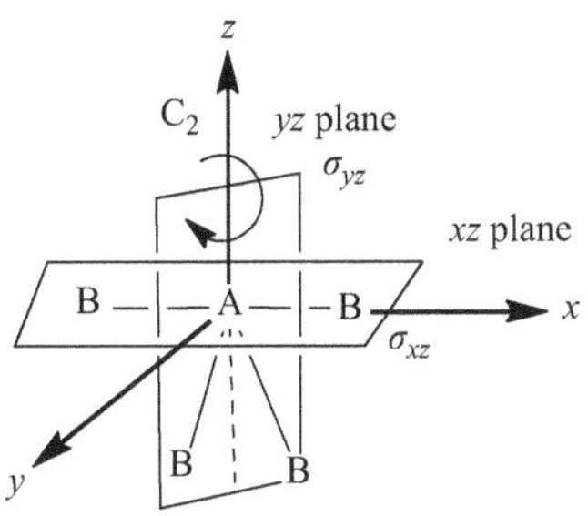

Cartesian coordinate method or 3N vectors method

We will consider here Cartesian coordinate method only. The symmetry operations in such case are two proper rotations through 0°(E) and 180°(C_2) and two improper rotations through 0°[$\sigma_v(xz)$] and 0°[$\sigma_v(yz)$].The following table provides the calculations of the total characters.

C_{2v}	E	C_2	σ_{yz}	σ_{xz}
β	0	180	0	0
Cos β	1	−1	1	1
±1 +2Cosβ	3	−1	1	1
n_R	5	1	3	3
Γ_{3N}	15	−1	3	3

The vertical dashed line separates the proper and improper rotations. The corresponding angleβ, cosβ and also (±1 + 2cosβ) values are tabulated for each operation. The proper rotations have (+1 + 2cosβ) and the improper rotations have (−1 + 2cosβ) values. The number of atoms retaining position (n_R) for each operation is also given in the table.

The above reducible representation (Γ_{3N}) can be reduced to irreducible representations using the reduction formula and character table of C_{2v} point group as follows:

$$N_i = \frac{1}{h}\sum_R \chi(R) . n . \chi_i (R)$$

where N_i is the number of times the *i*th irreducible representation occurs in a reducible representation, h is the order of group (number of symmetry operations), $\chi(R)$ is the character of a particular operation in the reducible representation, n is the number of operation of that type and $\chi_i(R)$ is the character of the same operation in the irreducible representation.

Reducible representation in question

C_{2v}	E	C_2	$\sigma_{v(yz)}$	$\sigma_{v(xz)}$
Γ_{3N}	15	−1	3	3

Character table of C_{2v} point symmetry

C_{2v}	E	C_2	$\sigma_{v(yz)}$	$\sigma_{v(xz)}$		
A_1	1	1	1	1	z	x^2, y^2, z^2
A_2	1	1	−1	−1	R_z	xy
B_1	1	−1	1	−1	x, R_y	xz
B_2	1	−1	−1	1	y, R_x	yz

$$\begin{aligned}
NA_1 &= 1/4\,[(15.1.1) + (-1.1.1) + (3.1.1) + (3.1.1)\\
&= 1/4\,[15 - 1 + 3 + 3]\\
&= 1/4\,[20] = 5\\
NA_2 &= 1/4\,[(15.1.1) + (-1.1.1) + (3.1.-1) + (3.1.-1)\\
&= 1/4\,[15 - 1 - 3 - 3]\\
&= 1/4\,[8] = 2\\
NB_1 &= 1/4\,[(15.1.1) + (-1.1.-1) + (3.1.1) + (3.1.-1)\\
&= 1/4\,[15 + 1 + 3 - 3]\\
&= 1/4\,[16] = 4\\
NB_2 &= 1/4\,[(15.1.1) + (-1.1.-1) + (3.1.-1) + (3.1.1)\\
&= 1/4\,[15 + 1 - 3 + 3]\\
&= 1/4\,[16] = 4
\end{aligned}$$

Thus, the total representation Γ_{3N} reduces to:

$$\Gamma_{3N} = 5A_1 + 2A_2 + 4B_1 + 4B_2$$

Thus, there are 15 modes (5 + 2 + 4 + 4 = 15) of degrees of freedom with different symmetries, A_1, A_2, B_1 and B_2. These include the vibrational, rotational and translational degrees of freedom.

The C_{2v} character table indicates that the translational modes have the same symmetry as the coordinates *x*, *y* and *z*, that is, B_1, B_2 and A_1. So also the rotational modes have the same symmetry as the rotation axes, R_x, R_y and R_z, that is, B_1, B_2 and A_2. Subtracting these translational and rotational degrees of freedom, we get the vibrational degrees of freedom only of appropriate symmetry.

$$\begin{aligned}
\Gamma_{3N} &= 5A_1 + 2A_2 + 4B_1 + 4B_2 \\
\Gamma_{tr} &= A_1 + \qquad\quad B_1 + B_2 \\
\Gamma_{rot} &= \qquad\quad A_2 + B_1 + B_2 \\
\hline
\Gamma_{vib} &= 4A_1 + A_2 + 2B_1 + 2B_2
\end{aligned}$$

Thus, out of nine vibrational modes $[4(A_1) + 1(A_2) + 2(B_1) + 2(B_2) = 9]$, one belongs to A_1, one belongs to A_2 and two belong to B_1 and two belong to B_2 irreducible representation.

Identification of IR active vibrations

IR active vibrations

From the character table of C_{2v} point group, it is clear that *x*-coordinate belongs to B_1, *y*-coordinate belongs to B_2 and *z*-coordinate belongs to A_1 irreducible representation. Hence, $4A_1$, $2B_1$ and $2B_2$ vibrational modes will be IR active. As these fundamental vibrations are non-degenerate vibrations, one can expect only eight frequencies in the IR spectra.

Raman active vibrations

An observation of the character table (C_{2V} point group) indicates that the binary products of the Cartesian coordinates belong to all the four irreducible representations (A_1, A_2, B_1 and B_2). Hence, all these vibrational modes are Raman active and we expect nine frequencies in the Raman spectrum.

IR and Raman active vibrational modes may be summarized as follows:

Mode	Infrared	Raman
$4A_1$	Active	Active
A_2	Inactive	Active
$2B_1$	Active	Active
$2B_2$	Active	Active

(d) Octahedral AB_6 Type of Molecules

Octahedral molecules such as SF_6, ML_6, etc. belong to O_h point symmetry and the possible symmetry elements/operations in such molecules are: ***E*, $6C_4$, $6S_4$, $3C_2$, $6C_2'$, $8C_3$, $8S_6$, $3\sigma_h$, $6\sigma_d$, *i*.**

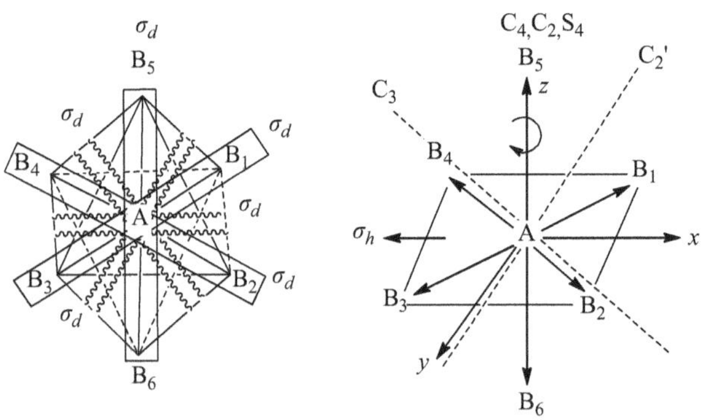

Cartesian coordinate method or 3N vector method

The symmetry operations in such molecules are **five** proper rotations through 0° (*E*), 90°(C_4), 120° (C_3), 180°(C_2) and 180°(C_2') and **five** improper rotations through 0°(σ_h), 0°(σ_d), 60°(S_6), 90°(S_4) and 180°(*i*). The following table provides the calculations of the total characters.

O_h	E	$8C_3$	$3C_2$	$6C_4$	$6C_2'$	i	$6S_4$	$8S_6$	$3\sigma_h$	$6\sigma_d$
β	0	120°	180°	90°	180°	180°	90°	60°	0°	0°
cosβ	1	−1/2	−1	0	−1	−1	0	1/2	1	1
±1+2cosβ	3	0	−1	1	−1	−3	−1	0	1	1
n_R	7	1	3	3	1	1	1	1	5	3
Γ_{3N}	21	0	−3	3	−1	−3	−1	0	5	3

The vertical dashed line separates the proper and improper rotations. The corresponding angleβ, cosβ and also (± 1 + 2cosβ) values are tabulated for each operation. The proper rotations have (+1 + 2cosβ) and the improper rotations have (-1 + 2cosβ) values. The number of atoms retaining position (n_R) for each operation is also given in the table.

The above reducible representation (Γ_{3N}) can be reduced to irreducible representations using the standard reduction formula and the character table of O_h point group as follows:

$$N_i = \frac{1}{h}\sum_R \chi(R) . n . \chi_i (R)$$

O_h	E	$8C_3$	$3C_2$	$6C_4$	$6C_2'$	i	$6S_4$	$8S_6$	$3\sigma_h$	$6\sigma_d$
Γ_{3N}	21	0	−3	3	−1	−3	−1	0	5	3

Character Table of O_h Point Group

O_h	E	$8C_3$	$3C_2$	$6C_4$	$6C_2'$	i	$6S_4$	$8S_6$	$3\sigma_h$	$6\sigma_d$		
A_{1g}	1	1	1	1	1	1	1	1	1	1		$x^2+y^2+z^2$
A_{2g}	1	1	1	−1	−1	1	−1	1	1	−1		
E_g	2	−1	2	0	0	2	0	−1	2	0		$(2z^2-x^2-y^2, x^2-y^2)$
T_{1g}	3	0	−1	1	−1	3	1	0	−1	−1	(R_x, R_y, R_z)	
T_{2g}	3	0	−1	−1	1	3	−1	0	−1	1		(xz, yz, xy)
A_{1u}	1	1	1	1	1	−1	−1	−1	−1	−1		
A_{2u}	1	1	1	−1	−1	−1	1	−1	−1	1		
E_u	2	−1	2	0	0	−2	0	1	−2	0		
T_{1u}	3	0	−1	1	−1	−3	−1	0	1	1	(x, y, z)	
T_{2u}	3	0	−1	−1	1	−3	1	0	1	−1		

$$NA_{1g} = 1/48\,[(21.1.1) + 0 + (-3.3.1) + (3.6.1) + (-1.6.1) + (-3.1.1) + (-1.6.1)$$
$$+ 0 + (5.3.1) + (3.6.1)$$
$$= 1/48\,[(21 - 9 + 18 - 6 - 3 - 6 + 15 + 18)]$$
$$= 1/48\,[72 - 24 = 48] = 1$$

$$NA_{2g} = 1/48\,[(21.1.1) + 0 + (-3.3.1) + (3.6.-1) + (-1.6.-1) + (-3.1.1) + (-1.6.-1)$$
$$+ 0 + (5.3.1) + (3.6.-1)$$
$$= 1/48\,[(21 - 9 - 18 + 6 - 3 + 6 + 15 - 18)]$$
$$= 1/48\,[48 - 48 = 0] = 0$$

$$E_g = 1/48\,[(21.1.2) + 0 + (-3.3.2) + 0 + 0 + (-3.1.2) + 0 + 0 + (5.3.2) + 0$$
$$= 1/48\,[42 - 18 - 6 + 30 = 48] = 1$$

$$NT_{1g} = 1/48\,[(21.1.3) + 0 + (-3.3.-1) + (3.6.1) + (-1.6.-1) + (-3.1.3) + (-1.6.1) + 0 + (5.3.-1) + (3.6.-1)$$

$$= 1/48\,[(63 + 9 + 0 + 18 + 6 - 9 - 6 + 0 - 15 - 18)]$$

$$= 1/48\,[96 - 48 = 48] = 1$$

$$NT_{2g} = 1/48\,[(21.1.3) + 0 + (-3.3.-1) + (3.6.-1) + (-1.6.1) + (-3.1.3) + (-1.6.-1) + 0 + (5.3.-1) + (3.6.-1)$$

$$= 1/48\,[(63 + 9 + 0 - 18 - 6 - 9 + 6 + 0 - 15 + 18)]$$

$$= 1/48\,[96 - 48 = 48] = 1$$

$$NA_{1u} = 1/48\,[(21.1.1) + 0 + (-3.3.1) + (3.6.1) + (-1.6.1) + (-3.1.-1) + (-1.6.-1) + 0 + (5.3.-1) + (3.6.-1)$$

$$= 1/48\,[(21 - 9 + 18 - 6 + 3 + 6 - 15 - 18)]$$

$$= 1/48\,[48 - 48 = 0] = 0$$

$$NA_{2u} = 1/48\,[(21.1.1) + 0 + (-3.3.1) + (3.6.-1) + (-1.6.-1) + (-3.1.-1) + (-1.6.1) + 0 + (5.3.-1) + (3.6.-1)$$

$$= 1/48\,[(21 - 9 - 18 + 6 + 3 - 6 - 15 + 18)]$$

$$= 1/48\,[48 - 48 = 0] = 0$$

$$E_u = 1/48\,[(21.1.2) + 0 + (-3.3.2) + 0 + 0 + (-3.1.-2) + 0 + 0 + (5.3.-2) + 0$$

$$= 1/48\,[42 - 18 + 6 - 30 = 0] = 0$$

$$NT_{1u} = 1/48\,[(21.1.3) + 0 + (-3.3.-1) + (3.6.1) + (-1.6.-1) + (-3.1.-3) + (-1.6.-1) + 0 + (5.3.1) + (3.6.1)$$

$$= 1/48\,[(63 + 9 + 0 + 18 + 6 + 9 + 6 + 0 + 15 + 18)]$$

$$= 1/48\,[144] = 3$$

$$NT_{2u} = 1/48\,[(21.1.3) + 0 + (-3.3.-1) + (3.6.-1) + (-1.6.1) + (-3.1.-3) + (-1.6.1) + 0 + (5.3.1) + (3.6.-1)$$

$$= 1/48\,[(63 + 9 + 0 - 18 - 6 + 9 - 6 + 0 + 15 - 18)]$$

$$= 1/48\,[96 - 48 = 48] = 1$$

Thus, the total representation Γ_{3N} reduces to:

$$T = A_{1g} + E_g + T_{1g} + T_{2g} + 3T_{1u} + T_{2u}(3N)$$

Thus, there are 21 modes (1 + 2 + 3 + 3 + 9 + 3 = 21) of degrees of freedom with different symmetry. These include the vibrational, rotational and translational degrees of freedom.

The O_h character table indicates that the translational modes have the same symmetry as the coordinates *x*, *y* and *z*, that is, T_{1u}. So also the rotational modes have the same symmetry as the rotation axes, R_x, R_y and R_z, that is, T_{1g}. Subtracting these translational and rotational degrees of freedom, we get the vibrational degrees of freedom only of appropriate symmetry.

$$\begin{aligned}
\Gamma_{3N} &= A_{1g} + E_g + T_{1g} + T_{2g} + 3T_{1u} + T_{2u} \\
\Gamma_{tr} &= \qquad\qquad\qquad\qquad\quad + T_{1u} \\
\Gamma_{rot} &= \qquad\quad + T_{1g} \\
&\cdots\cdots\cdots\cdots\cdots\cdots\cdots\cdots \\
\Gamma_{vib} &= A_{1g} + E_g + T_{2g} + 2T_{1u} + T_{2u}
\end{aligned}$$

Internal coordinate method

In the previous section, using internal coordinate method, we have worked out that

$$\begin{aligned}
\Gamma_{A-B} &= A_{1g} + E_g + T_{1u} \\
\Gamma_{ipd+opd} &= T_{2g} + T_{1u} + T_{2u}
\end{aligned}$$

Now, total irreducible representation $\Gamma_{vib} = \Gamma_{A-B} + \Gamma_{ipd} + \Gamma_{opd}$

$$\begin{aligned}
&= (A_{1g} + E_g + T_{1u}) + (T_{2g} + T_{1u} + T_{2u}) \\
&= A_{1g} + E_g + T_{2g} + 2T_{1u} + T_{2u}
\end{aligned}$$

This is equivalent to Γ_{vib} obtained from Cartesian coordinate or 3N vector method.

The vibrational modes of different symmetry species for AB_6 octahedral molecule belonging to O_h point group are given in the following Fig.

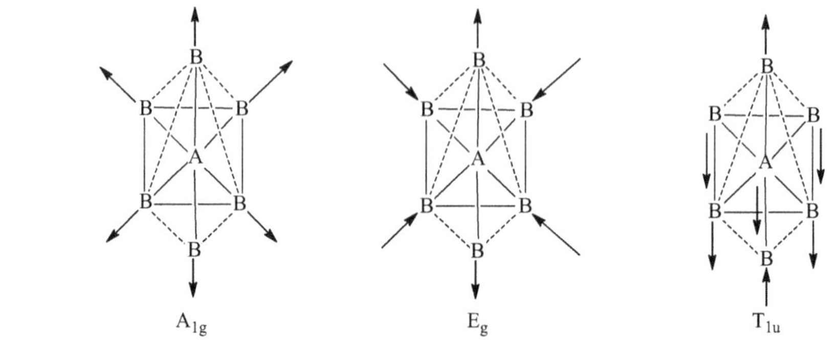

Stretchingmodes(A_{1g}, E_g,T_{1u})

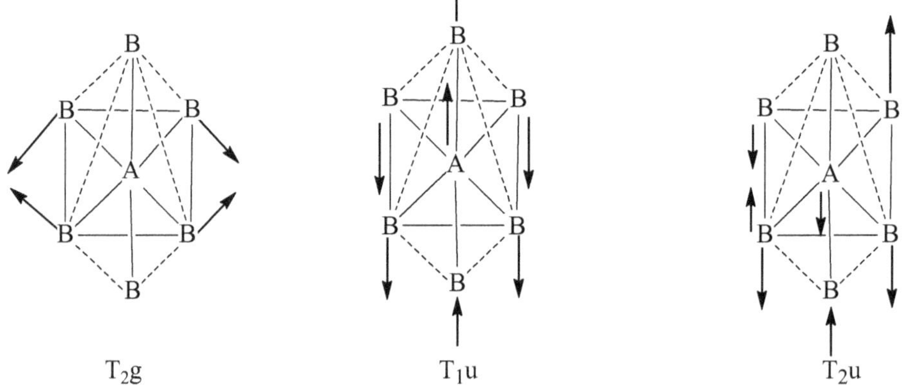

Bending modes (T_{2g},T_{1u},T_{2u})

Identification of IR active vibrations

IR active vibrations

From the character table of O_h point group, it is clear that *x*, *y*, *z*-coordinates belong to T_{1u} irreducible representation. Hence, $2T_{1u}$ vibrational modes will be IR active. As $2T_{1u}$ fundamental vibrations are two sets of triply degenerate vibrations, one can expect only two frequencies in the IR spectra.

Raman active vibrations

An observation of the character table of O_h point group indicates that the binary products of the Cartesian coordinates belong to irreducible representations A_{1g}, E_g, and T_{2g}. Hence A_{1g}, E_g, and T_{2g} vibrational modes are Raman active. As E_g fundamental vibrations are a set of doubly degenerate vibrations, T_{2g} is a set of triply degenerate vibrations and A_{1g} is a non-degenerate vibration, one can expect only 3 (1 + 1 + 1 = 3) frequencies in the Raman spectra.

Since AB_6 is an octahedral molecule with a center of symmetry, all those vibrational modes which are "*g*" type are Raman active and the others with "*u*" symmetry are infrared active. T_{2u} vibrations are both Raman and IR inactive.

IR and Raman active vibrational modes may be summarized as:

Mode	Infrared	Raman	Nature of vibration
A_{1g}	Inactive	Active	Pure stretching
E_g	Inactive	Active	Pure stretching
T_{2g}	Inactive	Active	Pure bending (angle of deformation)
$2T_{1u}$	Active	Inactive	Mixed (stretching and bending)
T_{2u}	Inactive	Inactive	Pure bending (angle of deformation)

Form the above table, it is notable that the mutual exclusion rule is satisfied and hence molecule possesses a center of symmetry. Also, this is an example where a vibrational mode is inactive in both IR and Raman.

(e) *trans*-N_2F_2 Molecule of C_{2h} Point Group

This molecule has a planar nonlinear structure and it belongs to C_{2h} point group. The possible number of symmetry operations is: E, C_2, i, σ_h.

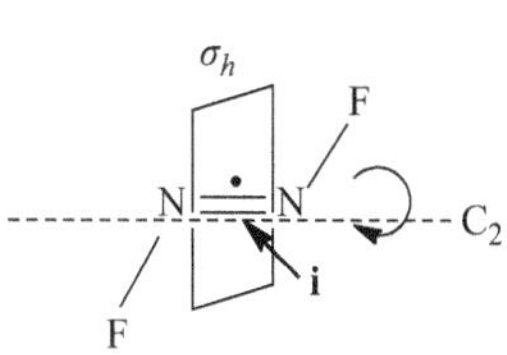

Cartesian coordinate method or 3N vector method

Out of four symmetry operations, E and C_2 are proper rotations through 0° and 180°, respectively, while i and σ_h are improper rotations through 180° and 0° (σ_h), respectively. The following table provides the calculations of the total characters.

Total character of *trans*-N_2F_2 molecule:

C_{2h}	E	C_2	i	σ_h
β	0	180	180	0
cos β	1	−1	−1	1
±1+2cos β	3	−1	−3	1
n_R	4	0	0	4
Γ_{3N}	**12**	**0**	**0**	**4**

The vertical dashed line separates the proper and improper rotations. The corresponding angleβ, cosβ and also (±1 + 2cosβ) values are tabulated for each operation. The proper rotations have (+1 + 2cosβ) and the improper rotations have (-1 + 2cosβ) values. The number of atoms retaining position (n_R) for each operation is also given in the table.

The above reducible representation (Γ_{3N}) can be reduced to irreducible representations using the standard reduction formula and character table of C_{2v} point group as follows:

$$N_i = \frac{1}{h}\sum_R \chi(R).n.\chi_i(R)$$

Reducible representation

C_{2h}	E	C_2	i	σ_h
Γ_{3N}	12	0	0	4

Character table of C_{2v} point symmetry

C_{2h}	E	C_2	i	σ_h		
A_g	1	1	1	1	R_z	x^2,y^2,z^2, xy
B_g	1	−1	1	−1	R_x, R_y	xz,yz
Au	1	1	−1	−1	z	
B_u	1	−1	−1	1	x, y	

$$\begin{aligned} NA_g &= 1/4\ [(12.1.1) + 0 + 0 + (4.1.1)] \\ &= 1/4\ [12 + 4 = 16] = 4 \end{aligned}$$

$$\begin{aligned} NB_g &= 1/4[(12.1.1) + 0 + 0 + (4.1.-1)] \\ &= 1/4\ [12 - 4 = 8] = 2 \end{aligned}$$

$$\begin{aligned} NA_u &= 1/4\ [(12.1.1) + 0 + 0 + (4.1.-1)] \\ &= 1/4\ [12 - 4 = 8] = 2 \end{aligned}$$

$$\begin{aligned} NB_u &= 1/4\ [(12.1.1) + 0 + 0 + (4.1.1)] \\ &= 1/4\ [12 + 4 = 16] = 4 \end{aligned}$$

Thus, the reducible representation reduces to:

$$T = 4A_g + 2B_g + 2A_u + 4B_u$$

Thus, there are 12 modes (4 + 2 + 2 + 4 = 12) of degrees of freedom with different symmetries, A_g, B_g, A_u and B_u. These include the vibrational, rotational and translational degrees of freedom.

The character table indicates that the translational modes have the same symmetry as the coordinates x, y and z, that is, A_u and $2B_u$. So also the rotational modes have the same symmetry as the rotation axes, R_x, R_y and R_z, that is, A_g and $2B_g$. Subtracting these translational and rotational degrees of freedom, we get the vibrational degrees of freedom only of appropriate symmetry.

$$\begin{aligned} \Gamma_{3N} &= 4A_g + 2B_g + 2A_u + 4B_u \\ \Gamma_{tr} &= \qquad\qquad\qquad A_u + 2B_u \\ \Gamma_{rot} &= A_g + 2B_g \\ &\cdots\cdots\cdots\cdots\cdots\cdots \\ T_{vib} &= 3A_g + A_u + 2B_u \end{aligned}$$

Thus, out of six vibrational modes [3 (A_g) + 1(A_u) + 2 (B_u)] = 6], thee belongs to A_g, one belongs to A_u and two belong to B_u irreducible representation.

Internal coordinate method

There are two N-F stretches and one N-N stretch and two angle deformations as shown in the Fig. We can use these stretches and bending vectors as the basis for representation to develop $\Gamma_{vib,}$ that is, vectors N-N, N-F and bending are used to find $\Gamma_{N\text{-}N}$, $\Gamma_{N\text{-}F}$ and Γ_{bend} as follows:

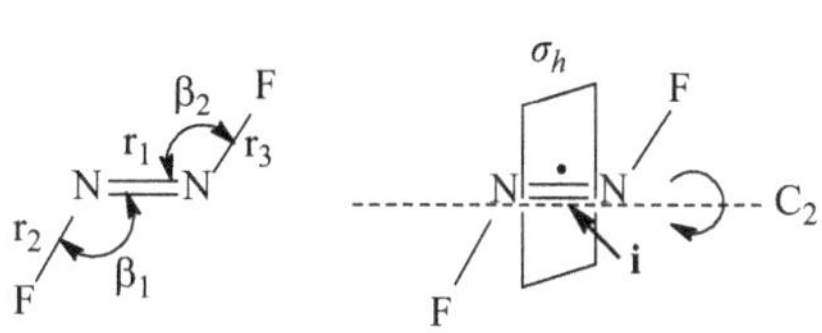

Γ_{N-N}

C_{2h}	E	C_2	i	σ_h
No. of unshifted N-N stretch (*n*)	1	1	1	1
$\chi(R)$	+ 1	+ 1	+ 1	+ 1
Γ_{N-N} $[n\times\chi(R)]$	1	1	1	1

The reducible representation $\Gamma_{N\text{-}N}$ can be reduced to irreducible representation applying standard reduction formula and using character table of C_{2h} point group.

Reducible representation in question

C_{2h}	E	C_2	i	σ_h
Γ_{N-N}	1	1	1	1

Character table of C_{2h} point symmetry

C_{2h}	E	C_2	i	σ_h		
A_g	1	1	1	1	R_z	x^2, y^2, z^2, xy
B_g	1	−1	1	−1	R_x, R_y	xz, yz
Au	1	1	−1	−1	z	
B_u	1	−1	−1	1	x, y	

$$NA_g = 1/4[(1.1.1) + (1.1.1) + (1.1.1) + (1.1.1)]$$
$$= 1/4[1 + 1 + 1 + 1] = 1$$
$$NB_g = 1/4[(1.1.1) + (1.1. - 1) + (1.1.1) + (1.1. - 1)]$$
$$= 1/4[1 - 1 + 1 - 1] = 0$$
$$NA_u = 1/4[(1.1.1) + (1.1.1) + (1.1. - 1) + (1.1. - 1)]$$
$$= 1/4[1 + 1 - 1 - 1] = 0$$
$$NB_u = 1/4[(1.1.1) + (1.1. - 1) + (1.1. - 1) + (1.1.1)]$$
$$= 1/4[1 - 1 - 1 + 1] = 0$$

Thus, the reducible representation reduces to:

$$\Gamma_{N-N} = A_g$$

$\Gamma_{N\text{-}F}$

C_{2h}	E	C_2	i	σ_h
No. of unshifted N-F stretch (*n*)	2	0	0	2
$\chi(R)$	+ 1	+ 1	+ 1	+ 1
$\Gamma_{N\text{-}N}$ $[n\times\chi(R)]$	2	0	0	2

The reducible representation $\Gamma_{N\text{-}F}$ can be reduced to irreducible representation applying standard reduction formula and using character table of C_{2h} point group.

Reducible representation in question

C_{2h}	E	C_2	i	σ_h
$\Gamma_{N\text{-}N}$	2	0	0	2

Character table of C_{2h} point symmetry

C_{2h}	E	C_2	i	σ_h		
A_g	1	1	1	1	R_z	x^2, y^2, z^2, xy
B_g	1	−1	1	−1	R_x, R_y	xz, yz
A_u	1	1	−1	−1	z	
B_u	1	−1	−1	1	x, y	

$$
\begin{aligned}
NA_g &= 1/4[(2.1.1) + (0) + (0) + (2.1.1)]\\
&= 1/4[2 + 0 + 0 + 2] = 1\\
&= 1/4[4] = 1\\
NB_g &= 1/4[(2.1.1) + (0) + (0) + (2.1. - 1)]\\
&= 1/4[2 + 0 + 0 - 2]\\
&= 1/4[0] = 0
\end{aligned}
$$

$$
\begin{aligned}
NA_u &= 1/4[(2.1.1) + (0) + (0) + (2.1. - 1)]\\
&= 1/4[2 + 0 + 0 - 2]\\
&= 1/4[0] = 0\\
NB_u &= 1/4[(2.1.1) + (0) + (0) + (2.1.1)]\\
&= 1/4[2 + 0 + 0 + 2]\\
&= 1/4[4] = 1
\end{aligned}
$$

Thus, the reducible representation reduces to:

$$\Gamma_{N-F} = A_g + B_u$$

Γ_{bend}

C_{2h}	E	C_2	i	σ_h
No. of unshifted double arrow headed vectors (n)	2	0	0	2
$\chi(R)$	+ 1	+ 1	+ 1	+ 1
Γ_{bend} $[n\times\chi(R)]$	2	0	0	2

This reducible representation Γ_{bend} is similar to $\Gamma_{N\text{-}F}$, and so it reduces to

$$\Gamma_{bend} = A_g + B_u$$

Now, the $\Gamma_{inter\ coord.} = \Gamma_{vib.} = \Gamma_{N-N} + \Gamma_{N-F} + \Gamma_{bend} = (A_g) + (A_g + B_u) + (A_g + B_u)$

$$= 3A_g + 2B_u(3 + 2 = 5)$$

When we compare this result with that of Γ_{vib} obtained by Cartesian coordinate method ($\Gamma_{vib} = 3A_g + A_u + 2B_u$), we find that the missing sixth mode is A_u.

This missing sixth mode is obviously out of plane (OOP) deformation mode. The symmetry species of the opd can be worked out as:

C_{2h}	E	C_2	i	σ_h
Γ_{oop}	1	1	−1	−1

In observation of character table of C_{2h} point group, we can identify that this representation belongs to A_u irreducible representation.

With the addition of Γ_{opp}, $\Gamma_{internal\ coord.} = \Gamma_{vib}$ now becomes

$$\Gamma_{vib} = 3A_g + 2B_u + A_u$$

Identification of IR and Raman active vibrations

IR active vibrations

From the character table of C_{2h} point group, we can see that there is no Cartesian coordinate transforming as A_g, and therefore the A_g is infrared inactive. But the z-coordinate transforms as A_u while the x- and y-coordinate transform as B_u. Hence, both the vibrational modes A_u and B_u will be IR active. As A_u is a non-degenerate vibration and $2B_u$ fundamental vibrations are two sets of non-degenerate vibrations, only three frequencies will be IR spectra.

Raman active vibrations

An observation of the character table (C_{2h} point group) indicates that the binary products x^2, y^2, z^2 and xy of the Cartesian coordinates transform as A_g irreducible representation. Hence, $3A_g$ vibrational modes will be Raman active. As $3A_g$ are three sets of non-degenerate vibrations, these three frequencies will be Raman active.

The IR and Raman activity of various vibrations of *cis*-N_2F_2 may be summarized as:

Mode	Infrared	Raman	Nature of vibration
$3A_g$	Inactive	Active	Mixed
$2B_u$	Active	Inactive	Mixed
$1A_u$	Active	Inactive	Pure out of plane (OPP) deformation

It is notable here that of the six normal vibrations, three are IR and three are Raman active. However, all the IR active vibrations are Raman inactive and vice versa in this case. This is true for all the molecules having a center of symmetry. This is stated as the mutual exclusion principle, which can be put as for molecules having a center of symmetry, infrared active vibrations are Raman inactive and vice versa.

(f) *cis*-N_2F_2 molecule of C_{2v} point group

This molecule has a planar non-linear structure and has C_{2v} point group. The possible number of symmetry operations is: E, C_2, σ_{xz}, σ_{yz}.

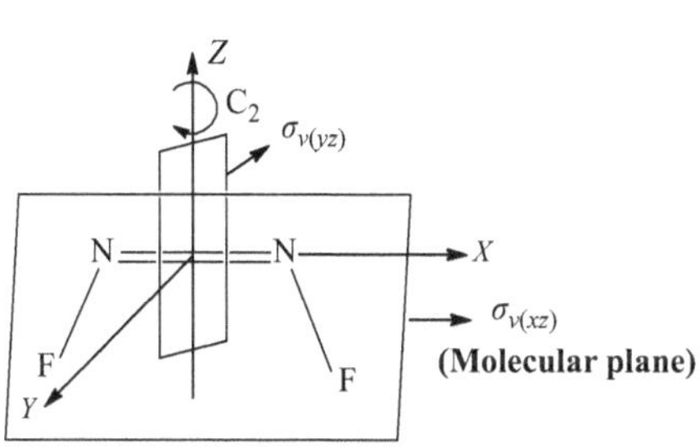

Cartesian coordinate method or 3N-vector method

Out of four symmetry operations, E and C_2 are proper rotations through 0° and 180°, respectively, while σ_{xz}, σ_{yz} are improper rotations through 0°, respectively. The following table provides the calculations of the total characters.

Table showing total character of cis-N_2F_2 molecule:

C_{2v}	E	C_2	σ_{xz}	σ_{yz}
β	0	180	0	0
cosβ	1	−1	1	1
±1 + 2cosβ	3	−1	1	1
n_R	4	0	4	0
Γ_{3N}	**12**	**0**	**4**	**0**

The vertical dashed line separates the proper and improper rotations. The corresponding angleβ, cosβ and also (± 1 + 2cosβ) values are tabulated for each operation. The proper rotations have (+ 1 + 2cosβ) and the improper rotations have (-1 + 2cosβ) values. The number of atoms retaining position (n_R) for each operation is also given in the table.

From the reducible representation Γ_{3N} in question, the irreducible representation can be worked out by referring to the character table of C_{2v} point group using standard reduction formula.

Reducible representation in question

C_{2v}	E	C_2	σ_{yz}	σ_{xz}
Γ_{3N}	12	0	0	4

Character table of C_{2v} point symmetry

C_{2v}	E	C_2	$\sigma_{v(yz)}$	$\sigma_{v(xz)}$		
A_1	1	1	1	1	z	x^2, y^2, z^2
A_2	1	1	−1	−1	R_z	xy
B_1	1	−1	1	−1	x, R_y	xz
B_2	1	−1	−1	1	y, R_x	yz

$$NA_1 = 1/4[(12.1.1) + 0 + 0 + (4.1.1)]$$
$$= 1/4[12 + 4 = 16] = 4$$

$$NA_2 = 1/4[(12.1.1) + 0 + 0 + (4.1. - 1)]$$
$$= 1/4[12 - 4 = 8] = 2$$

$$NB_1 = 1/4[(12.1.1) + 0 + 0 + (4.1. - 1)]$$
$$= 1/4[12 - 4 = 8] = 2$$

$$NB_2 = 1/4[(12.1.1) + 0 + 0 + (4.1.1)]$$
$$= 1/4[12 + 4 = 16] = 4$$

Thus, the reducible representation Γ_{3N} reduces to:

$$\Gamma_{3N} = 4A_1 + 2A_2 + 2B_1 + 4B_2$$

Thus, there are 12 modes (4 + 2 + 2 + 4 = 12) of degrees of freedom with different symmetries, A_1, A_2, B_1 and B_2. These include the vibrational, rotational and translational degrees of freedom.

The character table indicates that the translational modes have the same symmetry as the coordinates *x*, *y* and *z*, that is, A_1, B_1 and B_2. So also the rotational modes have the same symmetry as the rotation axes, R_x, R_y and R_z, that is, A_2, B_1 and B_2. Subtracting these translational and rotational degrees of freedom, we get the vibrational degrees of freedom only of appropriate symmetry.

$$\begin{aligned} \Gamma_{3N} &= 4A_1 + 2A_2 + 2B_1 + 4B_2 \\ \Gamma_{tr} &= A_1 \qquad\qquad B_1 + B_2 \\ \Gamma_{rot} &= \qquad\quad A_2 + B_1 + B_2 \\ &\cdots\cdots\cdots\cdots\cdots\cdots \\ \Gamma_{vib} &= 3A_1 + A_2 + 2B_2 \end{aligned}$$

Thus, out of six vibrational modes [3 (A_1) + 1(A_2) + 2(B_2)] = 6], thee belong to A_1, one belongs to A_2 and two belong to B_2 irreducible representation.

Internal coordinate method

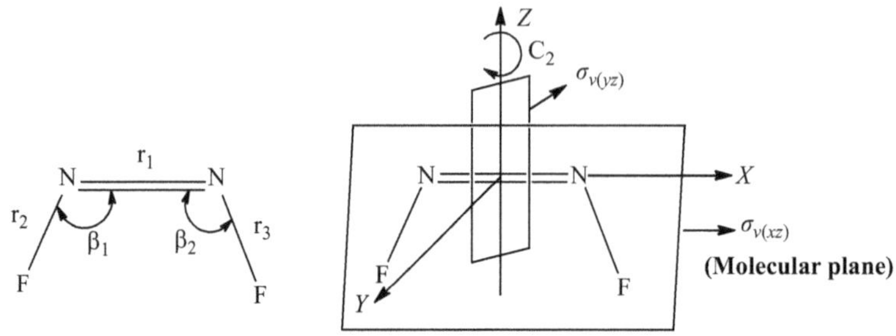

There is one N-N stretch (r_1) and two N-F stretches (r_2 and r_3) and two angle deformations (β_1 and β_2) as shown in the Fig. We can use these stretches and bending vectors as the basis for representation to develop $\Gamma_{vib,}$ that is, vectors N-N, N-F and bending are used to find $\Gamma_{N\text{-}N}$, $\Gamma_{N\text{-}F}$ and Γ_{bend} as follows:

$\Gamma_{N\text{-}N}$:

C_{2v}	E	C_2	σ_{yz}	σ_{xz}
No. of unshifted N-N stretch (n)	1	1	1	1
$\chi(R)$	+ 1	+ 1	+ 1	+ 1
$\Gamma_{N\text{-}N}$ $[n\times\chi(R)]$	1	1	1	1

This reducible representation $\Gamma_{N\text{-}N}$ can be reduced to irreducible representation applying standard reduction formula and using character table of C_{2v} point group.

Reducible representation in question

C_{2v}	E	C_2	σ_{yz}	σ_{xz}
$\Gamma_{N\text{-}N}$	1	1	1	1

Character table of C_{2v} point symmetry

C_{2v}	E	C_2	$\sigma_{v(yz)}$	$\sigma_{v(xz)}$		
A_1	1	1	1	1	z	x^2, y^2, z^2
A_2	1	1	−1	−1	R_z	xy
B_1	1	−1	1	−1	x, R_y	xz
B_2	1	−1	−1	1	y, R_x	yz

$$NA_1 = 1/4[(1.1.1) + (1.1.1) + (1.1.1) + (1.1.1)]$$
$$= 1/4[1 + 1 + 1 + 1 = 4] = 1$$
$$NA_2 = 1/4[(1.1.1) + (1.1.1) + (1.1. - 1) + (1.1. - 1)]$$
$$= 1/4[1 + 1 - 1 - 1 = 0] = 0$$
$$NB_1 = 1/4[(1.1.1) + (1.1. - 1) + (1.1.1) + (1.1. - 1)]$$
$$= 1/4[1 - 1 + 1 - 1 = 0] = 0$$
$$NB_2 = 1/4[(1.1.1) + (1.1. - 1) + (1.1. - 1) + (1.1.1)]$$
$$= 1/4[1 - 1 - 1 + 1 = 0] = 0$$

Thus, the reducible representation $\Gamma_{N\text{-}N}$ is reduced to:

$$\Gamma_{N-N} = A_1$$

$\Gamma_{N\text{-}F}$:

C_{2v}	E	C_2	σ_{yz}	σ_{xz}
No. of unshifted N-F stretch (*n*)	2	0	0	2
χ(*R*)	+ 1	+ 1	+ 1	+ 1
$\Gamma_{N\text{-}F}$ [*n*×χ(*R*)]	2	0	0	2

Now, the reducible representation $\Gamma_{N\text{-}F}$ can be reduced to irreducible representation applying standard reduction formula and using character table of C_{2v} point group.

Reducible representation in question

C_{2v}	*E*	C_2	σ_{yz}	σ_{xz}
Γ_{NF}	2	0	0	2

Character table of C_{2v} point symmetry

C_{2v}	E	C_2	$\sigma_{v(yz)}$	$\sigma_{v(xz)}$		
A_1	1	1	1	1	z	x^2, y^2, z^2
A_2	1	1	−1	−1	R_z	xy
B_1	1	−1	1	−1	x, R_y	xz
B_2	1	−1	−1	1	y, R_x	yz

$$NA_1 = 1/4[(2.1.1) + 0 + 0 + (2.1.1)]$$
$$= 1/4[2 + 2 = 4] = 1$$
$$NA_2 = 1/4[(2.1.1) + 0 + 0 + (2.1. - 1)]$$
$$= 1/4[2 - 2 = 0] = 1$$
$$NB_1 = 1/4[(2.1.1) + 0 + 0 + (2.1. - 1)]$$
$$= 1/4[2 - 2 = 0] = 0$$
$$NB_2 = 1/4[(2.1.1) + 0 + 0 + (2.1.1)]$$
$$= 1/4[2 + 2 = 4] = 1$$

Thus, the reducible representation $\Gamma_{N\text{-}F}$ is reduced to:

$$\Gamma_{N-F} = A_1 + B_2$$

Γ_{bend}:

C_{2v}	E	C_2	σ_{yz}	σ_{xz}
No. of unshifted double arrow headed vectors (n)	2	0	0	2
$\chi(R)$	+ 1	+ 1	+ 1	+ 1
Γ_{bend} $[n \times \chi(R)]$	2	0	0	2

This reducible representation Γ_{bend} is equal to the reducible representation $\Gamma_{N\text{-}F}$. Hence, it reduces to:

$$\Gamma_{bend} = A_1 + B_2$$

Now, the $\Gamma_{inter\ coord.} = \Gamma_{vib.} = \Gamma_{N-N} + \Gamma_{N-F} + \Gamma_{bend} = (A_1) + (A_1 + B_2) + (A_1 + B_2)$

$$= 3A_1 + 2B_2 (3 + 2 = 5)$$

When we compare this result with that of Γ_{vib} obtained by Cartesian coordinate method ($3A_1 + A_2 + 2B_2$), we see that the missing sixth mode is A_2.

This missing mode is obviously an OOP mode. This kind of vibration on the molecule results in a non-planar configuration for *cis*-N_2F_2 with the two N-N-F planes making a dihedral angle. The symmetry species of the opd can be worked out as:

C_{2v}	E	C_2	$\sigma_{v(yz)}$	$\sigma_{v(xz)}$
Γ_{oop}	1	1	−1	−1

In observation of character table of C_{2v} point group, we can identify that this representation belongs to A_2 irreducible representation.

With the addition of Γ_{oop}, the

$\Gamma_{internal\ coord.} = \Gamma_{vib}$ now becomes

$$\Gamma_{vib} = 3A_1 + 2B_2 + A_2$$

Identification of IR and Raman active vibrations

IR active vibrations

From the character table of C_{2v} point group, it is notable that *z*, *y*-coordinates belong to A_1 and B_2 irreducible representations. Hence, $3A_1$ and $2B_2$ vibrational modes will be IR active. As $2B_2$ vibrations are two sets of non-degenerate vibrations and $3A_1$ vibrations are three sets of non-degenerate vibrations, only five frequencies will be observed IR spectra.

Raman active vibrations

An observation of the character table (C_{2v} point group) indicates that the binary products of the Cartesian coordinates belong A_1, A_2, B_2 irreducible representation. Hence, all these vibrational modes will be Raman active. As $3A_1$ are three sets of non-degenerate vibrations, A_2 is a non-degenerate vibration and $2B_2$ are two sets of non-degenerate vibrations, we can expect six frequencies in Raman spectrum.

The IR and Raman activity of various vibrations of *cis*-N_2F_2 may be summarized as:

Mode	Infrared	Raman	Nature of vibration
$3A_1$	Active	Active	Mixed
$2B_2$	Active	Active	Mixed
A_2	Inactive	Active	Pure out of plane (OPP)deformation

(g) T-shaped ClF_3 molecule

This molecule belongs to C_{2v} point group. Various symmetry operations in this molecule are: *E*, C_2, σ_{xz}, σ_{yz}.

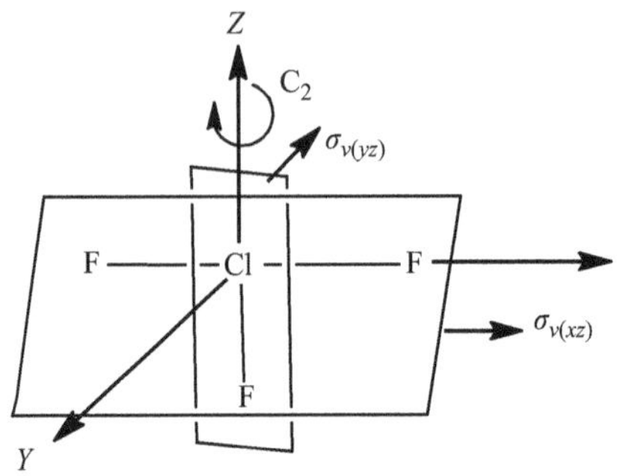

Cartesian coordinate method or 3N-vector method

Out of four symmetry operations, *E* and C_2 are proper rotations through 0° and 180°, respectively while σ_{xz}, σ_{yz} are improper rotations through 0°, respectively. The following table provides the calculations of the total characters

Table showing total character of ClF_3 molecule.

C_{2v}	E	C_2	σ_{yz}	σ_{xz}
β	0	180	0	0
cosβ	1	−1	1	1
±1+ 2cosβ	3	−1	1	1
n_R	4	2	2	4
Γ_{3N}	**12**	**−2**	**2**	**4**

The vertical dashed line separates the proper and improper rotations. The corresponding angle*β*, cos*β* and also (±1 + 2cos*β*) values are tabulated for each operation. The proper rotations have (+1 + 2cos*β*) and the improper rotations have (-1 + 2cos*β*) values. The number of atoms retaining position (n_R) for each operation is also given in the table.

From the reducible representation in Γ_{3N} question, the irreducible representation can be worked out by referring to the character table of C_{2v} point group using the standard reduction formula,

$$N_i = \frac{1}{h}\sum_R \chi(R).n.\chi_i(R)$$

Reducible representation in question

C_{2v}	E	C_2	σ_{yz}	σ_{xz}
Γ_{3N}	12	−2	2	4

Character table of C_{2v} point symmetry

C_{2v}	E	C_2	$\sigma_{v(yz)}$	$\sigma_{v(xz)}$		
A_1	1	1	1	1	z	x^2, y^2, z^2
A_2	1	1	−1	−1	R_z	xy
B_1	1	−1	1	−1	x, R_y	xz
B_2	1	−1	−1	1	y, R_x	yz

$$\begin{aligned}
NA_1 &= 1/4[(12.1.1) + (-2.1.1) + (2.1.1) + (4.1.1)] \\
&= 1/4[12 - 2 + 2 + 4 = 16] = 4 \\
NA_2 &= 1/4[(12.1.1) + (-2.1.1) + (2.1.-1) + (4.1.-1)] \\
&= 1/4[12 - 2 - 2 - 4 = 4] = 1 \\
NB_1 &= 1/4[(12.1.1) + (-2.1.-1) + (2.1.1) + (4.1.-1)] \\
&= 1/4[12 + 2 + 2 - 4 = 12] = 3 \\
NB_2 &= 1/4[(12.1.1) + (-2.1.-1) + (2.1.-1) + (4.1.1)] \\
&= 1/4[12 + 2 - 2 + 4 = 16] = 4
\end{aligned}$$

Thus, the reducible representation reduces to:

$$\Gamma_{3N} = 4A_1 + A_2 + 3B_1 + 4B_2 (3N = 12)$$

Thus, there are 12 modes (4 + 1 + 3 + 4 = 12) of degrees of freedom with different symmetries, A_1, A_2, B_1, and B_2. These include the vibrational, rotational and translational degrees of freedom.

The character table shows that the translational modes have the same symmetry as the coordinates x, y and z, that is, A_1, B_1 and B_2. So also the rotational modes have the same symmetry as the rotation axes, R_x, R_y and R_z, that is, A_2, B_1 and B_2. Subtracting these translational and rotational degrees of freedom, we get the vibrational degrees of freedom only of appropriate symmetry.

$$\begin{aligned}
\Gamma_{3N} &= 4A_1 + A_2 + 3B_1 + 4B_2 \\
\Gamma_{tr} &= A_1 \qquad\quad + B_1 + B_2 \\
\Gamma_{rot} &= \qquad\quad A_2 + B_1 + B_2 \\
&\cdots\cdots\cdots\cdots\cdots\cdots \\
\Gamma_{vib} &= 3A_1 + B_1 + 2B_2 (3N-6)
\end{aligned}$$

Thus, out of 6 vibrational modes [3 (A_1) + 1(B_1) + 2(B_2)] = 6], thee belong to A_1, one belongs to B_1 and two belong to B_2 irreducible representation.

Internal coordinate method

There are three Cl-F stretches (r_1, r_2 and r_3) and two angle deformations (β_1 and β_2) as shown in the Fig. We can use these stretches and bending vectors as the basis for representation to develop $\Gamma_{vib,}$ that is, vectors Cl-F and bending are used to find $\Gamma_{Cl\text{-}F}$ (r_1, r_2), $\Gamma_{Cl\text{-}F}$ (r_3) and Γ_{bend} as follows:

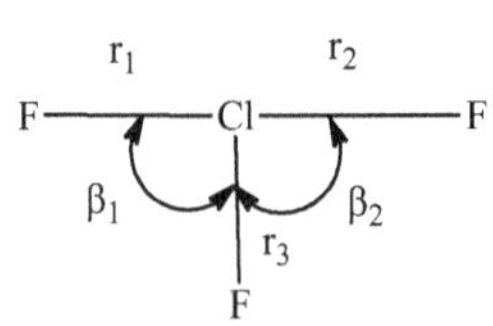

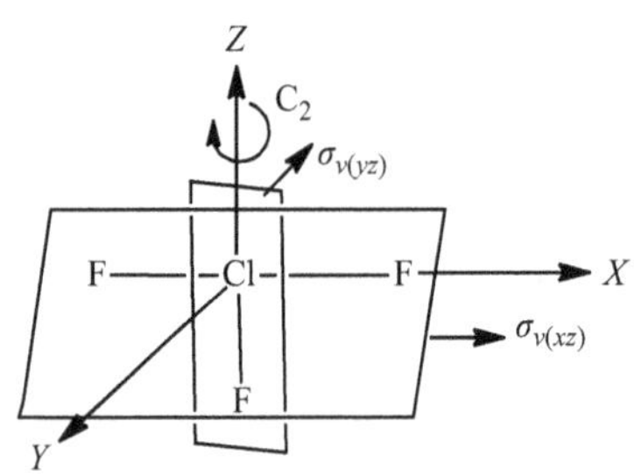

$\Gamma_{Cl\text{-}F}$ (r_1, r_2)

C_{2v}	E	C_2	σ_{yz}	σ_{xz}
No. of unshifted Cl-F stretch (*n*)	2	0	0	2
χ(*R*)	+ 1	+ 1	+ 1	+ 1
$\Gamma_{Cl\text{-}F}$ (r_1, r_2) [*n*×χ(*R*)]	2	0	0	2

This reducible representation $\Gamma_{Cl\text{-}F}$ (r_1,r_2) can be reduced to irreducible representation applying standard reduction formula and using character table of C_{2v} point group.

Reducible representation in question

C_{2v}	*E*	C_2	σ_{yz}	σ_{xz}
$\Gamma_{Cl\text{-}F}$ (r_1, r_2)	2	0	0	2

Character table of C_{2v} point symmetry

C_{2v}	E	C_2	$\sigma_{v(yz)}$	$\sigma_{v(xz)}$		
A_1	1	1	1	1	z	x^2, y^2, z^2
A_2	1	1	−1	−1	R_z	xy
B_1	1	−1	1	−1	x, R_y	xz
B_2	1	−1	−1	1	y, R_x	yz

$$NA_1 = 1/4[(2.1.1) + (0) + (0) + (2.1.1)]$$
$$= 1/4[2 + 2 = 4] = 1$$
$$NA_2 = 1/4[(2.1.1) + (0) + (0) + (2.1. - 1)]$$
$$= 1/4[2 - 2 = 0] = 0$$
$$NB_1 = 1/4[(2.1.1) + (0) + (0) + (2.1. - 1)]$$
$$= 1/4[2 - 2 = 0] = 0$$
$$NB_2 = 1/4[(2.1.1) + (0) + (0) + (2.1.1)]$$
$$= 1/4[2 + 2 = 4] = 1$$

Thus, the reducible representation reduces to:

$$\Gamma_{Cl-F}(r_1, r_2) = A_1 + B_2$$

$\Gamma_{Cl\text{-}F}$ (r_3)

C_{2v}	E	C_2	σ_{yz}	σ_{xz}
No. of unshifted Cl-F stretch (n)	1	1	1	1
$\chi(R)$	+ 1	+ 1	+ 1	+ 1
$\Gamma_{Cl\text{-}F}$ (r_3) [$n \times \chi(R)$]	1	1	1	1

Again, this reducible representation $\Gamma_{Cl\text{-}F}$ (r_3) can be reduced to irreducible representation applying standard reduction formula and using character table of C_{2v} point group.

$$NA_1 = 1/4[(1.1.1) + (1.1.1) + (1.1.1) + (1.1.1)]$$
$$= 1/4[1 + 1 + 1 + 1 = 4] = 1$$
$$NA_2 = 1/4[(1.1.1) + (1.1.1) + (1.1. - 1) + (1.1. - 1)]$$
$$= 1/4[1 + 1 - 1 - 1 = 0] = 0$$

$$NB_1 = 1/4[(1.1.1) + (1.1. - 1) + (1.1.1) + (1.1. - 1)]$$
$$= 1/4[1 - 1 + 1 - 1 = 0] = 0$$
$$NB_2 = 1/4[(1.1.1) + (1.1. - 1) + (1.1. - 1) + (1.1.1)]$$
$$= 1/4[1 - 1 - 1 + 1 = 0] = 0$$

Thus, the reducible representation reduces to:

$$\Gamma_{Cl-F}(r_3) = A_1$$

Γ_{bend}

C_{2v}	E	C_2	σ_{yz}	σ_{xz}
No. of unshifted double arrow headed vectors (*n*)	2	0	0	2
$\chi(R)$	+ 1	+ 1	+ 1	+ 1
Γ_{bend} $[n \times \chi(R)]$	2	0	0	2

This reducible representation Γ_{bend} is equal to the reducible representation $\Gamma_{N\text{-}F}$ (r_1, r_2). Hence, it reduces to:

$$\Gamma_{bend} = A_1 + B_2$$

Now, the $\Gamma_{inter\ coord.} = \Gamma_{vib.} = \Gamma_{Cl-F}(r_1, r_2) + \Gamma_{Cl-F}(r_3) + \Gamma_{bend} = (A_1 + B_2) + A_1 + (A_1 + B_2)$

$$= 3A_1 + 2B_2 (3 + 2 = 5)$$

On comparing this result with that of Γ_{vib} obtained by Cartesian coordinate method $(3A_1 + B_1 + 2B_2)$, we see that the missing sixth mode is B_1.

This missing sixth mode is obviously the vibration corresponding to the OOP mode of this planar molecule. The symmetry species of the opd can be worked out as:

C_{2v}	*E*	C_2	$\sigma_{v(yz)}$	$\sigma_{v(xz)}$
Γ_{oop}	1	−1	1	−1

Looking over the character table of C_{2v} point group, we can identify that this representation belongs to B_1 irreducible representation.

With the addition of Γ_{oop}, the $\Gamma_{internal\ coord.} = \Gamma_{vib}$ now becomes

$$\Gamma_{vib} = 3A_1 + 2B_2 + B_1$$

Identification of IR and Raman active vibrations

IR active vibrations

From the character table of C_{2v} point group, it is notable that *x*, *y*, *z*-coordinates belong to A_1 B_1 and B_2 irreducible representations. Hence, $3A_1$, B_1 and $2B_2$ vibrational modes will be IR active. As $3A_1$ are three sets of non-degenerate vibrations, B_1 is a non-degenerate vibration and $2B_2$ are two sets of non-degenerate vibrations, we can expect six frequencies in Raman spectrum.

Raman active vibrations

An observation of the character table (C_{2v} point group) indicates that the binary products of the Cartesian coordinates belong to $A_{1,}$ $B_{1,}$ and B_2 irreducible representations. Hence, $3A_{1,}$ $B_{1,}$ and $2B_2$ vibrational modes will be Raman active. One can expect six frequencies in Raman spectrum.

The IR and Raman activity of various vibrations of T-shaped ClF_3 molecule may be summarized as:

Mode	Infrared	Raman	Nature of vibration
$3A_1$	Active	Active	Mixed
$2B_2$	Active	Active	Mixed
B_1	Active	Active	Pure out of plane (OPP) deformation

(h) *cis*-$PtCl_2L_2$ molecule of C_{2v} point group

This compound exhibits geometrical isomerism. In the *cis*-form, it belongs to C_{2v} point group and in the *trans*-form to D_{2h} point group. The possible symmetry operations in the *cis*-form are: *E*, C_2, σ_{xz} and σ_{yz}.

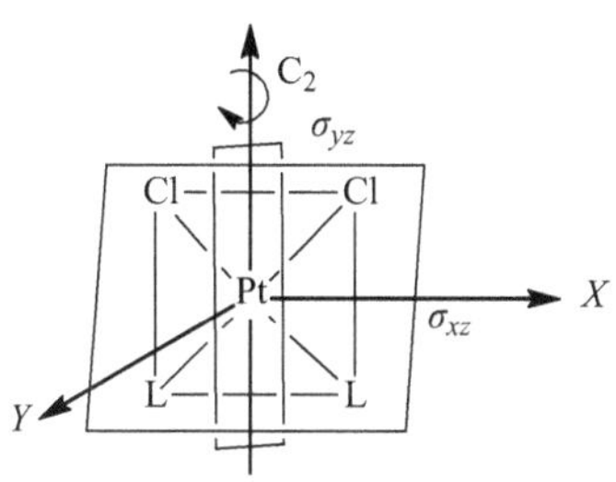

Cartesian coordinate method or 3N-vector method

Out of four symmetry operations, *E* and C_2 are proper rotations through 0° and 180°, respectively while σ_{xz}, σ_{yz} are improper rotations through 0°, respectively. The following table provides the calculations of the total characters.

Table showing total character of *cis*-$PtCl_2L_2$

C_{2v}	E	C_2	σ_{yz}	σ_{xz}
β	0	180	0	0
cosβ	1	−1	1	1
±1 + 2cosβ	3	−1	1	1
n_R	5	1	1	5
Γ_{3N}	**15**	**−1**	**1**	**5**

The vertical dashed line separates the proper and improper rotations. The corresponding angle*β*, cos*β* and also (±1 + 2cos*β*) values are tabulated for each operation. The proper rotations have (+1 + 2cos*β*) and the improper rotations have (-1 + 2cos*β*) values. The number of atoms retaining position (n_R) for each operation is also given in the table.

The above reducible representation (Γ_{3N}) can be reduced to irreducible representations using the standard reduction formula and character table of C_{2v} point group as follows:

Reducible representation in question

C_{2v}	*E*	C_2	$\sigma_{v(yz)}$	$\sigma_{v(xz)}$
Γ_{3N}	15	−1	1	5

Character table of C_{2v} point symmetry

C_{2v}	E	C_2	$\sigma_{v(yz)}$	$\sigma_{v(xz)}$		
A_1	1	1	1	1	z	x^2, y^2, z^2
A_2	1	1	−1	−1	R_z	xy
B_1	1	−1	1	−1	x, R_y	xz
B_2	1	−1	−1	1	y, R_x	yz

$$NA_1 = 1/4[(15.1.1) + (-1.1.1) + (1.1.1) + (5.1.1)]$$
$$= 1/4[15 - 1 + 1 + 5 = 20]$$
$$= 5$$

$$\begin{aligned} NA_2 &= 1/4[(15.1.1) + (-1.1.1) + (1.1.-1) + (5.1.-1)] \\ &= 1/4[15 - 1 - 1 - 5 = 8] \\ &= 2 \end{aligned}$$

$$\begin{aligned} NB_1 &= 1/4[(15.1.1) + (-1.1.-1) + (1.1.1) + (5.1.-1)] \\ &= 1/4[15 + 1 + 1 - 5 = 12] \\ &= 3 \end{aligned}$$

$$\begin{aligned} NB_2 &= 1/4[(15.1.1) + (-1.1.-1) + (1.1.-1) + (5.1.1)] \\ &= 1/4[15 + 1 - 1 + 5 = 20] \\ &= 5 \end{aligned}$$

Thus, the reducible representation reduces to:

$$\Gamma_{3N} = 5A_1 + 2A_2 + 3B_1 + 5B_2 (3N = 15)$$

Thus, there are 15 modes (5 + 2 + 3 + 5 = 15) of degrees of freedom with different symmetries, A_1, A_2, B_1, B_2. These include the vibrational, rotational and translational degrees of freedom.

The character table shows that the translational modes have the same symmetry as the coordinates *x*, *y* and *z*, that is, A_1, B_1 and B_2. So also the rotational modes have the same symmetry as the rotation axes, R_x, R_y and R_z, that is, A_2, B_1 and B_2. Subtracting these translational and rotational degrees of freedom, we get the vibrational degrees of freedom of appropriate symmetry.

$$\begin{aligned} \Gamma_{3N} &= 5A_1 + 2A_2 + 3B_1 + 5B_2 \\ \Gamma_{tr} &= A_1 \qquad\qquad B_1 + B_2 \\ \Gamma_{rot} &= \qquad\quad A_2 + B_1 + B_2 \\ &\cdots\cdots\cdots\cdots\cdots\cdots \\ \Gamma_{vib} &= 4A_1 + A_2 + B_1 + 3B_2 (3N - 6 = 9) \end{aligned}$$

Thus, out of nine vibrational modes [4(A_1) + 1(A_2) + 1(B_1) + 3(B_2) = 9], four belong to A_1, one belongs to A_2, one belongs to B_1 and three belong to B_2 irreducible representation.

Internal coordinate method

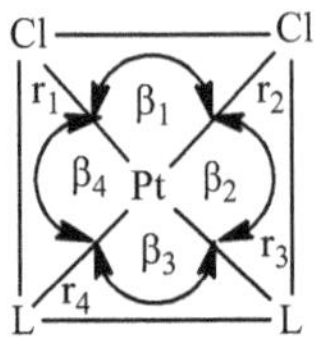

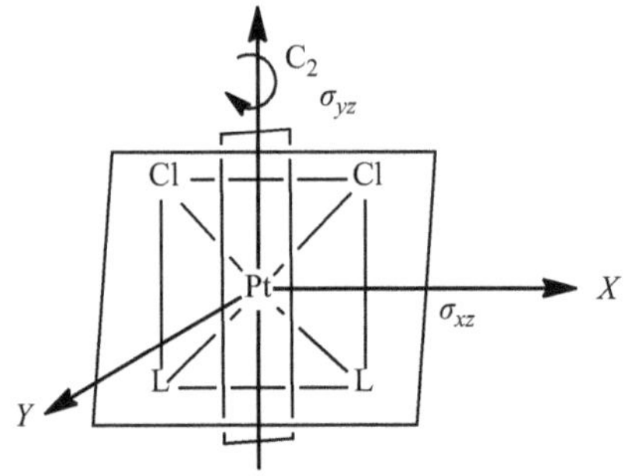

This molecule contains two Pt-Cl and two Pt-L stretches (r_1, r_2, r_3 and r_4) and four angle deformations (β_1 to β_4) as shown in the Fig. We can use these stretches and bending vectors as the basis for representation to develop $\Gamma_{vib,}$ that is, vectors Pt-Cl/ Pt-L and bending are used to find $\Gamma_{Pt\text{-}Cl/Pt\text{-}L}$ and Γ_{bend} as follows:

$\Gamma_{Pt\text{-}Cl/\ Pt\text{-}L}$

C_{2v}	E	C_2	σ_{yz}	σ_{xz}
No. of unshifted Pt-Cl/Pt-L stretch (n)	4	0	0	4
$\chi(R)$	+ 1	+ 1	+ 1	+ 1
$\Gamma_{Pt\text{-}Cl/\ Pt\text{-}L}$ $[n\times\chi(R)]$	4	0	0	4

This reducible representation $\Gamma_{Pt\text{-}Cl/\ Pt\text{-}L}$ can be reduced to irreducible representation applying standard reduction formula and using character table of C_{2v} point group.

Reducible representation in question

C_{2v}	E	C_2	$\sigma_{v(yz)}$	$\sigma_{v(xz)}$
$\Gamma_{Pt\text{-}Cl/Pt\text{-}L}$	4	0	0	4

Character table of C_{2v} point symmetry

C_{2v}	E	C_2	$\sigma_{v(yz)}$	$\sigma_{v(xz)}$		
A_1	1	1	1	1	z	x^2, y^2, z^2
A_2	1	1	−1	−1	R_z	xy
B_1	1	−1	1	−1	x, R_y	xz
B_2	1	−1	−1	1	y, R_x	yz

$$NA_1 = 1/4[(4.1.1) + (0) + (0) + (4.1.1)]$$
$$= 1/4[4 + 4 = 8] = 2$$
$$NA_2 = 1/4[(4.1.1) + (0) + (0) + (4.1.-1)]$$
$$= 1/4[4 - 4 = 0] = 0$$
$$NB_1 = 1/4[(4.1.1) + (0) + (0) + (4.1.-1)]$$
$$= 1/4[4 - 4 = 0] = 0$$
$$NB_2 = 1/4[(4.1.1) + (0) + (0) + (4.1.1)]$$
$$= 1/4[4 + 4 = 8] = 2$$

Thus, the reducible representation reduces to:

$$\Gamma_{Pt-Cl/Pt-L} = 2A_1 + 2B_2$$

Γ_{bend}

C_{2v}	E	C_2	σ_{yz}	σ_{xz}
No. of unshifted double arrow headed vectors (n)	4	2	2	4
$\chi(R)$	+ 1	+ 1	+ 1	+ 1
Γ_{bend} $[n \times \chi(R)]$	4	2	2	4

This reducible representation Γ_{bend} can be reduced to irreducible representation applying standard reduction formula and using character table of C_{2v} point group.

Reducible representation in question

C_{2v}	E	C_2	σ_{yz}	σ_{xz}
Γ_{bend}	4	2	2	4

Character table of C_{2v} point symmetry

C_{2v}	E	C_2	$\sigma_{v(yz)}$	$\sigma_{v(xz)}$		
A_1	1	1	1	1	z	x^2, y^2, z^2
A_2	1	1	−1	−1	R_z	xy
B_1	1	−1	1	−1	x, R_y	xz
B_2	1	−1	−1	1	y, R_x	yz

$$
\begin{aligned}
NA_1 &= 1/4[(4.1.1) + (2.1.1) + (2.1.1) + (4.1.1)] \\
&= 1/4[4+2+2+4=12] = 3 \\
NA_2 &= 1/4[(4.1.1) + (2.1.1) + (2.1.-1) + (4.1.-1)] \\
&= 1/4[4+2-2-4=0] = 0 \\
NB_1 &= 1/4[(4.1.1) + (2.1.-1) + (2.1.1) + (4.1.-1)] \\
&= 1/4[4-2+2-4=0] = 0 \\
NB_2 &= 1/4[(4.1.1) + (2.1.-1) + (2.1.-1) + (4.1.1)] \\
&= 1/4[4-2-2+4=4] = 1
\end{aligned}
$$

Thus, the reducible representation reduces to:

$$\Gamma_{bend} = 3A_1 + B_2$$

Now,

$$\Gamma_{int.\ coord} = \Gamma_{Pt-Cl/Pt-L} + \Gamma_{bend} = (2A_1 + 2B_2) + (3A_1 + B_2) = 5A_1 + 3B_2$$

On comparing the results $\Gamma_{int.\ coord}$ so obtained with Γ_{vib} worked out through Cartesian coordinate method, we see that A_1 mode is spurious and A_2 and B_1modes are missing. So, neglecting A_1 mode, we get

$$\Gamma_{int.coord} = 4A_1 + 3B_2$$

The missing modes are obviously the OOP type and such a deformity results in the square plane changing into square pyramid or tetrahedral. Thus,

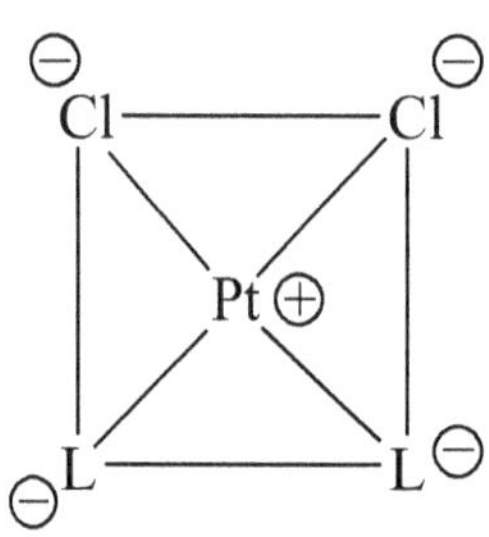

$\Gamma_{oop}(1)$

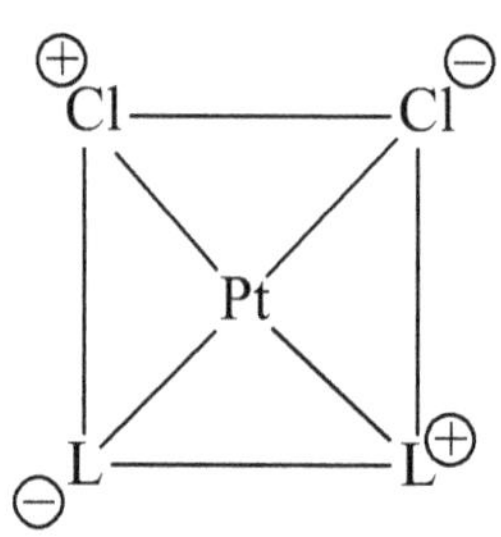

$\Gamma_{oop}(2)$

The symmetry species of the opd can be worked out as:

C_{2v}	E	C_2	$\sigma_{v(yz)}$	$\sigma_{v(xz)}$
$\Gamma_{oop\,(1)}$	1	−1	1	−1
$\Gamma_{oop\,(2)}$	1	1	−1	−1

Looking over the character table of C_{2v} point group, we can identify that the representations $\Gamma_{oop}(1)$ and $\Gamma_{oop}(2)$ belong to B_1 and A_2 irreducible representation, respectively.

With addition of $\Gamma_{oop}(1)$ and $\Gamma_{oop}(2)$ to $\Gamma_{int.\ coord} = \Gamma_{vib}$, we get

$$\begin{aligned}\Gamma_{int.\ coord} = \Gamma_{vib} &= (4A_1 + 3B_2) + (A_2 + B_1) \\ &= 4A_1 + A_2 + B_1 + 3B_2\end{aligned}$$

Identification of IR and Raman active vibrations

IR active vibrations

Looking over the character table of C_{2v} point group, it is notable that x, y, z-coordinates belong to A_1, B_1, and B_2 irreducible representations. Hence, $4A_1$, B_1, and $3B_2$ vibrational modes will be IR active. Thus, we can expect only eight frequencies in the IR spectra.

Raman active vibrations

An observation of the character table (C_{2v} point group) indicates that the binary products of the Cartesian coordinates belong to A_1, A_2, B_1 and B_2 irreducible representations. Hence, $4A_1$, A_2, B_1 and $3B_2$ vibrational modes are Raman active. Thus, we can expect nine frequencies in the IR spectra.

The IR and Raman activity of various vibrations of *cis*-$PtCl_2L_2$ molecule may be summarized as:

Mode	Infrared	Raman	Nature of vibration
$4A_1$	Active	Active	Mixed
A_2	Inactive	Active	Pure out of plane (OOP)deformation
B_1	Active	Active	Pure out of plane (OOP)deformation
$3B_2$	Active	Active	Mixed

(i) *trans*-1,2-Dichloroethylene molecule of C_{2h} point group

This molecule belongs to C_{2h} point group with symmetry operations E, $C_{2(x)}$, i and $\sigma_{h(xz)}$.

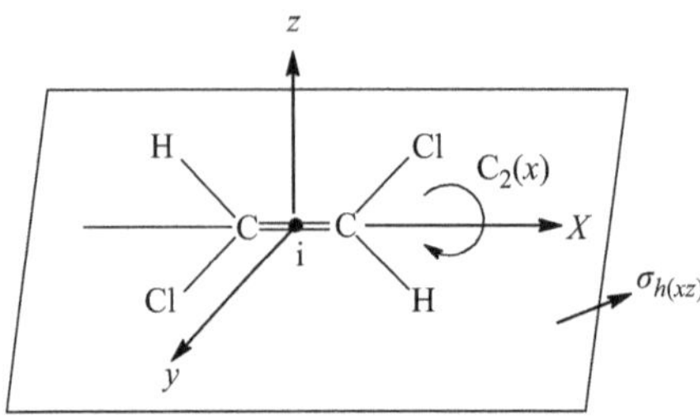

(a) Cartesian coordinate or 3N vector method

Out of four symmetry operations, E and $C_{2(x)}$ are proper rotations through 0° and 180°, respectively, while i and $\sigma h_{(xz)}$ are improper rotations through 180° and 0°, respectively. The following table provides the calculations of the total characters.

C_{2v}	E	$C_{2(x)}$	i	$\sigma_{h(xz)}$
β	0	180	180	0
cosβ	1	−1	1	1
±1 + 2cosβ	3	−1	−3	1
n_R	6	0	0	6
Γ_{3N}	**18**	**0**	**0**	**6**

The vertical dashed line separates the proper and improper rotations. The corresponding angleβ, cosβ and also (±1 + 2cosβ) values are tabulated for each operation. The proper rotations have (+1 + 2cosβ) and the improper rotations have (-1 + 2cosβ) values. The number of atoms retaining position (n_R) for each operation is also given in the table.

The above reducible representation (Γ_{3N}) can be reduced to irreducible representations using the reduction formula and character table of C_{2h} point group as follows:

Reducible representation in question

C_{2h}	E	C_2	i	σ_h
Γ_{3N}	18	0	0	6

Character table of C_{2v} point symmetry

C_{2h}	E	C_2	i	σ_h		
A_g	1	1	1	1	R_z	x^2, y^2, z^2, xy
B_g	1	−1	1	−1	R_x, R_y	xz, yz
Au	1	1	−1	−1	z	
B_u	1	−1	−1	1	x, y	

$$
\begin{aligned}
NA_g &= 1/4[(18.1.1) + (0) + (0) + (6.1.1)]\\
&= 1/4[18 + 0 + 0 + 6]\\
&= 1/4[24] = 6\\
NB_g &= 1/4[(18.1.1) + (0) + (0) + (6.1.-1)]\\
&= 1/4[18 + 0 + 0 - 6]\\
&= 1/4[12] = 3\\
NA_u &= 1/4[(18.1.1) + (0) + (0) + (6.1.-1)]\\
&= 1/4[18 + 0 + 0 - 6]\\
&= 1/4[12] = 3\\
NB_u &= 1/4[(18.1.1) + (0) + (0) + (6.1.1)]\\
&= 1/4[18 + 0 + 0 + 6]\\
&= 1/4[24] = 6
\end{aligned}
$$

Thus, the reducible representation reduces to:

$$6A_g + 3B_g + 3A_u + 6B_u$$

Thus, there are 18 modes (6 + 3 + 3 + 6 = 18) of degrees of freedom with different symmetries, A_g, B_g, A_u and B_u. These include the vibrational, rotational and translational degrees of freedom.

The character table of C_{2h} point group indicates that the translational modes have the same symmetry as the coordinates *x*, *y* and *z*, that is, A_u and $2B_u$. So also the rotational modes have the same symmetry as the rotation axes, R_x, R_y and R_z, that is, A_g and $2B_g$. Subtracting these translational and rotational degrees of freedom, we get the vibrational degrees of freedom only of appropriate symmetry.

$$
\begin{aligned}
\Gamma_{3N} &= 6A_g + 3B_g + 3A_u + 6B_u\\
\Gamma_{tr} &= \qquad\qquad\qquad A_u + 2B_u\\
\Gamma_{rot} &= A_g + 2B_g\\
&\cdots\cdots\cdots\cdots\cdots\cdots\\
\Gamma_{vib} &= 5A_g + B_g + 2A_u + 4B_u
\end{aligned}
$$

Thus, out of 12 vibrational modes $[5(A_g) + 1(B_g) + 2(A_u) + 4(B_u)] = 12]$, five belong to A_g, one belongs to B_g, two belong to A_u and four belong to B_u irreducible representation.

Internal coordinate method

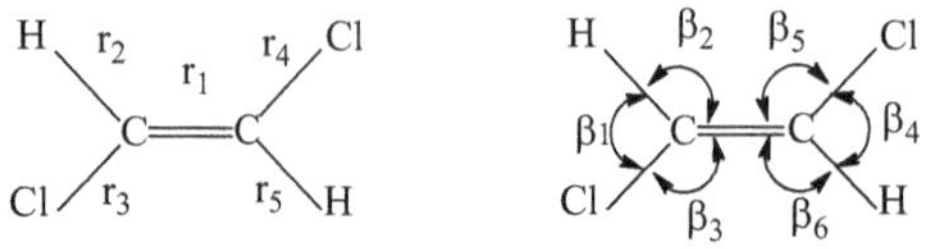

This molecule contains five stretches (r_1, r_2, r_3, r_4 and r_5) and six angle deformations (β_1 to β_6) as shown in the Fig. We can use these stretches and bending vectors as the basis for representation.

Γ_r

C_{2h}	E	C_2	i	σ_h
No. of unshifted stretches (n)	5	1	1	5
$\chi(R)$	+ 1	+ 1	+ 1	+ 1
Γ_r	5	1	1	5
$[n\times\chi(R)]$				

This reducible representation Γ_r can be reduced applying the reduction formula and using the character table of C_{2h} point group follows:

Reducible representation in question

C_{2h}	E	C_2	i	σ_h
Γ_r	5	1	1	5

Character table of C_{2h} point symmetry

C_{2h}	E	C_2	i	σ_h		
A_g	1	1	1	1	R_z	x^2, y^2, z^2, xy
B_g	1	−1	1	−1	R_x, R_y	xz, yz
A_u	1	1	−1	−1	z	
B_u	1	−1	−1	1	x, y	

$$NA_g = 1/4[(5.1.1) + (1.1.1) + (1.1.1) + (5.1.1)]$$
$$= 1/4[12]$$
$$= 3$$

$$
\begin{aligned}
NB_g &= 1/4[(5.1.1) + (1.1.-1) + (1.1.1) + (5.1.-1)] \\
&= 1/4[0] \\
&= 0 \\
NA_u &= 1/4[(5.1.1) + (1.1.1) + (1.1.-1) + (5.1.-1)] \\
&= 1/4[0] \\
&= 0 \\
NB_u &= 1/4[(5.1.1) + (1.1.-1) + (1.1.-1) + (5.1.1)] \\
&= 1/4[8] \\
&= 2
\end{aligned}
$$

Thus, the reducible representation reduces to:

$$3A_g + 2B_u$$

Γ_{bend}

C_{2h}	E	C_2	i	σ_h
No. of unshifted double arrow headed vectors (n)	6	0	0	6
$\chi(R)$	+ 1	+ 1	+ 1	+ 1
Γ_{bend} [i×$\chi(R)$]	6	0	0	6

This reducible representation Γ_{bend} can be reduced to irreducible representation applying standard reduction formula and using character table of C_{2h} point group.

Reducible representation in question

C_{2h}	E	C_2	i	σ_h
Γ_{bend}	6	0	0	6

Character table of C_{2h} point symmetry

C_{2h}	E	C_2	i	σ_h		
A_g	1	1	1	1	R_z	x^2, y^2, z^2, xy
B_g	1	−1	1	−1	R_x, R_y	xz, yz
A_u	1	1	−1	−1	z	
B_u	1	−1	−1	1	x, y	

$$\begin{aligned}NA_g &= 1/4[(6.1.1) + (0.1.1) + (0.1.1) + (6.1.1)]\\ &= 1/4[12]\\ &= 3\\ NB_g &= 1/4[(6.1.1) + (0.1.-1) + (0.1.1) + (6.1.-1)]\\ &= 1/4[0]\\ &= 0\\ NA_u &= 1/4[(6.1.1) + (0.1.1) + (0.1.-1) + (6.1.-1)]\\ &= 1/4[0]\\ &= 0\\ NB_u &= 1/4[(6.1.1) + (0.1.-1) + (0.1.-1) + (6.1.1)]\\ &= 1/4[12]\\ &= 3\end{aligned}$$

Thus, the reducible representation reduces to:

$$\Gamma_{bend} = 3A_g + 3B_u$$

Now,

$$\begin{aligned}\Gamma_{int\ coord} = \Gamma_r + \Gamma_{bend} &= (3A_g + 2B_u) + (3A_g + 3B_u)\\ &= 6A_g + 5B_u\end{aligned}$$

On comparison with Γ_{vib} of Γ_{3N} it can be seen that one A_g and one B_u of $\Gamma_{int\ coord}$, particularly Γ_{bend} occur as extra and spurious modes. Removing these modes, we get,

$$\Gamma_{int\ coord} = 5A_g + 4B_u$$

This requires three more modes to be found. As we have not considered OOP modes while considering the internal coordinates, the missing mode can be found from them as given below:

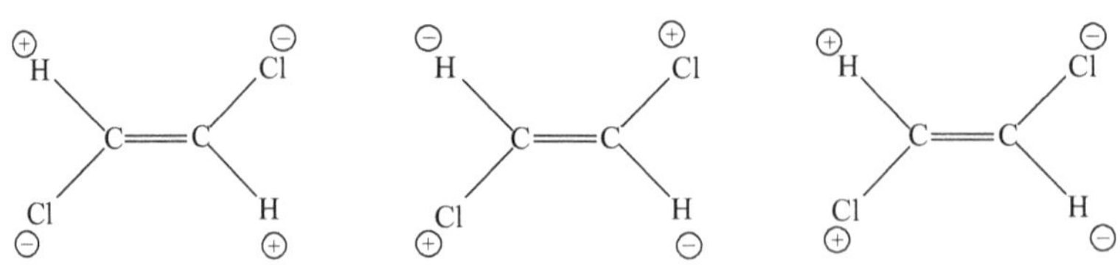

$\Gamma_{OOP}(1)$ $\Gamma_{OOP}(2)$ $\Gamma_{OOP}(3)$

The character system of the above configurations can be worked out as:

C_{2h}	E	C_2	i	$\sigma_{(xy)}$
Γ_{oop} (1)	1	1	−1	−1
Γ_{oop} (2)	1	1	−1	−1
Γ_{oop} (3)	1	−1	1	−1

Looking over the character table of C_{2h} point group, we can identify that the representations $\Gamma_{oop}(1)$ and $\Gamma_{oop}(2)$ and $\Gamma_{oop}(3)$ belong to A_u, A_u and B_g irreducible representation, respectively.

With addition of these OOP modes to $\Gamma_{int.\ coord} = \Gamma_{vib}$, we get

$$\Gamma_{int.\ coord} = \Gamma_{vib} = \left(5A_g + 4B_u\right) + \left(2A_u + B_g\right)$$
$$= 5A_g + 2A_u + B_g + 4B_u$$

Identification of IR and Raman active vibrations

Here, in the present case, modes are either *g* or *u* type. So, all the *g* modes are active in Raman, whereas all the *u* modes are IR active. Thus, we can expect only six frequencies ($5A_g + B_g$) in the Raman spectrum and only six frequencies ($2A_u + 4B_u$) in IR spectrum.

The IR and Raman activity of various vibrations of *trans*- 1,2-Dichloroethylene molecule may be summarized as:

Mode	Infrared	Raman	Nature of vibration
$5A_g$	Inactive	Active	Mixed
$2A_u$	Active	Inactive	Pure out of plane (OOP)deformation
B_g	Inactive	Active	Pure out of plane (OOP)deformation
$4B_u$	Active	Inactive	Mixed

(j) Ethylene molecule of D_{2h} point group

This molecule belongs to C_{2h} point group with symmetry operations: E, $C_{2(z)}$, $C_{2(y)}$, $C_{2(x)}$, i, $\sigma_{(xy)}$, $\sigma_{(xz)}$ and $\sigma_{(yz)}$.

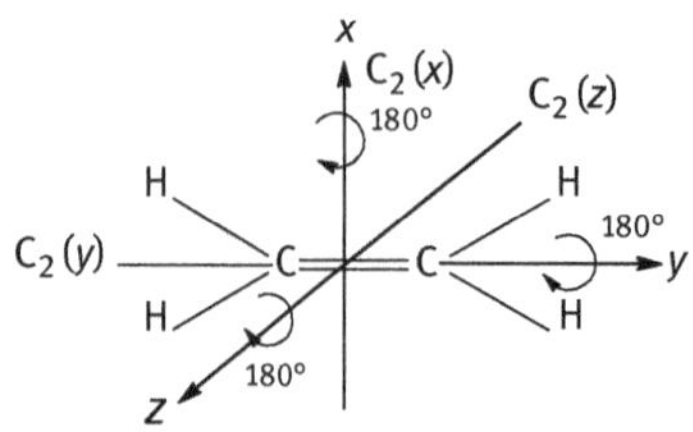

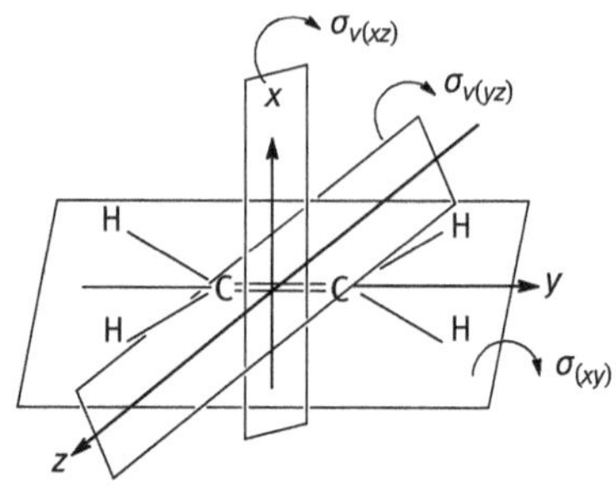

Cartesian coordinate method or 3N vector method

The symmetry operations in this case are four proper rotations through 0°(E), 180°[$C_{2(x)}$], 180°[$C_{2(y)}$], 180°[$C_{2(z)}$] and four improper rotations through 180°(*i*), 0°(σ_{xy}), 0°(σ_{xz}) and 0°(σ_{yz}). The following table provides the calculations of the total characters.

Table: Determination of total character for C_2H_4 molecule

D_{2h}	E	$C_2(z)$	$C_2(y)$	$C_2(x)$	i	σ_{xy}	σ_{xz}	σ_{yz}
β	0	180	180	180	180	0	0	0
cosβ	1	−1	−1	−1	−1	1	1	1
±1 + 2cosβ	3	−1	−1	−1	−3	1	1	1
n_R	6	0	2	0	0	6	0	2
Γ_{3N}	**18**	**0**	**−2**	**0**	0	**6**	**0**	**2**

The vertical dashed line separates the proper and improper rotations. The corresponding angleβ, cosβ and also (±1 + 2cosβ) values are tabulated for each operation. The proper rotations have (+1 + 2cosβ) and the improper rotations have (-1 + 2cosβ) values. The number of atoms retaining position (n_R) for each operation is also given in the table.

The above reducible representation (Γ_{3N}) can be reduced to irreducible representations using the reduction formula and character table of D_{2h} point group as follows:

Reducible representation in question

D_{2h}	E	$C_2(z)$	$C_2(y)$	$C_2(x)$	i	$\sigma_{(xy)}$	$\sigma_{(xz)}$	$\sigma_{(yz)}$
Γ_{3N}	18	0	−2	0	0	6	0	2

Character table of D_{2h} point group

D_{2h}	E	$C_2(z)$	$C_2(y)$	$C_2(x)$	i	$\sigma_{(xy)}$	$\sigma_{(xz)}$	$\sigma_{(yz)}$		
A_g	1	1	1	1	1	1	1	1		x^2, y^2, z^2
B_{1g}	1	1	−1	−1	1	1	−1	−1	R_z	xy
B_{2g}	1	−1	1	−1	1	−1	1	−1	R_y	xz
B_{3g}	1	−1	−1	1	1	−1	−1	1	R_x	yz
A_u	1	1	1	1	−1	−1	−1	−1		
B_{1u}	1	1	−1	−1	−1	−1	1	1	z	
B_{2u}	1	−1	1	−1	−1	1	−1	1	y	
B_{3u}	1	−1	−1	1	−1	1	1	−1	x	

$$NA_g = 1/8[(18.1.1) + (0.1.1)) + (-2.1.1) + (0.1.1) + (0.1.1) + (6.1.1) + (0.1.1) + (2.1.1)]$$
$$= 1/8[(18 + 0 - 2 + 0 + 0 + 6 + 0 + 2]$$
$$= 1/8[24] = 3$$
$$NB_{1g} = 1/8[(18.1.1) + (0.1.1)) + (-2.1.-1) + (0.1.1) + (0.1.1) + (6.1.1) + (0.1.1) + (2.1.-1)]$$
$$= 1/8[(18 + 0 + 2 + 0 + 0 + 6 + 0 - 2]$$
$$= 1/8[24] = 3$$
$$NB_{2g} = 1/8[(18.1.1) + (0.1.-1)) + (-2.1.1) + (0.1.-1) + (0.1.1) + (6.1.-1) + (0.1.1) + (2.1.-1)]$$
$$= 1/8[(18 + 0 - 2 + 0 + 0 - 6 + 0 - 2]$$
$$= 1/8[8] = 1$$

$$
\begin{aligned}
NB_{3g} &= 1/8[(18.1.1) + (0.1.-1)) + (-2.1.-1) + (0.1.1) + (0.1.1) + (6.1.-1) + (0.1.-1) \\
&\quad + (2.1.1)] \\
&= 1/8[(18 + 0 + 2 + 0 + 0 - 6 + 0 + 2] \\
&= 1/8[16] = 2
\end{aligned}
$$

$$
\begin{aligned}
NA_{u} &= 1/8[(18.1.1) + (0.1.1)) + (-2.1.1) + (0.1.1) + (0.1.-1) + (6.1.-1) + (0.1.-1) \\
&\quad + (2.1.-1)] \\
&= 1/8[(18 + 0 - 2 + 0 + 0 - 6 + 0 - 2] \\
&= 1/8[8] = 1
\end{aligned}
$$

$$
\begin{aligned}
NB_{1u} &= 1/8[(18.1.1) + (0.1.1)) + (-2.1.-1) + (0.1.-1) + (0.1.-1) + (6.1.-1) + (0.1.1) \\
&\quad + (2.1.1)] \\
&= 1/8[(18 + 0 + 2 + 0 + 0 - 6 + 0 + 2] \\
&= 1/8[16] = 2
\end{aligned}
$$

$$
\begin{aligned}
NB_{2u} &= 1/8[(18.1.1) + (0.1.-1)) + (-2.1.1) + (0.1.-1) + (0.1.-1) + (6.1.1) + (0.1.-1) \\
&\quad + (2.1.1)] \\
&= 1/8[(18 + 0 - 2 + 0 + 0 + 6 + 0 + 2] \\
&= 1/8[24] = 3
\end{aligned}
$$

$$
\begin{aligned}
NB_{3u} &= 1/8[(18.1.1) + (0.1.-1)) + (-2.1.-1) + (0.1.1) + (0.1.-1) + (6.1.1) + (0.1.1) \\
&\quad + (2.1.-1)] \\
&= 1/8[(18 + 0 + 2 + 0 + 0 + 6 + 0 - 2] \\
&= 1/8[24] = 3
\end{aligned}
$$

Thus, the reducible representation (Γ_{3N}) is reduced to

$$\Gamma_{3N} = 3A_g + 3B_{1g} + B_{2g} + 2B_{3g} + A_u + 2B_{1u} + 3B_{2u} + 3B_{3u}$$

Thus, there are 18 modes (3 + 3 + 1 + 2 + 1 + 2 + 3 + 3 = 18) of degrees of freedom with different symmetries given above. These include the vibrational, rotational and translational degrees of freedom.

The character table indicates that the translational modes have the same symmetry as the coordinates *x*, *y* and *z*, that is, B_{3u}, B_{2u} and B_{1u}. So also the rotational modes have the same symmetry as the rotation axes, R_x, R_y and R_z, that is, B_{3g}, B_{2g} and B_{1g} . Subtracting these translational and rotational degrees of freedom, we get the vibrational degrees of freedom only of appropriate symmetry.

$$\begin{aligned}
\Gamma_{3N} &= 3A_g + 3B_{1g} + B_{2g} + 2B_{3g} + A_u + 2B_{1u} + 3B_{2u} + 3B_{3u} \\
\Gamma_{tr} &= \qquad\qquad\qquad\qquad\qquad\qquad B_{1u} \quad B_{2u} \quad B_{3u} \\
\Gamma_{rot} &= \qquad B_{1g}, + B_{2g} + B_{3g} \\
&\cdots\cdots\cdots\cdots\cdots\cdots \\
\Gamma_{vib} &= 3A_g + 2B_{1g} + B_{3g} + A_u + B_{1u} + 2B_{2u} + 2B_{3u}
\end{aligned}$$

The number of vibrational mode corresponds to 3N-6 = 12.

Internal coordinate method

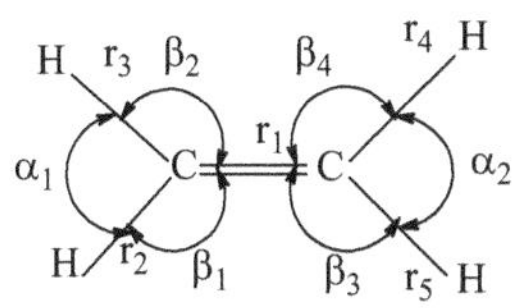

Possible internal coordinates of ethylene molecule

This molecule contains five stretches [r_1(C-C), r_2-r_5(C-H)] and six angle deformations (σ_1, σ_2; β_1 to β_4) as shown in the Fig. We can use these stretches and bending vectors as the basis for representation.

Γ_{r1}

D_{2h}	E	$C_{2(z)}$	$C_{2(y)}$	$C_{2(x)}$	i	$\sigma_{(xy)}$	$\sigma_{(xz)}$	$\sigma_{(yz)}$
No. of unshifted stretches (*n*)	1	1	1	1	1	1	1	1
$\chi(R)$	+ 1	+ 1	+ 1	+ 1	+ 1	+ 1	+ 1	+ 1
Γ_{r1} [$n\times\chi(R)$]	1	1	1	1	1	1	1	1

Looking over the character table of D_{2h} point group, we can observe that the representation Γ_{r1} belong to A_g irreducible representation.

Γ_{r2-r5}

D_{2h}	E	$C_{2(z)}$	$C_{2(y)}$	$C_{2(x)}$	i	$\sigma_{(xy)}$	$\sigma_{(xz)}$	$\sigma_{(yz)}$
No. of unshifted stretches (*n*)	4	0	0	0	0	4	0	0
$\chi(R)$	+ 1	+ 1	+ 1	+ 1	+ 1	+ 1	+ 1	+ 1
Γ_{r2-r5} [$n\times\chi(R)$]	4	0	0	0	0	4	0	0

This representation Γ_{r2-r5} reduces to:

$$\Gamma_{r2-r5} = A_g + B_{1g} + B_{2u} + B_{3u}$$

Γ_α

D_{2h}	E	$C_{2(z)}$	$C_{2(y)}$	$C_{2(x)}$	i	$\sigma_{(xy)}$	$\sigma_{(xz)}$	$\sigma_{(yz)}$
No. of unshifted double arrow headed vectors (*n*)	2	0	2	0	0	2	0	2
$\chi(R)$	+ 1	+ 1	+ 1	+ 1	+ 1	+ 1	+ 1	+ 1
Γ_α $[n\times\chi(R)]$	2	0	2	0	0	2	0	2

This reducible representation Γ_α reduces to:

$$\Gamma_\alpha = A_g + B_{2u}$$

Γ_β

D_{2h}	E	$C_{2(z)}$	$C_{2(y)}$	$C_{2(x)}$	i	$\sigma_{(xy)}$	$\sigma_{(xz)}$	$\sigma_{(yz)}$
No. of unshifted double arrow headed vectors (*n*)	4	0	0	0	0	4	0	0
$\chi(R)$	+ 1	+ 1	+ 1	+ 1	+ 1	+ 1	+ 1	+ 1
Γ_β $[n\times\chi(R)]$	4	0	0	0	0	4	0	0

Similar to reducible representation $\Gamma_{r2\text{-}r5}$, this reducible representation Γ_β is reduced to:

$$\Gamma_\beta = A_g + B_{1g} + B_{2u} + B_{3u}$$

Thus, the total reducible representation $\Gamma_{int\ coord}$ becomes

$$\begin{aligned}\Gamma_{int\,coord} &= \Gamma_{r1} + \Gamma_{r2-r5} + \Gamma_\alpha + \Gamma_\beta \\ &= A_g + \left(A_g + B_{1g} + B_{2u} + B_{3u}\right) + \left(A_g + B_{2u}\right)\ \left(A_g + B_{1g} + B_{2u} + B_{3u}\right) \\ &= 4A_g + 2B_{1g} + 3B_{2u} + 2B_{3u}\ \text{(11 coordinates)}\end{aligned}$$

When the results of $\Gamma_{int\ coord}$ and Γ_{vib} (= $3A_g + 2B_{1g} + B_{3g} + A_u + B_{1u} + 2B_{2u} + 2B_{3u}$) obtained through Cartesian coordinate method are compared, it is observed that one A_g and one B_{2u} of $\Gamma_{int\ coord}$ occur as extra modes and B_{3g}, A_u and B_{1u} as missing modes in $\Gamma_{int\ coord}$. The extra modes present in $\Gamma_{int\ coord}$ are called spurious modes. When these modes are subtracted from $\Gamma_{int\ coord}$, we get the actual modes

$$\Gamma_{\text{int coord}} = (4A_g + 2B_{1g} + 3B_{2u} + 2B_{3u}) - (A_g + B_{2u})$$
$$= 3A_g + 2B_{1g} + 2B_{2u} + 2B_{3u}(9 \text{ modes})$$

This requires three more modes A_u, B_{1u} and B_{3g},. As we have not considered OOP modes, while considering the internal coordinates, the missing mode can be found from them as given below:

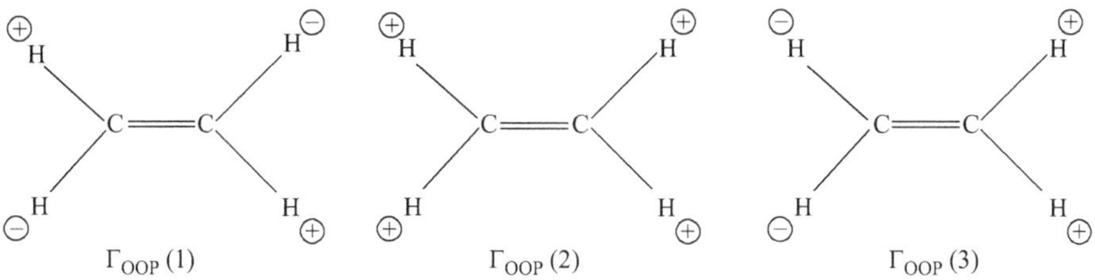

The character system of the above deformations can be worked out as:

D_{2h}	E	$C_{2(z)}$	$C_{2(y)}$	$C_{2(x)}$	i	$\sigma_{(xy)}$	$\sigma_{(xz)}$	$\sigma_{(yz)}$
Γ_{OOP} (1)	1	1	1	1	−1	−1	−1	−1
Γ_{OOP} (2)	1	1	−1	−1	−1	−1	1	1
Γ_{OOP} (3)	1	−1	−1	1	1	−1	−1	1

Looking over the character table of D_{2h} point group, we can identify that the representations $\Gamma_{oop}(1)$ and $\Gamma_{oop}(2)$ and $\Gamma_{oop}(3)$ belong to A_u, B_{1u} and B_{3g} irreducible representation, respectively.

With addition of these OOP modes to $\Gamma_{\text{int. coord}} = \Gamma_{\text{vib}}$, we get

$$\Gamma_{\text{int. coord}} = \Gamma_{\text{vib}} = (3A_g + 2B_{1g} + 2B_{2u} + 2B_{3u}) + (A_u + B_{1u} + B_{3g})$$
$$= 3A_g + 2B_{1g} + B_{3g} + A_u + B_{1u} + 2B_u + 2B_{3u} \ (12 \text{ modes})$$

Identification of IR and Raman active vibrations

From the character table of D_{2h} point group, it is notable that *x*, *y*, *z*-coordinates belong to B_{1u}, $2B_u$, $2B_{3u}$ irreducible representations. Hence, B_{1u}, $2B_u$ and $2B_{3u}$ vibrational modes will be IR active. Thus, we can expect only five frequencies in the IR spectra.

An observation of the character table indicates that the binary products of the Cartesian coordinates belong to A_g, B_{1g} and B_{3g} irreducible representations. Hence, $3A_g$, $2B_{1g}$, and B_{3g} vibrational modes are Raman active. Thus, we can expect six frequencies in the Raman spectrum. A_u vibrational mode is neither IR active nor Raman active.

The IR and Raman activity of various vibrations of C_2H_4 molecule may be summarized as:

Mode	Infrared	Raman	Nature of vibration
$3A_g$	Inactive	Active	Mixed
$2B_{1g}$	Inactive	Active	Mixed
B_{3g}	Inactive	Active	Pure out of plane (OOP)deformation
A_u	Inactive	Inactive	Pure out of plane (OOP)deformation
B_{1u}	Active	Inactive	Pure out of plane (OOP)deformation
$2B_u$	Active	Inactive	Mixed
$2B_{3u}$	Active	Inactive	Mixed

C_2H_4 is a centrosymmetric molecule, its IR and Raman spectra are mutually exclusive.

(k) Fullerene C_{60} of I_h point group

One of the best examples of a molecule belonging to I_h point group is fullerene C_{60}. It has 12 pentagonal and 20 hexagonal rings. Molecular symmetry and group theory have simplified the problem of finding and predicting IR and Raman active modes in this molecule. The symmetry operations in this point group are: E, $12C_5$, $12C_5^{\ 2}$, $20C_3$, $15C_2$, i, $12S_{10}$, $12S_{10}^{\ 3}$, $20S_6$ and 15σ. Thus, the total number of symmetry operations in this point group is 120 and these belong to 10 different classes.

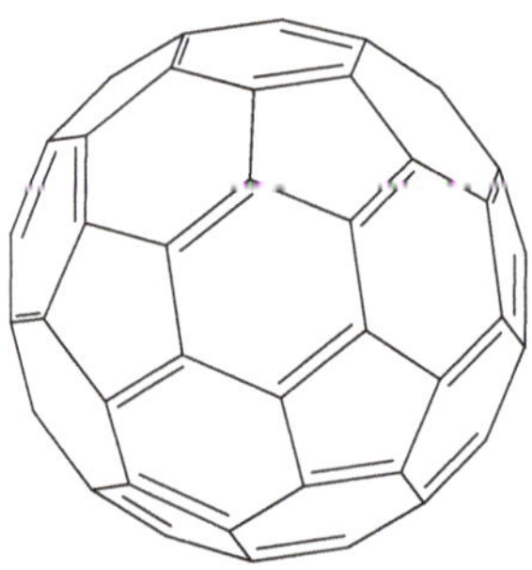

Fullerene C_{60}

Fullerene C_{60}

The character table of I_h point group is given below. This shows that characters for various rotational symmetry operations are of the type ½(1 + √5) and ½(1 − √5). There are only 10 irreducible representations in this point group. These irreducible representations are one-dimensional (A type), three-dimensional (T-type), four-dimensional (G-type) and five-dimensional (H-type).

If one carefully looks at the structure of C_{60}, it can be seen that no rotation axis pass through any of the atom of the molecule. It means that when rotational symmetry operations are carried out over these axes, then all atoms will be shifted. Hence, contribution by the basis set towards χ of the reducible representation will be zero. Likewise symmetry operation *i* will shift all the atoms, and so χ will be zero in this case. Each symmetry plane in the molecule contains only four atoms, and so they are not shifted by these planes will contribute to χ.

Character Table of I_h Point group:

I_h	E (1)	$12C_5$ (2)	$12C_5^2$ (3)	$20C_3$ (4)	$15C_2$ (5)	*i* (6)	$12S_{10}$ (7)
A_g	1	1	1	1	1	1	1
T_{1g}	3	½(1 + √5)	½(1 − √5)	0	−1	3	½(1 − √5)
T_{2g}	3	½(1 − √5)	½(1 + √5)	0	−1	3	½(1 + √5)
G_g	4	−1	−1	1	0	4	−1
H_g	5	0	0	−1	1	5	0
A_u	1	1	1	1	1	−1	−1
T_{1u}	3	½(1 + √5)	½(1 − √5)	0	−1	−3	−½(1 − √5)
T_{2u}	3	½(1 − √5)	½(1 + √5)	0	−1	−3	−½(1 + √5)
G_u	4	−1	−1	1	0	−4	1
H_u	5	0	0	−1	1	−5	0

I_h	$12S_{10}^3$ (8)	$20S_6$ (9)	15σ (10)	Linear function,rotation (11)	Quadratic functions (12)
A_g	1	1	1	–	$x^2 + y^2 + z^2$
T_{1g}	½(1 + √5)	0	−1	R_x, R_y, R_z	–
T_{2g}	½(1 − √5)	0	−1	–	–
G_g	−1	1	0	–	–
H_g	0	−1	1	–	$[2z^2-x^2-y^2, x^2-y^2, xy, xz, yz]$
A_u	−1	−1	−1	–	–
T_{1u}	−½(1 + √5)	0	1	(x, y, z)	–
T_{2u}	−½(1 − √5)	0	1	–	–
G_u	1	−1	0	–	–
H_u	0	1	−1	–	–

Cartesian coordinate method or 3N vector method

Looking over the complexity of the molecule, only Cartesian coordinate method will be applied here to get fundamental vibrational modes.

The symmetry operations in this case are five proper rotations through 0°(E), 72°(C_5) 144°(C_5^2), 120°(C_3) and 180°(C_2), and five improper rotations through 180°(i), 36°(S_{10}), 108°(S_{10}^3), 60°(S_6), and 0°(σ). The following table provides the calculations of the total characters.

Table: Determination of Total character of C_{60} fullerene molecule

I_h	E	$12C_5$	$12C_5^2$	$20C_3$	$15C_2$	i	$12S_{10}$	$12S_{10}^3$	$20S_6$	15σ
β	0	72	144	120	180	180	36	108	60	0
cosβ	1	$\frac{(\sqrt{5}-1)}{4}$	$\frac{(\sqrt{5}+1)}{4}$	$-\frac{1}{2}$	−1	−1	$\frac{(\sqrt{5}+1)}{4}$	$-\frac{(\sqrt{5}-1)}{4}$	$\frac{1}{2}$	1
±1 + 2cosβ	3	$\frac{\sqrt{5}+1)}{2}$	$\frac{\sqrt{5}+3)}{2}$	0	−1	−1	$-1+\frac{(\sqrt{5}+1)}{2}$	$-1-\frac{(\sqrt{5}-1)}{2}$	0	1
n_R	60	0	0	0	0	0	0	0	0	4
Γ_{3N}	180	0	0	0	0	0	0	0	0	4

The vertical dashed line separates the proper and improper rotations. The corresponding angleβ, cosβ and also (±1 + 2cosβ) values are tabulated for each operation. The proper rotations have (+1 + 2cosβ) and the improper rotations have (−1 + 2cosβ) values. The number of atoms retaining position (n_R) for each operation is also given in the table.

The above reducible representation (Γ_{3N}) can be reduced to irreducible representations applying the standard reduction formula and using the character table of I_h point group as follows:

$$N_i = \frac{1}{h}\sum_R \chi(R).n.\chi_i(R)$$

$$NA_g = 1/120[(180.1.1) + 0 + 0 + 0 + 0 + 0 + 0 + 0 + 0 + (4.15.1)$$
$$= 1/120[(180 + 60 = 240)] = 2$$
$$NT_{1g} = 1/120[(180.1.3) + 0 + 0 + 0 + 0 + 0 + 0 + 0 + 0 + (4.15.-1)$$
$$= 1/120[(540 - 60 = 480)] = 4$$
$$NT_{2g} = 1/120[(180.1.3) + 0 + 0 + 0 + 0 + 0 + 0 + 0 + 0 + (4.15.-1)$$
$$= 1/120[(540 - 60 = 480)] = 4$$

$$NA_g = 1/120[(180.1.1) + 0 + 0 + 0 + 0 + 0 + 0 + 0 + 0 + (4.15.1)$$
$$= 1/120[(180 + 60 = 240)] = 2$$
$$NT_{1g} = 1/120[(180.1.3) + 0 + 0 + 0 + 0 + 0 + 0 + 0 + 0 + (4.15. - 1)$$
$$= 1/120[(540 - 60 = 480)] = 4$$
$$NT_{2g} = 1/120[(180.1.3) + 0 + 0 + 0 + 0 + 0 + 0 + 0 + 0 + (4.15. - 1)$$
$$= 1/120[(540 - 60 = 480)] = 4$$
$$NG_g = 1/120[(180.1.4) + 0 + 0 + 0 + 0 + 0 + 0 + 0 + 0 + (4.15.0)$$
$$= 1/120[(720 + 0 = 720)] = 6$$
$$NH_g = 1/120[(180.1.5) + 0 + 0 + 0 + 0 + 0 + 0 + 0 + 0 + (4.15.1)$$
$$= 1/120[(900 + 60 = 960)] = 8$$
$$NA_u = 1/120[(180.1.1) + 0 + 0 + 0 + 0 + 0 + 0 + 0 + 0 + (4.15. - 1)$$
$$= 1/120[(180 - 60 = 120)] = 1$$
$$NT_{1u} = 1/120[(180.1.3) + 0 + 0 + 0 + 0 + 0 + 0 + 0 + 0 + (4.15.1)$$
$$= 1/120[(540 + 60 = 600)] = 5$$
$$NT_{2u} = 1/120[(180.1.3) + 0 + 0 + 0 + 0 + 0 + 0 + 0 + 0 + (4.15.1)$$
$$= 1/120[(540 + 60 = 600)] = 5$$
$$NG_u = 1/120[(180.1.4) + 0 + 0 + 0 + 0 + 0 + 0 + 0 + 0 + (4.15.0)$$
$$= 1/120[(720 + 0 = 720)] = 6$$
$$NH_u = 1/120[(180.1.5) + 0 + 0 + 0 + 0 + 0 + 0 + 0 + 0 + (4.15. - 1)$$
$$= 1/120[(900 - 60 = 840)] = 7$$

Thus, the reducible representation Γ_{3N} is reduced to:

$$\Gamma_{3N} = 2A_g + 4T_{1g} + 4T_{2g} + 6G_g + 8H_g + A_u + 5T_{1u} + 5T_{2u} + 6G_u + 7H_u$$

Thus, there are 180 modes (2 + 12 + 12 + 24 + 40 + 1 + 15 + 15 + 24 + 35 = 180) of degrees of freedom with different symmetry. These include the vibrational, rotational and translational degrees of freedom.

The I_h character table point out that the translational modes have the same symmetry as the coordinates x, y and z, that is, T_{1u}. So also the rotational modes have the same symmetry as the rotation axes, R_x, R_y and R_z, that is, T_{1g}. Subtracting these translational and rotational degrees of freedom, we get the vibrational degrees of freedom only of appropriate symmetry

$$\Gamma_{3N} = 2A_g + 4T_{1g} + 4T_{2g} + 6G_g + 8H_g + A_u + 5T_{1u} + 5T_{2u} + 6G_u + 7H_u$$
$$\Gamma_{tr} = T_{1u}$$
$$\Gamma_{rot} = T_{1g}$$

....

$$\Gamma_{vib} = 2A_g + 3T_{1g} + 4T_{2g} + 6G_g + 8H_g + A_u + 4T_{1u} + 5T_{2u} + 6G_u + 7H_u$$

Now, the total dimensionality number is 180−3(T_{1u}) − 3(T_{1g}) = 174

Identification of IR active vibrations

From the character table of I_h point group, it is notable that *x, y, z*-coordinates belong to T_{1u} irreducible representations. Hence, 4T_{1u} vibrational modes are IR active.

An observation of the I_h point group character table indicates that the binary products of the Cartesian coordinates belong to A_g and H_g irreducible representations. Hence 10 vibrational modes (2A_g and 8H_g) will be Raman active.

By taking examples of so many molecules discussed above, we have seen that how molecular symmetry and group theory is helpful in predicting IR and Raman active modes. We can analyze spectra of any molecule using Cartesian coordinate method and internal coordinate method and specific frequency to every vibrational mode and symmetry species can be ascertained.

3.8 Complications in IR and Raman spectra and difficulties in assignments

Because of certain factors, namely, overtones, combination bands, etc., sometimes the spectra become complicated, and so assignments also become very difficult. For larger molecules that have low symmetry, and for molecules that have more than one likely structures, the above procedure becomes very complex. Molecules may have different conformations and for such molecules different structures are assumed. Another difficulty encountered is that all fundamental modes may not appear in the spectrum with sufficient energy. Sometimes, frequency with similar symmetry may couple with each other and the resulting vibrations are mixtures of the original vibrations. Let us now discuss some of these factors that complicate the original spectrum, and where symmetry and group theory can help to resolve these complications.

3.8.1 Overtones, combination band, hot bands and Fermi resonance

By now, we have discussed the fundamental modes of vibrations for which basic selection rule, $\Delta v = \pm 1$ for a vibrational transition (fundamental transition) from quantum mechanics, is followed. If all the transitions are simple harmonics, then all other transitions (except fundamental transitions) will be forbidden. In fact, molecular vibrations are generally anharmonic. In this situation, it is possible to excite two or more vibrational modes at a time and the excitation can be from ν_0 to ν_2 or ν_0 to $(\nu_1 + \nu_2)$ or ν_0 to $(\nu_1 - \nu_2)$ or ν_0 to $(2\nu_1 + \nu_2)$. This gives rise overtones, combination bands, etc. Consequently, a fundamental mode that is forbidden in IR may be allowed as an overtone or combination band.

Appearance of overtones or combination bands is determined by symmetry selection rules similar to one we have considered earlier, that is, we have to consider the direct product and vanishing integrals involving ground-state symmetry (ν_0) and excited-state symmetry ($\nu_2, \nu_3, \ldots$). For non-degenerate vibrations, as per selection rule, the symmetry of the upper level should be the same as that of at least one of the components of the transitional vectors, T_x, T_y, T_z or x, y, z-Cartesian coordinates. In order to determine the overtone band that are allowed in IR, it is necessary to take the direct product of irreducible representation of the fundamental that make up the overtone.

Now, we will take up detailed discussion each type of such bands one by one.

3.8.2 Overtones

If a molecule absorbs two or more quanta of energy in one and the same vibrational mode say ν_1, an overtone band arises with a frequency close to $n\nu_1$ (where $n = 2,3,4,\ldots\ldots$). So, bands at $2\nu_1$, $3\nu_1$, $4\nu_1$ may occur with $\Delta\nu = \pm 2, \pm 3, \pm 4 \ldots \pm n$. Consequently, band at $2\nu_1$ is called first overtone, at $3\nu_1$ is second overtone and so on. The observed frequencies of overtone are always slightly lower than the whole

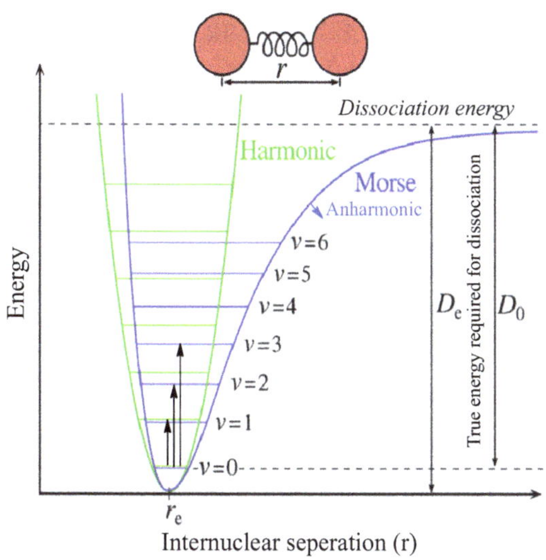

Fig. 3.8: Potential energy curve for diatomic molecules having anharmonic vibrations.

number multiple values ($2\nu_1$, $3\nu_1$,... etc.) because the spacing between successive vibrational levels decreases due to anharmonicity (Fig. 3.8). Hence, the difference between actual overtone frequency and integral value is a measure of anharmonicity of the vibration.

Let us explain the formation of overtones taking few examples.

(i) Molecule of C_{3v} point group

Let us consider NH_3 molecule belonging to C_{3v} point group and a fundamental of A_2 species as shown in the character table of this point group. Let us observe whether there will be overtone for this fundamental or not. Now, we take here the direct product of ($A_2 \times A_2$) and reduce it. If this product contains symmetric mode, then overtone will be IR active.

Character table of C_{3v} point group

C_{3v}	E	$2C_3$	3σ		
A_1	1	1	1	z, T_z	x^2+y^2, z^2
A_2	1	1	−1	R_z	
E	2	−1	0	$(T_x, T_y) + (R_x, R_y)$	(x^2-y^2, xy) (xz, yz)

So,

$$(A_2 \times A_2) = (1 \times 1), (1 \times 1), (-1 \times -1) = 1, 1, 1$$

Here, the character of each symmetry operation in ($A_2 \times A_2$) = 1, 1, 1, that is, A_1 mode. Thus, the direct product of ($A_2 \times A_2$) on reduction contains A_1 mode, which is symmetric. This suggests that the first overtone of IR inactive A_2 mode is IR active

Let us consider whether the second overtone of A_2 mode is IR active or not. For this, we have to take the direct product of ($A_2 \times A_2 \times A_2$).

$$(A_2 \times A_2 \times A_2) = (1 \times 1 \times 1), (1 \times 1 \times 1), (-1 \times -1 \times -1) = 1, 1, -1 (A_2 \text{ mode, not symmetric})$$

This indicates that the second overtone of A_2 mode is IR inactive.

Similarly, such direct product method may be used to find overtones of fundamental modes in other point group.

3.8.3 Method for finding overtones for degenerate vibrational modes

(a) Doubly degenerate vibrational mode (E)

In case of non-degenerate vibrations, we have just used the direct product method for finding the characters of symmetry operations for reducible representation. For degenerate vibration mode, the method is quite complicated. For doubly degenerate vibrational mode (E), the following equation is used for finding the character of symmetry operations for reducible representation corresponding to *n*th overtones.

$$^{n}\chi_E(R) = 1/2\,[^{n-1}\chi_E(R)\cdot\chi_E(R) + \chi_E(R^n)] \tag{i}$$

Here, $^{n}\chi_E(R)$ is the character of nth overtones for symmetry operation R and $^{n-1}\chi_E(R)$ is the character of (n-1)th overtone. $\chi_E(R)$ is the character of the fundamental for which activity of overtone is to be worked out. For the first overtone, n = 2 and for second overtone, n = 3. The $\chi_E(R^n)$ is the character corresponding to operation R performed n times in succession.

For E mode of vibration and for n = 2 (first overtone), the above equation may be written as:

$$^{2}\chi_E(R) = 1/2[^{1}\chi_E(R).\chi_E(R) + \chi_E(R^2)]$$

where $\chi_E(R^2)$ is the character of operation R performed two times in succession.

Let us find $\chi_E(R^2)$ for each symmetry operation for C_{3v} point group.

For E representation $\chi_E(E^2) = \chi_E(E) = 2$, that is, character E (identity) in E mode. $\chi_E(C_3^2) = \chi_E(C_3^3.C_3^{-1}) = \chi_E(E.C_3^{-1}) = \chi_E(C_3^{-1}) = -1$ and $\chi_E(\sigma_v^2) = \chi_E(E) = 2$ [values taken from C_{3v} character table].

Hence, $\chi_E(R^2)$ for E, $2C_3$, $3\sigma_v$ are 2, −1, 2, respectively. The values of $\chi_E(R^n)$ for n up to 5 are summarized as:

C_{3v}	E	$2C_3$	$3\sigma_v$
$\chi_E(R^2)$	2	−1	2
$\chi_E(R^3)$	2	2	0
$\chi_E(R^4)$	2	−1	2
$\chi_E(R^5)$	2	−1	0

Thus, using equation (i) given above, for n = 2, that is, for first overtone, $^{2}\chi_E(R)$ is obtained as follows:

C_{3v}	E	$2C_3$	$3\sigma_v$
$^{1}\chi_E(R)$	2	−1	0
$\chi_E(R)$	2	−1	0
$^{1}\chi_E(R).\ \chi_E(R)$	4	1	0 ......(a)
$\chi_E(R^2)$	2	−1	2 ......(b)
$^{1}\chi_E(R).\ \chi_E(R) + \chi_E(R^2)$	6	0	2 (a + b)
$^{2}\chi_E(R) = 1/2[^{1}\chi_E(R).\ \chi_E(R) + \chi_E(R^2)]$	3	0	1 [½(a + b)]

The above manipulation shows that $^{2}\chi_E(R)$ is (E = 3, $2C_3$ = 0, $3\sigma_v$ = 1). This reducible representation can be reduced applying usual reduction formula and using character table of C_{3v} point group.

$$N_i = \frac{1}{h}\sum_R \chi(R).n.\chi_i(R)$$

Reducible representation in question

C_{3v}	E	$2C_3$	3σ
$^2\chi_E$ (R)	3	0	1

Character table of C_{3v} point group

C_{3v}	E	$2C_3$	3σ		
A_1	1	1	1	z	x^2+y^2, z^2
A_2	1	1	−1	R_z	
E	2	−1	0	(x,y) (R_x, R_y)	(x^2-y^2, xy) (xz,yz)

$$\begin{aligned}
NA_1 &= 1/6[(3.1.1) + 0 + (1.3.1)] \\
&= 1/6[3 + 0 + 3] \\
&= 1/6[6] = 1 \\
NA_2 &= 1/6[(3.1.1) + 0 + (1.3.-1)] \\
&= 1/6[3 - 3] \\
&= 1/6[0] = 0 \\
NE &= 1/6[(3.1.2) + 0 + 0] \\
&= 1/6[6 + 0 + 0] \\
&= 1/6[6] - 1
\end{aligned}$$

Thus, the reducible representation $^2\chi_E(R)$ is reduced to:

$$A_1 + E$$

This reduction result contains A_1 mode which is totally symmetric irreducible representation and E representation is also IR active (because x, y belong to E; see character table). Hence, both the components are IR active. So, the first overtone of E mode in C_{3v} point group is IR active.

For the second overtone of E mode ($n = 3$), the equation (i) may be written as:

$$^3\chi_E(R) = 1/2[^2\chi_E(R).\chi_E(R) + \chi_E(R^3)]$$

The different components for getting the value of $^3\chi_E(R)$ are given in the following table.

C_{3v}	E	$2C_3$	$3\sigma_v$
$^2\chi_E(R)$ (First overtone) (obtained above)	3	0	1
$\chi_E(R)$	2	−1	0
$^2\chi_E(R).\ \chi_E(R)$	6	0	0 ......(a)
$\chi_E(R^3)$ (taken from above table)	2	2	0 ......(b)
$^2\chi_E(R).\ \chi_E(R) + \chi_E(R^3)$	8	2	0 (a + b)
$^3\chi_E(R) = 1/2[^1\chi_E(R).\ \chi_E(R) + \chi_E(R^2)]$	4	1	0 [½(a + b)]

From the above manipulation we see that $^3\chi_E(R)$ is ($E = 4$, $2C_3 = 1$, $3\sigma_v = 0$). This reducible representation can be reduced applying usual reduction formula and using character table of C_{3v} point group.

$$N_i = \frac{1}{h}\sum_R \chi(R)\,.\,n\,.\,\chi_i\,(R)$$

Reducible representation in question

C_{3v}	E	$2C_3$	3σ
$^3\chi_E$ (R)	4	1	0

Character table of C_{3v} point group

C_{3v}	E	$2C_3$	3σ		
A_1	1	1	1	z	x^2+y^2, z^2
A_2	1	1	−1	R_z	
E	2	−1	0	$(x,y)\ (R_x, R_y)$	$(x^2-y^2, xy)\ (xz,yz)$

$$\begin{aligned} NA_1 &= 1/6[(4.1.1) + (1.2.1) + 0] \\ &= 1/6[4 + 2 + 0] \\ &= 1/6[6] = 1 \end{aligned}$$

$$\begin{aligned} NA_2 &= 1/6[(4.1.1) + (1.2.1) + 0] \\ &= 1/6[4 + 2 + 0] \\ &= 1/6[6] = 1 \end{aligned}$$

$$\begin{aligned} NE &= 1/6[(4.1.2) + (1.2. - 1) + 0] \\ &= 1/6[8 - 2 + 0] \\ &= 1/6[6] = 1 \end{aligned}$$

Thus, the reducible representation $^3\chi_E(R)$ is reduced to:

$$A_1 + A_2 + E$$

Of the three components, A_1 and E are IR active (because z belongs to $A_1$and x, y belong to E; see character table) and A_2 is not IR active. As A_1 and E (we require at least one active mode) are IR active, second overtone of E-mode is also IR active).

(b) Triply degenerate vibrational mode (T or F)

For triply degenerate (T or F) vibrational modes, the equation for working our character for reducible representation is somewhat more complicated as shown below:

$$^n\chi_T(R) = 1/3[2\chi_T{}^{n-1}(R).\chi_T(R) - 1/2\{\chi_T{}^{n-2}(R)\ (\chi_T(R))^2\} + 1/2\chi_T{}^{n-2}(R).\chi_T(R^2) + \chi_T(R^n)] \quad \ldots(ii)$$

Let us consider T_2 mode in T_d point group and find the IR activity of first overtone ($n = 2$).

Character table of T_d point group

T_d	E	$8C_3$	$3C_2$	$6S_4$	$6\sigma_d$		
A_1	1	1	1	1	1		$x^2+y^2+z^2$
A_2	1	1	1	−1	−1		
E	2	−1	2	0	0		$(2z^2-x^2-y^2, x^2-y^2)$
T_1	3	0	−1	1	−1	(Rx, Ry, Rz)	
T_2	3	0	−1	−1	1	(x, y, z)	(xy, yz, xz)

For the first overtone of T_2 mode in T_d point group, the equation (ii) may be simplified as:

$$\begin{aligned} 2\chi_{T2}(R) &= 1/3[2\chi_{T2}{}^{2-1}(R).\chi_{T2}(R) - 1/2\{\chi_{T2}{}^{2-2}(R)(\chi_{T2}(R)))^2\} + 1/2\chi_{T2}{}^{2-2}(R).\chi_{T2}(R^2) \\ &\quad + \chi_{T2}(R^2)] \\ &= 1/3[2\{\chi_{T2}(R)\}^2 - 1/2\{(\chi_{T2}(R))^2\} + 1/2\chi_{T2}(R^2) + \chi_{T2}(R^2) \\ &= 1/3[2\{\chi_{T2}(R)\}^2 - 1/2\{(\chi_{T2}(R))^2\} + 3/2\chi_{T2}(R^2) \\ &= 1/3[3/2\{\chi_{T2}(R)\}^2 + 3/2\chi_{T2}(R^2)] \\ &= 1/2[\{\chi_{T2}(R)\}^2 + \{\chi_{T2}(R^2)\}] \end{aligned}$$

Now, we can find the values of terms $\chi_{T2}(R)$ and $\chi_{T2}(R^2)$ in this simplified equation. $\chi_{T2}(R)$ is given as:

T_d	E	$8C_3$	$3C_2$	$6S_4$	$6\sigma_d$
$\chi_{T2}(R)$	3	0	−1	−1	1

The next step is now to find $\chi_T(R^2)$ for various symmetry operations of this point group (T_d) and the results obtained are as follows:

$$[\chi_{T2}(E^2)] = \chi_{T2}(E) = 3$$
$$[\chi_{T2}(C_3^2)] = \chi_{T2}(C_3^{-1}) = 0$$
$$[\chi_{T2}(C_2^2)] = \chi_{T2}(E) = 3$$
$$[\chi_{T2}(S_4^2)] = \chi_{T2}(C_2) = -1$$
$$[\chi_{T2}(\sigma_d^2)] = \chi_{T2}(E) = 3$$

Hence,

T_d	E	$8C_3$	$3C_2$	$6S_4$	$6\sigma_d$
$\chi_{T2}(R)$	3	0	−1	−1	1
$\chi_T(R^2)$	3	0	3	−1	3

Now, the values for $\chi_{T2}(R^n)$ for various values of n are tabulated below:

T_d	E	$8C_3$	$3C_2$	$6S_4$	$6\sigma_d$
$\chi_{T2}(R)$	3	0	−1	−1	1
$\chi_T(R^2)$	3	0	3	−1	3
$\chi_T(R^3)$	3	3	−1	−1	1
$\chi_T(R^4)$	3	0	3	3	3

Let us find ${}^2\chi_{T2}(R)$ using the data collected above as follows:

T_d	E	$8C_3$	$3C_2$	$6S_4$	$6\sigma_d$
$\chi_{T2}(R)$	3	0	−1	−1	1
$\chi_T(R)\times\chi_T(R) = \{\chi_T(R)\}^2$	9	0	1	1	1...... (a)
$\chi_T(R^2)$	3	0	3	−1	3......(b)
$\{\chi_T(R)\}^2 + \chi_T(R^2)$	12	0	4	0	4 (a + b)
$1/2[\{\chi_T(R)\}^2 + \chi_T(R^2)]$	6	0	2	0	2
${}^2\chi_{T2}(R)$	6	0	2	0	2

From the above table we see that ${}^2\chi_{T2}(R)$ is [$E = 6$, $8C_3 = 0$, $3C_2 = 2$, $6S_4 = 0$, $6\sigma_d = 2$]. This reducible representation can be reduced applying usual reduction formula and using character table of C_{3v} point group.

Reducible representation in question

T_d	E	$8C_3$	$3C_2$	$6S_4$	$6\sigma_d$
$^2\chi_{T2}(R)$	6	0	2	0	2

Character table of T_d point group

T_d	E	$8C_3$	$3C_2$	$6S_4$	$6\sigma_d$		
A_1	1	1	1	1	1		$x^2+y^2+z^2$
A_2	1	1	1	−1	−1		
E	2	−1	2	0	0		$(2z^2-x^2-y^2, x^2-y^2)$
T_1	3	0	−1	1	−1	(Rx, Ry, Rz)	
T_2	3	0	−1	−1	1	(x, y, z)	(xy, yz, xz)

$$\begin{aligned}
NA_1 &= 1/24[(6.1.1)+0+(2.3.1)+0+(2.6.1)] \\
&= 1/24[(6+0+6+0+12] \\
&= 1/24[24] = 1 \\
NA_2 &= 1/24[(6.1.1)+0+(2.3.1)+0+(2.6.-1)] \\
&= 1/24[(6+0+6+0-12] \\
&= 1/24[0] = 0 \\
NE &= 1/24[(6.1.2)+0+(2.3.2)+0+0] \\
&= 1/24[(12+0+12+0+0] \\
&= 1/24[24] = 1 \\
NT_1 &= 1/24[(6.1.3)+0+(2.3.-1)+0+(2.6.-1)] \\
&= 1/24[(18+0-6+0-12] \\
&= 1/24[0] = 0 \\
NT_2 &= 1/24[(6.1.3)+0+(2.3.-1)+0+(2.6.1)] \\
&= 1/24[(18+0-6+0+12] \\
&= 1/24[24] = 1
\end{aligned}$$

Thus, the reducible representation $^2\chi_{T2}(R)$ is reduced to:

$$A_1 + E + T_2$$

Out of these three components, T_2 is IR active (we need at least one active mode). So, the first overtone of T_2 mode in T_d point group is IR active.

Similarly, we can work out the activity of overtones of non-degenerate and degenerate vibrational modes of any point group.

3.8.4 Combination bands

These types of bands occur from transitions to an excited state where the molecule has gained two or more vibrational quanta of energy that is distributed among two or more modes. Thus, if two fundamental bands are at ν_i and ν_j, then a binary combination will generally be found at ~(ν_i+ν_j). The transition energy will be slightly less than the sum of fundamentals.

Sometimes, difference bands may also be observed. In such a case, a molecule which is already in the excited state ν_n absorbs other quanta of energy ν_m and looses the earlier quanta, ν_n of energy. Then a difference band may be observed at (ν_n−ν_m). Such bands are rare as only few molecules exist initially in the excited state except at high temperature. It is, therefore, necessary to find the symmetries of these combinations/difference bands and check these are vibrationally active or not.

The occurrence of **combination/difference** bands may be explained by taking example of CH_4, a molecule belonging to T_d point group.

In the previous section, we have already observed that for this molecule,

$$\Gamma_{\text{vib}} = A_1 + E + 2T_2$$
$$\Gamma_{\text{str}} = A_1 + T_2$$
$$\Gamma_{\text{bend}} = E + T_2$$

and various IR/Raman frequencies observed are:

$$\nu_1(A_1):2900\ \text{cm}^{-1} \rightarrow \text{Raman active}$$
$$\nu_2(E):1525\ \text{cm}^{-1} \rightarrow \text{Raman active}$$
$$\nu_3(T_2):3020\ \text{cm}^{-1} \rightarrow \text{IR/Raman active}$$
$$\nu_4(T_2):1300\ \text{cm}^{-1} \rightarrow \text{IR/Raman active}$$

Based on the above data, the possible combination bands are:

$$\nu_1 + \nu_2 = 4425\ \text{cm}^{-1} \qquad \nu_1 + \nu_3 = 5920\ \text{cm}^{-1}$$
$$\nu_1 + \nu_4 = 4200\ \text{cm}^{-1} \qquad \nu_2 + \nu_3 = 4545\ \text{cm}^{-1}$$
$$\nu_2 + \nu_4 = 2825\ \text{cm}^{-1} \qquad \nu_3 + \nu_4 = 4320\ \text{cm}^{-1}$$
$$2\nu_1 = 5800\ \text{cm}^{-1} \qquad 2\nu_2 = 3050\ \text{cm}^{-1}$$

Now, we will see whether any of these combinations will appear in IR/ Raman spectra. For this, we have to take the direct product of the symmetry species of these two states to which these frequencies belong and reduce it. If the direct product contains symmetry species that is, IR or Raman active, then the combination band will be vibrationally active.

Some of the above examples are discussed here with regard to vibrational activity.

(i) Let us take $2\nu_1 = 5800\ \text{cm}^{-1}$. ν_1 has A_1 symmetry. So, we will obtain $\Gamma_{A1.\ A1}$ using part of the character table of T_d point group as shown below:

T_d	E	$8C_3$	$3C_2$	$6S_4$	$6\sigma_d$
A_1	1	1	1	1	1
A_1	1	1	1	1	1
$\Gamma_{A1.\ A1}$	1	1	1	1	1

Inspection of character table reveals that $\Gamma_{A1.\ A1}$ contains A_1 symmetry, which is Raman active. Hence, $2\nu_1$ combination band is Raman active.

(ii) $2\nu_2 = 3050\ \text{cm}^{-1}$. Here, ν_2 has *E* symmetry. So, we will obtain $\Gamma_{E.\ E}$ using part of the character table of T_d point group as shown below:

T_d	E	$8C_3$	$3C_2$	$6S_4$	$6\sigma_d$
E	2	−1	2	0	0
E	2	−1	2	0	0
$\Gamma_{E.\ E}$	4	1	4	0	0

This reducible representation $\Gamma_{E.\ E}$ can be reduced to irreducible representation as follows:

Reducible representation in question

T_d	E	$8C_3$	$3C_2$	$6S_4$	$6\sigma_d$
$\Gamma_{E.E}$	4	1	4	0	0

Character table of T_d point group

T_d	E	$8C_3$	$3C_2$	$6S_4$	$6\sigma_d$		
A_1	1	1	1	1	1		$x^2+y^2+z^2$
A_2	1	1	1	−1	−1		
E	2	−1	2	0	0		$(2z^2-x^2-y^2, x^2-y^2)$
T_1	3	0	−1	1	−1	(Rx, Ry, Rz)	
T_2	3	0	−1	−1	1	(x, y, z)	(xy, yz, xz)

$$NA_1 = 1/24[(4.1.1) + (1.8.1) + (4.3.1) + 0 + 0]$$
$$= 1/24[(4 + 8 + 12 + 0 + 0]$$
$$= 1/24[24] = 1$$
$$NA_2 = 1/24[(4.1.1) + (1.8.1) + (4.3.1) + 0 + 0]$$
$$= 1/24[(4 + 8 + 12 + 0 + 0]$$
$$= 1/24[24] = 1$$
$$NE = 1/24[(4.1.2) + (1.8. - 1) + (4.3.2) + 0 + 0]$$
$$= 1/24[(8 - 8 + 24 + 0 + 0]$$
$$= 1/24[24] = 1$$
$$NT_1 = 1/24[(4.1.3) + 0 + (4.3. - 1) + 0 + 0]$$
$$= 1/24[(12 + 0 - 12 + 0 + 0]$$
$$= 1/24[0] = 0$$
$$NT_2 = 1/24[(4.1.3) + 0 + (4.3. - 1) + 0 + 0]$$
$$= 1/24[(12 + 0 - 12 + 0 + 0]$$
$$= 1/24[0] = 0$$

Thus, the reducible representation $\Gamma_{E.\ E}$ is reduced to:

$$A_1 + A_2 + E$$

Inspection of character table reveals that A_1 and E is Raman active. Hence, the combination $2\nu_2$ is vibrationally active.

(iii) $\nu_3 + \nu_4 = 4320\ cm^{-1}$. As ν_3 and ν_4 have T_2 symmetry, so, we will obtain $\Gamma_{T2.\ T2}$ using part of the character table of T_d point group as shown below:

T_d	E	$8C_3$	$3C_2$	$6S_4$	$6\sigma_d$
T_2	3	0	−1	−1	1
T_2	3	0	−1	−1	1
$\Gamma_{T2.\ T2}$	9	0	1	1	1

The reducible representation $\Gamma_{T2.T2}$ can be reduced to irreducible representation as follows:

Reducible representation in question

T_d	E	$8C_3$	$3C_2$	$6S_4$	$6\sigma_d$
$\Gamma_{T2.T2}$	9	0	1	1	1

Character table of T_d point group

T_d	E	$8C_3$	$3C_2$	$6S_4$	$6\sigma_d$		
A_1	1	1	1	1	1		$x^2+y^2+z^2$
A_2	1	1	1	−1	−1		
E	2	−1	2	0	0		$(2z^2-x^2-y^2, x^2-y^2)$
T_1	3	0	−1	1	−1	(Rx, Ry, Rz)	
T_2	3	0	−1	−1	1	(x, y, z)	(xy, yz, xz)

$$NA_1 = 1/24[(9.1.1) + 0 + (1.3.1) + (1.6.1) + (1.6.1)]$$
$$= 1/24[(9 + 0 + 3 + 6 + 6]$$
$$= 1/24[24] = 1$$
$$NA_2 = 1/24[(9.1.1) + 0 + (1.3.1) + (1.6. - 1) + (1.6. - 1)]$$
$$= 1/24[(9 + 0 + 3 - 6 - 6]$$
$$= 1/24[0] = 0$$
$$NE = 1/24[(9.1.2) + 0 + (1.3.2) + 0 + 0]$$
$$= 1/24[(18 + 0 + 6 + 0 + 0]$$
$$= 1/24[24] = 1$$
$$NT_1 = 1/24[(9.1.3) + 0 + (1.3. - 1) + (1.6.1) + (1.6. - 1)]$$
$$= 1/24[(27 + 0 - 3 + 6 - 6]$$
$$= 1/24[24] = 1$$
$$NT_2 = 1/24[(9.1.3) + 0 + (1.3. - 1) + (1.6. - 1) + (1.6.1)]$$
$$= 1/24[(27 + 0 - 3 - 6 + 6]$$
$$= 1/24[24] = 1$$

Thus, the reducible representation $\Gamma_{T2.T2}$ is reduced to:

$$A_1 + E + T_1 + T_2$$

Inspection of character table reveals that A_1 and E is Raman active and T_2 is IR and Raman active. Hence, the combination $\nu_3 + \nu_4$ is both IR and Raman active.

(iv) $\nu_3 - \nu_4 = 1720\ \text{cm}^{-1}$. As ν_3 and ν_4 belong to T_2 symmetry, so, we will obtain $\Gamma_{T2.\ T2}$ using part of the character table of T_d point group.

Again, $\Gamma_{T2.\ T2}$ gives $A_1 + E + T_1 + T_2$ irreducible representation (as derived above). Hence, the $(\nu_3 - \nu_4)$ difference band is both IR and Raman active.

3.8.5 Hot bands

When a molecule already present in the excited state absorbs one quanta of energy and excited to higher vibrational state, hot band arise. These bands are usually found when the spectrum is recorded at high temperature. These bands vanish when the spectrum is recorded at lower temperature. The activity of such bands can be checked in a similar manner as discussed above. The hot band corresponding to the transition $\nu_1 \rightarrow \nu_2$ is nearly one-tenth as intense as $\nu_0 \rightarrow \nu_1$ fundamental band. The hot bands along with fundamental and overtone bands are shown in the following Fig. 3.9. The transition 1 is fundamental transition, 2 and 3 transitions are first and second overtones, respectively and transitions 4 and 5 are hot bands.

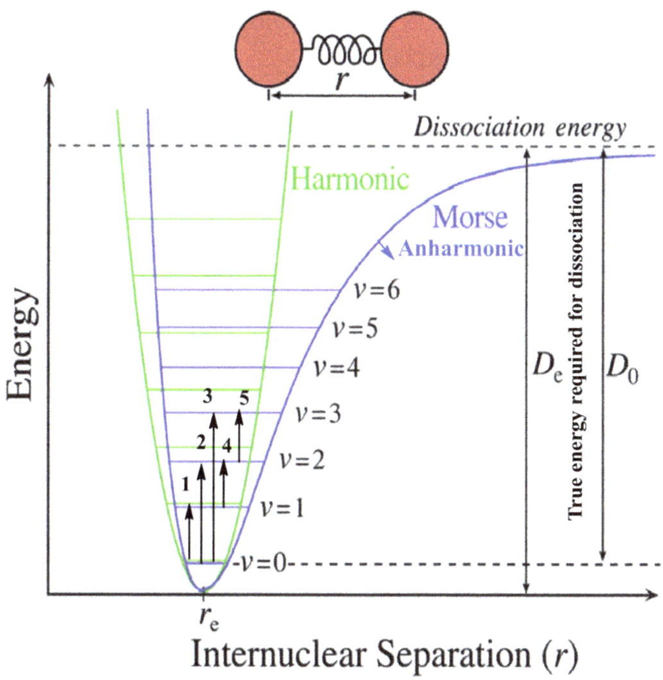

Fig.3.9: Fundamental, overtones and hot bands in vibrational spectrum.

3.8.6 Fermi resonance

Usually, the overtones and combination bands have low intensities but in some cases, they have surprisingly high intensities. This occurs when they appear near to a fundamental with the same symmetry. In such cases, the overtones or combination bands borrow intensity from the fundamental. The two bands (fundamental band and the overtone or combination band) mix (quantum mechanical interaction or mixing) and split, losing their individual identity and give rise to a pair of bands of comparable intensity. This interaction of fundamental and binary bands (overtones or combination) of same symmetry and approximately same energy is called Fermi resonance. It is notable here that symmetry of fundamental bands and the binary bands should be the same; otherwise, there will be no interaction or quantum mechanical mixing. Although, the interaction is in between different kinds of bands but due to approximately same energy and same symmetry we may call them accidently degenerate.

The phenomenon of Fermi resonance has two noticeable effects:

(i) The overtone can gain intensity from nearby fundamental of the same symmetry.

(ii) Both energy levels are shifted from each other, therefore, higher energy band shifts to higher energy side and lower energy band shifts to lower energy side and both have nearly equal intensity.

So, when we expect a single band and actually if it is a doublet, then the possibility of Fermi resonance is considered for its interpretation.

(a) Let us explain the Raman spectrum of CCl_4 molecule considering the occurrence of Fermi resonance. In the previous section, we have already observed that Raman spectrum of CCl_4 should show four bands for:

(i) $A_1 + T_2$: C–Cl stretches and (ii) $E + T_2$: angle of deformation bands.

In reality, five Raman bands are observed with the data shown below:

$$791\text{cm}^{-1}\ (T_2);\ 762\ \text{cm}^{-1}\ (T_2);\ 459\ \text{cm}^{-1}\ (A_1);\ 315\ \text{cm}^{-1}\ (T_2)\ \text{and}\ 217\ \text{cm}^{-1}\ (E),$$

Since frequencies 791cm^{-1} and 762 cm^{-1} correspond to C-Cl stretching, these have T_2 symmetries. 491 cm^{-1} is A_1 stretching mode. So, we need only one stretching frequency corresponding to T_2 symmetry but we are getting two frequencies (doublet) of T_2 symmetry. Now the possibility of Fermi resonance is the only way to account for the doublet as follows:

The C-Cl antisymmetric stretches come in the range 770–780 cm^{-1}. The combination band of 459 (A_1) and 315 cm^{-1}(T_2) comes at 774 cm^{-1}. The combination band at 774 cm^{-1} is due to the direct product of A_1 and T_2 mode, that is, $A_1 \times T_2 = T_2$ which is both IR and Raman active. The combination band at 774 cm^{-1} lies between the expected C-Cl stretches range770–780 cm^{-1}, and both have the same symmetry T_2. So, Fermi resonance will occur between fundamental and combination bands. Due to this Fermi resonance, one frequency will be lowered to 762 cm^{-1} and other will be raised to 791cm^{-1}. Consequently, one will get five lines instead of four in the Raman spectrum of CCl_4.

(b) Another interesting molecule exhibiting Fermi resonance is CO_2, a linear molecule. As CO_2 is a linear molecule, there should be $3n-5 = 3 \times 3-5 = 4$ vibrations possible. These are ν_s, ν_{as} and bending vibrations in plane (δ) and out of plane (π).

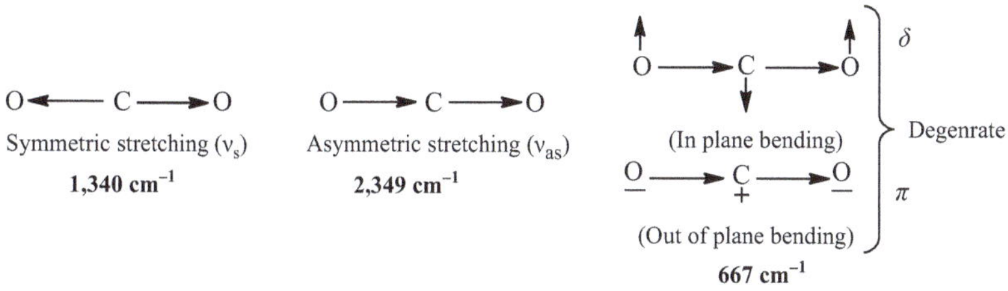

Out of four vibrations, ν_s does not bring change in the dipole moment and should be IR inactive. The other three vibrations bring change in dipole moment, and hence are IR active. Though δ and π are different vibration modes, but for a linear molecule, they are equivalent and hence are degenerate.

The IR bands occurring at 2349 cm^{-1} and 667 cm^{-1} correspond to ν_{as} and δ or π, respectively. The band at 1340 cm^{-1} corresponding to ν_s does not occur because it is IR inactive. However, two bands are observed at 1388 and 1286 cm^{-1}. This is due to the phenomena called Fermi resonance. The first overtone δ ($2 \times 667 = 1334$) occurs very near to fundamental ν_s (1340 cm^{-1}). These two vibrations are very close in energy and are of same symmetry. Hence, they undergo Fermi resonance resulting in two bands at 1385 and 1286 cm^{-1}. In fact, the ν_s which is IR inactive is Raman active and appears at 1340 cm^{-1}.

(c) In benzene, a combination band (606 cm^{-1} + 992 cm^{-1}) = 1598 cm^{-1} interacts with a fundamental band at 1595 cm^{-1} to give two bands at 1585 cm^{-1} and at 1606 cm^{-1} in the Raman spectrum due to Fermi resonance.

(d) Benzoyl chloride shows two intense peaks for carbonyl group instead of one due to Fermi resonance.

3.9 Ascent-descent or group–subgroup in symmetry: interpretation of spectral data

When the structure of the molecule changes due to atomic substitution or deformation, this affects the change in chemical properties of the molecule. This happens due to ascent or descent in symmetry of the molecule. In fact, such changes may lower or raise the symmetry of the molecule. If the change is from lower symmetry to higher symmetry, then it is called ascent in symmetry. On the other hand if change is from higher symmetry to lower symmetry, then it is called descent in symmetry. In both the cases, there is group–subgroup relationship.

In descent in symmetry, the degeneracy is lifted while in ascent in symmetry some of the properties which were distinguishable earlier may become indistinguishable, that is, become degenerate. Knowledge of the relationship between irreducible representation of group and irreducible representation of subgroup can be helpful in predicting the changes in properties of the molecule as the structure of the molecule changes. This type of knowledge is very important for the appropriate interpretation of the spectral data.

The change in symmetry of a molecule due to change in its structure may be understood taking an example of a tetrahedral molecule MA_4 of T_d symmetry. When we substitute atom A by B one by one, we get the structures as, MA_3B and MA_2B_2. This changes the geometry and hence point group of the molecule (Fig. 3.10)

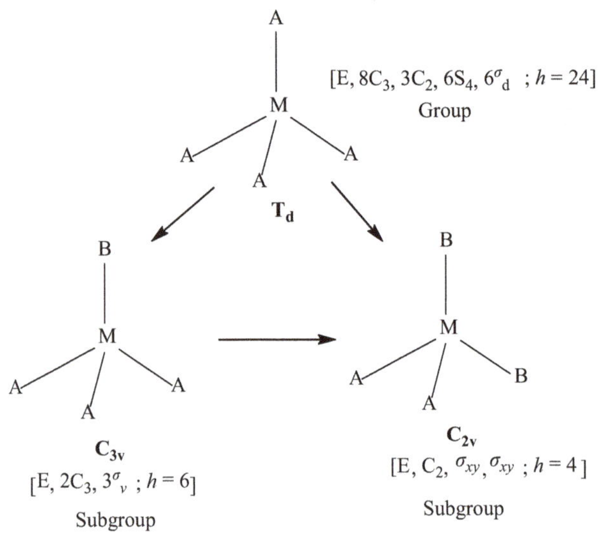

Fig. 3.10: The change in symmetry of a tetrahedral molecule due to change in its structure.

Here, the change in symmetry from $T_d \rightarrow C_{3v} \rightarrow C_{2v}$ is descent in symmetry. As evident from the Fig. 3.10 given above that in descent symmetry some of the symmetry elements are lost and so also the symmetry operations. Hence, the order of the group is also lowered. Thus, in descending symmetry, there is sudden change in symmetry, that is, basic geometry is lost.

There may not be continuous relationship between groups before and after. For example, if there is change in geometry from trigonal bipyramidal (D_{3h}) to square pyramidal (C_{4v}) or vice versa for MA_5 molecule, there will be no group –subgroup relationship. In D_{3h} point group (with symmetry operations: E, $2C_3$, $3C_2$, $2S_3$, σ_h, $3\sigma_v$), $h = 12$ and C_{4v} (with symmetry operations: E, $2C_4$, C_2, $2\sigma_v$, $2\sigma_d$), $h = 8$ is not an integral divisor of group with $h = 12$. When group–subgroup relation exists, then there also exists relationship between the irreducible representation of group and subgroup.

Let us consider the MA_4 molecule of T_d point group. The character table of T_d point group is given below.

Character table of T_d point group

T_d	E	$8C_3$	$3C_2$	$6S_4$	$6\sigma_d$		
A_1	1	1	1	1	1		$x^2 + y^2 + z^2$
A_2	1	1	1	−1	−1		
E	2	−1	2	0	0		$(2z^2 - x^2 - y^2, x^2 - y^2)$
T_1	3	0	−1	1	−1	(Rx, Ry, Rz)	
T_2	3	0	−1	−1	1	(x, y, z)	(xy, yz, xz)

In this table, symmetry operations enclosed in boxes are shared by subgroup C_{3V}. In C_{3v} point group, there are symmetry operations E, $2C_3$ and $3\sigma_v$. So, during descending in symmetry, the symmetry operations lost are: $6C_3$, $3C_2$, $6S_4$ and $3\sigma_d$. Descent in symmetry can be used to predict the vibrational modes as these are related to each other.

In MA_4 (T_d), $\Gamma_{vib} = A_1 + E + 2T_2$. The character of A_1, E and T_2 are shown in the following Table.

T_d	E	$8C_3$	$3C_2$	$6S_4$	$6\sigma_d$
A_1	1	1	1	1	1
E	2	−1	2	0	0
T_2	3	0	−1	−1	1

If we go from T_d to C_{3v} (that is from MA_4 to MA_3B), then the character of the remaining operations in the C_{3V} (that is, E, $2C_3$, $3\sigma_v$) for the irreducible representations (A_1, E, T_2) will remain same as it is there in T_d point group and these are given below:

C_{3v}	E	$2C_3$	$3\sigma_v$
A_1	1	1	1
E	2	−1	0
T_2	3	0	1

If we look over the irreducible representations in the T_d point group (A_1, A_2, E, T_1, T_2) and in the C_{3V} point group (A_1, A_2, E), then the irreducible representations A_1 and E are still there in C_{3V} point group. But T_2 irreducible representation in T_d is reducible in C_{3v} as shown below:

C_{3v}	E	$2C_3$	$3\sigma_v$
T_2	3	0	1

The reducible representation T_2 can be reduced by applying standard reduction formula and using character table of C_{3v} point group as follows:

$$N_i = \frac{1}{h}\sum_R \chi(R).n.\chi_i(R)$$

Reducible representation in question

C_{3v}	E	$2C_3$	$3\sigma_v$
T_2	3	0	1

Character table of C_{3v} point group

C_{3v}	E	$2C_3$	$3\sigma_v$		
A_1	1	1	1	z	x^2+y^2, z^2
A_2	1	1	−1	R_z	
E	2	−1	0	(x,y) $R_x,R_y)$	(x^2-y^2, x_y) (xz,yz)

$$\begin{aligned}
NA_1 &= 1/6[(3.1.1) + 0 + (1.3.1)] \\
&= 1/6[3 + 0 + 3] \\
&= 1/6[6] = 1 \\
NA_2 &= 1/6[(3.1.1) + 0 + (1.3. - 1)] \\
&= 1/6[3 + 0 - 3] \\
&= 1/6[6] = 0 \\
NE &= 1/6[(3.1.2) + 0 + 0] \\
&= 1/6[6 + 0 + 0] \\
&= 1/6[6] = 1
\end{aligned}$$

Thus, the reducible representation T_2 is reduced to: $A_1 + E$.

So, in going from T_d to C_{3v}, the relationship between irreducible representations is as follows:

$T_d \rightarrow C_{3v}$

....

$A_1 \rightarrow A_1$

$E \rightarrow E$

$T_2 \rightarrow A_1 + E.$

This shows that A_1 and E modes of vibrations in MA_4 retains the same symmetry labels on descent in symmetry to C_{3v} but each of T_2 modes splits into $A_1 + E$ modes in C_{3v} point group. It is now simple to work out the IR/Raman activity of these vibrations in C_{3v} point group.

If we go from $T_d \rightarrow C_{2v}$, then the symmetry operations retained are: E, C_2, $\sigma_{v(yz)}$, $\sigma_{v(xz)}$. Let us take the character of A_1, E, T_2 modes as given in the Table below.

C_{2v}	E	C_2	$\sigma_{v(yz)}$	$\sigma_{v(xz)}$
A_1	1	1	1	1
E	2	2	0	0
T_2	3	−1	1	1

We know that C_{2v} point group have A_1, A_2, B_1, B_2 irreducible representations. It means that E and T_2 modes can be reduced. Again, the reducible representation E and T_2 can be done one by one applying the standard reduction formula and using character table of C_{2v} point group.

$$N_i = \frac{1}{h}\sum_R \chi(R).n.\chi_i(R)$$

(A) E-representation

Reducible representation in question

C_{2v}	E	C_2	$\sigma_{v(yz)}$	$\sigma_{v(xz)}$
E	2	2	0	0

Character table of C_{2v} point group

C_{2v}	E	C_2	$\sigma_{v(yz)}$	$\sigma_{v(xz)}$		
A_1	1	1	1	1	z	x^2, y^2, z^2
A_2	1	1	−1	−1	R_z	xy
B_1	1	−1	1	−1	x, R_y	xz
B_2	1	−1	−1	1	y, R_x	yz

$$\begin{aligned} NA_1 &= 1/4[(2.1.1) + (2.1.1) + 0 + 0] \\ &= 1/4[2 + 2 + 0 + 0] \\ &= 1/4[4] = 1 \end{aligned}$$

$$\begin{aligned} NA_2 &= 1/4[(2.1.1) + (2.1.1) + 0 + 0] \\ &= 1/4[2+2+0+0] \\ &= 1/4[4] = 1 \\ NB_1 &= 1/4[(2.1.1) + (2.1.-1) + 0 + 0] \\ &= 1/4[2-2+0+0] \\ &= 1/4[0] = 0 \\ NB_2 &= 1/4[(2.1.1) + (2.1.-1) + 0 + 0] \\ &= 1/4[2-2+0+0] \\ &= 1/4[0] = 0 \end{aligned}$$

Thus, the reducible representation E is reduced to: $A_1 + A_2$

(B) T_2-representation

Reducible representation in question

C_{2v}	E	C_2	$\sigma_{v(yz)}$	$\sigma_{v(xz)}$
T_2	3	−1	1	1

Character table of C_{2v} point group

C_{2v}	E	C_2	$\sigma_{v(yz)}$	$\sigma_{v(xz)}$		
A_1	1	1	1	1	z	x^2, y^2, z^2
A_2	1	1	−1	−1	R_z	xy
B_1	1	−1	1	−1	x, R_y	xz
B_2	1	−1	−1	1	y, R_x	yz

$$\begin{aligned} NA_1 &= 1/4[(3.1.1) + (-1.1.1) + (1.1.1) + (1.1.1)] \\ &= 1/4[3-1+1+1] \\ &= 1/4[4] = 1 \\ NA_2 &= 1/4[(3.1.1) + (-1.1.1) + (1.1.-1) + (1.1.-1)] \\ &= 1/4[3-1-1-1] \\ &= 1/4[0] = 0 \\ NB_1 &= 1/4[(3.1.1) + (-1.1.-1) + (1.1.1) + (1.1.-1)] \\ &= 1/4[3+1+1-1] \\ &= 1/4[4] = 1 \end{aligned}$$

$$\begin{aligned} NB_2 &= 1/4[(3.1.1) + (-1.1.-1) + (1.1.-1) + (1.1.1)] \\ &= 1/4[3+1-1+1] \\ &= 1/4[4] = 1 \end{aligned}$$

Thus, the reducible representation T_2 is reduced to: $A_1 + B_1 + B_2$

Hence, the correlation between irreducible representation of T_d and C_{2v} point group is as follows:

$$T_d \rightarrow C_{2v}$$

....

$$\begin{aligned} &A_1 \rightarrow A_1 \\ &E \rightarrow A_1 + A_2 \qquad (\text{E splits}) \\ &T_2 \rightarrow A_1 + B_1 + B_2 \qquad (T_2\,\text{splits}) \end{aligned}$$

Such relations between irreducible representation of group and subgroups are common. These types of relations are summarized in Correlation Tables in appendix II. This principle is applicable to each and every area of applied group theory.

3.10 IR and Raman spectra of linear molecules

The linear molecules belong to either of the two infinite groups namely, $C_{\infty v}$ and $D_{\infty h}$ depending on the absence or presence of center of symmetry, respectively. All linear molecules possess the molecular axis which is the line connecting the atoms. This axis is called C_∞ axis since there are innumerable angle of rotation,θ, carrying the molecule into an equivalent configuration.

$C_{\infty v}$

The point group $C_{\infty v}$ is characterized by the presence of a C_∞ axis which is the molecular axis itself, and an infinite vertical planes (i.e., $\infty\sigma_v$) containing the C_∞ axis (Fig.). Examples of molecules belonging to this group are the hetero nuclear diatomic molecules and unsymmetrical linear tri or tetra atomic molecules, which lack center of symmetry, such as, HX (X = F, Cl, Br, I), CO, NO, COS, SCN^-, HCN, N_2O (NNO), HC = CX, etc.

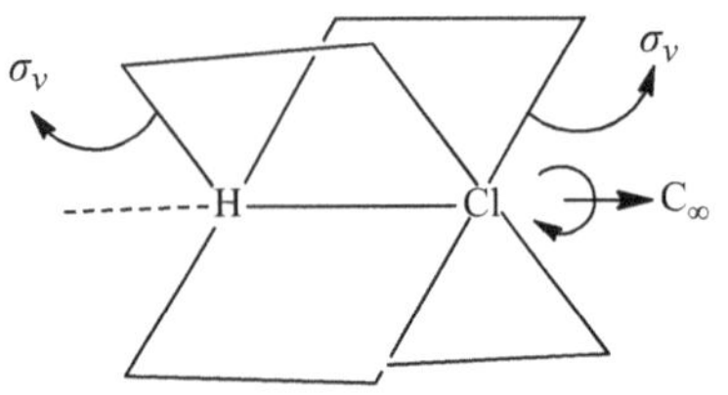

Symmetry element in HCl molecule (only two σ_v are shown, but are ∞ in number)

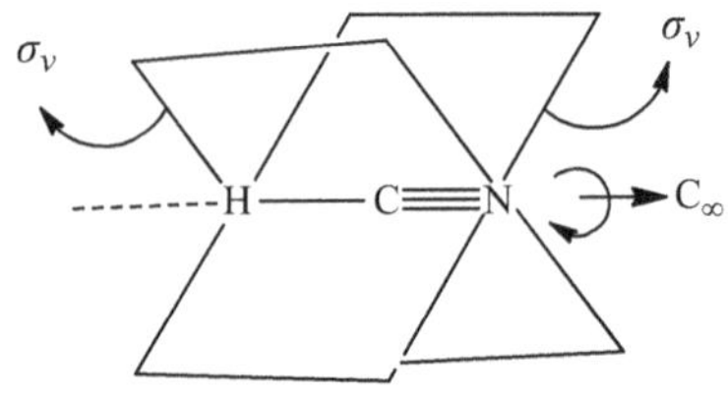

Symmetry element in HCN molecule (only two σ_v are shown, but are ∞ in number)

D∞h

Homonuclear diatomic molecules, such as N_2, H_2, F_2, Cl_2, O_2, etc. are the best examples of this point group. Hg_2^{2+} (mercurous ions) also belong to this point group. Among the linear polyatomics, HC≡CH, CO_2, $HgCl_2$, N_3^-, NO_2^+, XeF_2, etc. are some of the examples that belong to $D_{\infty h}$ point group. These molecules have a C_∞ axis which is the molecular axis itself. The equivalence of the two halves in these molecules means that any line which is the perpendicular bisector of the molecular axis is C_2 axis. Since, there is infinity of rotations about the molecular axis, C_∞, there is also an infinity of C_2 axis, that is, ∞C_2. These molecules also have infinity number of vertical planes ($\infty\sigma_v$), i, S_∞ in addition to one σ_h perpendicular to C_∞ axis.

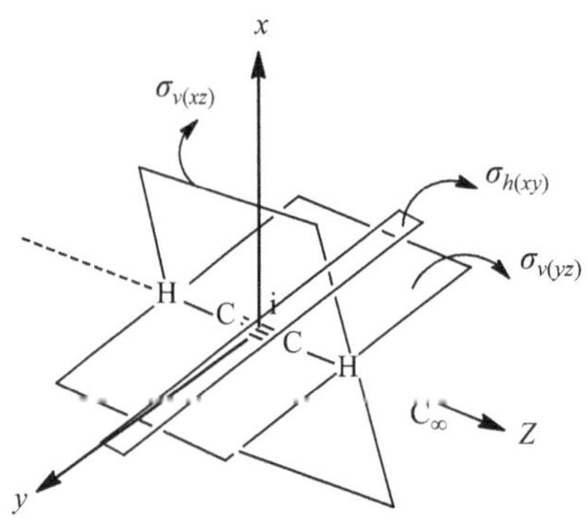

Symmetry element in HC≡CHmolecule (only two σ_v are shown, but are ∞ in number)

The procedure used for vibrational spectroscopy for linear molecules is similar to procedure used for non-linear molecules. The reducible representation Γ_{3N} is constructed by considering three Cartesian coordinate vectors on each atom in similar manner and Γ_{vib} (3N-5) is obtained by subtracting three translational and two rotational modes from Γ_{3N}. But for linear molecules, after obtaining the reducible representation, we cannot use the reduction formula,

$$N_i = \frac{1}{h}\sum_R \chi(R)\,.\,n\,.\,\chi_i\,(R)$$

This is because the order of the group h is ∞ in linear molecules, and the term 1/h becomes zero.

In many cases, reduction can be done by inspection method but for representation of higher dimensions, alternative methods have to be devised to solve the problem of linear molecules. The following alternative methods can be used for the reduction of reducible representation into the various irreducible representations in linear molecules.

(i) Subgroup method
(ii) Integration method

3.10.1 Inspection method

This method may be explained by taking example of a linear molecule CO_2 in deciding the type and number of irreducible representations. For CO_2, two C = O bonds are taken as basis for generating reducible representation as shown below:

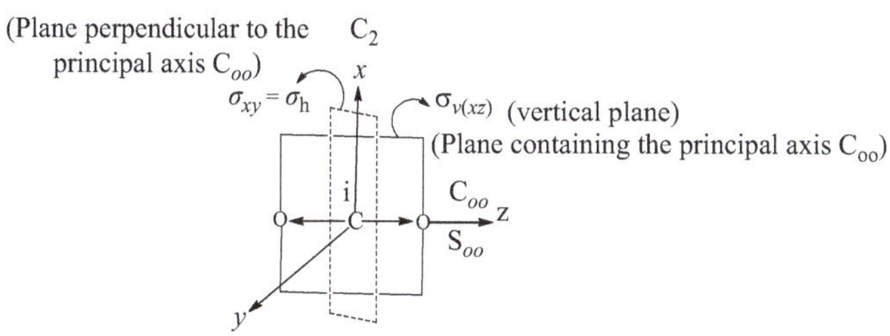

The reducible representation $\Gamma_{C=O}$ using C = O bond vectors as the basis is shown below.

	E	C_∞	$\sigma_{v(xz)}$	i	S_∞	$C_{2(x)}$
No. of unshifted C = O bond vectors (n)	2	2	2	0	0	0
$\chi(R)$	+ 1	+ 1	+ 1	+ 1	+ 1	+ 1
$\Gamma_{C=O}$ [$n \times \chi(R)$]	2	2	2	0	0	0

Now, we will make use of the partial character table of $D_{\infty h}$ point group given below to reduce $\Gamma_{C=O}$.

$D_{\infty h}$	E	C_∞	$\sigma_{v(xz)}$	i	S_∞	$C_{2(x)}$		
Σ_g^+	1	1	1	1	1	1		x^2+y^2, z^2
Σ_g^-	1	1	−1	1	1	−1	R_z	
π_g	2	$2\cos\phi$	0	2	$-2\cos\phi$	0	R_x, R_y	(xz,yz)
Δ_g	2	$2\cos2\phi$	0	2	$2\cos2\phi$	0		(x^2-y^2, xy)
...	...	...	...	...	...	...		
Σ_u^+	1	1	1	−1	−1	−1	z	
Σ_u^-	1	1	−1	−1	−1	1		
π_u	2	$2\cos\phi$	0	−2	$2\cos\phi$	0	(x, y)	
Δ_u	2	$2\cos2\phi$	0	−2	$-2\cos2\phi$	0		
...	...	...	...	...	...	...		

Now, let us have a look on the irreducible representations of the character table, combinations of which of these will give the added character as that of $\Gamma_{C=O}$. There must be two such irreducible representations of one-dimension or one irreducible representation of dimension two. As in $\Gamma_{C=O}$, there are no cosine terms in the characters, hence π and Δ irreducible representations (having cosine terms) will not be there in the desired combinations. We are now left with Σ type of one-dimensional irreducible representation.

Based on the above, the possible combinations are: (i) $(\Sigma_g^+ + \Sigma_g^+)$ (ii) $(\Sigma_g^+ + \Sigma_g^-)$ (iii) $(\Sigma_g^- + \Sigma_g^-)$ (iv) $(\Sigma_g^+ + \Sigma_u^+)$ (v) $(\Sigma_g^+ + \Sigma_u^-)$ (vi) $(\Sigma_u^- + \Sigma_u^-)$. The combination that gives added character as that of $\Gamma_{C=O}$ will be the right combination of irreducible representations. If we look on the characters of $\sigma_{v(xz)}$ and i in the reducible representation $\Gamma_{C=O}$, these are 2 and 0, respectively. The characters of $\sigma_{v(xz)}$ and i in these six combinations are:

Combination	$\chi[\sigma_{v(xz)}]$	$\chi[i]$
(i) $(\Sigma_{g^+} + \Sigma_{g^+})$	2	2
(ii) $(\Sigma_{g^+} + \Sigma_{g^-})$	0	2
(iii) $(\Sigma_{g^-} + \Sigma_{g^-})$	−2	2
(iv) $(\Sigma_{g^+} + \Sigma_{u^+})$	2	0
(v) $(\Sigma_{g^+} + \Sigma_{u^-})$	0	0
(vi) $(\Sigma_{u^-} + \Sigma_{u^-})$	−2	−2

This shows that the combination (iv) gives the correct added character [$\sigma_{v(xz)} = 2$ and $i = 0$] for $\Gamma_{C=O}$. So, the two irreducible representations are Σ_g^+ and Σ_u^+ or in other words, the reducible representation $\Gamma_{C=O}$ is reduced to Σ_g^+ and Σ_u^+.

From the character table of $D_{\infty h}$ point group, it is notable that z-coordinate belongs to Σ_u^+. Therefore, this vibrational mode will be IR active. Again, the 4th area of character table indicates that $x^2 + y^2$ and z^2 transform as Σ_{g+}. Therefore, vibration, Σ_{g+} is Raman active.

This procedure is time-consuming and difficult too for reducible representation of higher dimensions. Therefore, one requires a workable technique.

3.10.2 Subgroup method

This group–subgroup correlation technique or descent in symmetry technique was given by D. P. Strammen and E. R. Lippincott in 1972 [J. Chem. Educ., 49, 341 (1972)] to obtain the reducible representations of vibrational modes in linear molecules. This method consists of the following steps.

(i) Presume a lower molecular symmetry which corresponds to a subgroup of the infinite point group. For linear molecules, the subgroups generally used are C_{2v} and D_{2h}.

(ii) Workout the reducible representation Γ_{3N} using the symmetry operations of the subgroup. The character of each operation in the reducible representation $\chi(R)$, is obtained by using the relationship,

$$\chi(R) = \text{NUA} \,.\, \chi_{xyx}(R)$$

where NUA denotes the number of unshifted atoms by the symmetry operation R and $\chi_{x\,y\,x}$ (R) refers to the character of the matrix for the operation R.

(iii) The reducible representation is then reduced to irreducible representation using the standard reduction formula

$$N_i = \frac{1}{h}\sum_R \chi(R) \,.\, n \,.\, \chi_i(R)$$

(iv) Find the symmetry species for rotational and translational modes form the character table of the subgroup and subtract them from irreducible representations obtained for the reducible representation Γ_{3N}.

(v) Finally, we compare the basis vectors of the irreducible representations of the infinite group with those obtained for the molecule presuming subgroup. The resulting irreducible representations in the infinite group correspond to the symmetry species of the vibrational modes in the linear molecule.

Let us explain this method my taking some examples:

(i) Acetylene (HC≡CH) molecule

This molecule belongs to $D_{\infty h}$ point group. The following Fig. gives some of the symmetry elements present in this molecule.

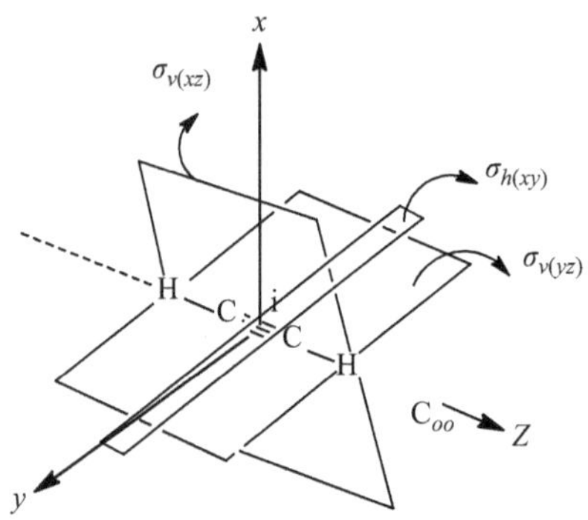

The subgroup chosen is D_{2h} and the reducible representation Γ_{3N} is obtained for HC≡CH is given below:

D_{2h}	E	$C_2(z)$	$C_2(y)$	$C_2(x)$	i	$\sigma_{v(xy)}$	$\sigma_{v(xz)}$	$\sigma_{v(yz)}$
NUA (n)	4	4	0	0	0	0	4	4
$\chi(R)$/UA	3	−1	−1	−1	−3	1	1	1
Γ_{3N} [n×$\chi(R)$]	12	−4	0	0	0	0	4	4

The reducible representation Γ_{3N} can be reduced by applying standard reduction formula and using character table of D_{2h} point group.

Reducible representation in question

D_{2h}	E	$C_2(z)$	$C_2(y)$	$C_2(x)$	i	$\sigma_{(xy)}$	$\sigma_{(xz)}$	$\sigma_{(yz)}$
Γ_{3N}	12	−4	0	0	0	0	4	4

Character table of D_{2h} point group

D_{2h}	E	$C_2(z)$	$C_2(y)$	$C_2(x)$	i	$\sigma_{(xy)}$	$\sigma_{(xz)}$	$\sigma_{(yz)}$		
A_g	1	1	1	1	1	1	1	1		x^2, y^2, z^2
B_{1g}	1	1	−1	−1	1	1	−1	−1	R_z	xy
B_{2g}	1	−1	1	−1	1	−1	1	−1	R_y	xz
B_{3g}	1	−1	−1	1	1	−1	−1	1	R_x	yz
A_u	1	1	1	1	−1	−1	−1	−1		
B_{1u}	1	1	−1	−1	−1	−1	1	1	z	
B_{2u}	1	−1	1	−1	−1	1	−1	1	y	
B_{3u}	1	−1	−1	1	−1	1	1	−1	x	

$$\begin{aligned}
NA_{1g} &= 1/8[(12.1.1) + (-4.1.1) + 0 + 0 + 0 + 0 + (4.1.1) + (4.1.1)] \\
&= 1/8[12 - 4 + 0 + 0 + 0 + 0 + 4 + 4] \\
&= 1/8[16] = 2 \\
NB_{1g} &= 1/8[(12.1.1) + (-4.1.1) + 0 + 0 + 0 + 0 + (4.1.-1) + (4.1.-1)] \\
&= 1/8[12 - 4 + 0 + 0 + 0 + 0 - 4 - 4] \\
&= 1/8[0] = 0 \\
NB_{2g} &= 1/8[(12.1.1) + (-4.1.-1) + 0 + 0 + 0 + 0 + (4.1.1) + (4.1.-1)] \\
&= 1/8[12 + 4 + 0 + 0 + 0 + 0 + 4 - 4] \\
&= 1/8[16] = 2 \\
NB_{3g} &= 1/8[(12.1.1) + (-4.1.-1) + 0 + 0 + 0 + 0 + (4.1.-1) + (4.1.1)] \\
&= 1/8[12 + 4 + 0 + 0 + 0 + 0 - 4 + 4] \\
&= 1/8[16] = 2
\end{aligned}$$

$$
\begin{aligned}
NA_{1u} &= 1/8[(12.1.1) + (-4.1.1) + 0 + 0 + 0 + 0 + (4.1.-1) + (4.1.-1)] \\
&= 1/8[12 - 4 + 0 + 0 + 0 + 0 - 4 - 4] \\
&= 1/8[0] = 0 \\
NB_{1u} &= 1/8[(12.1.1) + (-4.1.1) + 0 + 0 + 0 + 0 + (4.1.1) + (4.1.1)] \\
&= 1/8[12 - 4 + 0 + 0 + 0 + 0 + 4 + 4] \\
&= 1/8[16] = 2 \\
NB_{2u} &= 1/8[(12.1.1) + (-4.1.-1) + 0 + 0 + 0 + 0 + (4.1.-1) + (4.1.1)] \\
&= 1/8[12 + 4 + 0 + 0 + 0 + 0 - 4 + 4] \\
&= 1/8[16] = 2 \\
NB_{3u} &= 1/8[(12.1.1) + (-4.1.-1) + 0 + 0 + 0 + 0 + (4.1.1) + (4.1.-1)] \\
&= 1/8[12 + 4 + 0 + 0 + 0 + 0 + 4 - 4] \\
&= 1/8[16] = 2
\end{aligned}
$$

Thus, the irreducible representation Γ_{3N} is reduced to:

$$\Gamma_{3N} = 2A_{1g} + 2B_{2g} + 2B_{3g} + 2B_{1u} + 2B_{2u} + 2B_{3u}$$

Thus, there are 12 modes of degrees of freedom with different symmetries, A_{1g}, B_{2g}, B_{3g}, B_{1u}, B_{2u}, B_{3u}. These include the vibrational, rotational and translational degrees of freedom.

The character table shows that the translational modes have the same symmetry as the coordinates *x*, *y* and *z*, that is, B_{1u}, B_{2u} and B_{3u}. So also the rotational modes have the same symmetry as the rotation axes, R_x, R_y and R_z, that is, B_{3g}, B_{2g} and B_{1g}. Subtracting these translational and rotational degrees of freedom (excluding B_{1g} because it is not present in Γ_{3N}; in linear molecule rotation about the molecular axis *z* is not possible), we get the vibrational degrees of freedom only of appropriate symmetry.

$$
\begin{aligned}
\Gamma_{3N} &= 2A_{1g} + 2B_{2g} + 2B_{3g} + 2B_{1u} + 2B_{2u} + 2B_{3u} \\
\Gamma_{tr} &= \qquad\qquad\qquad\qquad\qquad\; B_{1u} + B_{2u} + B_{3u} \\
\Gamma_{rot} &= \qquad\quad B_{2g} + B_{3g} \\
&\cdots\cdots\cdots\cdots\cdots\cdots\cdots\cdots\cdots\cdots \\
T_{vib} &= 2A_{1g} + B_{2g} + B_{3g} + B_{1u} + B_{2u} + B_{3u}
\end{aligned}
$$

The basis vectors corresponding to these irreducible representations are as follows:

Symmetry species:	$2A_{1g}$	B_{2g}	B_{3g}	B_{1u}	B_{2u}	B_{3u}
Basis vector:	z^2	xz	yz	z	y	x (From D_{2h} char. Tab.)

These basis vectors correspond to the following representations in $D_{\infty h}$ point group.

Character table of $D_{\infty h}$ point group

$D_{\infty h}$	E	$2C_\infty^\phi$	…	$\infty\sigma_v$	i	$2S_\infty^\phi$	…	∞C_2		
$A_{1g} = \Sigma_g^+$	1	1	…	1	1	1	…	1		x^2+y^2, z^2
$A_{2g} = \Sigma_g^-$	1	1	…	−1	1	1	…	−1	R_z	
$E_{1g} = \Pi_g$	2	$2\cos\phi$	…	0	2	$-2\cos\phi$	…	0	(R_x, R_y)	(xz, yz)
$E_{2g} = \Delta_g$	2	$2\cos 2\phi$	…	0	2	$2\cos 2\phi$	…	0		(x^2-y^2, xy)
…	…	…	…	…	…	…	…	…		
$A_{1u} = \Sigma_u^+$	1	1	…	1	−1	−1	…	−1	z	
$A_{2u} = \Sigma_u^-$	1	1	…	−1	−1	−1	…	1		
$E_{1u} = \Pi_u$	2	$2\cos\phi$	…	0	−2	$2\cos 2\phi$	…	0	(x, y)	
$E_{2u} = \Delta_u$	2	$2\cos 2\phi$	…	0	−2	$-2\cos\phi$	…	0		
…	…	…	…	…	…	…	…	…		

Basis vector:	z^2	(xz, yz)	z	(x, y)
Symmetry species:	A_{1g}	E_{1g}	A_{1u}	E_{1u}

Hence, Γ_{vib} for acetylene can be written as:

$$\Gamma_{vib} = 2A_{1g} + E_{1g} + A_{1u} + E_{1u}$$

From the character table of $D_{\infty h}$ point group, it is notable that *x*, *y*, *z*-coordinates belong to E_{1u} and A_{1u} irreducible representations. Hence, E_{1u} and A_{1u} vibrational modes will be IR active. Thus, we can expect only **three** frequencies in the IR spectrum. An observation of the character table ($D_{\infty h}$ point group) indicates that the binary products of the Cartesian coordinates belong to $A_{1g,}$ and E_{1g} irreducible representations. Hence, $2A_{1g}$ and E_{1g} vibrational modes will be Raman active. Thus, we can expect **four** frequencies in the Raman spectrum. In all, seven vibrational modes are expected in IR and Raman spectra.

(ii) CO_2 molecule/ NO_2^+ ion

These species belong to $D_{\infty h}$ point group. The following Fig. 3.11 gives some of the symmetry elements present in this molecule/ion.

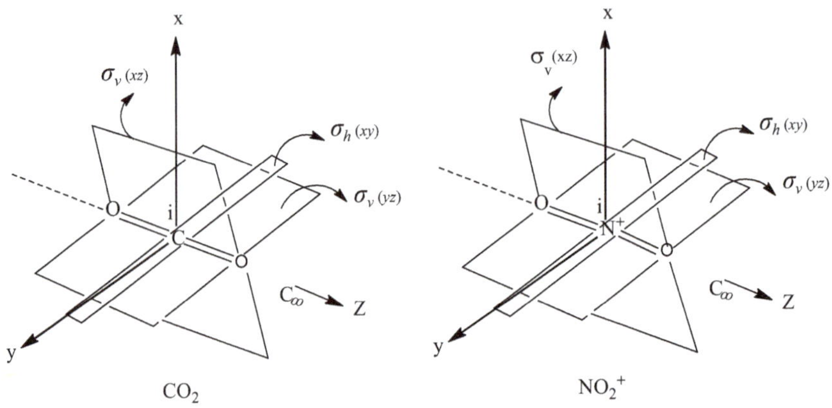

Fig. 3.11: Some of the symmetry elements present in CO_2 and NO_2^+.

Here, the subgroup chosen is D_{2h} and the reducible representation Γ_{3N} is obtained for O = C = O /O = N$^+$ = O is given below:

D_{2h}	E	$C_2(z)$	$C_2(y)$	$C_2(x)$	i	$\sigma_{v(xy)}$	$\sigma_{v(xz)}$	$\sigma_{v(yz)}$
NUA (n)	3	3	1	1	1	1	3	3
χ(R)/UA	3	−1	−1	−1	−3	1	1	1
Γ_{3N} [n×χ(R)]	9	−3	−1	−1	−3	1	3	3

Now, the reducible representation Γ_{3N} can be reduced by applying standard reduction formula and using character table of D_{2h} point group.

Reducible representation in question

D_{2h}	E	$C_2(z)$	$C_2(y)$	$C_2(x)$	i	$\sigma(xy)$	$\sigma(xz)$	$\sigma(yz)$
Γ_{3N}	9	−3	−1	−1	−3	1	3	3

Character table of D_{2h} point group

D_{2h}	E	$C_2(z)$	$C_2(y)$	$C_2(z)$	i	$\sigma(xy)$	$\sigma(xz)$	$\sigma(yz)$		
A_{1g}	1	1	1	1	1	1	1	1		x^2, y^2, z^2
B_{1g}	1	1	−1	−1	1	1	−1	−1	R_z	xy
B_{2g}	1	−1	1	−1	1	−1	1	−1	R_y	xz
B_{3g}	1	−1	−1	1	1	−1	−1	1	R_x	yz
A_{1u}	1	1	1	1	−1	−1	−1	−1		
B_{1u}	1	1	−1	−1	−1	−1	1	1	z	
B_{2u}	1	−1	1	−1	−1	1	−1	1	y	
B_{3u}	1	−1	−1	1	−1	1	1	−1	x	

$$\begin{aligned}NA_{1g} &= 1/8[(9.1.1) + (-3.1.1) + (-1.1.1) + (-1.1.1) + (-3.1.1) + (1.1.1) + (3.1.1) + (3.1.1)]\\ &= 1/8[9 - 3 - 1 - 1 - 3 + 1 + 3 + 3]\\ &= 1/8[8] =\end{aligned}$$

$$\begin{aligned}NB_{1g} &= 1/8[(9.1.1) + (-3.1.1) + (-1.1.-1) + (-1.1.-1) + (-3.1.1) + (1.1.1) + (3.1.-1)\\ &\quad + (3.1.-1)]\\ &= 1/8[9 - 3 + 1 + 1 - 3 + 1 - 3 - 3]\\ &= 1/8[0] = 0\end{aligned}$$

$$\begin{aligned}NB_{2g} &= 1/8[(9.1.1) + (-3.1.-1) + (-1.1.1) + (-1.1.-1) + (-3.1.1) + (1.1.-1) + (3.1.1)\\ &\quad + (3.1.-1)]\\ &= 1/8[9 + 3 - 1 + 1 - 3 - 1 + 3 - 3]\\ &= 1/8[8] = 1\end{aligned}$$

$$\begin{aligned}NB_{3g} &= 1/8[(9.1.1) + (-3.1.-1) + (-1.1.-1) + (-1.1.1) + (-3.1.1) + (1.1.-1) + (3.1.-1)\\ &\quad + (3.1.1)]\\ &= 1/8[9 + 3 + 1 - 1 - 3 - 1 - 3 + 3]\\ &= 1/8[8] = 1\end{aligned}$$

$$
\begin{aligned}
NA_{1u} &= 1/8[(9.1.1) + (-3.1.1) + (-1.1.1) + (-1.1.1) + (-3.1.-1) + (1.1.-1) + (3.1.-1) \\
&\quad + (3.1.-1)] \\
&= 1/8[9 - 3 - 1 - 1 + 3 - 1 - 3 - 3] \\
&= 1/8[0] = 0 \\
NB_{1u} &= 1/8[(9.1.1) + (-3.1.1) + (-1.1.-1) + (-1.1.-1) + (-3.1.-1) + (1.1.-1) + (3.1.1) \\
&\quad + (3.1.1)] \\
&= 1/8[9 - 3 + 1 + 1 + 3 - 1 + 3 + 3] \\
&= 1/8[16] = 2 \\
NB_{2u} &= 1/8[(9.1.1) + (-3.1.-1) + (-1.1.1) + (-1.1.-1) + (-3.1.-1) + (1.1.1) + (3.1.-1) \\
&\quad + (3.1.1)] \\
&= 1/8[9 + 3 - 1 + 1 + 3 + 1 - 3 + 3] \\
&= 1/8[16] = 2 \\
NB_{3u} &= 1/8[(9.1.1) + (-3.1.-1) + (-1.1.-1) + (-1.1.1) + (-3.1.-1) + (1.1.1) + (3.1.1) \\
&\quad + (3.1.-1)] \\
&= 1/8[9 + 3 + 1 - 1 + 3 + 1 + 3 - 3] \\
&= 1/8[16] = 2
\end{aligned}
$$

The above calculations show that the reducible representation Γ_{3N} is reduced to:

$$A_{1g} + B_{2g} + B_{3g} + 2B_{1u} + 2B_{2u} + 2B_{3u}(3N = 9)$$

Now, from character Table of D_{2h} pint group:

$\Gamma_{tr} = B_{1u} + B_{2u} + B_{3u}$
$\Gamma_{rot} = B_{1g} + B_{2g} + B_{3g}$ (excluding B_{1g} because it is not present in Γ_{3N};
Rotation about molecular axis z is not possible in linear molecule)

Hence, $\Gamma_{vib} = \Gamma_{3N} - (\Gamma_{tr} + \Gamma_{rot})$

$$
\begin{aligned}
&= A_{1g} + B_{2g} + B_{3g} + 2B_{1u} + 2B_{2u} + 2B_{3u} - (B_{1u} + B_{2u} + B_{3u}) - (B_{2g} + B_{3g}) \\
&= A_{1g} + B_{1u} + B_{2u} + B_{3u}
\end{aligned}
$$

The basis vectors corresponding to these irreducible representations are:

Symmetry species:	A_{1g}	B_{1u}	B_{2u}	B_{3u}
Basis vector:	z^2	z	y	x

These basis vectors correspond to the following representations in $D_{\infty h}$ point group.

Character table of $D_{\infty h}$ point group

$D_{\infty h}$	E	$2C_\infty^\phi$	…	$\infty\sigma_v$	i	$2S_\infty^\phi$	…	∞C_2		
$A_{1g}=\Sigma_g^+$	1	1	…	1	1	1	…	1		x^2+y^2, z^2
$A_{2g}=\Sigma_g^-$	1	1	…	−1	1	1	…	−1	R_z	
$E_{1g}=\Pi_g$	2	$2\cos\phi$	…	0	2	$-2\cos\phi$	…	0	(R_x, R_y)	(xz, yz)
$E_{2g}=\Delta_g$	2	$2\cos 2\phi$	…	0	2	$2\cos 2\phi$	…	0		(x^2-y^2, xy)
…	…	…	…	…	…	…	…	…		
$A_{1u}=\Sigma_u^+$	1	1	…	1	−1	−1	…	−1	z	
$A_{2u}=\Sigma_u^-$	1	1	…	−1	−1	−1	…	1		
$E_{1u}=\Pi_u$	2	$2\cos\phi$	…	0	−2	$2\cos\phi$	…	0	(x, y)	
$E_{2u}=\Delta_u$	2	$2\cos 2\phi$	…	0	−2	$-2\cos 2\phi$	…	0		
…	…	…	…	…	…	…	…	…		

Basis vector:	z^2	z	(x, y)	
Symmetry species:	A_{1g}	A_{1u}	E_{1u}	or
	Σ_g^+	Σ_u^+	π_u	

Here, the Σ_g^+, Σ_u^+ and π_u designations are based on angular momentum and A_{1g}, A_{1u} and E_{1u} are Mulliken symbols (see character table of $D_{\infty h}$ point group).

Hence, Γ_{vib} for CO_2 can be written as:

$$\Gamma_{vib} = A_{1g} + A_{1u} + E_{1u} \quad \text{or} \quad \Sigma_g^+ + \Sigma_u^+ + \pi_u$$

The symmetric stretching and antisymmetric vibrations of CO_2 belong to A_{1g} and A_{1u} symmetry species, respectively. The doubly degenerate bending modes of CO_2 belong to the symmetry species E_{1u}.

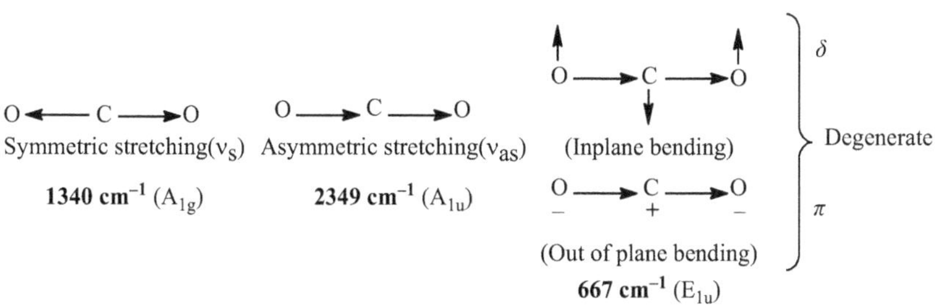

Out of four vibrations, ν_s of A_{1g} symmetry is IR inactive but Raman active and appears at 1340 cm^{-1}. The other three vibrations of A_{1u} and E_u symmetry are IR active. Though δ and π are different bending vibration modes, but for a linear molecule, they are equivalent and hence are degenerate. The IR bands occurring at 2349 cm^{-1}

and 667 cm^{-1} correspond to ν_{as} (A_{1u}) and δ or π (E_{1u}), respectively. In the IR spectrum of CO_2, we observe two additional bands at 1385 and 1286 cm^{-1}. This can be explained on the basis of Fermi resonance.

According to Fermi resonance, if there are two vibrations very close in energy and are of same symmetry, they combine by a quantum mechanical resonance resulting in two new vibrations, one having higher energy **and the other having lower energy** than the combining vibrations. In case of CO_2, ν_s is expected at 1340 cm^{-1} but is IR inactive. First overtone δ (2 × 667 = 1334) occurs very near to fundamental ν_s (1340 cm^{-1}). These two vibrations are very close in energy and are of same symmetry. Hence they undergo Fermi resonance resulting in two bands at 1385 and 1286 cm^{-1}. Hence, the two bands due to Fermi resonance are observed as intense bands.

CO_2/NO_2^+ is centrosymmetric molecule so mutual exclusion rule applies.

Alternative method for Γ_{3N} in CO_2 molecule

The reducible representation Γ_{3N} in the linear molecule CO_2 belonging to $D_{\infty h}$ point group can also be directly worked out using the character table of $D_{\infty h}$ instead of a subgroup D_{2h}. The character table of $D_{\infty h}$ and the reducible representation Γ_{3N} derived are given below:

Character table of $D_{\infty h}$ point group

	$D_{\infty h}$	E	$2C_{\infty}^{\theta}$	...	$\infty\sigma_v$	i	$2S_{\infty}^{\theta}$	...	∞C_2		
A_{1g} =	Σ_g^+	1	1	...	1	1	1	...	1		x^2+y^2, z^2
A_{2g} =	Σ_g^-	1	1	...	−1	1	1	...	−1	R_z	
E_{1g} =	Π_g	2	2 cos θ	...	0	2	−2 cos θ	...	0	(R_x, R_y)	(xz, yz)
E_{2g} =	Δ_g	2	2 cos 2θ	...	0	2	2 cos 2θ	...	0		(x^2-y^2, xy)
	...	...	...	...	...	...	...	...	...		
A_{1u} =	Σ_u^+	1	1	...	1	−1	−1	...	−1	z	
A_{2u} =	Σ_u^-	1	1	...	−1	−1	−1	...	1		
E_{1u} =	Π_u	2	2 cos θ	...	0	−2	2 cos θ	...	0	(x, y)	
E_{2u} =	Δ_u	2	2 cos 2θ	...	0	−2	−2 cos 2θ	...	0		
	...	...	...	...	...	...	...	...	...		
χ (R)	$\Gamma_{x,y,z}$	3	1+2cosθ		1	−3	−1+2cosθ		−1		
	NUA	3	3		3	1	1		1		
	Γ_{3N}	9	3+6cosθ		3	−3	−1+2cosθ		−1		

The $\chi(R)$ for $\Gamma_{x,y,z}$ is obtained by the summation of diagonal elements in the 3×3 general matrices for E, C_n, σ_v, S_n and i, which we have already derived in previous section (p-152–156).

Now, in order to get Γ_{vib}, we will subtract Γ_{tr} and Γ_{rot} from Γ_{3N}. It is notable here that for Γ_{rot} representation, we will use only (R_x, R_y) because there is no rotation about z-axis in linear molecule. This is given in the following table.

$D_{\infty h}$	E	$2C^{\theta}_{\infty}$	$\infty\sigma_v$	i	$2S^{\theta}_{\infty}$	∞C_2
Γ_{3N}	9	3 + 6cos θ	3	−3	−1 + 2cos θ	−1
Γ_{tr} [(x,y) + z] ($\pi_u+\Sigma_u^+$)	3	1 + 2cos θ	1	−3	−1 + 2cos θ	−1
Γ_{rot} (R_x, R_y) (π_g)	2	2cos θ	0	2	−2cos θ	0
$\Gamma_{vib} = \Gamma_{3N}-\Gamma_{tr}-\Gamma_{rot}$	4	2 + 2cos θ	2	−2	2cos θ	0

It is noticeable from the result of Γ_{vib} that it contains factors of 2cosθ for C^{θ}_{∞} and S^{θ}_{∞}. Now, from character table of $D_{\infty h}$, we have doubt that π_u representation may be involved in Γ_{vib} because π_u also has character of 2cos θ for C^{θ}_{∞} and S^{θ}_{∞}. Hence, if we subtract π_u from Γ_{vib}, we get, Γ_{vib} -π_u equals to:

	E	$2C^{\theta}_{\infty}$	$\infty\sigma_v$	i	$2S^{\theta}_{\infty}$	∞C_2
Γ_{vib} -π_u =	2	2	2	0	0	0

We see that this representation is just obtained by adding Σ^+_g and Σ^+_u representations of $D_{\infty h}$ point group. This concludes that the normal modes of vibrations of CO_2 transform as $\Gamma_{vib} = \Sigma^+_g + \Sigma^+_u + \pi_u$. The π_u is counted as two representations, and so we get the four vibrational degrees of freedom. When two modes transform as a single degenerate irreducible representation (π or E), they must be degenerate. The symmetry species of the vibrational modes ν_1 (ν_s), ν_2 (δ_d) and ν_3 (ν_{as}) can be determined by performing symmetry operations of the group on these vibrational modes and whatever character representation we get is then matched with the irreducible representation. The results so obtained are summarized below:

$D_{\infty h}$	E	$2C^{\theta}_{\infty}$	$\infty\sigma_v$	i	$2S^{\theta}_{\infty}$	∞C_2
ν_1	1	1	1	1	1	1
ν_2	2	$2\cos\theta$	0	−2	$2\cos\theta$	0
ν_3	1	1	1	−1	−1	−1

The above results show that the symmetry species of vibrational modes are as follows:

$$\nu_1 \rightarrow \Sigma^+{}_g \equiv A_{1g};\quad \nu_2 \rightarrow \pi_u \equiv E_{1u};\quad \nu_3 \rightarrow \Sigma^+{}_u \equiv A_{1u}$$

The two bending vibrational modes (ν_2) transform together as the degenerate representation π_u. These are taken together because the two bending vibrations in the xz and yz planes are mixed by $C^{\theta}{}_{\infty}$ and $S^{\theta}{}_{\infty}$ operations.

Looking over the 3rd and 4th columns of the $D_{\infty h}$ character table it is inferred that A_{1g} symmetry vibrational mode is IR inactive and Raman active while the other modes of A_{1u} and E_{1u} vibrational modes are IR active.

3.10.3 Integration method

The reducible representation in a linear molecule is split into various symmetry species using the following reduction formula

$$\mathrm{Ni} = 1/\mathrm{h} \sum_{R} \int_{0}^{2\pi} \chi(R) \,.\, \chi i\, (R)\, d\theta \tag{i}$$

where N_i is the number of times the ith irreducible representation occurs in a reducible representation, h is the order of group (number of symmetry operations), $\chi(R)$ is the character of a particular operation in the reducible representation, $\chi_i(R)$ is the character of the same operation in the irreducible representation and θ denotes the angle of rotation.

This formula can be applied to obtain the symmetries of the 3N degrees of freedom in HCN molecule belonging to $C_{\infty v}$ point group with symmetry operations, E, $C^{\theta}{}_{\infty}$ and $\infty\sigma_v$.

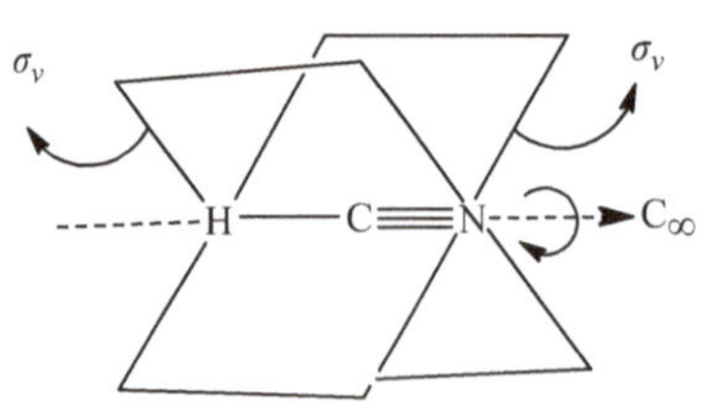

Symmetry eelemente in HCN molecule (only two σ_v are shown, but are ∞ in number)

The reducible representation Γ_{3N} for the HCN molecule can be obtained as follows:

(i) The number of unshifted atoms (NUA) for each symmetry operation is found out.

(ii) The character of each operation in the reducible representation $\chi(R)$ is obtained by using the relationship,

$$\chi(R) = NUA.\chi_{xyx}(R)$$

where NUA denotes the number of unshifted atoms by the symmetry operation R and $\chi_{x\,y\,x}$ (R) refers to the character of the matrix for the operation R.

The reducible representation Γ_{3N} obtained for HCN is given below:

$C_{\infty v}$	E	C^{θ}_{∞}	σ_v
NUA (n)	3	3	3
$\chi(R)$/UA	3	$2\text{Cos}\theta + 1$	1
Γ_{3N} [$n\times\chi(R)$]	9	$6\text{Cos}\theta + 3$	3

This reducible representation Γ_{3N} can now be reduced to various irreducible representations using the reduction formula

$$Ni = 1/h \sum_R \int_0^{2\pi} \chi(R).\chi i(R)\, d\theta \qquad \text{(i)}$$

Since C^{θ}_{∞} is equivalent to E, the summation in the above equation is performed using characters of C^{θ}_{∞} and σ_v operations only. Hence, the above equation reduces to

$$Ni = 1/h \int_0^{2\pi} \chi(C^{\theta}_{\infty})\chi_i(C^{\theta}_{\infty})\, d\theta + 1/h \int_0^{2\pi} \chi(\sigma_v)\chi_i(\sigma_v)\, d\theta \qquad \text{(ii)}$$

The symbol h in this equation is assumed as 4π and is equal to the sum of 2π C^{θ}_{∞} and $2\pi\sigma_v$ operations. The χ_i (C^{θ}_{∞}) and χ_i (σ_v) can be obtained from the character table of $C_{\infty v}$ point group given below:

$C_{\infty v}$	E	$2C_{\infty}^{\theta}$	...	$\infty\sigma_v$		
$A_1 \equiv \Sigma^+$	1	1	...	1	z	x^2+y^2, z^2
$A_2 \equiv \Sigma^-$	1	1	...	−1	R_z	
$E_1 \equiv \Pi$	2	$2\cos\theta$	...	0	$(x, y); (R_x, R_y)$	(xz, yz)
$E_2 \equiv \Delta$	2	$2\cos 2\theta$	...	0		(x^2-y^2, xy)
$E_3 \equiv \varphi$	2	$2\cos 3\theta$	...	0		
...	...	...	...	...		

Substituting the values of $\chi(C^{\theta}_{\infty})$, $\chi_i(C^{\theta}_{\infty})$, $\chi(\sigma_v)$ and $\chi i(\sigma_v)$ in equation (ii), we get

$$NA_1 = 1/4\pi\left[\int_0^{2\pi}(6cos\theta+3)\times 1d\theta + \int_0^{2\pi} 3\times 1d\theta\right]$$

$$= 1/4\pi\left[\int_0^{2\pi} 6cos\theta\, d\theta + \int_0^{2\pi} 3d\theta + \int_0^{2\pi} 3d\theta\right]$$

$$= 1/4\pi\left[\int_0^{2\pi} 6cos\theta\, d\theta. + \int_0^{2\pi} 6d\theta\right]$$

$$= 1/4\pi\,[6\sin\theta + 6\theta]_0^{2\pi}$$

$$= 1/4\pi\,[6\sin 360^\circ + 6\times 2\pi] \quad [\text{Sin}360^\circ = 0]$$

$$= 1/4\pi\,[6\times 0 + 12\pi]$$

$$= 3$$

$$Ni = 1/h\int_0^{2\pi}\chi(C_{\infty}^{\theta})\chi_i(C_{\infty}^{\theta})\,d\theta + 1/h\int_0^{2\pi}\chi(\sigma_v)\chi_i(\sigma_v)\,d\theta \qquad \text{(ii)}$$

$$NA_2 = 1/4\pi\left[\int_0^{2\pi}(6cos\theta+3)\times 1d\theta + \int_0^{2\pi} 3\times -1d\theta\right]$$

$$NA_2 = 1/4\pi\left[\int_0^{2\pi}(6cos\theta + 3 - 3)d\theta\right]$$

$$NA_2 = 1/4\pi\left[\int_0^{2\pi} 6cos\theta\, d\theta\right]$$

$$NA_2 = 1/4\pi\,[6sin\theta]_0^{2\pi}$$

$$= 1/4\pi\,[6sin360^\circ] = 0$$

By similar calculations we can find that

$$NE_1 = 3$$

All the remaining irreducible representations of $C_{\infty v}$ point group do not occur in the reducible representation Γ_{3N}. Thus, the reducible representation Γ_{3N} is reduced to

$$\Gamma_{3N} = 3A_1 + 3E_1$$

Thus, there are nine of degrees of freedom of HCN with different symmetries. These include the vibrational, rotational and translational degrees of freedom. The symmetries of translational and rotational degrees of freedom from character table given below are subtracted from Γ_{3N} to obtain Γ_{vib}.

$$\Gamma_{tr} = A_1 + E_1 \text{ and } \Gamma_{rot} = E_1$$

Hence,

$$\begin{aligned}\Gamma_{vib} &= \Gamma_{3N} - (\Gamma_{tr} + \Gamma_{rot}) = 3A_1 + 3E_1 - A_1 - E_1 - E_1 \\ &= 2A_1 + E_1\end{aligned}$$

Thus, the normal modes of HCN belong to A_1 and E_1 symmetry. Looking over the third and fourth columns of the character table, it is expected that both A_1 and E modes are infrared and Raman active. This is in favour of non-centrosymmetric HCN molecule.

3.11 Structural diagnosis: application of infrared and Raman spectra

The following points are worth considering in this regard:

(i) Vibrational spectroscopy can be used to predict the structure of a molecule. The full assignment of bands can be made by the complete normal mode analysis that helps in predicting the number and symmetry of the fundamentals. In some cases, it is possible to assign the appearance of extra bands as combination, overtone bands and Fermi resonance.

(ii) Assignment based on the reduction in the "local symmetry" of a ligand species can be made when it becomes a part of the huge molecular structure. This type of partial analysis helps in differentiating and identifying such popular phenomenon as linkage isomerism. This treatment also helps in interpreting the "denticity" of those ligands which exhibit flexi dentate character.

(iii) The occurrence of "geometrical isomerism" is another important phenomenon in coordination compounds. The information from the vibrational spectrum can be discerned with a lot of advantage for identifying the individual isomers as well as distinguishing them from each other. The coordination compounds that are known to present with this isomerism are predominantly square planar and octahedral complexes.

(iv) Metal carbonyls are special class of compounds. Group theoretical analysis helps in predicting the number of fundamentals that might occur only due to coordinated carbonyl group. Their number and position in the spectrum is of great use in knowing the type of carbonyl (terminal or bridging) and the number of such groups.

(v) Coordinate bond vibrations and the new bands activated by coordination can also be used for structural diagnosis.

3.11.1 Predicting /fitting structure/geometry of molecule

Results and discussion so far made can be utilized to predict the geometry of the molecule provided its vibrational spectral data is available in gaseous/liquid state correctly. For finding Γ_{vib}, we can use internal coordinate/Cartesian coordinate method. Let us take some examples of molecules to predict their geometry.

(i) SF_4 Molecule

The possible geometries of SF_4 molecule using *S* as the central atom is:

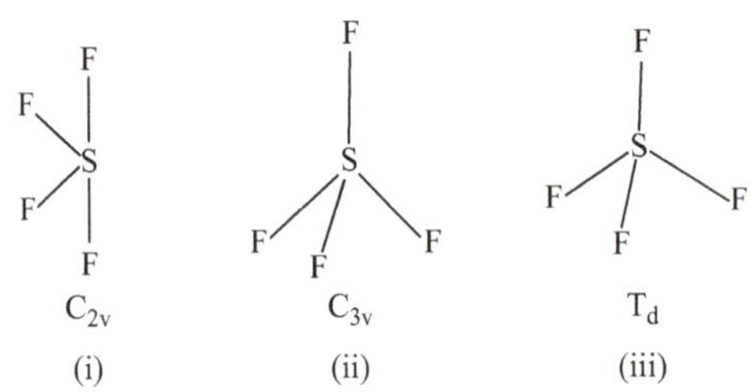

Let us consider the structure (i), (ii) and (iii) separately.

(a) The structure (i) belongs to C_{2v} point group with symmetry operations, *E*, C_2, σ_{xy}, σ_{yz}. Based on the Cartesian coordinate or 3N-vector method, the total character of the reducible representation Γ_{3N} obtained is as follows:

C_{2v}	E	C_2	σ_{xz}	σ_{yz}
β	0	180	0	0
Cos β	1	−1	1	1
±1+2Cos β	3	−1	1	1
n_R	5	1	3	3
T_{3N}	15	−1	3	3

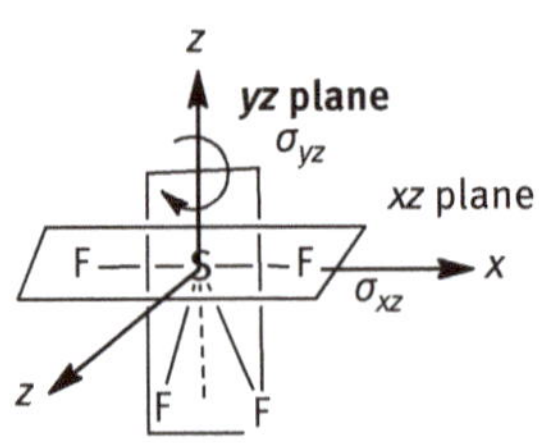

On applying reduction formula and using character table of C_{2v} point group, the reducible representation Γ_{3N} is reduced to:

$$\Gamma_{3N} = 5A_1 + 2A_2 + 4B_1 + 4B_2$$

Now, from the character table of C_{2v} point group, one can identify the translational (Γ_{tr}) and rotational modes (Γ_{rot}) as

Character table of C_{2v} point group

C_{2v}	E	C_2	$\sigma_{v(yz)}$	$\sigma_{v(xz)}$		
A_1	1	1	1	1	z	x^2, y^2, z^2
A_2	1	1	−1	−1	R_z	xy
B_1	1	−1	1	−1	x, R_y	xz
B_2	1	−1	−1	1	y, R_x	yz

$$\Gamma_{tr} = A_1 + B_1 + B_2 \text{ and } \Gamma_{rot} = A_2 + B_1 + B_2$$

Hence, the representation for vibrational modes (Γ_{vib}) is:

$$\begin{aligned}\Gamma_{vib} &= \Gamma_{3N} - (\Gamma_{tr} + \Gamma_{rot})\\ &= 5A_1 + 2A_2 + 4B_1 + 4B_2 - (A_1 + B_1 + B_2) - (A_2 + B_1 + B_2)\\ &= 4A_1 + A_2 + 2B_1 + 2B_2 (9 \text{ modes})\end{aligned}$$

From the C_{2v} character table, it is clear that

IR active modes are: $4A_1 + 2B_1 + 2B_2$ and

Raman active modes are: $4A_1 + A_2 + 2B_1 + 2B_2$

(b) The structure (ii) belongs to C_{3v} point group with symmetry operations, E, C_3 and $3\sigma_v$. Again, based on the Cartesian coordinate or 3N-vector method, the total character of the reducible representation Γ_{3N} obtained is as follows:

C_{2v}	E	$2C_3$	$3\sigma_v$
β	0	120	0
cosβ	1	$\frac{-1}{2}$	1
±1 + 2cosβ	3	0	1
n_R	5	2	3
T_{3N}	**15**	**0**	**3**

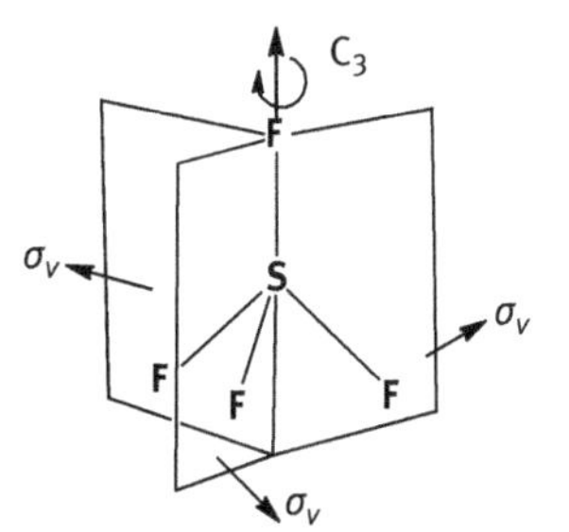

On applying reduction formula and using character table of C_{3v} point group, the reducible representation Γ_{3N} reduces to:

$$\Gamma_{3N} = 4A_1 + A_2 + 5E \quad (3N = 15)$$

Again, from the character table of C_{3v} point group, we can conclude the Translational (Γ_{tr}) and rotational modes (Γ_{rot}) as:

Character table of C_{3v} point group

C_{3v}	E	$2C_2$	3s		
A_1	1	1	1	z	x^2+y^2, z^2
A_2	1	1	−1	R_z	
E	2	−1	0	(x, y) $R_x, R_y)$	(x^2-y^2, xy) (xz, yz)

$$\Gamma_{tr} = A_1 + E \text{ and } \Gamma_{rot} = A_2 + E$$

$$\text{Now, } \Gamma_{vib} = \Gamma_{3N} - (\Gamma_{tr} + \Gamma_{rot})$$

$$= 4A_1 + A_2 + 5E - (A_1 + E) - (A_2 + E)$$

$$= 3A_1 + 3E(9 \text{ modes})$$

From the C_{3v} character table, it is notable that-
IR active modes are: $3A_1 + 3E$ and
Raman active modes are: $3A_1 + 3E$

(c) The structure (iii) belongs to T_d point group with symmetry operations, E, $8C_3$, $3C_2$, $6S_4$, $6\sigma_d$. Again, based on the Cartesian coordinate or 3N-vector method, the total character of the reducible representation Γ_{3N} obtained is as follows:

T_d	E	$8C_3$	$3C_2$	$6S_4$	$6\sigma_d$
β	0	120	180	90	0
cos β	1	$-\frac{1}{2}$	−1	0	1
±1+2cos β	3	0	−1	−1	1
n_R	5	2	1	1	3
T_{3N}	**15**	**0**	**−1**	**−1**	**3**

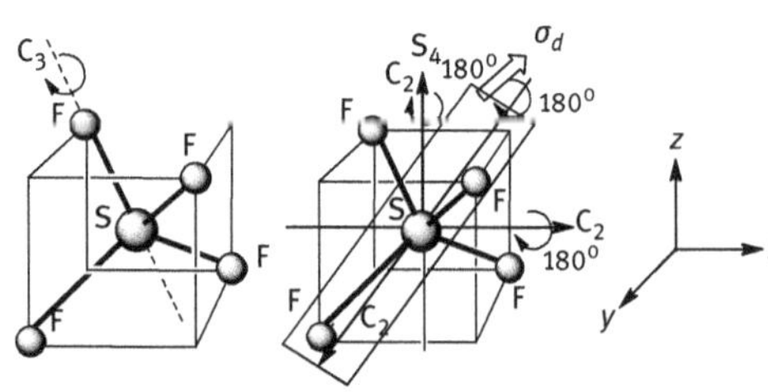

On applying reduction formula and using character table of T_d point group, the reducible representation Γ_{3N} reduces to:

$$\Gamma_{3N} = A_1 + E + T_1 + 3T_2(3N = 15)$$

Again, from the character table of T_d point group, we can identify the translational (Γ_{tr}) and rotational modes (Γ_{rot}) as:

Character Table of T_d point group

T_d	E	$8C_3$	$3C_2$	$6S_4$	$6\sigma_d$		
A_1	1	1	1	1	1		$x^2+y^2+z^2$
A_2	1	1	1	−1	−1		
E	2	−1	2	0	0		$(2z^2-x^2-y^2, x^2-y^2)$
T_1	3	0	−1	1	−1	(Rx, Ry, Rz)	
T_2	3	0	−1	−1	1	(x, y, z)	(xy, yz, xz)

$\Gamma_{tr} = T_2$ and $\Gamma_{rot} = T_1$

Now, $\Gamma_{vib} = \Gamma_{3N} - (\Gamma_{tr} + \Gamma_{rot})$

$$= A_1 + E + T_1 + 3T_2 - (T_2) - (T_1)$$
$$= A_1 + E + 2T_2 (9\text{ modes})$$

From the T_d character table, it is notable that-

IR active modes are: $2T_2$ and
Raman active modes are: $A_1 + E + 2T_2$

The following Table enlists the predicted IR and Raman active modes and experimentally observed bands of SF_4 molecule in different geometries.

Point group	IR active modes	Raman active modes	Raman polarized bands
C_{2v}	$8(4A_1 + 2B_1 + 2B_2)$	$9(4A_1 + A_2 + 2B_1 + 2B_2)$	4 $[4A_1]$
C_{3v}	6 $(3A_1 + 3E)$	6 $(3A_1 + 3E)$	3 $[3A_1]$
T_d	2 $(2T_2)$	4 $(A_1 + E + 2T_2)$	1 $[A_1]$
Experimentally observed bands	5 (Five)	5 (Five)	1(One)

Now, we will match the predicted results based on different geometry with the experimentally observed spectrum of SF_4. As given in the Table, the actual spectrum of SF_4 contains five Infrared and five Raman active bands and one Raman polarized. For T_d symmetry of SF_4, less vibrational modes are predicted in IR and Raman than experimentally observed, which is unlikely. So, T_d symmetry for SF_4 is ruled out. Thus, we have remaining two structures C_{3v} and C_{2v} of this compound.

In C_{3v} symmetry, one S-F bond is supposed to be different from the other three S-F bonds, which also seems to be unlikely. By detailed analysis of the band contours and assignment of frequency, C_{3v} point symmetry is also ruled out. The comparative data clearly show that there is close agreement between predicted and observed bands in case of C_{2v} point symmetry. This example also demonstrates the important point that although there cannot be more fundamental vibrations than allowed by a symmetry species, we see that generally all vibrations are not detected in the spectrum. Some fundamentals have low intensity, so they are not observed in the spectrum, and in the application of group theoretical concept here, the problem of separating overtones and combination bands from fundamentals is usually encountered. So, by a detailed analysis and separating the overtones and combinations from fundamentals, it was concluded that the structure C_{2v} (i) of SF_4 is the correct structure.

(ii) AB_3 Molecule

Let us consider molecules of AB_3 type which might be proposed to have any one of the following structures:

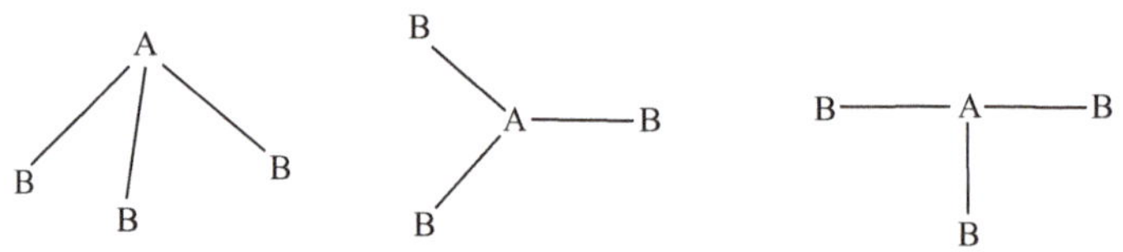

Pyramidal (C_{3v}) Planar(D_{3h}) T-Shaped (C_{2v})

Following the procedure similar to that described above, theoretically predicted vibrational modes along with the number of coincidences for each of the above geometry are given in the following Table.

Table (a): IR/ Raman predicted modes of AB_3 molecules in various point groups.

Point group	Number and symmetry of IR modes	Number and symmetry of Raman modes	Number of coincidences	Polarized Taman lines
C_{3v}	4 $(2A_1 + 2E)$	4 $(2A_1 + 2E)$	4 $(2A_1 + 2E)$	2 $(2A_1)$
D_{3h}	3 $(A_2'' + 2E')$	3 $(A_1' + 2E')$	2$(2E')$	1 (A_1')
C_{2v}	6 $(3A_1 + B_1 + 2B_2)$	6 $(3A_1 + B_1 + 2B_2)$	6$(3A_1 + B_1 + 2B_2)$	3 $(3A_1)$

Let us illustrate the structural elucidation of PCl_3, BF_3 and ClF_3 molecules which are of AB_3 type. Their complete vibrational spectral data are given in the following Table.

Table (b): IR and Raman spectral data of PCl_3, BF_3 and ClF_3.

Molecule	IR (cm^{-1})	Raman (cm^{-1})
PCl_3 (C_{3v})	4 [192, 260, 485, 510]	4 [190, 261, 486, 515]
BF_3 (D_{3h})	3 [480, 690, 1450]	3 [480, 688, 1454]
ClF_3 (C_{2v})	6[320, 365, 435, 525,700, 750]	6[322, 365, 435, 525, 702, 750]

The data can be analyzed for the possible structures of the molecules. The information given in Table (a) will help in fitting this data appropriately.

(i) The fact that PCl_3 exhibits four Raman lines, two of which are strongly polarized, rules out planar and T-shaped structures. The appearance of all the four bands in both IR and Raman confirms that the molecule possesses structure with C_{3v} point symmetry.

(ii) The BF_3 molecule shows three IR active modes and also three Raman active modes with one polarized. This data fits in very well with the planar structure of BF_3 molecule having D_{3h} point symmetry.

(iii) ClF_3 molecule exhibits six IR active modes and six Raman active modes with three polarized. This well fits in for C_{2V} geometry for T-shaped ClF_3 molecule.

3.12 Prediction of coordination sites and linkage isomerism

There are several anionic ligands which exhibit linkage isomerism in their metal complexes. Some of them are: SO_3^{2-}, SCN^-, NO_2^- and NH_2CONH_2

(i) SO_3^{2-}

It has been observed that sulfite ion (SO_3^{2-}) displays two coordination modes during complex formation as given below:

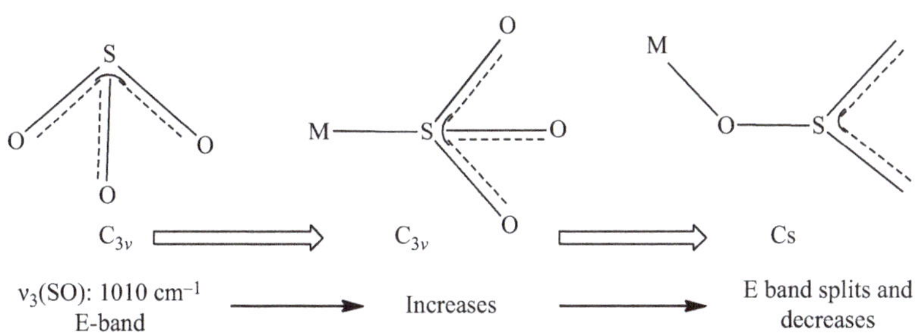

It is expected that ν_3(SO) increases when it is S-bonded and decreases when O-bonded relative to free ν(SO) at 1010 cm^{-1}. It is notable here that ν_1, ν_2, and ν_4 bands are not diagnostic. The details of these four bands in some S-and O-bonded complexes of sulfite ion are given in the following Table.

SO_3^{2-}/Complex	Coordination mode	$\nu_3(E)(cm^{-1})$	$\nu_1(A_1)$ (cm^{-1})	$\nu_2(A_1)$ (cm^{-1})	$\nu_4(E)$ (cm^{-1})
SO_3^{2-}	Free	1010	961	633	496
$K_6[Pt(SO_3)_4]$	S-bonded	1082–1057	964	660	540
$[Co(NH_3)_5(SO_3)]Cl$	S-bonded	1110	985	633	519
$[Co(en)_2(SO_3)\ Cl]$	S-bonded	1117–1075	984	625	–
$Tl_2[Cu((SO_3)_4]$	O-bonded	902–862	989	673	506 and 450

(ii) SCN^-

This ligand can coordinate to the metal either through the S-atom or through N-atom. Soft metal ions, such as, Pd^{2+} and Hg^{2+} generally form S-bonded thiocyanato complexes, whereas relatively hard metal ions, such as, Cr^{3+} and Fe^{3+} form N-bonded isothiocyanato complexes. The following results based on IR spectroscopy are used to distinguish the mode of coordination.

Mode of Coordination	ν(CN) (cm^{-1})	ν(CS) (cm^{-1})	δ(NCS) (cm^{-1})
M-SCN	2100 (sharp)	~700	440–400
M-NCS	< 2100 (broad)	860–780	490–450

Of the three bands given above, the band due to ν(CS) can be used as more diagnostic for assigning M-SCN or M-NCS coordination mode. A suitable metal ion for the study of linkage isomerism with thiocyanato ion is Pd^{2+} ion. The following data give evidence of this fact.

Mode of Coordination	ν(CS) (cm^{-1})	ν(CN) (cm^{-1})
trans-$[Pd(SCN)_2(AsPh_3)_2]$	~700	2119 (sharp)
trans-$[Pd(NCS)_2(AsPh_3)_2]$	854	2089 (broad)

(iii) NO_2^-

The nitrite ion in the complex may be present in the three forms:

(i) Anionic nitrite, $[ML_n](NO_2)_m$

(ii) Coordination of nitrite to metals through nitrogen atom, $[M(NO_2)_6]$, called nitro complexes

(iii) Nitrite ion coordinated to metals through one oxygen atom, $[M(ONO)_6]$, called nitrito complexes

Out of these three types of attachment of nitrite ion with a metal in a particular complex can be diagnosed by the IR spectrum of the complex as follows:

(a) Free nitrite ion belongs to C_{2v} point symmetry, and so it exhibits three IR bands at 1335 (ν_3), 1250 (ν_1) and 850 (ν_2) cm^{-1}.

(b) When NO_2^- coordinates to the metal ion through N-atom, the symmetry of NO_2 group remains C_{2v}. Therefore, three bands as given in (a) will also appear in the IR spectrum of the complex with some change in the positions. However, a new band at ~625 cm^{-1} corresponding to the out- of- plane bending vibration (wagging) also appear in the spectrum of the complex. This mode (~ 625 cm^{-1}) was a part of rotation in the free nitrite ion but it becomes IR active on coordination.

(c) When NO_2^- coordinates to the metal ion through O-atom, the symmetry of NO_2^- is lowered to C_S. Consequently, the two ν(NO) vibrations occur at ~ 1468 and ~ 1065 cm^{-1}. The ν_2 band occurs at 825 cm^{-1}.

In nitro complexes, both N–O bonds are equivalent and therefore ν_{as} and ν_s have a separation of ~ 100 cm^{-1}. But in nitrito complexes, there is one N–O single bond and another is N = O double bond, so their stretching frequencies, that is, ν(N = O) (1468 cm^{-1}) and ν(N–O) (1065 cm^{-1}) have more separation. The structures of both types of complexes are shown in the following Fig.

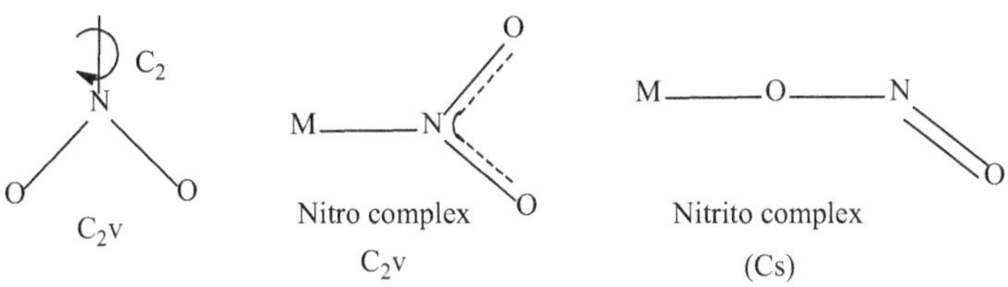

Some specific examples

It has been reported that NO_2^- displays two coordination modes in the complex $[Co(NH_3)_5(NO_2)]^{2+}$. When NO_2^- coordinates to cobalt through one of its oxygen atoms, the resulting red complex is called *nitrito* isomer, whereas the yellow complex is *nitro* isomer where nitrogen is the coordinating atom in NO_2^-.

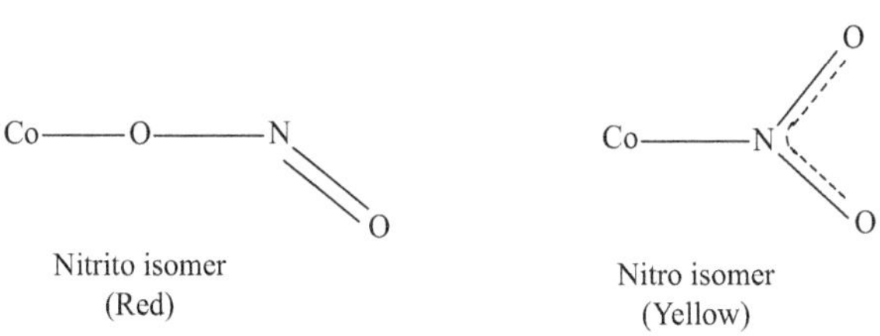

The IR spectrum of *nitro* complex shows the symmetric and asymmetric NO_2 stretching bands at 1315 and 1430 cm^{-1} (with a smaller separation of bands, 115 cm^{-1}), respectively. The *nitrito* complex exhibits two ONO stretching bands at 1065 and 1460 cm^{-1}, with larger separation (~ 400 cm^{-1}) of the bands. The larger separation is indicative of the inequality of the two N-O bonds in nitrito mode of linkage. The rocking mode at 600 cm^{-1} observed in *nitro* complex is absent in *nitrito* complex.

The NO_2^- can also coordinate in three other different ways as given in Fig shown below:

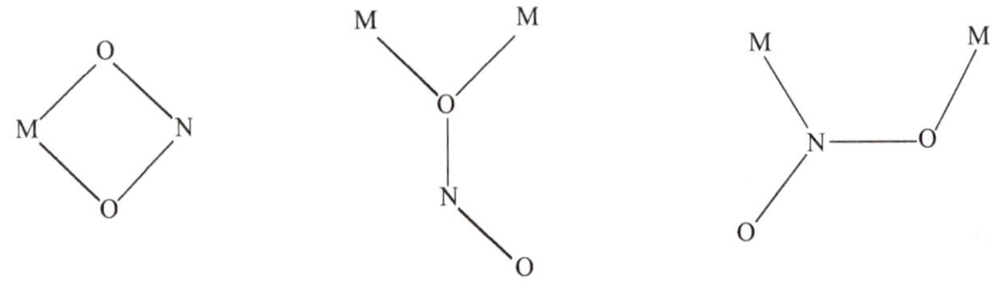

(i) η^2, monomeric bidentate (ii) η^1, bimetallic bridging (iii) η^2, bimetallic bridging bidentate

All these three types of coordination were realized in the complex [Ni_3(3-methylpyridine)$_6$(NO_2)$_6$] (Fig. 3.12). The bridging NO_2 ligand absorbs around 1515 and 1200 cm^{-1} in the complex $[(NH_3)_3Co(OH)(NO_2)Co(NH_3)_3]^{3+}$ and the observed frequencies correspond to structure (iii) given above.

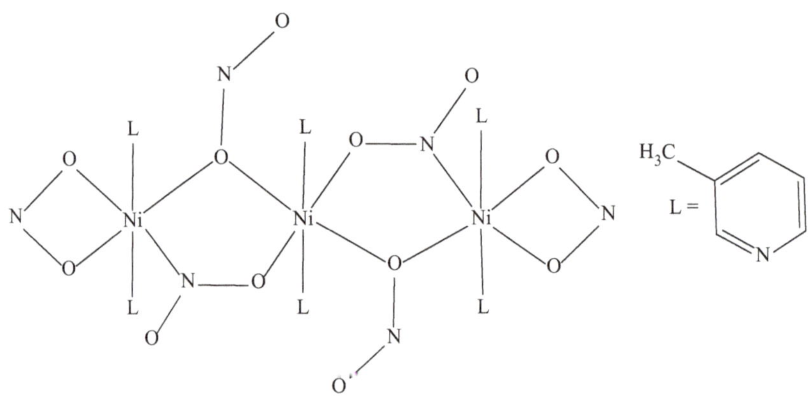

Fig. 3.12: Structure of [Ni_3(3-methylpyridine)$_6$(NO_2)$_6$] having three types of coordination modes.

(iv) NH_2CONH_2

We have seen by now, the ambidentate character of only anionic ligands that was diagnosable from absorption spectrum. There are several complex ligand molecules, particularly having multi donor atoms and separated by spacers also exhibit ambidentate character. Urea molecule is one of such ligand which has N and O-donor atoms. The donor character of this molecule may be understood by its resonating structures shown below:

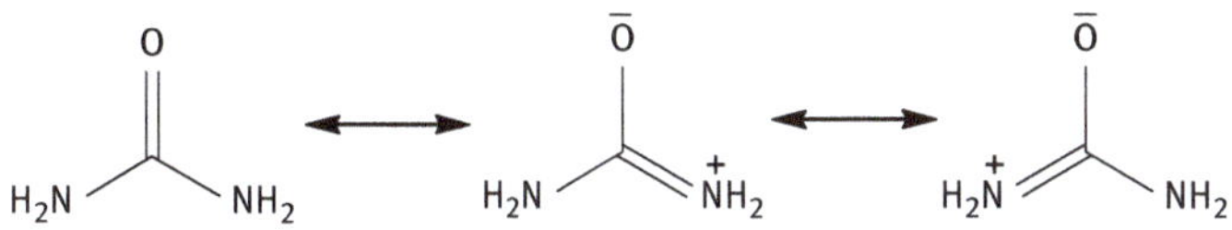

(i) (ii) (iii)

Coordination of urea through amino *N* is expected to increase the ν(CO) and decrease ν(CN) [structure (i)]. On the other hand coordination through carbonyl O is expected to decrease the ν(CO) and increase ν(CN) [structure (ii) and (iii)].

The data presented below with regard to the two complexes, $[Pt(urea)_2Cl_2]$ and $[Cr(urea)_6]Cl_3$ make it clear.

Urea/Complex	ν(CO) (cm^{-1})	ν(CN) (cm^{-1})
Urea	1683	1471
$[Pt(urea)_2Cl_2]$	1725	1395
$[Cr(urea)_6]Cl_3$	1505	1505

The above results suggest that in Pt(II) complex, Pt-N and in Cr(III) complex, Cr-O bonds are presents.

3.13 Denticity assignment for anionic ligands

Many simple anionic ligands may coordinate to metal ion in several modes resulting in simple or chelated monometallic or bimetallic complexes. Vibrational spectroscopy can distinguish between these coordination modes in metal complexes. Let us illustrate this taking trigonal planar, namely, CO_3^{2-} (carbonate) and NO_3^- (nitrate) and tetrahedral, such as, SO_4^{2-}(sulfate), ClO_4^- (perchlorate) and PO_4^{3-} (phosphate) anions.

(a) Mode of coordination of trigonal planar CO_3^{2-} and NO_3^- ions

The possible modes of coordination of these anions are given in the Fig. 3.13.

In order to know the symmetries of fundamental modes in these ions, we have to look at the character table of D_{3h} point group (given below) to which these ions belong.

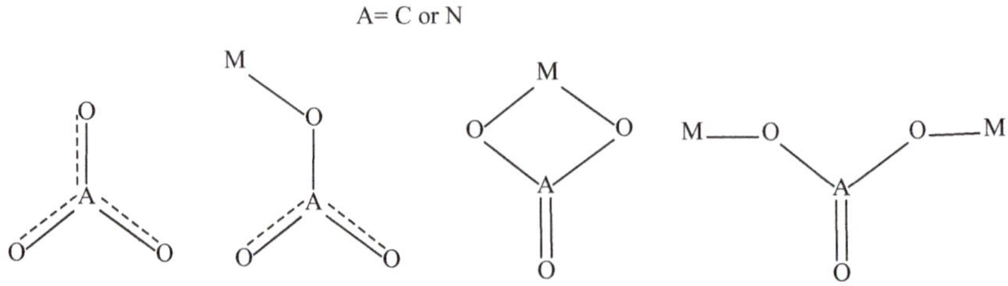

Fig. 3.13: Possible modes of coordination of CO_3^{2} and NO_3 ions.

D_{3h}	E	$2C_3$	$3C_2$	σ_h	$2S_3$	$3\sigma_v$		
A_1'	1	1	1	1	1	1		x^2+y^2, z^2
A_2'	1	1	−1	1	1	−1	R_z	
E'	2	−1	0	2	−1	0	(x, y)	(x^2-y^2, xy)
A_1''	1	1	1	−1	−1	−1		
A_2''	1	1	−1	−1	−1	1	z	
E''	2	−1	0	−2	1	0	(R_x, R_y)	(xz, yz)

The CO_3^{2-}, for example, has four normal/fundamental modes of A_1', A_2'' and 2E′ type. Of these A_1' is pure stretching and $A_2''$is pure bending. The two E′ modes are mixed type. A_1' (ν_1) is Raman active, $A_2''(\nu_2)$ is IR active and E′ (ν_3 and ν_4) are active in both IR and Raman.

The local symmetry of the CO_3^{2-} ion is reduced from D_{3h} to C_s and C_{2v} as shown in the Fig. 3.13 above depending on the mode of coordination. In each of these cases, there are changes in the spectral properties as well as the degeneracy. The correlation table given below summarizes all these changes.

Point group	ν_1	ν_2	ν_3	ν_4
D_{3h}	A_1'(Raman)	A_2''(IR)	E′(Raman, IR)	E′(Raman, IR)
C_{2v}	A_1(Raman, IR)	B_1(Raman, IR)	A_1(Raman, IR) B_2(Raman, IR)	A_1(Raman, IR) B_2(Raman, IR)
C_s	A′(Raman, IR)	A″(Raman, IR)	2A′(Raman, IR)	2A′(Raman, IR)

From this correlation table, it is well clear that $A_1'(\nu_1)$ which was inactive in IR becomes active in both lower symmetries C_{2v} and C_s. A_2 mode (see character table of

C_{2v} point group) in C_{2v} changes its symmetry and E′ modes lift their degeneracies as the symmetry is lowered, each of this giving rise two IR bands. The splitting of ν_3 is much larger in C_{2v} case. The character tables of C_s and C_{2v} point symmetry are given below for convenience of readers.

Character table of C_s

C_s	E	σ_h		
A'	1	1	x, y, R_z	x^2, y^2, z^2, xy
A"	1	−1	z, R_x, R_y	yz, xz

Character table of C_{2v}

C_{2v}	E	C_2	$\sigma_{v(xz)}$	$\sigma_{v(yz)}$		
A_1	1	1	1	1	z	x^2, y^2, z^2
A_2	1	1	−1	−1	R_z	xy
B_1	1	−1	1	−1	x, R_y	xz
B_2	1	−1	−1	1	y, R_x	yz

It is notable from the correlation table that carbonate ion in both unidentate and bidentate coordination displays two separate bands each for ν_3 and ν_4. However, it is possible to distinguish between these two types of coordination from the magnitude of separation or splitting of these bands and ν_3 band splitting can be monitored in this regard. The separation is larger in bidentate mode (C_{2v}) than in monodentate (C_s).

The two examples of unidentate and bidentate carbonato complexes give evidence of the above facts/results. The IR spectrum of $[Co(NH_3)_5CO_3]Br$ (1) exhibits two separate bands for ν_3 at 1450 and 1370 cm^{-1}, whereas that of $[Co(NH_3)_4CO_3]Cl$ (2)exhibit two separate bands for ν_3 at 1592 and 1255 cm^{-1}. It can now be concluded that the smaller the separation ($\Delta\nu = 1450–1370 = 80\ cm^{-1}$) correspond to unidentate coordination (Complex 1) and the larger separation ($\Delta\nu = 1592–1255 = 337\ cm^{-1}$) to the bidentate coordination (Complex 2). The bridging bidentate may give a similar spectrum to that of bidentate coordination, but distinction between these two can be made from the metal oxygen stretching mode in the far IR region.

The results obtained in case of nitrato complexes (NO_3^-, D_{3h}) can be similarly worked out. The ν_3 band of free NO_3^- splits in $[Ni(en)_2(NO_3)_2]$ (unidentate, C_s) to give two bands at 1420 and 1305 cm^{-1}($\Delta\nu = 1420–1305 = 115\ cm^{-1}$). In $[Ni(dien)(NO_3)_2]ClO_4$, the split bands of ν_3 are observed at 1440 and 1315 cm^{-1} ($\Delta\nu = 1440–1315 = 125\ cm^{-1}$) for unidentate and 1480 and 1300 ($\Delta v = 180$ cm^{-1}) for bidentate NO_3^- ligand, suggesting a mixed coordination.

(b) Mode of coordination of SO_4^{2-} and ClO_4^- ions

The sulfate and perchlorate ions may present in the complex in the following ways:

(i) In free ionic from with T_d point symmetry (ii) as a monodentate ligand through one of the oxygen atoms with C_{3v} point symmetry and (iii) as bidentate chelating ligand with C_{2v} symmetry and (iv) as a bridging ligand involving C_{2v} symmetry. Taking example of SO_4^{2-} ion, these situations are presented in the following Fig. 3.14.

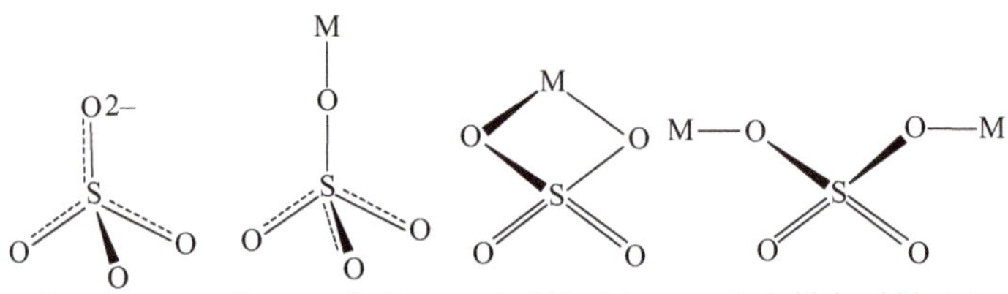

Fig. 3.14: Point symmetries of free sulfate ion and its various coordination modes.

As shown in the Fig. 3.14, the local symmetry of sulfate ion is reduced to C_{3v} and C_{2v} depending on the mode of coordination. The correlation table given below summarizes all these changes.

Point group	v_1	v_2	v_3	v_4
T_d (Free ion)	A_1(R*)	E (R)	T_2(R, IR)	T_2 (R, IR)
C_{3v} (Monodentate)	A_1(R, IR)	E(R, IR)	(A_1, E) (R, IR)	(A_1, E) (R, IR)
C_{2v} (Bidentate)	A_1(R, IR)	A_1(R, IR), A_2(R)	(A_1, B_1, B_2) (R, IR)	(A_1, B_1, B_2) (R, IR)
C_{2v} (Bridging bidentate)	A_1(R, IR)	A_1(R, IR), A_2(R)	(A_1, B_1, B_2) (R, IR)	(A_1, B_1, B_2) (R, IR)

*R = Raman

The SO_4^{2-} ion has four fundamental modes of vibration and only two (v_3 and v_4) of them are IR active. After coordination, the local symmetry of SO_4^{2-} ion changes to C_{3v} (monodentate) or C_{2v} (bidentate or bridging bidentate) and this is accompanied by changes in the spectral activity of v_1 and v_2 (which become active in IR and Raman) and splitting of v_3 and v_4 bands into two ($A_1 + E$; monodentate) or three ($A_1 + B_1 + B_2$; bidentate) components.

The IR spectra of simple sulfates exhibit only two bands (v_3 and v_4) around 1104 and 613 cm^{-1}. The other two bands (v_1 and v_2), being active in Raman are found at

~ 983 and 450 cm^{-1}, respectively. Similar to CO_3^{2-}/NO_3^- ion complexes, the splitting of ν_3 bands is much greater than that of ν_4 bands in SO_4^{2-} ion complexes.

The IR spectral data of SO_4^{2-} ion and some of its complexes are given in the following Table.

Ion/Complex	Point group	ν_1	ν_2	ν_3	ν_4
SO_4^{2-} ion	T_d	–	–	1104	613
$[Co(NH_3)_5(SO_4)]Br$	C_{3v}	970	438	1032–1044 br 1117–1143 br	613 604
$[(NH_3)_4Co(NH_2)(SO_4)Co(NH_3)_4]^{3+}$	C_{2v}	995	462	1050–1060 br 1105 s 1170 s	641 610 571

The ν_1 and ν_2 bands got active in IR and appear at 970 and 438 cm^{-1}, respectively, when SO_4^{2-} ion is coordinated as a monodentate ligand in $[Co(NH_3)_5(SO_4)]Br$, while ν_3 and ν_4 split into two bands as shown in the table . These features suggest that the local symmetry of SO_4^{2-} is lowered from T_d to C_{3v}.

As bidentate and bridging ligand, the SO_4^{2-} has C_{2v} symmetry and this further lowering of symmetry causes the appearance of ν_1 and ν_2 bands with medium intensity and ν_3 and ν_4 split up into three bands each. For example, the complex $[(NH_3)_4Co(NH_2)(SO_4)Co(NH_3)_4]^{3+}$ exhibits three spit up bands of ν_3 at 1050–1060, 1105 and 1170 cm^{-1}.

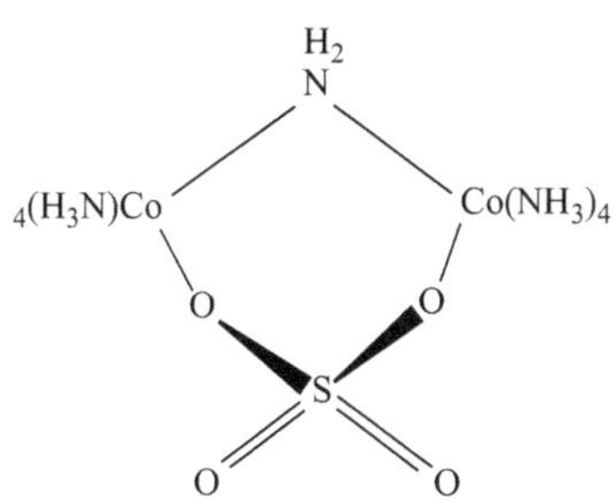

A similar pattern of bands has been observed in the IR spectrum of $[Co(en)_2(SO_4)]Br$ in which sulfate is linked to the cobalt ion in a bidentate fashion. In the present case, the three split up bands of ν_3 vibration occur at 1211, 1176, 1075 cm^{-1}, and ν_1 band at 993 cm^{-1}.

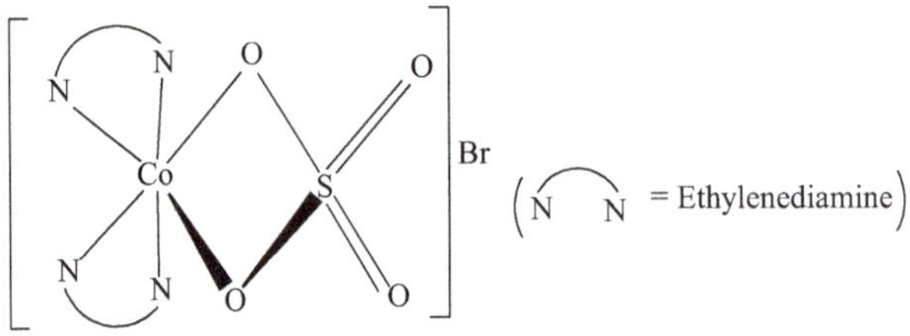

Thus, it is clear that both chelated and bridged sulfate ligand show same number of bands in the IR spectrum. But the only distinction between chelated and bridged sulfate ligand is that ν_3 band in chelated sulfate occurs at higher energy than that of bridged sulfate.

The perchlorate (ClO_4^-) ion has fewer tendencies to coordinate. So, it is present in complex as an anion or as unidentate or rarely as bidentate ligand. If it is present as anionic perchlorate, it shows only IR band corresponding to asymmetric stretching vibration (ν_3) at 1170 cm^{-1}. A low intensity band at 935 cm^{-1} is also observed for symmetrical stretching (ν_1) vibration. However, in monodentate perchlorate, as in $[Ni(CH_3CN)_4(ClO_4)_2]$, the ν_1 vibration becomes IR active and gives rise to an intense band at 972 cm^{-1}. The band corresponding to ν_3 vibrations also spilt up into two bands. As a bidentate perchlotate (C_{2v} symmetry) in the complex, $[Ni(CH_3CN)_4(ClO_4)]^+$, the ν_1 is observed at 920 cm^{-1} and ν_3 also split up into three bands due to lowering of symmetry (from T_d to C_{2v}) of perchlorate (bidentate).

So, the results presented above on the splitting of ν_3 and ν_4 bands and the appearance of ν_1 band in the IR spectra of some complexes give good evidence for the flexi dentate behavior of XO_3^{n-} and XO_4^{n-} anionic ligands.

(c) Mode of Coordination of Ethylenediamine: Chelating or Bridging

Two coordination modes are expected in ethylenediamine when coordinated with metal ions and these are: (i) bridged bidentate and (ii) chelated bidentate. These coordination modes in ethylenediamine can be distinguished by IR spectroscopy. In general, coordination of ethylenediamine with metal ion causes lowering in N-H frequency ($\nu_{N\text{-}H}$).

It can exist in three conformations as shown in the following Fig. 3.15.

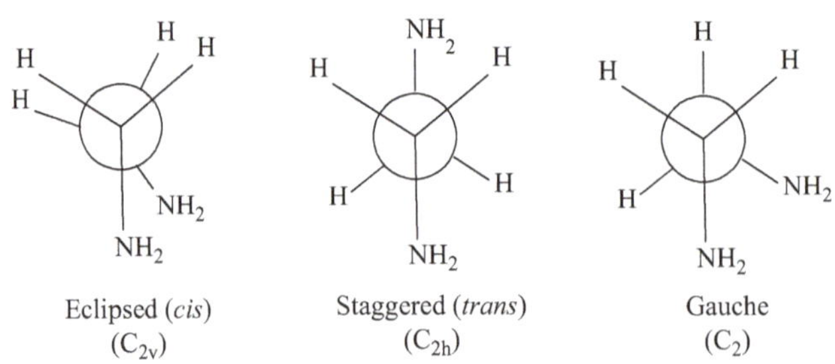

Fig. 3.15: Different conformations of ethylenediamine.

The eclipsed form of ethylenediamine has C_{2v} point symmetry, the staggered from has C_{2h} and the Gauche has C_2 symmetry. Hence, it can be expected that the staggered and gauche forms will exhibit more IR active bands because of lowering in symmetry.

It is notable here that when ethylenediamine is in the bidentate chelated form in the complex, it acquires gauche conformation and in the bridged form it exists as staggered conformation. Hence, in the IR spectrum of two complexes having ethylenediamine as ligand, the complex which shows lesser number of bands may contain ethylenediamine in staggered from (C_{2h}), and therefore it will serve as a bidentate bridging ligand in that complex, and if the complex shows relatively more number of IR bands corresponding to ligand (ethylenediamine) frequencies, then it may indicate that ethylenediamine is present as bidentate chelated form (C_2 symmetry). This is because as the symmetry is lowered, the number of IR bands increases. Here, the order of symmetry of the ligand is $C_{2v} > C_{2h} > C_2$. The order of symmetry is based on the number of symmetry elements present in the point groups).

(d) Mode of coordination of $C_2O_4^{2-}$ ion

Free oxalate ion has D_{2h} symmetry. When it is coordinated to metal ion as a bidentate ligand, its symmetry is reduced from D_{2h} to C_{2v}. Therefore, IR inactive vibrations in the free oxalate become IR active when it is coordinated. Consequently, the number of IR bands is more in oxalate complex. The free oxalate and coordinated oxalate in bidentate fashion along with symmetry are given in Fig. 3.16 is shown below.

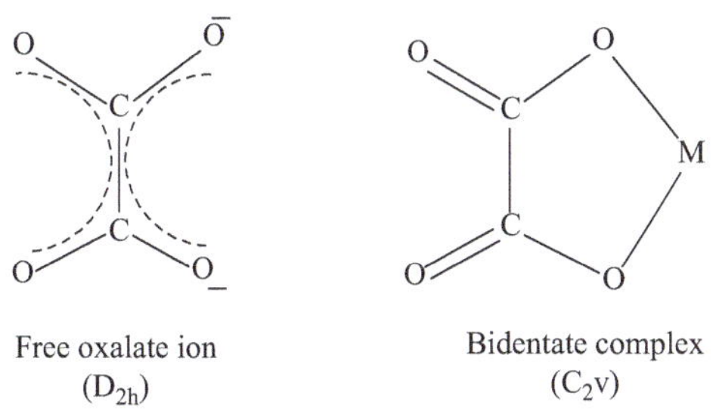

Fig. 3.16: Structures of Free oxalate and coordinated oxalate (bidentate) and their point symmetries.

In the free/ionic oxalate ion, all the four C–O bonds are equivalent as shown in the Fig. On coordination, two bonds become C–O single bonds (two oxygen coordinated to M) and two free bonds are C = O. So, the two different stretching vibrations corresponding to ν(C–O) and ν(C = O) may be observed in the spectrum. However, this is an empirical interpretation based on group frequency concept. In the chelated complex, coupling between various ligand vibrations and the new M–O bending vibration may occur and a quantitative interpretation of IR spectra can be made only by normal coordinate analysis. However, in general, as ν(M-O) increases, ν(C-O) decreases and ν(C = O) and ν(C-C) increases.

3.14 Geometrical isomers: distinction

One of the most useful applications of vibrational spectroscopy, particularly IR spectroscopy is to distinguish the geometrical isomers of complexes.

(i) *cis-trans* isomerism

This type of isomerism is shown by square planar and octahedral complexes. Each of this will be discussed one by one.

(a) Square planar *cis-* and *trans-*$[Pt(NH_3)_2Cl_2]$

The two isomers of $[Pt(NH_3)_2Cl_2]$ are shown in the Fig. 3.17. The *trans*-isomer has D_{2h} symmetry while the *cis*-isomer has C_{2v} symmetry.

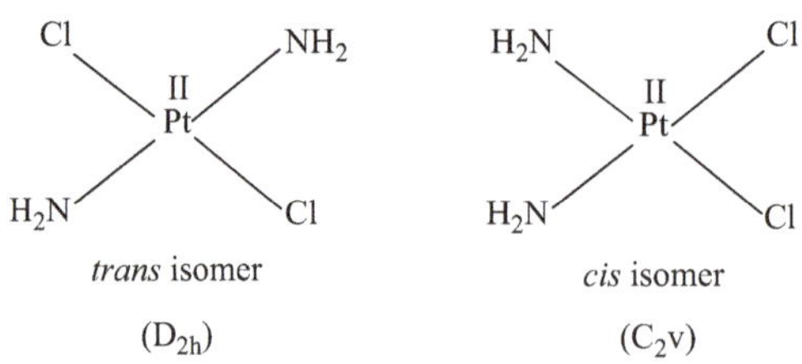

Fig. 3.17: Point symmetry of *cis*-and *trans*-isomers of $[Pt(NH_3)_2Cl_2]$.

From group theoretical consideration, for *trans*-isomer of D_{2h} point symmetry, there should be one ν(Pt-N) stretching vibration of B_{2u} symmetry, and one ν(Pt-Cl) stretching vibration of also B_{2u} symmetry, and both are IR active. In *trans*-isomer of [Pt $(NH_3)_2Cl_2$], the bands corresponding to Pt-N and Pt-Cl vibrational modes of B_{2u} symmetry are observed at 504 and 206 cm^{-1} in the IR spectrum.

As in the *cis*-isomer of $[Pt(NH_3)_2Cl_2]$, symmetry is reduced to C_{2v}, there should be two IR active stretching vibrations of A_1 and B_2 symmetry for both Pt-N and Pt-Cl vibrations. In the actual practice, the IR spectrum of *cis*-$[Pt(NH_3)_2Cl_2]$ exhibits three bands, one at 510 cm^{-1} corresponding to ν(Pt-N) and two bands at 330 and 323 cm^{-1} corresponding to ν(Pt-Cl) stretching vibration.

(b) Octahedral *cis-* and *trans-*$[ML_4X_2]$

In octahedral complexes, symmetry is reduced from the regular octahedral O_h. For instance, in $[Co(NH_3)_4Cl_2]$, it has two isomers with different symmetry as shown in the Fig. given below.

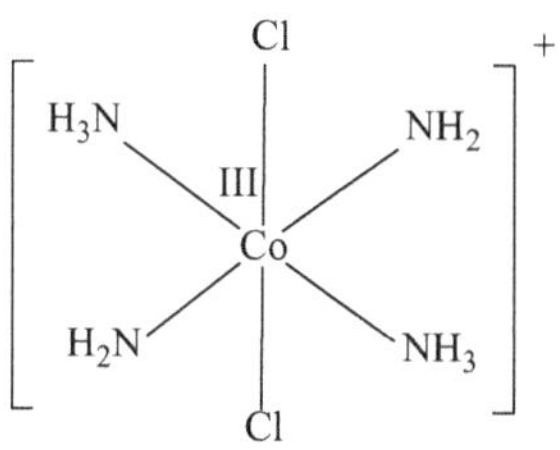

trans isomer (D_{4h})

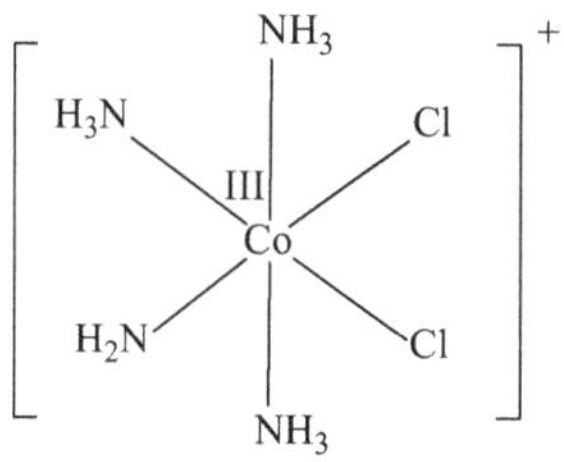

cis isomer (C_{2v})

As the *cis*-form has lower symmetry (C_{2v}) compared to the *trans*–form (D_{4h}), it is expected to show more IR active bands. In the *trans* form of this complex with D_{4h} symmetry, there should be one M-N stretching of E_u symmetry and one M-Cl stretching of A_{2u} symmetry. These are observed at 501 and 353 cm^{-1}, respectively. In the *cis*-form, there should be four M-N stretching modes (two of A_1, one of B_1 and one of B_2 symmetry) and two M-Cl stretching modes of A_1 and B_1 symmetry.

(c) Octahedral *cis*- and *trans*-[ML_2X_4]

The normal mode of analysis for MX_4 unit in [ML_2X_4] suggests four bands ($2A_1$ + B_1 + B_2) for *cis* complex of C_{2v} symmetry, and one band (E_u) for the *trans* complex of D_{4h} symmetry in their respective IR spectra. The observed spectral bands in case of *cis*-[$Pt(NH_3)_2Cl_4$] and *trans*-[$Pt(NH_3)_2Cl_4$] are presented below.

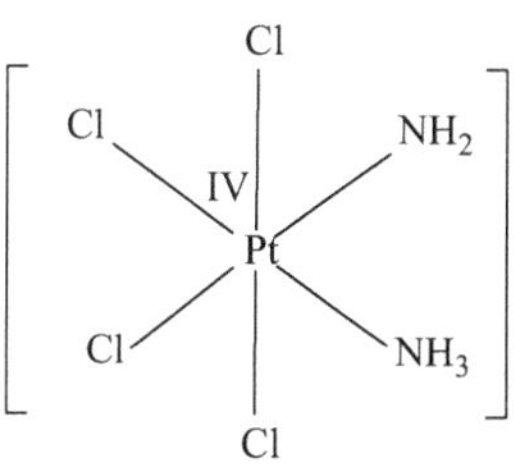

cis isomer (C_{2v})

cis-[$Pt(NH_3)_2Cl_4$]
353, 344, 330, 206 cm^{-1}

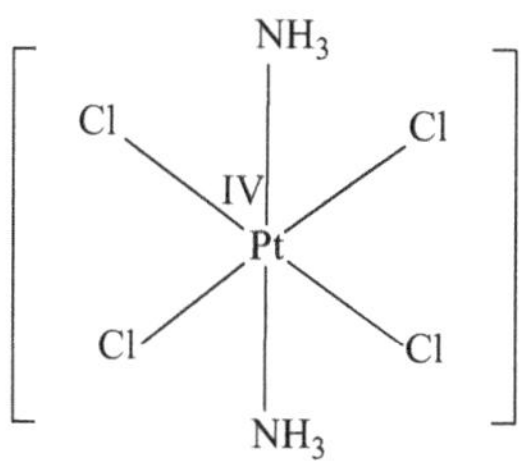

trans isomer (D_{4h})

trans-[$Pt(NH_3)_2Cl_4$]
352 (346sh) cm^{-1}

(d) Octahedral *fac*- and *mer*-[ML_3X_3]

The normal mode of analysis suggests that the *mer*-isomer of C_{2v} symmetry gives three ν(M-X) bands of $2A_1$ and B_2 symmetry and the *fac*-isomer (C_{3v}) gives two bands of A_1 and E symmetry in their respective IR spectra.

Complex	ν(M–Cl), *mer*-isomer (cm^{-1})	ν(M–Cl), *fac*–isomer (cm^{-1})
$[Ru(H_2O)_3Cl_3]$	315, 311, 291	314, 291
$[Rh(py)_3Cl_3]$	355, 322, 295	341, 325

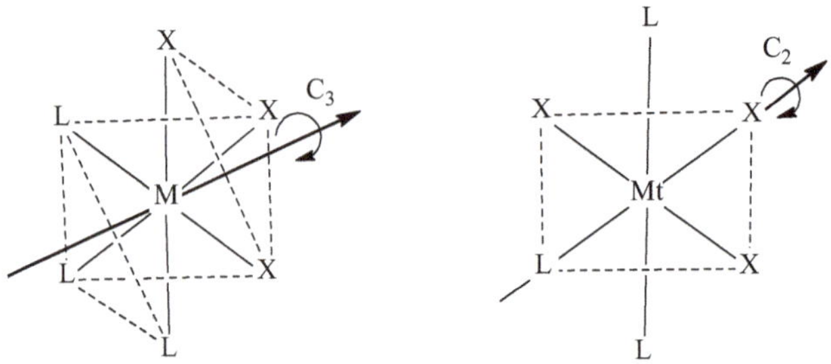

fac- isomer(C_{3v}) *mer*- isomer (C_{2v})

The IR spectral results of $[Ru(H_2O)_3Cl_3]$ and $[Rh(py)_3Cl_3]$ support the above predictions.

3.15 Metal carbonyls: structural elucidation

IR spectroscopy has been the subject of interest in the structural elucidation of metal carbonyls having different structures. Uncoordinated CO absorbs at 2141 cm^{-1} and in complexes, it absorbs at lower values than this value depending on the mode of coordination. Two coordination modes: (i) terminal and (ii) bridging are known for coordinated CO in metal carbonyls. In either mode, the CO bond order is reduced that results in lower energy absorption band. In bridging mode, it can coordinate to two or three metal atoms resulting in doubly bridging of triply bridging metal carbonyls.

The stretching frequencies of CO ligand in various coordination modes generally follow the order as given below.

Coordination mode →	Free CO >	MCO >	M_2CO >	M_3CO
Frequency(cm^{-1}) →	2141	2100–1900	1900–1700	1700–1550

Back donation in metal carbonyls can also be monitored conveniently by IR spectroscopy. Vibrational bands corresponding CO stretching vibrations are usually quite intense and fairly well isolated from other type of vibrations.

Besides IR, ^{13}C-NMR spectroscopy has been used to determine the types of carbonyls present in metal carbonyls. This is because separate ^{13}C-signals are expected for inequivalent CO ligands when the molecule is not fluxional. The terminal CO is most shielded. Though NMR gives more structural information than IR when the molecule is not fluxional, but the IR is the simplest and least expensive method to use. Both the range of absorption and the number of CO bands are important for the structural elucidation.

It has been observed that metal carbonyls which do not have a center of inversion or a C_n (n ≥3) axis of symmetry, such molecules will have a band for each CO ligand. On the other hand, metal carbonyls with a center of inversion or a C_n (≥3) axis of symmetry will have only one CO stretching band in their IR spectra. For example, metal carbonyls, such as, *mer*-$M(CO)_3L_3$, *cis*-$M(CO)_2L_4$ and *cis*-$M(CO)_4L_2$ type which do not have C_n-axis ((n ≥3), have given rise IR bands equal to the number of CO groups (**Table given below).** Metal carbonyls like $M(CO)_6$, *trans*-$M(CO)_4L_2$, *trans*-$M(CO)_2L_4$ and a linear $M(CO)_2$ with centre of inversion have only one CO stretching band observed in their IR spectra (**Table**). This is an important observation for the structural elucidation of metal carbonyls.

Table: The number of CO stretching band(s) with symmetry in the IR and Raman spectra of Some Monomeric Metal Carbonyls of different compositions.

Metal carbonyl	Point group	No. of IR band(s) with symmetry	No. of Raman band(s) with symmetry
$[M(CO)_6]$	O_h	$1(T_{1u})$	$2(A_{1g}, E_g)$
$[M(CO)_5L]$	C_{4v}	$3(2A_1, E)$	$4(2A_1, B_1, E)$
cis-$[M(CO)_4L_2]$	C_{2v}	$4(2A_1, B_1, B_2)$	$4(2A_1, B_1, B_2)$
trans-$[M(CO)_4L_2]$	D_{4h}	$1(E_u)$	$2(A_{1g}, B_{1g})$
mer-$[M(CO)_3L_3]$	C_{2v}	$3(2A_1, B_1)$	$3(2A_1, B_1)$
fac-$[M(CO)_3L_3]$	C_{3v}	$2(A_1, E)$	$2(A_1, E)$
cis-$[M(CO)_2L_4]$	C_{2v}	$2(A_1, B_1)$	$2(A_1, B_1)$
trans-$[M(CO)_2L_4]$	D_{4h}	$1(A_{2u})$	$1(A_{1g})$
$[M(CO)L_5]$	C_{4v}	$1(A_1)$	$1(A_1)$
$[M(CO)_5]$(SP)	C_{4v}	$3(2A_1, E)$	$4(2A_1, B_1, E)$
$[M(CO)_5]$(TBP)	D_{3h}	$2(A_2'', E')$	$3(2A_1', E')$
$[M(CO)_4L]$(ax)	C_{3v}	$3(2A_1, E)$	$3(2A_1, E)$
$[M(CO)_4L]$(eq)	C_{2v}	$4(2A_1, B_1, B_2)$	$4(2A_1, B_1, B_2)$
$[M(CO)_3L_2]$(ax)	D_{3h}	$1(E')$	$2(A_1', E')$
$[M(CO)_3L_2]$	C_s	$3(2A', A'')$	$3(2A', A'')$

Table: (continued)

Metal carbonyl	Point group	No. of IR band(s) with symmetry	No. of Raman band(s) with symmetry
$[M(CO)_4]$	T_d	$1(T_2)$	$1(A_1)$
$[M(CO)_3L]$	C_{3v}	$2(A_1, E)$	$2(A_1, E)$
$[M(CO)_2L_2]$	C_{2v}	$2(A_1, B_1)$	$2(A_1, B_1)$
$[M(CO)L_3]$	C_{3v}	$1(A_1)$	$1(A_1)$

mer = meridional, that is, similar ligands are arranged on a meridian of the octahedral;
fac = facial, that is, similar ligands are arranged on the same face of the octahedral;
SP = square pyramidal, TBP = trigonal bipyramidal, ax = axial, eq = equatorial

The use of ν(M–C) in combination with ν(CO) is worth considering for structural elucidation. A smooth trend in these parameters is revealed from the IR data for some metal carbonyls given in the following Table.

Table: The M—C and CO stretching frequency for some metal carbonyl.

Metal carbonyl	ν(CO) (cm^{-1})	ν(M–C) (cm^{-1})
$[Mn(CO)_6]^+$	2096	416
$[Cr(CO)_6]$	2213	441
$[V(CO)_6]^-$	1859	460
$[Ni(CO)_4]$	2058	–
$[Co(CO)_4]^-$	1883	–
$[Fe(CO)_4]^{2-}$	1729	–
$[Mn(CO)_4]^{3-}$	1670	–

There is regular decrease in CO bond order and increase in M–C bond order in both O_h and T_d complexes. Due to the increase in negative charge on the metal carbonyl, the π-back donation from the metal to the π* orbitals of CO weakens the CO bond and hence reduces ν(CO). Greater is the –ve charge, greater is the lowering in ν(CO). The increase in the ease of electron transfer also increases M–C bond order.

The structure of $[Fe(CO)_5]$ complex had been a subject of long debate because it has two structures: (a) trigonal bipyramidal (D_{3h}) and (b) square pyramidal (C_{4v}) as shown in the Fig. 3.18. The dipole moment (0.64 Debye) of this compound supports the square pyramidal structure while the electron diffraction study shows that trigonal bipyramidal is the correct structure.

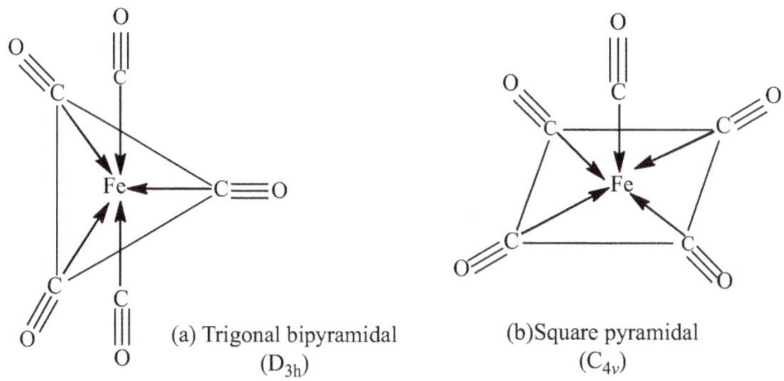

Fig. 3.18: Two possible structures of $[F_e(CO)_5]$.

Group theory predicts that for trigonal bipyramidal structure of $[Fe(CO)_5]$ (D_{3h}), there should be two IR active (A_2'', E') and three Raman active ($2A_1'$, E') normal modes of vibrations for both CO and Fe–C stretching. But for square pyramidal structure (C_{4v}), it should exhibit three IR active ($2A_1$, E) and four Raman active ($2A_1$, B_1, E) CO stretching vibrations (see Table above).

The observation of two IR and three Raman bands (results shown in Table given below) are consistent with trigonal bipyramidal (D_{3h}) structure of $[Fe(CO)_5]$. The totally symmetric band (A_1') at 2114 cm^{-1} is absent in IR spectrum but strong in Raman spectrum (Fig. 3.19).

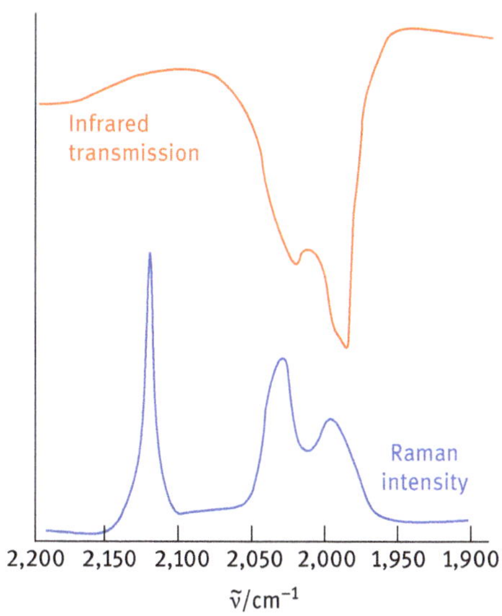

Fig. 3.19: IR and Raman spectra of liquid $[Fe(CO)_5]$ shown only in the CO stretching region.

Vibrational spectrum	TBP	SP	Experimental ν(CO) bands (cm^{-1})
IR	A_2'', E'	$2A_1$, E	2088, 1994
Raman	$2A_1'$, E'	$2A_1$, B, E	2114, 2031, 1984

In polynuclear metal carbonyls, the CO may be present as terminal as well as bridged ligand. The IR spectra of polynuclear carbonyls have been found very helpful in identifying two types of carbonyl ligands, terminal as well as bridging. Terminal CO ligands in neutral molecules have stretching frequency in the region 2100–1900 cm^{-1}. If the two COs are bridged between two metal atoms then its stretching frequency occurs in the region 1900–1700 cm^{-1} and if the three COs are bridged then its stretching frequency is observed below 1700 cm^{-1}. The progressive decrease in energy with the increase in the number of bridging CO indicates progressive decrease in C—O bond order.

Let us see how the IR spectroscopy is helpful in identifying bridging and terminal carbonyl groups in metal carbonyls with some examples.

(i) The polynuclear golden yellow carbonyl compound $Mn_2(CO)_{10}$ has two different structures as shown in Fig. 3.20.

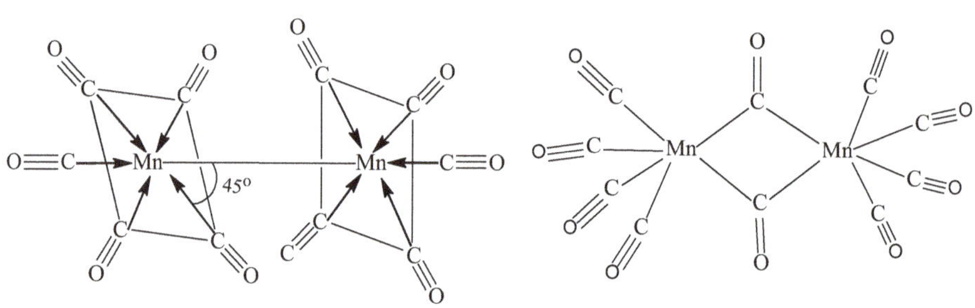

(a) Two $Mn(CO)_5$ units are staggered to each other

(b) Structure having two bridging CO group

Fig. 3.20: Two possible structures of $Mn_2(CO)_{10}$.

The IR spectrum of $Mn_2(CO)_{10}$ exhibits no band corresponding to the bridging CO group and shows only bands corresponding to CO stretching vibration in the region 2100–2214 cm^{-1} for terminal COs. So, the bridging structure (b) is ruled out.

(ii) IR spectral studies of $Fe_2(CO)_9$ and $Fe_3(CO)_{12}$ suggest that both contain the terminal and the bridging CO ligand, whereas the IR spectrum of $Ru_3(CO)_{12}$ or $Os_3(CO)_{12}$ display only one band type, that, the terminal CO bands in its IR spectrum. Accordingly, their structures are shown below.

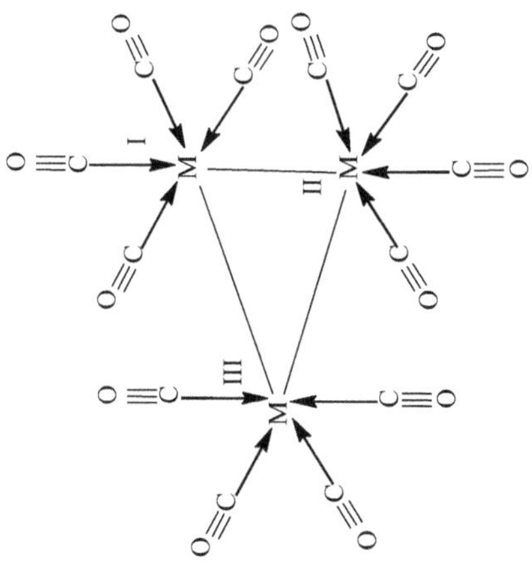

Structure of $M_3(CO)_{12}$ (M = Ru or Os)

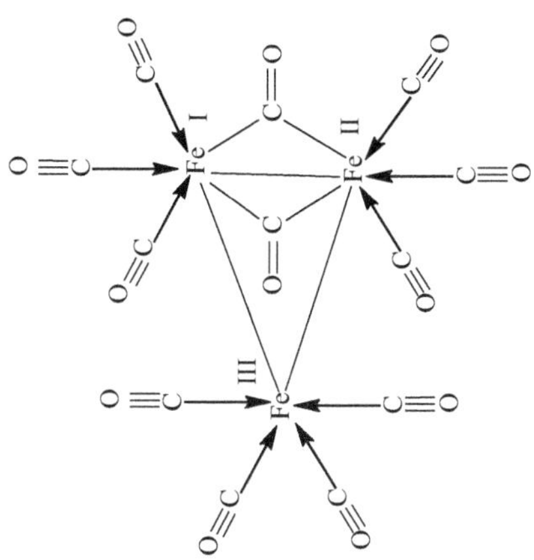

Structure of $Fe_3(CO)_{12}$

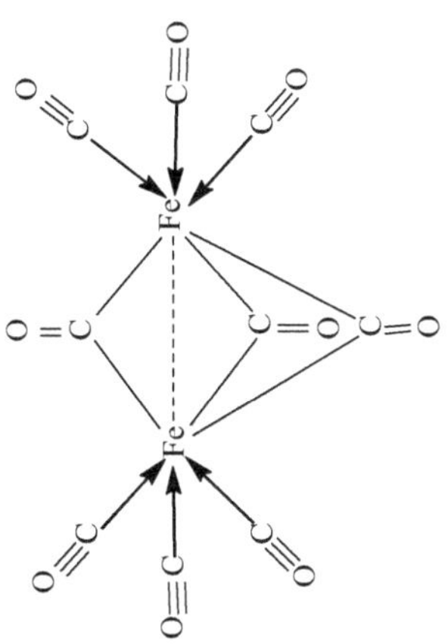

Structure of $Fe_2(CO)_9$

The CO stretching frequencies [ν(CO)] are also indicative of the electron donor abilities of the metal and other ligands in metal carbonyls containing terminal COs. The greater is the electron density supplied by the metal and other ligands, the greater would be the back donation, and therefore, the lower will be the CO bond order. Consequently, the observed stretching frequency of CO is lowered.

Exercises

Multiple choice questions/fill in the blank

1. A linear molecule will have rotational degrees of freedom in number:
   (a) 1 (b) 2 (c) 3 (d) none of these

2. A non-linear molecule containing N number of atoms will have stretching mode in number:
   (a) N-2 (b) 2N-2 (c) N-1 (d) 2N-1

3. The fundamental difference between IR and Raman techniques is that in the former …………… of photon is studied while in the latter ………… of photons is studied.

4. In case of Antistokes lines in Raman spectra, the energy of the scattered photons is:
   (a) more than the incident photons (b) less than the incident photons
   (c) equal to the incident photons (d) any one of these

5. For finding reducible representation for C_6H_6 by usual method, one has to deal with matrices of the type:
   (a) 12×12 (b) 24×24 (c) 36×36 (d) none of these

6. The character $\chi(S_n)$ for improper rotation using 3N unit vectors as the basis is
   (a) $2\cos\theta + 1$ (b) $2\cos\theta + 2$ (c) $2\cos\theta - 1$ (d) $2\cos\theta - 2$

7. Square pyramidal AB_5 molecules belong to point symmetry:
   (a) D_{4h} (b) D_{5h} (c) C_{4v} (d) C_{5v}

8. Considering five d-orbitals as the basis function for representation of any point group, the character of proper rotation is given by:
   (a) $4\cos^2 (360°/n) - 2\cos (360°/n) - 1$
   (b) $4\cos^2 (360°/n) + 2\cos (360°/n) - 1$
   (c) $4\cos^2 (360°/n) - 2\cos (360°/n) - 2$
   (d) $4\cos^2 (360°/n) + 2\cos (360°/n) - 2$

9. In AB_6 type molecule of O_h point group, the number of internal coordinates to be used to as bases to generate reducible representation is:
   (a) 12 (b) 14 (c) 18 (d) 20

10. How many vibrational modes is Raman active in case of H_2O molecule of C_{2v} point symmetry?
    (a) 2 (b) 3 (c) 4 (d) none of these

Short answer type questions

1. How many translational, rotational and vibrational motions are possible in linear and non-linear polyatomic molecules? Explain.
2. Briefly explain the simplified procedure for determining reducible representation.
3. Tabulate the character $\chi(R)$ of various symmetry operations using 3N unit vectors as the basis.
4. Workout the reducible representation of a molecule belonging to D_{4h} point group using 3N unit vectors as the basis.
5. How reducible representation of a point group is generated using bond vectors as the basis?
6. Workout the reducible representation of PCl_5 molecule belonging to D_{3h} point group using bond vectors as the basis.
7. How reducible representation of H_2O molecule belonging to C_{2v} point group is generated using p-orbitals as the basis?
8. Explain how reducible representation of a point group is generated using five d-orbitals as the basis function.
9. Taking a suitable example of a point group explain how reducible representation is generated using internal coordinates as the basis.
10. What is mutual exclusion Principle? Explain with a suitable example.
11. What are the various steps to obtain the reducible representations of vibrational modes in linear molecules by group-subgroup correlation technique?
12. How the IR spectroscopy is helpful in identifying bridging and terminal carbonyl groups in metal carbonyls?

Long answer type questions

1. Explain the utility of vibrational spectroscopy in distinction of the geometrical isomers of complexes.
2. How is group theory helpful in the structural elucidation of metal carbonyls? Justify this with suitable examples.

3. Based on the concept of group theory explain the mode of coordination trigonal planar CO_3^{2-} and NO_3^- with the help of vibrational spectroscopy.
4. Explain the usefulness of IR spectroscopy in prediction of coordination sites in some anionic ligands and linkage isomerism in NH_2CONH_2.
5. Present a detailed account of IR and Raman spectra of linear molecules.
6. Explain the phenomenon of Fermi resonance with suitable examples.
7. What are combination bands? Explain the occurrence of combination bands taking example of CH_4 molecule belonging to T_d point group.
8. Explain the method for finding overtones for degenerate vibrational modes.
9. Explain how molecular symmetry and group theory is helpful in predicting IR and Raman active modes in fullerene C_{60} belonging to I_h point group.
10. How Cartesian coordinate and internal coordinate methods are helpful in prediction of IR and Raman active modes in molecules of different point groups?

4 Chemical reactions: orbital symmetry rules

4.1 Introduction

Symmetry arguments/properties of atomic orbitals, molecular orbitals (MOs) and electronic states can be applied to predict the course of certain inorganic and organic chemical reactions. Activation energy, because of its exponential effect, plays a dominant role in predicting the course of a chemical reaction. For this reason, any rule that can be used to predict whether a given reaction has large or small activation energy will be of utmost importance. A great stride in this direction was taken by R. B. Woodward (Nobel prize for Chemistry, 1965) and R. Hoffmann (Nobel prize for Chemistry, 1980), K. Fukui and R. G. Pearson. They emphasized on using the symmetry properties of MOs of reactants and products. The change in symmetry properties is simply the change in sign that occurs for the wave function (that is, MO) at different parts of the molecule during the course of chemical reaction. If the reactant molecule has symmetry elements (rotation axes, symmetry planes, etc.), then this information can be easily utilized for studying the course of chemical reaction of the molecule and its products using certain symmetry elements (s) that is/are conserved during the course of chemical reaction.

According to *Woodward and Hoffmann*, if filled MOs of the product are correlated with filled MOs of the reactant, then the reaction is said to be allowed by the orbital symmetry, that is, the reaction has *low activation barrier*. Contrary to this, if the filled MOs of the reactants and products do not correlate with each other, then the reaction is said to be forbidden by orbital symmetry, that is, has *high activation barrier*.

Pearson used quantum mechanical perturbation theory in which MOs of the reactants develop into MOs of the products as a result of the motion along reaction coordinates.

Breaking of certain bonds between atoms and making of new bonds take place in chemical reactions. All MOs correspond to bonding, anti-bonding and non-bonding MOs of the molecule. As the reaction coordinate changes in a chemical reaction, certain MOs vacated into MOs of products and thereby initiating electron density flow from one part to the another part of the molecule. *Fukui* proposed the concept of highest occupied molecular orbitals (HOMO) and lowest occupied molecular orbitals (LUMO) to explain the flow of electron density during the course of chemical reactions. Movement of electrons between the two orbitals cannot occur unless the orbitals meet symmetry requirements. The two types of orbitals (HOMO and LUMO) act as an essential part in wide range of chemical reactions of various saturated or unsaturated compounds. These two orbitals are coined as "Frontier Orbitals" by *Fukui*. For a bimolecular reaction, the requirement is simply that the orbitals of two reactants must have net overlap.

https://doi.org/10.1515/9783110635034-004

The salient features of above three approaches (i) one because of Woodward and Hoffmann, (ii) second because of Pearson and (iii) third one because of Fukui will be used in a mixed manner for obtaining the symmetry rules for predicting the course of a chemical reaction.

There are limitations of the applications of symmetry rules to chemical reactions. These rules are applicable to concerted reactions only. In these types of reactions, all relevant changes occur simultaneously, that is, transformations of reactants into product take place in one step only without formation of any intermediate.

4.2 Chemical reactions: symmetry rules

Symmetry rules derived by perturbation theory will be argued very briefly. MO theory will still be used. Here, chemical reaction is treated as a small perturbation on the reacting system. That is, small displacement along the reaction coordinate is taken as small perturbation. Using quantum mechanical method of second-order perturbation, the symmetry rule for a chemical reaction can be easily predicted and stated [36].

Let us consider a concerted reaction of any molecularity and see how the concept of symmetry enters into the system when variation of potential energy with reaction coordinate is investigated. Group theory will be used to answer this question.

The Fig. 4.1 gives the adiabatic (without change in heat) plot of potential energy (*E*) versus reaction coordinate (Q). Point A, B, C or any other point on the plot in the diagram refers to some arrangement of nuclei of the reactants during the course of the reaction. Here, the arrangement means that system at that point belongs to certain symmetry point group (C_{2v}, C_{3v}, T_d, etc.). Points A, B, C on the plot will be used

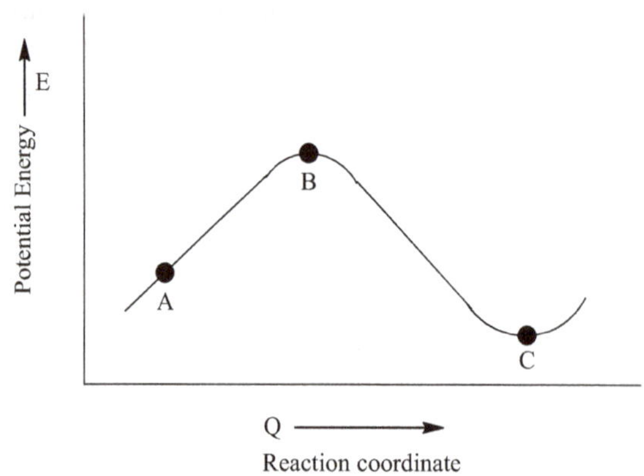

Fig. 4.1: Plot of potential energy (*E*) versus reaction coordinate (Q).

to derive the symmetry rules. It is well known that all the symmetry characteristics are contained in the irreducible representations of that point group to which the system belongs at a given point on the plot of the diagram.

The wave function for the system is supposed to be solved exactly. In other words, there is number of Eigen states ψ_0, ψ_1, ψ_2, ψ_3,.... ψ_k, and corresponding Eigen values E_0, E_1, E_2, E_3,.. E_k, where ψ_0 and E_0 refer to the ground electronic state. All the wave functions must span an irreducible representation of the type A, B (nondegenerate), E (doubly degenerate), T (triply degenerate) of the point group of the system at any point on the plot in the diagram, that is, each energy value has a symmetry label attached to it. Any small motion of the nuclear arrangement can now be discussed in terms of symmetry assuming the small displacements correspond to the normal modes of the pseudo molecular reactant system. Each of these normal modes belongs to certain symmetry species of the point group.

The eq. (4.1) given below represents the variation of energy, E, of the electronic state.

$$E = E_0 + Q\left\langle \psi_0 \left| \frac{\partial u}{\partial Q} \right| \psi_0 \right\rangle + \frac{Q^2}{2}\left\langle \psi_0 \left| \frac{\partial^2 u}{\partial Q^2} \right| \psi_0 \right\rangle + Q^2 \sum_k \left[\frac{< \psi_0 \left| \frac{\partial u}{\partial Q} \right| \psi_k >}{E_0 - E_k} \right]^2 \tag{4.1}$$

where E_0 is the energy at Q = 0.

As the Hamiltonian used is invariant (not changing) under all symmetry operations of point group of the pseudo molecule, it means Q and $\partial u/\partial Q$ must have the same symmetry because reaction coordinate Q is always symmetric. Q also represents a magnitude of small displacement from Q_0. Q^2 is also symmetric and $\partial^2 u/\partial Q^2$ is also symmetric. It is notable here that u represents nuclear–electronic and nuclear–nuclear potential energy. The symbol <....> represents integration over all electronic coordinates covering all space.

At maximum and minimum in the potential energy curve, $\partial u/\partial Q = 0$. Therefore, integral must be identically zero independent of symmetry. At all other points, this term (that is, $\partial u/\partial Q$) must be dominant one since Q is small. As the direct product of a nondegenerate state itself is totally symmetric, the first rule is "*except for maximum or minimum in potential energy curve, all reaction coordinates belong to totally symmetric representation.*" It (integral) must be nonzero for all the rising and falling parts of plot in Fig. 4.1. Thus, once the reaction starts on reaction path, it must stay in within same point group symmetry until it reaches maximum or minimum in the energy curve. The nuclear motion can change bond angles and bond distance, but it cannot change point group symmetry.

At point A on the curve (Fig. 4.1), the integral $< \psi_0 |\partial u/\partial Q| \psi_0>$ has nonzero value and is positive term since activation energy is positive. The term Q^2 in eq. (4.1) now becomes important. Their sum determines the curvature of the potential energy plot. The integral

$< \psi_O |\partial^2 u/\partial^2 Q| \psi_O>$ has a nonzero value and by symmetry $\partial^2 u/\partial^2 Q$ is also totally symmetric. The last term in eq. (4.1) represents the change in energy that occurs from changing the electron distribution. Its value is always negative as $E_O - E_k$ is a negative number. Each excited state wave function is mixed into ground state wave function. Group theory tells us that only excited state wave function ψ_k for a chemical reaction that have symmetry as that of ψ_O can mix in and that will lower the energy barrier. Thus, for a chemical reaction to occur with reasonable activation energy there must be low-lying excited states for the reacting system of the same symmetry as that of ground state.

Symmetry is related only to those symmetry elements that are not changed, that is, are conserved during the course of chemical reaction. Such a reaction is called *symmetry allowed.* A *symmetry forbidden* reaction is the one that has high activation energy because of the absence of suitable excited states of proper symmetry.

One drastic assumption made for the derivation of symmetry rules is that instead of exact wave functions ψ_O and ψ_k, LCAO-MO theory is used. This assumption works because we do not need exact overlap of wave function but we need overlap of some definite value. Now, we can replace ψ_O and ψ_k wave functions by ϕ_o and ϕ_k orbitals correspondingly. Special importance is paid to HOMO and LUMO.

At point B and C, the theory is changed to some extent. Again it is important that low-lying state ψ_k of correct symmetry should match with the symmetry of ψ_O. The ψ_O, ψ_k and $\partial u/\partial Q$ is still bound by the symmetry requirements so that the direct product ($\psi_O \times \psi_k \times \partial u/\partial Q$) contain the totally symmetric representation.

In LCAO-MO theory, the $\psi_O \times \psi_k$ is replaced by $\phi_o \times \phi_k$. HOMO and LUMO must be of the same molecule. The electron density moves from HOMO to LUMO. The electron density increases in the region where HOMO and LUMO have the same symmetry. In other words, positive overlap and electron density decreases when HOMO and LUMO do not match in symmetry, that is, negative overlap. The positive charge nuclei move in the direction of maximum electron density. The motion of the nuclei defines the reaction coordinate, that is, the reaction path.

For activated complex at point B on plot (Fig. 4.1), there must be low energy excited state. The symmetry of the excited state and ground state of the activated complex determines the mode of decomposition of the activated complex. At point C (low potential well), the symmetry of this state and the ground state will determine the type of reaction of the unstable molecule that it undergoes.

The exact treatment of the perturbation theory is not being given here. However, for the exact treatment of this theory, it is advised to consult some relevant book on quantum chemistry. Symmetry rules derived from perturbation theory may be summarized as follows:

(i) All reaction coordinates belong to totally symmetric representations. That is, Q and $\partial u/\partial Q$ must be totally symmetrical.

(ii) Critical MOs are HOMO and LUMO. When reactants approach to each other electrons flow from HOMO to LUMO.

(iii) HOMO and LUMO must be of right symmetry and close in energy.
(iv) For chemical reactions to occur with reasonable activation energy there must be low-lying excited states of the reacting system with the same symmetry as the symmetry of the ground state. In such cases, the reaction is *symmetry allowed*. A *symmetry forbidden* reaction is the one in which no such low-lying excited states with same symmetry are available.
(v) As the reacting molecules approach each other, they may be considered larger pseudo molecule, that is, transition state (TS) with definite point group. In pseudo molecule HOMO and LUMO must be close in energy and must be of right symmetry of the point group.
(vi) HOMO and LUMO of reacting species must approach to each other in such a manner that there is a net positive overlap.
(vii) HOMO must correspond to bond breaking and LUMO must correspond to bond making if MOs are bonding in nature. For antibonding MOs, reverse statement is true.
(viii) If no such MOs (HOMO, LUMO) of right symmetry are available, then reaction will be symmetry forbidden and will have high activation energy.
(ix) Free radical reactions or atomic reactions are always symmetry allowed.

Now, we will apply the above rules to some chemical reactions both inorganic and organic. Any suggested mechanism must follow the symmetry test and must be examined whether the symmetry properties of HOMO and LUMO involved will allow the reaction to proceed or not.

4.3 Inorganic/organic reactions: symmetry considerations

(a) Reaction between H_2 and I_2 and I_2 and Cl_2

These well-known reactions are:

$$H_2 + I_2 \rightleftharpoons 2HI \tag{4.2}$$

$$I_2 + Cl_2 \rightleftharpoons 2ICl \tag{4.3}$$

Since long, these reactions were supposed to proceed through bimolecular processes in which two reacting molecules collide broad sideways, that is, in C_{2v} symmetry as shown below:

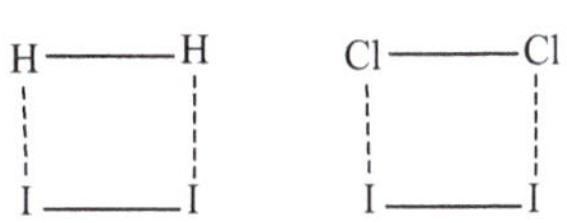

(i) We will now take up first the reaction between H_2 and I_2. The HOMO and LUMO of H_2 and I_2 molecules are shown below:

H_2 Molecule (H_1 = $1S^1$)

The MO electronic configuration of H_2 is $\sigma_{1g}{}^2\sigma_{1u}{}^\star$. The HOMO of H_2 is σ_{1g} and LUMO is $\sigma_{1u}{}^\star$. The HOMO and LUMO of H_2 are shown in Fig. 4.2.

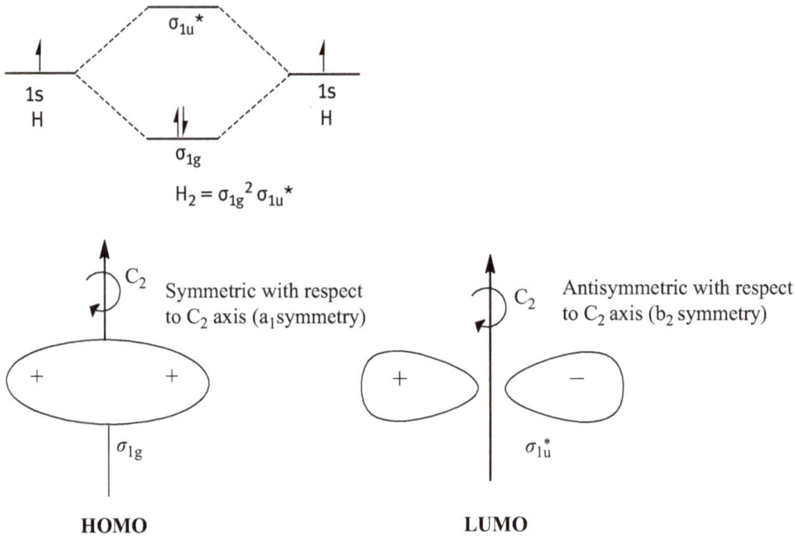

Fig. 4.2: M.O. Electronic configuration and HOMO and LUMO of H_2 molecule.

I_2 Molecule (I_{53}: $1s^2\ 2s^2\ 2p^6\ 3s^23p^63d^{10}\ 4s^24p^64d^{10}\ 5s^25p^5$)

The MO electronic configuration of I_2 may be written as $\sigma_g{}^2\ \pi_u{}^4\pi_g{}^{\star 4}\sigma_u{}^\star$ (considering only $5p^5$ electronic configuration of each I atom). The HOMO of I_2 is $\pi_g{}^\star$ and LUMO is $\sigma_u{}^\star$. The HOMO and LUMO of I_2 is shown in Fig. 4.3.

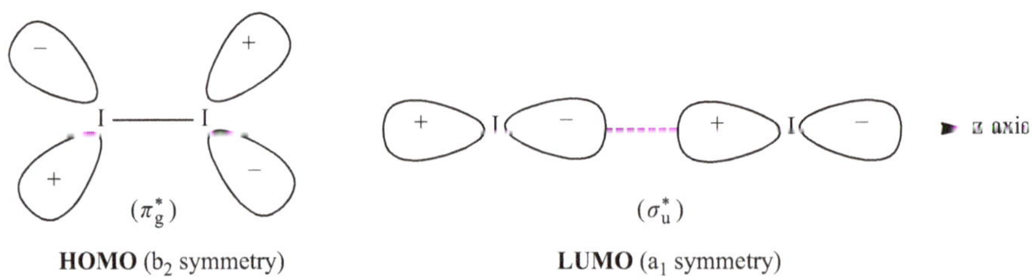

Fig.4.3: HOMO and LUMO of I_2 molecule in C_{2v} symmetry.

Now, let us take broad side ways collision of H_2 and I_2 to give rise to activated complex or large pseudo molecule (Fig. 4.4). This has C_{2v} symmetry.

It is notable here that σ_{1g} (HOMO) of H_2 and $\sigma_u{}^\star$ (LUMO) of I_2 belong to A_1 irreducible representation and $\sigma_{1u}{}^\star$ (LUMO) of H_2 and $\pi_g{}^\star$ (HOMO) in I_2 belong to B_2 irreducible in C_{2v} point group and these are shown in Table 4.1.

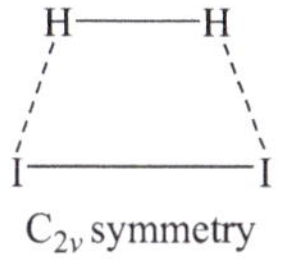

Fig. 4.4: Large pseudo molecule.

Table 4.1: Symmetry of HOMO and LUMO of H_2 and I_2.

C_{2v}	E	C_2	σ_{xz}	σ_{yz}	H_2	I_2
A_1	1	1	1	1	σ_{1g} (HOMO)	σ_u* (LUMO)
A_2	1	1	−1	−1	–	–
B_1	1	−1	1	−1	–	–
B_2	1	−1	−1	1	σ_{1u}* (LUMO)	π_g* (HOMO)

Thus, H_2 HOMO (σ_{1g}) and I_2 LUMO (σ_u*) belong to A_1 symmetry, that is, these are a_1 type orbitals while H_2 LUMO (σ_{1u}*) and I_2 HOMO (π_g*) belong to B_2 symmetry, that is, these are b_2 type orbitals.

Let us check the overlaps of these orbitals.

HOMO (H_2) and LUMO (I_2)

This type of overlap is shown (Fig. 4.5) below:

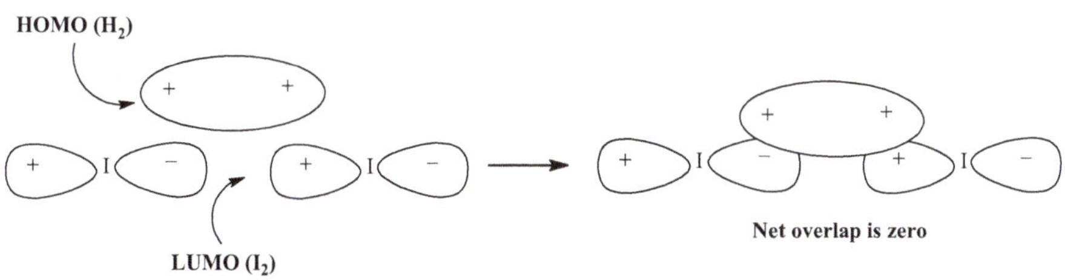

Fig. 4.5: Overlap of HOMO of H_2 and LUMO of I_2.

The above overlap indicates that this reaction is *symmetry forbidden.*

LUMO (H_2) and HOMO (I_2)

This type of overlap is shown in Fig. 4.6 given below.
The electron flow from HOMO (I_2) to LUMO (H_2) is allowed by symmetry rule. Here electrons flow from antibonding π_g* HOMO of I_2 to empty σ_{1u}* LUMO of H_2. Hence, this type of overlap does not lead to the desired reaction. As this type of flow of electron density from antibonding MO of I_2 will strengthen I–I bond and also it

makes no sense that electrons are transferred from more electronegative element iodine to more electropositive element (hydrogen). Thus, the reaction

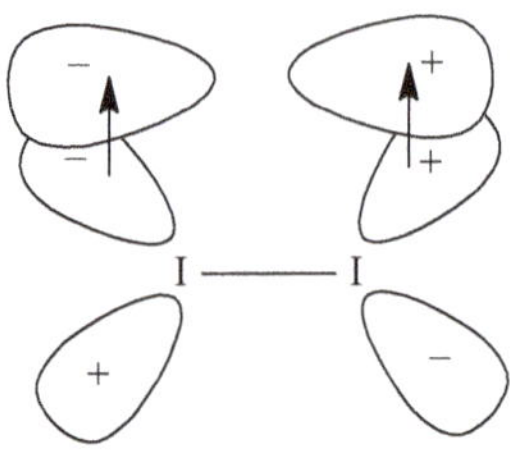

Fig. 4.6: Overlap of HOMO of I_2 and LUMO of H_2.

$$H_2 + I_2 \rightleftharpoons 2HI$$

by broad side ways molecular attack is *symmetry forbidden*. Hence, the possible reaction paths are as follows:

(i) Reaction can occur between H_2 and two atoms of iodine at low temperature.

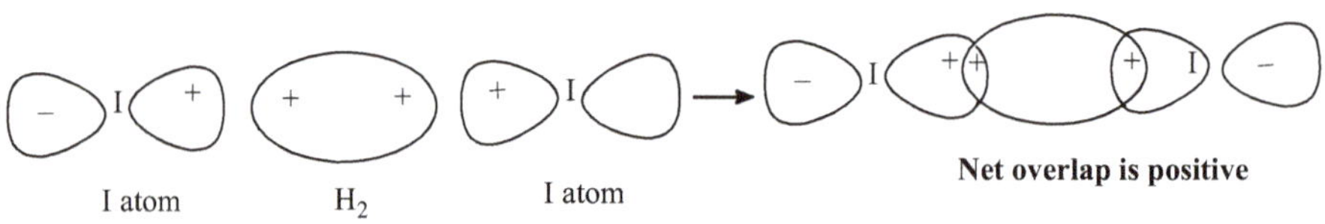

(ii) A free radical reaction at high temperature is symmetry allowed because of desired symmetry of HOMO and LUMO.

$$I_2 \rightarrow 2I^*$$
$$I^* + H_2 \rightarrow HI + H^*$$

Similarly other reactions like

$$H_2 + F_2 \rightleftharpoons 2HF$$
$$I_2 + Br_2 \rightleftharpoons 2IBr$$
$$I_2 + Cl_2 \rightleftharpoons 2ICl$$
$$H_2 + D_2 \rightleftharpoons 2HD$$

are not symmetry allowed by broad sideways attack of molecules. These reactions occur by atomic or free radical mechanism.

Let us consider the reaction

$$H_2 + D_2 \rightleftharpoons 2HD$$

just to reconfirm the above statement.

The overlapping HOMO (H_2) and LUMO (D_2) may be shown as follows:

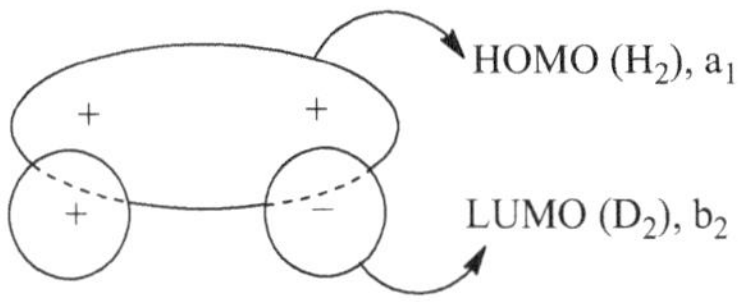

Net overlap is zero

This gives no net overlap, so sideways attack is symmetry forbidden. Similar to H_2 molecule, HOMO of D_2 is a_1 type and LUMO is b_2 type. The combined HOMO and LUMO of H_2 and D_2 are shown below:

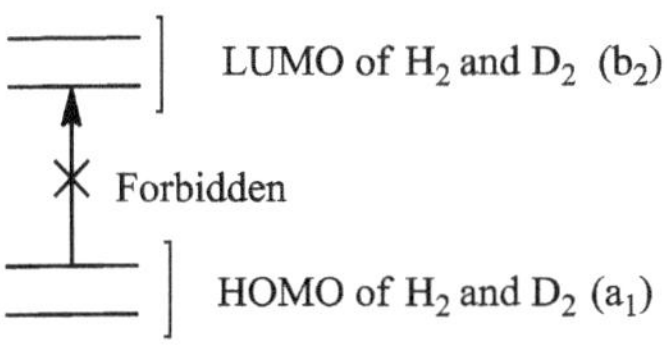

As there is no antibonding orbital of right symmetry, transfer of electron from a_1 to b_2 is symmetry forbidden.

Contrary to this, reactions of the types:

$$D_2 \rightarrow 2D$$
$$H_2 + 2D \rightarrow HD + D$$

are allowed and overlap between HOMO and LUMO is depicted below:

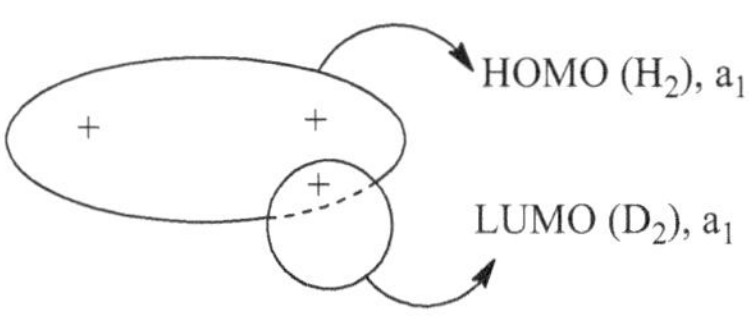

Net overlap is non zero

As HOMO of H_2 and LUMO of D_2 both have a_1 symmetry, this overlap is symmetry allowed.

(b) Reaction between N_2 and O_2

This is a bimolecular reaction resulting in the synthesis of nitric oxide.

$$N_2 + O_2 \rightleftharpoons 2NO \quad \Delta H = 43\,\text{kcal/mole}$$

For years, this reaction was thought to proceed through bimolecular processes in which two reacting molecules collide broad sideways, that is, in C_{2v} symmetry as shown below:

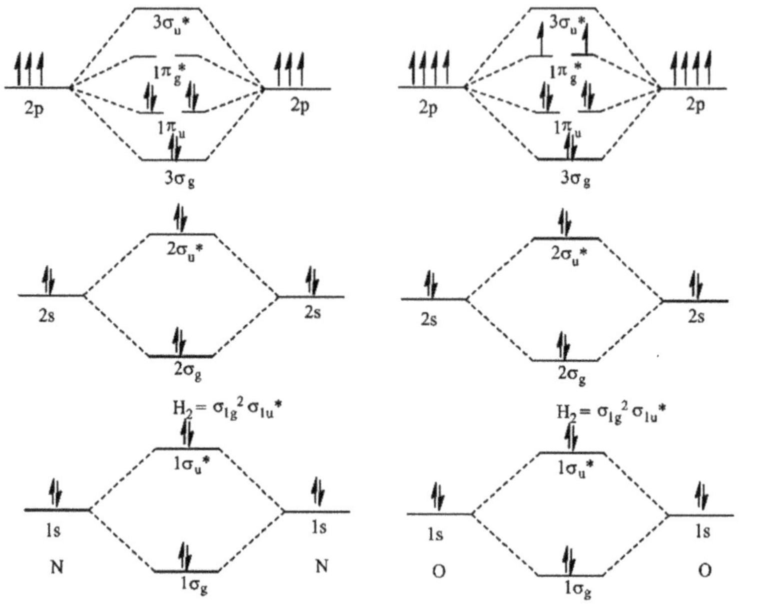

$^{14}N_2 = 1\sigma_g^2 1\sigma_u^{*2} 2\sigma_g^2 2\sigma_u^{*2} 3\sigma_g^2 1\pi_u^4 1\pi_g^*$ $\quad$ $^{16}O_2 = 1\sigma_g^2 1\sigma_u^{*2} 2\sigma_g^2 2\sigma_u^{*2} 3\sigma_g^2 1\pi_u^4 1\pi_g^{2*} 3\sigma_u^*$

In a simple way (removing prefix 1, 2, 3), the M. O. electronic configurations of N_2 and O_2 are:

$$N_2 = \sigma_g^2 \sigma_u^{*2} \sigma_g^2 \sigma_u^{*2} \sigma_g^2 \pi_u^4 \pi_g^*$$

$$O_2 = \sigma_g^2 \sigma_u^{*2} \sigma_g^2 \sigma_u^{*2} \sigma_g^2 \pi_u^4 \pi_g^{*2} \sigma_u^*$$

Thus, HOMO of N_2 is π_u^4 (a_1 type) and LUMO of N_2 is π_g^* (b_2 type). HOMO of O_2 is π_u^{*4} (a_1 type) and LUMO of O_2 is π_g^{*2} or σ_u^*(b_2 type). Here, π_g^{*2} of O_2 is half filled and this can be treated as HOMO and LUMO. These MOs are shown in Fig. 4.7. The results are given in Table 4.2 for C_{2v} point group of TS or pseudo molecule.

After having information about HOMO and LUMO of N_2 and O_2, let us take HOMO of N_2 and LUMO of O_2 and check their overlap with each other.

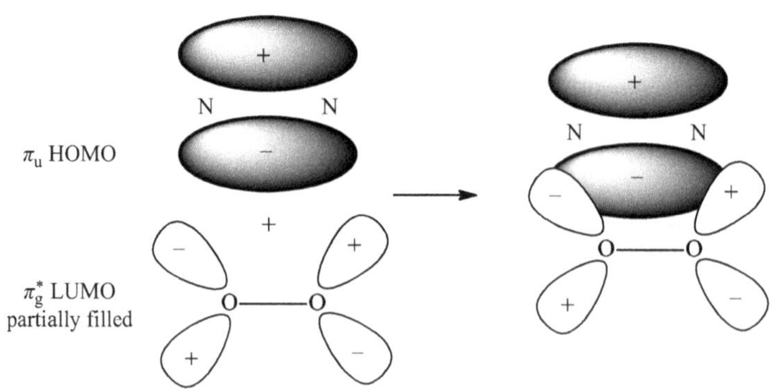

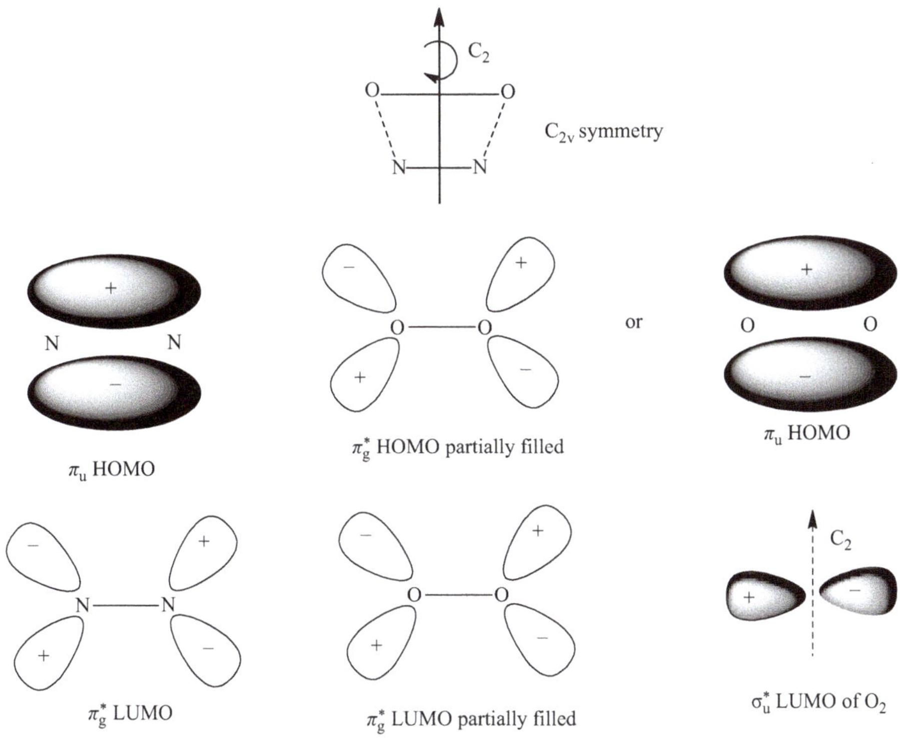

Fig.4.7: HOMO and LUMO of N_2 and O_2.

From the above overlapping scheme, it is very clear that the net overlap is zero and hence is symmetry forbidden.

Now, let us see the overlapping of HOMO of O_2 and LUMO of N_2 depicted below:

Table 4.2: Symmetry of HOMO and LUMO of O_2 and N_2.

C_{2v}	E	C_2	σ_{xz}	σ_{yz}	O_2	N_2
A_1	1	1	1	1	π_u HOMO (a_1)	π_uHOMO (a_1)
A_2	1	1	−1	−1		
B_1	1	−1	1	−1		
B_2	1	−1	−1	1	π_g* LUMO or π_u* LUMO (b_2)	π_g* LUMO (b_2)

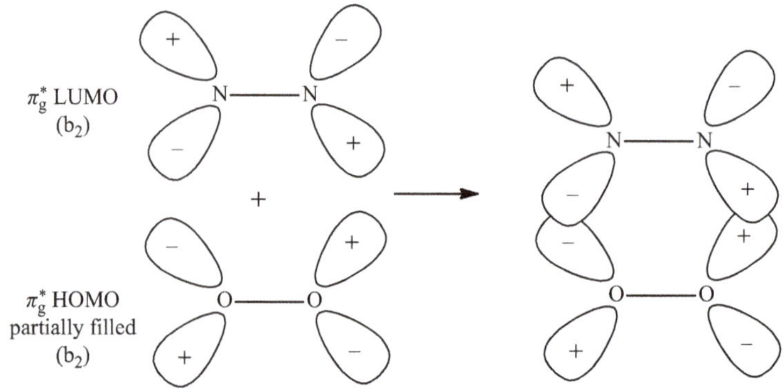

The overlap shown above is symmetry allowed but chemically incorrect because electrons cannot flow from more electronegative element (O) to less electronegative element (N). Moreover, flow of electron from antibonding $\pi_g{}^*$ MO of O_2 will strengthen the O–O bond rather than weakening it.

Reaction of N_2 + O_2 to give rise to NO takes place only at high temperature (E_a = 93 kcal) (Fig. 4.8).

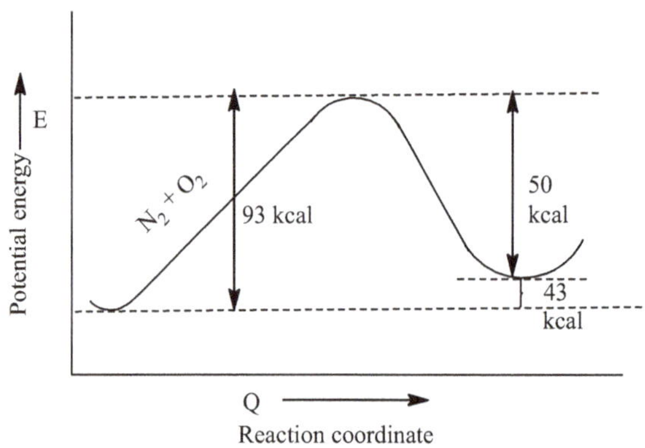

Fig. 4.8: Plot of potential energy curve for reaction between O_2 and N_2.

At high temperature, electron flow from HOMO of N_2 to excited LUMO of N_2 can occur. Now, from this excited LUMO of N_2 (partially filled because of excitation at high temperature), electron can flow to $\pi_g{}^*$ LUMO of O_2 (partially filled) as both are b_2 symmetry.

(c) Addition to H_2 or Cl_2 to ethylene ($H_2C = CH_2$)

As concerted processes, this type of reaction is forbidden as evident from the overlapping of HOMO of H_2/Cl_2 and LUMO of **$H_2C = CH_2$** shown below:

$$C_2H_4 + H_2 \rightarrow C_2H_6$$
$$C_2H_4 + Cl_2 \rightarrow C_2H_4Cl_2$$

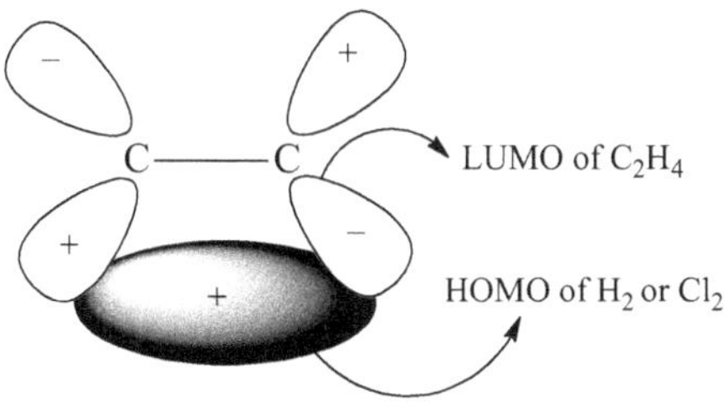

From the overlapping scheme given above, it is very clear that the net overlap is zero and hence is symmetry forbidden.

Let us consider this reaction in terms of symmetry of bonds.

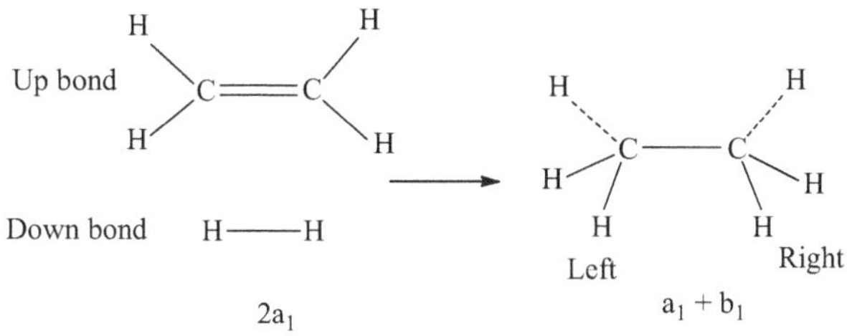

Thus, we see that in the above reaction, two $2a_1$ types of bonds have been converted to a_1 + b_1 type. Therefore, from symmetry point of view, above reaction is not allowed. It is thus concluded that all reactions are symmetry forbidden in which we take bonds that are initially written as up and down and are converted into left to write bonds.

Thus, concerted *cis*-addition of H_2 to ethylene is symmetry forbidden. However, concerted *trans* addition of H_2 to ethylene is symmetry allowed.

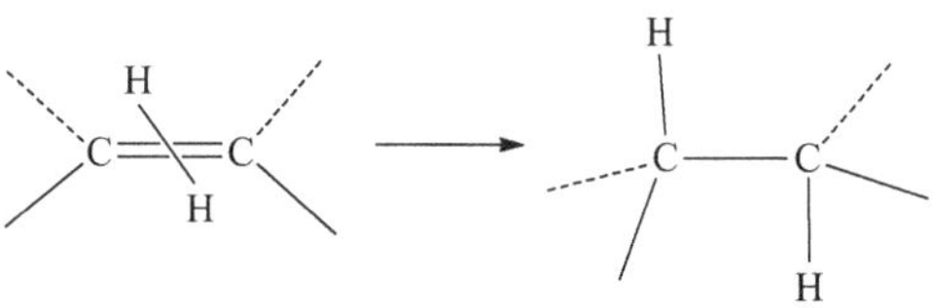

This type of *trans*-addition is called antarafacial and *cis*-addition is called suprafacial. The reaction path of *trans*-addition with the formation of TS is given in Fig. 4.9.

(i) Addition of HCl to ethylene (C_2H_4)

The reaction C_2H_4 + HCl→ C_2H_5Cl is partially symmetry forbidden. This is well evident from the following Fig. 4.10.

The overlap of HOMO and LUMO is not zero (slightly more than zero) because of unsymmetrical MO of HCl (unequal distribution of electron density in H–Cl bond).

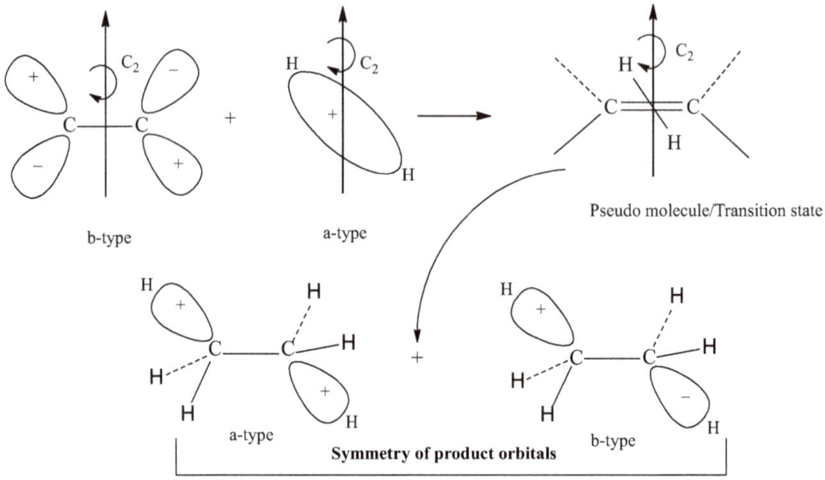

Fig. 4.9: Symmetries of C–C π-bond and H–H σ-bond in C_2 symmetry in antarafacial addition of H_2 to ethylene.

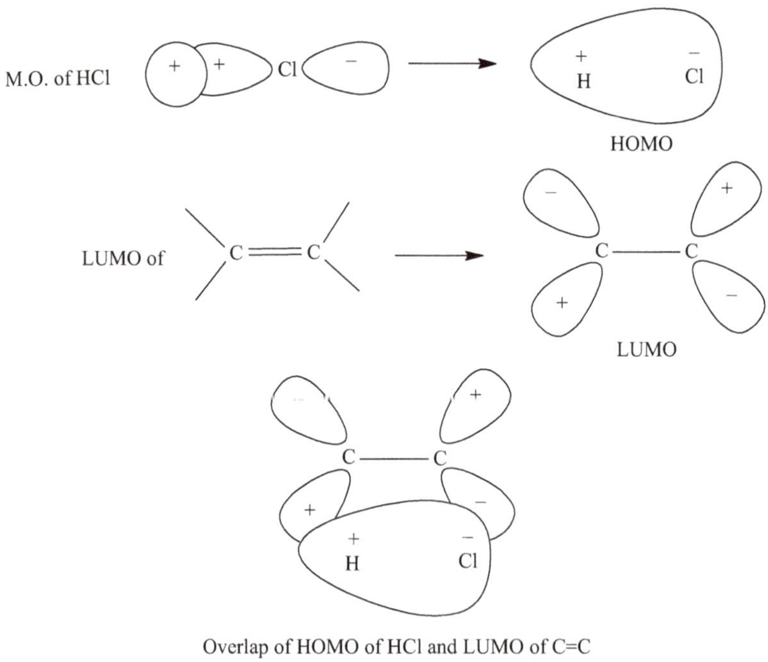

Fig. 4.10: MO of HCl and C=C and their overlap with each other.

(ii) Addition of SCl_2 and Cl_2 and SO_2 and Cl_2

The reaction of $SCl_2 + Cl_2 \rightarrow SCl_4$ is symmetry allowed. This can be checked as follows:

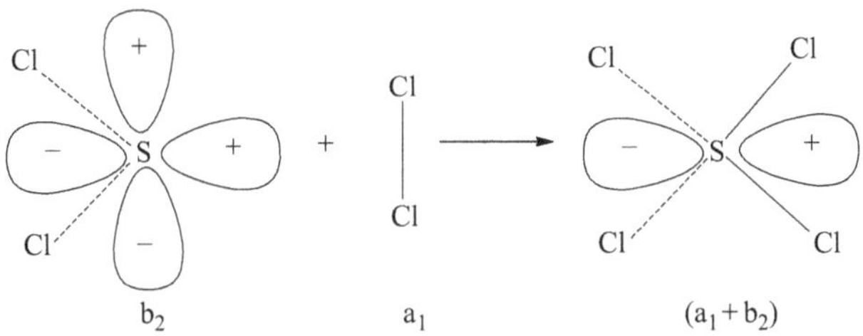

The additional lone pair of electron in SCl_2 occupies b_2 orbital. Thus, orbital symmetry at both side matches. Hence, the reaction is symmetry allowed and really occurs instantly at 0 °C.

The reaction of $SO_2 + Cl_2 \rightarrow SO_2Cl_2$ is not symmetry allowed. This can be shown as follows:

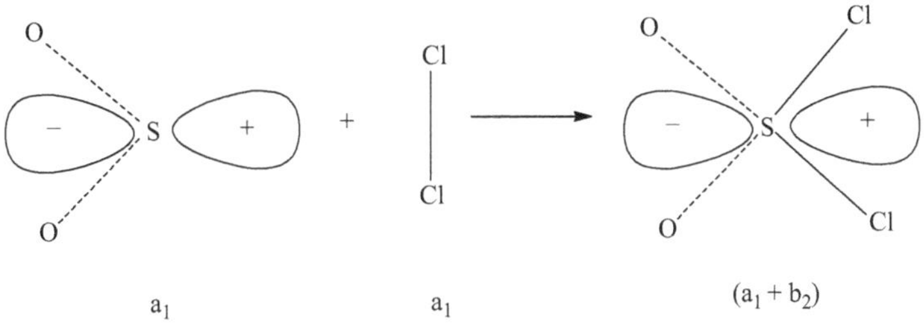

HOMO of SO_2 and Cl_2 are a_1 type. The resulting two bonds in SO_2Cl_2 are a_1 and b_2 types. Thus, orbital symmetry both side does not match. Therefore, from symmetry point of view, the reaction is not allowed.

(iii) Reaction of $PbCl_4$ or $SbCl_5$ and $ArICl_2$ to olefin

When $PbCl_4$ interacts with olefin, *cis*-addition of chlorine to olefinic double bond takes place as given in the following reaction scheme:

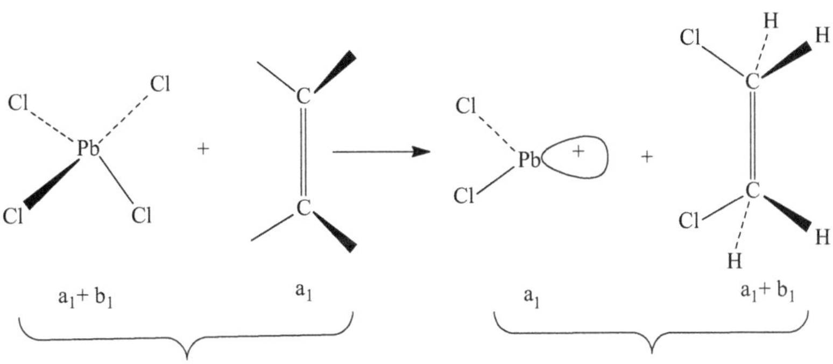

We see that orbital symmetry both side matches. Therefore, the reaction is symmetry allowed. When $SbCl_5$ is taken as a source of chlorination, one axial and one equatorial Cls are removed.

When $ArICl_2$ is used as a source of chlorination, it adds chlorine *cis* to olefins in concerted manner. However, it is symmetry forbidden reaction.

We see that orbitals on both sides do not match, and so reaction is not symmetry allowed.

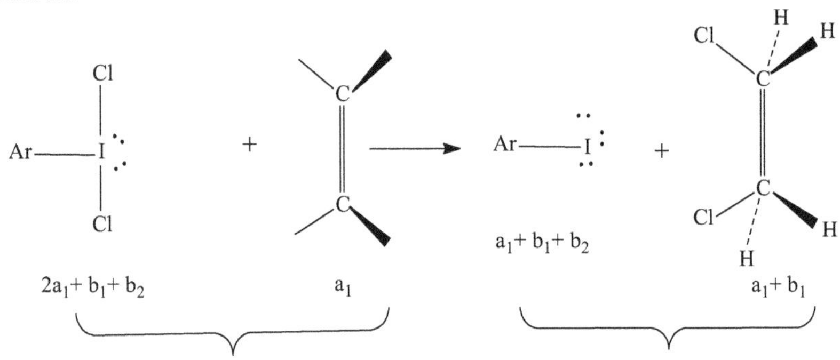

(iv) Addition of MnO_4^- or OsO_4^- to olefin

Concerted addition of MnO_4^- or OsO_4^- to olefinic double bond occurs as shown below.

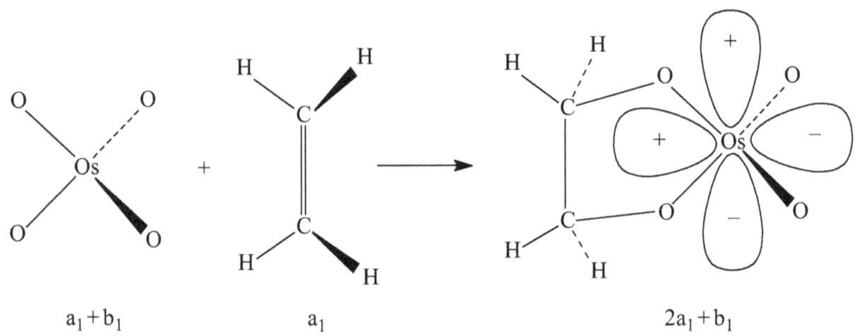

As the orbital symmetry match both sides, the reaction is symmetry allowed.

(v) Addition of hydrogen to olefin double bond

Though the direct addition of hydrogen to an olefin or acetylene is symmetry forbidden, still it is possible to add simultaneously two atoms of hydrogen to unsaturated linkage. All that necessary here is that hydrogen should be bonded to one or two other atoms. The following two schemes will make it possible:

(a) The *cis*-diimide can act as hydrogenating agent in one step concerted process of the type:

$$N_2H_2 + C_2H_4 \rightarrow N_2 + C_2H_6$$

The HOMO in diimide is one that is bonding for both N–H and antibonding for N–N bond. The electron flow from HOMO into π^* orbital of C_2H_4 will break C–C double bond and the two N–H bonds. At the same time, N–N bond will acquire triple bond character and two new C–H bonds will be formed.

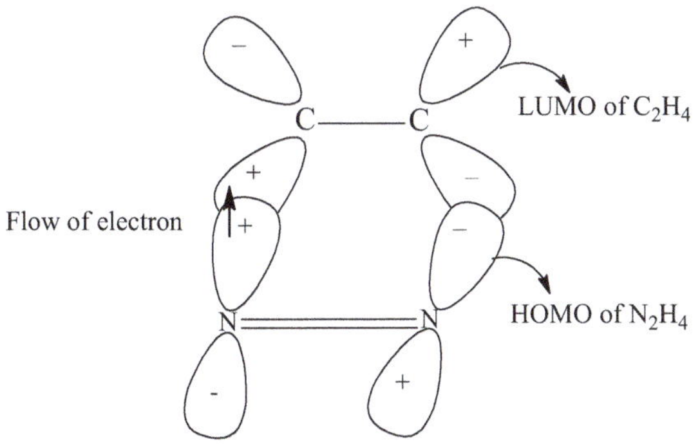

Because of presence of hydrogen atoms on two different nitrogen atoms (like two isolated atoms), symmetries of HOMO and LUMO are matched, and hence the reaction is symmetry allowed.

(b) Another possibility is that H_2 atom is first linked to a transition metal. The d-orbitals of transition metal have the right symmetry to react directly with H_2 molecule. Electrons flow from filled d-orbitals into empty σ^* orbital of H_2 will break the H–H bond and will create two new metal–hydrogen bonds. The metal hydride can now transfer two hydrogen atoms to olefin double bond in one step concerted process as shown in Fig. 4.11.

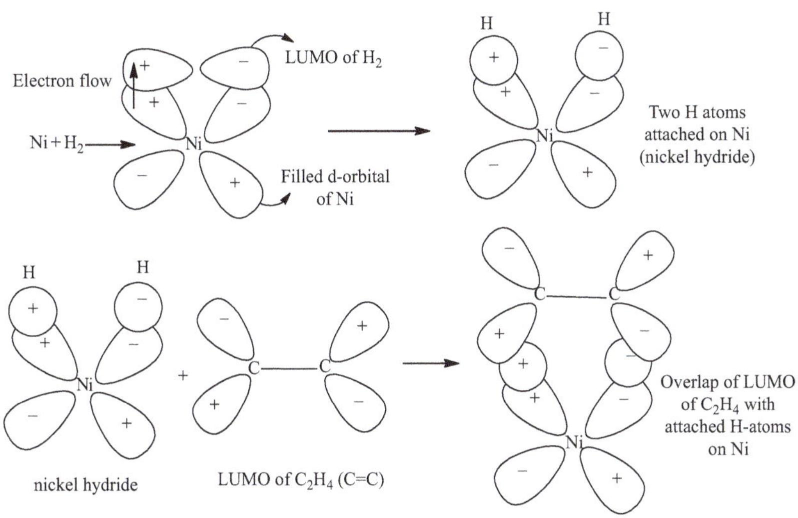

Fig. 4.11: Catalytic role of Ni on addition of hydrogens to olefin double bond.

(i) Oxidation of N_2O_2 in presence of O_2

The HOMO of N_2O_2 and LUMO of O_2 and their overlap (Fig. 4.12) unveil the plausible mechanism of oxidation of N_2O_2 by O_2.

It is noticeable from the above diagram that the flow of electrons take place from HOMO of N_2O_2 (b_2 symmetry) to LUMO of O_2 (b_2 symmetry). This helps in breaking of O–O bond (because of flow of electrons to antibonding LUMO of O_2), and thus strengthening of N–N bond in N_2O_2 and formation of two new N–O bonds by positive overlap resulting in N_2O_4. The N–N bond strength in N_2O_4 is much greater than N_2O_2.

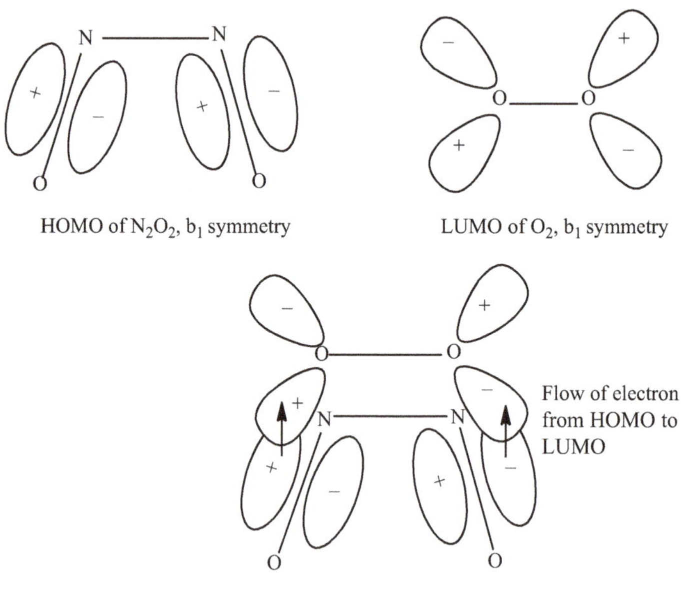

Fig. 4.12: HOMO and LUMO of N_2O_2 and O_2, and their overlap.

4.4 Nucleophilic displacement reactions

(a) Let us consider a molecule of the type A–B through which nuleophilic displacement occurs. The LUMO of A–B (R–X or C–X, where R = alkyl group and X= halide ion) is of σ^* type. The electron must flow from HOMO of nucleophile (Nu:) to this (σ^*) orbital to break A–B (R–X or C–X) bond or electron must be removed from filled σ orbital.

Let us take molecule A–B as alkyl halide (C–Cl) and observe the overlapping of HOMO and LUMO of the reactants.

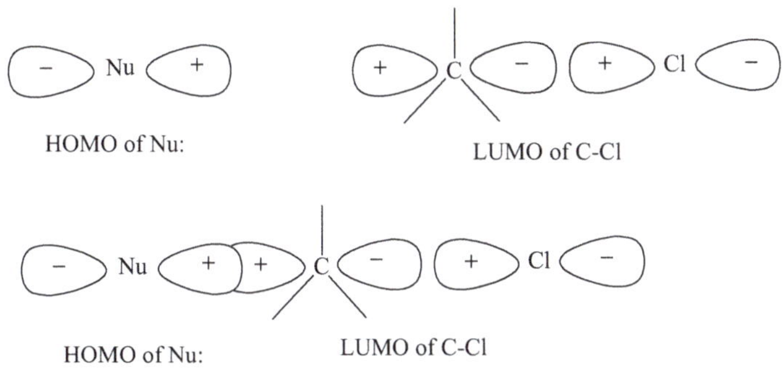

This is well clear that the overlap of HOMO and LUMO is symmetry allowed. The usual displacement mechanism is as follows:

$$\text{Nu:} + \text{C} - \text{Cl} \rightarrow \text{Nu} - \text{C} + \text{Cl}$$

This displacement takes place with retention in configuration at the carbon center, and is symmetry allowed for σ-type donor orbital.

When the nucleophile is π-donor, it can react at the front side of C–Cl bond in another way. Here, a d-orbital would be suitable for this type of interaction. However, p-orbital may also be effective. For π-donor, electron transfer from the HOMO of nucleophile to the LUMO of alkyl halide will result in breaking of C–Cl bond and formation of Nu–C bond. This overlap is pictorially shown in Fig. 4.13. Again, such an overlap results in the retention of configuration at the carbon atom.

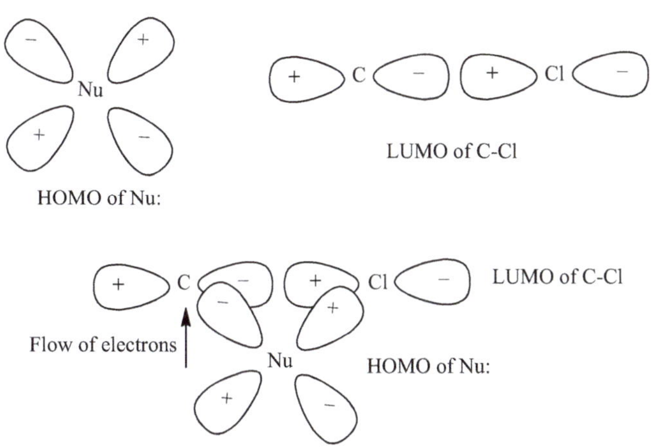

Fig. 4.13: Front side overlap of HOMO and LUMO.

(b) Let us consider another well-known reaction, the oxidative addition of A–B (alkyl halide) to square planar d^8 metal complexes in which carbon and halogen are attached to metal ion. For example, reaction of CH_3I with Vaska's compound $[Ir(Cl)(CO)(PPh_3)_2]$ as shown below:

$$[Ir^{I}(Cl)(CO)(PPh_3)_2] + CH_3I \rightarrow [[Ir^{III}(Cl)(CO)(PPh_3)_2(CH_3)(I)]]$$

Both *cis*- and *trans*-addition of A–B (CH_3I) can occur to this complex with retention in configuration at carbon as found in the product. Iridium atom can act as the electron donor (nucleophile) in these reactions. These are the two cases where carbon atom attaches directly to the metal atom. The following Fig. 4.14 shows these mechanisms from symmetry point of view.

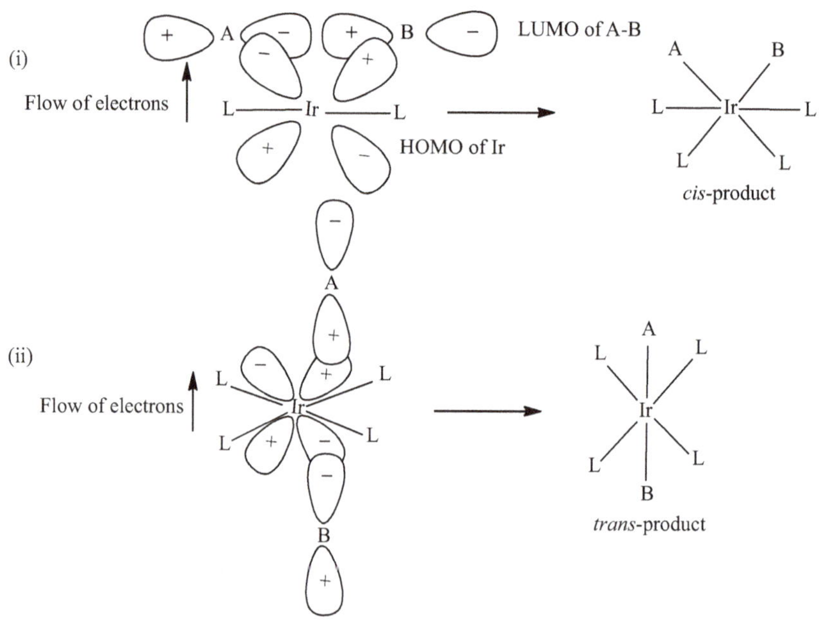

Fig. 4.14: (i) One step *cis*-addition and (ii) one step *trans*-addition of A and B components of A–B (CH_3I) molecule.

4.5 Berry's pseudorotation: orbital symmetry control

This type of rotation engrosses intramolecular rearrangement of fluxional trigonal bipyramidal (TBP) molecules. In such rearrangement, interchange of axial and equatorial substituents takes place without breaking of bonds with minimal structural distortion via square pyramidal (SP) TS as shown in Fig. 4.15.

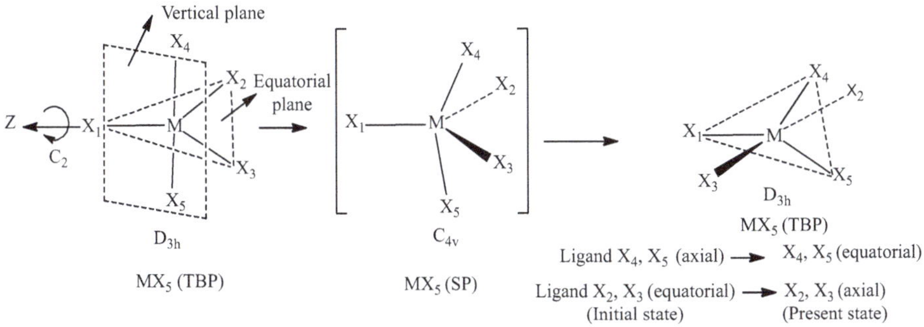

Fig. 4.15: Pseudorotation of MX_5.

A symmetry analysis of Berry's pseudorotation provides an insight to the electronic *allowedness/acceptability* of the fluxional process [35].

Figure 4.15 given above shows that the pentagonal bipyramidal molecule, MX_5, belonging to D_{3h} point group, changes into equivalent D_{3h} structure via a TS with C_{4v} symmetry point group. In order to do symmetry analysis only those symmetry elements that are preserved throughout the rearrangement path are considered. For Berry's pseudorotation of MX_5 molecule, these symmetry elements are C_2 axis and two mutually perpendicular planes containing this C_2 axis. These are shown in Fig. 4.14 [(i) C_2 along X_1 and bisecting $\angle X_2MX_3$, (ii) planes are (a) equatorial plane containing M, X_1, X_2 and X_3 and (b) vertical plane passing through $X_1...X_4MX_5$]. This figure also shows the geometry of pseudorotation with respect to Cartesian coordinates, *z*-axis coinciding with conserved C_2-axis.

A correlation diagram for various d-orbitals under pseudorotation, including the valence of d-orbitals of the initial, transition and final states is shown in Fig. 4.16.

The d-orbital energy has been estimated on the basis of crystal field theory (CFT).

In the reaction path of Berry's pseudorotation, it is assumed that M–X bond distances are equal and remain unchanged during the course of rearrangement. The only changes that take place are in bond angles. The two types of bond angles are θ and ϕ. The $\theta = \angle X_4M\,X_5 = 180°$ and $\phi = \angle X_2M\,X_3 = 120°$ for initial state (D_{3h}). In TS $\theta = \phi = 105°$. For the final state, these angles are $\theta = \angle X_4M\,X_5 = 120°$ and $\phi = \angle X_2M\,X_3 = 180°$.

In *case-I* [Fig. 4.16 **(a)**], in SP TS, the b_2 orbital is predicted to be of higher energy than the ($a_2 + b_1$) orbitals. In *case-II* [Fig. 4.16 **(b)**], the energy ordering obtained from extended Huckel method is qualitatively similar to *case-I*, except that the relative energies of b_2 and ($a_2 + b_1$) orbitals are reversed. Both the correlation diagrams also show the d-orbital composition of each energy level for the initial and final states of the TBP molecule (D_{3h}). In C_{2v} symmetry, both d_{z^2} and $d_{x^2-y^2}$ orbitals are of a_1 symmetry, and are mixed in the initial and final TBP state.

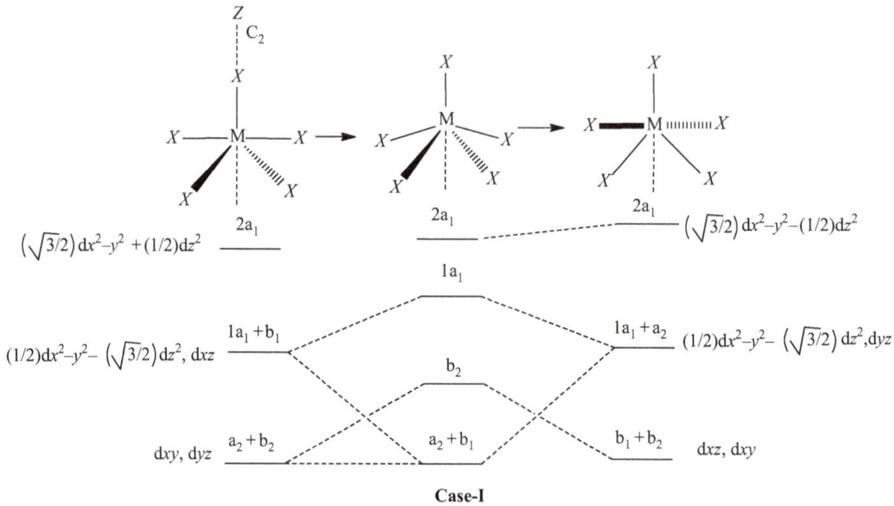

Fig. 4.16 (a): d-Orbital correlation diagram for Berry's pseudorotation when $Eb_2 > E\,(a_2 + b_1)$ in SP state.

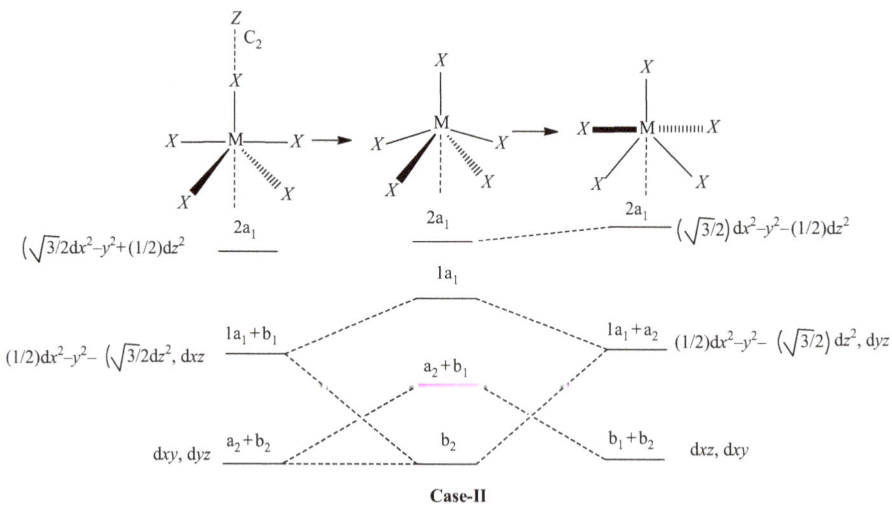

Fig. 4.16 (b): d-Orbital correlation diagram for Berry's pseudorotation when $Eb_2 < E\,(a_2 + b_1)$ in SP state.

The correlation diagrams for *case-I* and *case-II* may now be used to predict orbital symmetry allowedness of Berry's pseudorotation of d^n configuration. For an allowed rearrangement, the occupied orbitals of the initial state must correlate to the ground state occupied orbitals of both transition and final states. Generally, most of the complexes that undergo pseudorotation have a d^8 configuration in low-spin state. For such complexes or molecules, the electronic configuration of the initial TBP state is $a_2^2b_2^21a_1^2\ b_1^2\ 2a_1^0$ in C_{2v} point symmetry. The electronic configuration is same for SP TS in both the cases (*case-I* and *case-II*, Fig. 4.16) and also for the final TBP state. Thus, Berry's pseudorotation is predicted to be orbitally allowed for all low-spin configurations.

4.6 Correlation diagrams: prediction of orbital symmetry allowedness for Berry's pseudorotation

Let us consider a d^1 case. The electronic configuration may be either a_2^1 or b_2^1 initially in TBP state. Suppose the initial configuration is a_2^1. Then, the correlation diagram [Fig. 4.16 **(a)**] shows that this starting ground state a_2^1 configuration does not have any matching ground state orbital in final ground state. In fact a_2^1 matches in symmetry with the excited state configuration ($1a_1 + a_2$) of the final state. So, in this configuration it is orbitally forbidden.

If we take configuration b_2^1, then the initial ground state orbitals and final state ground state orbitals match in symmetry with ($b_1 + b_2$) but this again has to pass through high energy SP TS b_2. Thus, if we consider only the initial and final states, then the reaction is allowed but it has to pass through excited TS, and thus the process is kinetically inert to rearrangement by Berry's pseudorotation.

If we consider *case-II* [Fig. 4.16 (b)] and suppose the initial configuration b_2^1, then it matches in symmetry with ground state of final state ($b_1 + b_2$) and matches also with the TS orbital symmetry b_2 in ground state ($b_2 \rightarrow b_2 \rightarrow b_2$). Then, Berry's pseudorotation path is predicted to be orbitally allowed.

Parallel analysis can also be carried out with other d^n configuration in TBP symmetry. Some more examples of various d^n configuration and their allowedness are given below in Table 4.3.

Table 4.3: Orbital allowedness for Berry's pseudorotation for some d^n electronic configuration.

d^n configuration	Initial ground state orbital configuration	Allowed or not in per view of case-I	Allowed or not in per view of case-II
d^2	$a_2^1\ b_2^1$	Not allowed	Not allowed
d^3 (low-spin)	$a_2^2\ b_2^1$	Not allowed	Not allowed
d^4 (low-spin)	$a_2^2\ b_2^2$	Not allowed	Not allowed
d^9	$a_2^2b_2^21a_1^2\ b_1^2\ 2a_1^1$	Allowed	Allowed
d^{10}	$a_2^2b_2^21a_1^2\ b_1^2\ 2a_1^2$	Allowed	Allowed

4.7 Stable shape of the molecules: symmetry rules

Symmetry rules can be applied to predict the stable shape/structure of the molecules. Molecules of the type AB_n or A_2B_n generally have regular structures. It can be predicted by symmetry rules whose regular structure these molecules can assume.

The procedure involves in examining a molecule in possible structure, viz., octahedral, square planar, tetrahedral, linear or bent is to see the gap between HOMO and LUMO. In fact, stable structure of a molecule will have large gap between HOMO and LUMO that matches for electron transition. Unstable structures have small gap between HOMO and LUMO.

Let us explain the procedure by taking H_2AB molecule. Whether it is stable in planar form of C_{2v} symmetry or it will distort to pyramidal form belonging to C_s symmetry point group (Fig. 4.17).

Examples:

(i) **Formaldehyde (H_2CO)** The *abinitio* calculations in the present case give the ordering of MOs as $(4a_1)^2(1b_1)^2(5a_1)^2(1b_2)^2(2b_1)^2(2b_2)^0(6a_1)^0$. The lowest energy transition is $2b_1 \to 2b_2$ orbitals. This is also called n to π^* transition because $2b_1$ is non-bonding and localized on oxygen atom. The energy gap between $2b_1$and $2b_2$ is 3.5 eV, which is not a large gap. The symmetry of the transition is given by the direct product B_1 (corresponding to b_1 orbital) and B_2 (corresponding to b_2 orbital) in C_{2v} point group, $B_1 \times B_2 = A_2$. This is well clear from the character table (CT) of C_{2v} point group.

C_{2v}	E	C_2	$\sigma_{v(xz)}$	$\sigma_{v(yz)}$		
A_1	1	1	1	1	z	x^2, y^2, z^2
A_2	1	1	−1	−1	R_z	xy
B_1	1	−1	1	1	x, R_y	xz
B_2	1	−1	−1	1	y, R_x	yz

Here, $B_1 \times B_2 = 1 \times 1 \quad -1 \times -1 \quad 1 \times -1 \quad -1 \times 1$ (character of B_1 and B_2 taken from CT)

$= 1 \quad 1 \quad -1 \quad -1$

$= A_2$

There is no A_2 type of vibrational mode (vibrational modes are only $2A_1$ and B_2) in C_{2v} point group. So, the transition is not feasible. The next excited state ($6a_1$) lies very high in energy (>7.0 eV). Thus, the transition $2b_1 \to 2b_2$ is not feasible, and H_2CO or CH_2O remains in planar structure. In fact, it has planar structure.

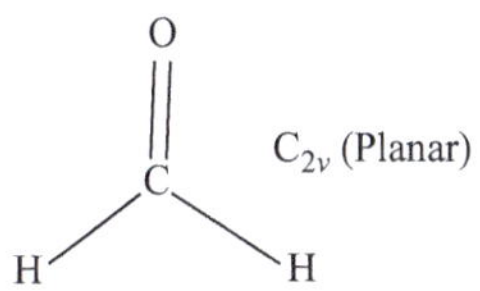

(ii) **NH_2O radical** Based on *abinitio* calculations, the MO ordering of this radical is as follows:

$$(1b_2)^2(2b_1)^2(2b_2)^1(6a_1)^0$$

So, the lowest energy transition is $2b_2 \rightarrow 6a_1$ and the symmetry of the transition is $B_2 \times A_1 = B_2$. This B_2 corresponds to normal vibrational mode in C_{2v} point group. Hence, the electronic transition is possible, and structure will distort from planar structure to pyramidal one.

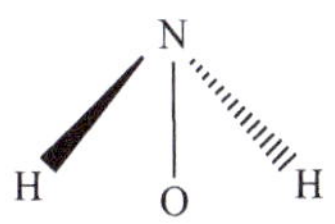

C_s (Pyramidal)

(iii) **NH_2F** The MO sequence for NH_2F is: ....$(1b_2)^2(2b_1)^2(2b_2)^1(6a_1)^0$. This is similar to that of NH_2O. So, it is predicted to be pyramidal in geometry.

(iv) **CH_2F** It is isoelectronic with NH_2O, and it is predicted to be pyramidal in shape. Unlike the CH_3 radical, CH_2F, CHF_2 and CF_3 are pyramidal.

(v) **O_3 and SO_2** We will see the stability of these two molecules on the basis of symmetry rules. SO_2 is colorless and O_3 is blue in color and is highly unstable. Let us examine whether the dissociation $O_3 \rightarrow O_2 + O$ is symmetrically feasible or not. The MO sequence for O_3 is: ....$(3b_1)^2(4b_1)^2(6a_1)^2(1a_2)^2(2b_2)^0$. The lowest energy transition is $a_2 \rightarrow b_2$ and the transition has the symmetry $A_2 \times B_2 = B_1$. This is asymmetrical vibration in C_{2v} point group, that is, one O–O bond is short and other bond is long. It means that transition is feasible and hence the dissociation $O_3 \rightarrow O_2 + O$ is feasible and will occur.

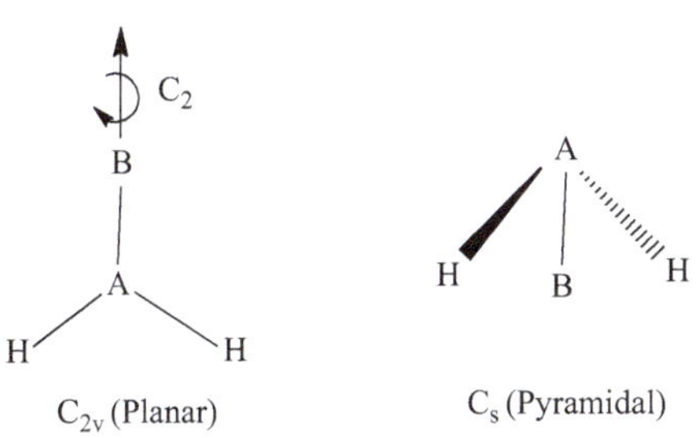

Fig. 4.17: Structures of H_2AB molecule.

The absorption bands for O_2 are at 1.5 and 2.1 eV and those for SO_2 are at 3.2 and 3.7 eV. For SO_2, it corresponds to triplet to singlet transition, that is, $6a_1 \rightarrow 2b_2$. From the view point of symmetry, this transition is spin forbidden and thus transition is not feasible. It means that SO_2 is stable with respect to asymmetric dissociation as shown below:

$$SO_2 \rightarrow SO + O$$

From the above discussion, we see that for unimolecular reactions, viz., isomerism and dissociation the following points are pertinent.

(a) In such reaction, HOMO and LUMO are on the same molecule.
(b) HOMO to LUMO transfer of electrons results in shift of electron density in the molecule.
(c) When HOMO to LUMO overlap is positive, the electron density increases and when overlap is negative, electron density decreases. In such situation, the positive nuclei will move in the direction of increasing electron density, that is, isomerization or dissociation starts.
(d) Small difference in energy in HOMO and LUMO means that the structure is unstable and rearrangement occurs in such a manner that HOMO and LUMO gap is increased.
(e) The gap between HOMO and LUMO determines the absorption of UV/visible light. Hence, colored compounds are generally more reactive because HOMO–LUMO gas is small and transition occurs in visible region.

(vi) **H_2X molecule** Let us consider H_2X molecule and see whether it acquires linear or bent structure or whether it rearranges or dissociates to HX and X. For MO sequencing, only s and p orbitals on X and s orbitals of two hydrogen atoms are considered. The MO sequence is . . .σ_g σ_u π_u $\sigma_g{}^{\star}\sigma_u{}^{\star}$. π_u is nonbonding The energy gap between $\pi_u \rightarrow \sigma_g{}^{\star}$ is much smaller than $\sigma_g \rightarrow \sigma_g{}^{\star}$ gap. The Fig. 4.18 displays MO sequence for such molecules (H_2X).

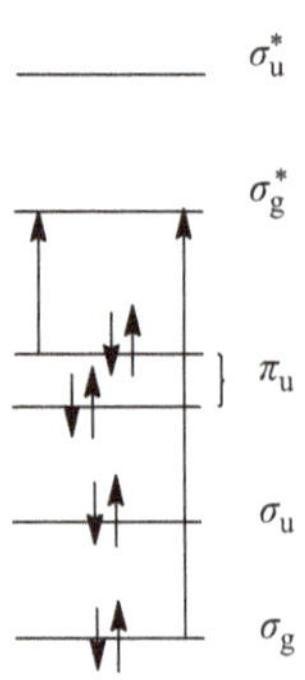

Fig. 4.18: Molecular orbital sequence of H_2X.

Molecules with 5, 6, 7 and 8 electrons, that is, $\sigma_g^2\ \sigma_u^2\ \pi_u^1\ \sigma_g{}^{\star}\sigma_u{}^{\star}$ to $\sigma_g^2\ \sigma_u^2\ \pi_u^4$ $\sigma_g{}^{\star}\sigma_u{}^{\star}$ configurations are unstable and $\pi_u \rightarrow \sigma_g{}^{\star}$ is possible because of the small gap. These structures rearrange to maximize the gap between HOMO and LUMO. For example, BH_2 (5 electrons), CH_2 (6 electrons), NH_2 (7 electrons) and H_2O (8 electrons) are predicted to be bent. BeH_2 (4 electrons) with electron configuration $\sigma_g^2\sigma_u^2$ will be linear because the transition ($\pi_u \rightarrow \sigma_g{}^{\star}$) energy is very high and the molecule is stable toward rearrangement. This is why it is a linear molecule. The electronic transition $\sigma_u \rightarrow \pi_u$ (bonding to nonbonding) simply causes molecule to rotate.

H_2F (9 electrons) and NeF_2 (10 electrons) will be stable to bending and will be unstable toward dissociation to HF + H or to NeF + F. This is because of energy gap between $\sigma_g{}^{\star} \rightarrow \sigma_u{}^{\star}$ is very small and both are antibonding.

4.8 Symmetry controlled pericyclic reactions

Pericyclic reactions are relatively important class of reactions in which a reactant is converted to a product through cyclic conjugated TS without giving any intermediate either thermally or photochemically. Such reactions differ from ionic or free radical reactions in number of respects. These reactions are insensitive toward solvents, catalyst, radical initiators or inhibitors and effected only under the influence of heat or light.

Pericyclic reactions occur in concerted fashion in which a reactant is converted to a product without intervention of an intermediate. The bond breakage and formation occur synchronously/simultaneously. These reactions are highly stereospecific and the results obtained under thermal and photochemical conditions may be markedly different.

4.9 Classes of pericyclic reactions

The main classes of pericyclic reactions are briefly presented below:

(i) Cycloaddition reaction

It is an intra- or intermolecular reaction usually thermally induced (exceptions are photoinduced dimerisation of simple olefin) concerted and reversible pericyclic reaction in which two terminals of two different π-systems within the same or different molecules combine to form two σ-bonds and an additional ring. It may be represented as follows:

$$(m+n)\pi \Leftrightarrow [(m+n)-2]\pi + 2\sigma^2 + \text{ring}$$

where m and n are number of π-bonds and in σ^2, 2 stands for the number of electrons in a σ-bond.

Examples:

(a)

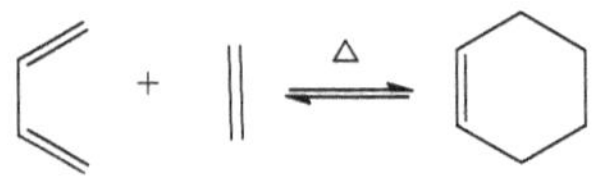

$$(2+1)\,\pi \Leftrightarrow [(2+1)-2]\,\pi + 2\sigma^2 + \text{ring}$$
$$\Leftrightarrow \pi + 2\sigma^2 + \text{ring}$$

(b)

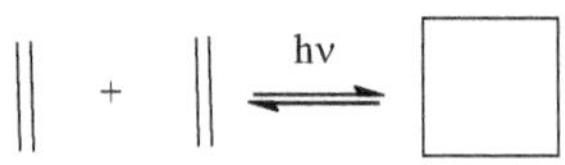

$$(1+1)\,\pi \Leftrightarrow [(1+1)-2]\,\pi + 2\sigma^2 + \text{ring}$$
$$\Leftrightarrow 2\sigma^2 + \text{ring}$$

(ii) Cheletropic reactions

These are thermally induced reversible concerted cycloaddition reactions in which two σ-bonds are formed on the same atom or two σ-bonds are broken on the same atom. A few examples are given below:

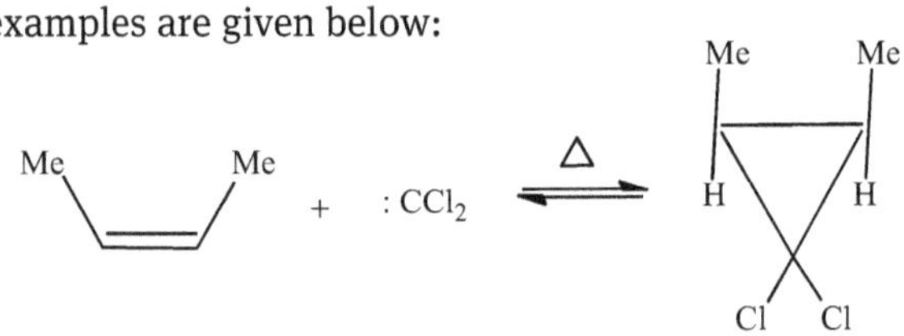

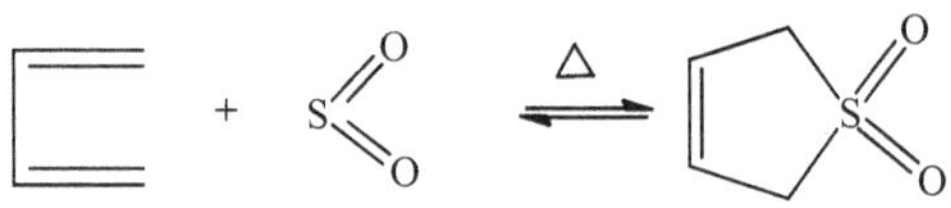

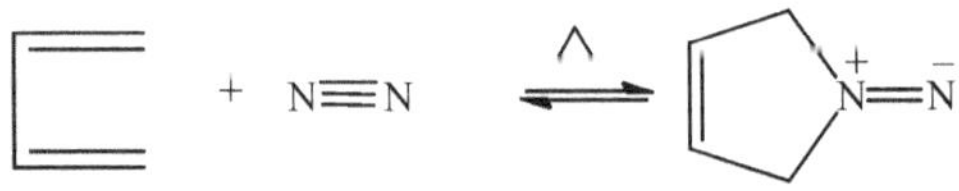

(iii) Electrocyclic reactions

An electrocyclic reaction is an intramolecular thermal or photoinduced concerted reversible pericyclic reaction in which two terminals of a conjugated polyene (acyclic or cyclic) are joined through a single bond, the net change being the conversion of a π-bond into a σ-bond with ring formation or reversibly the conversion of a σ-bond into a π-bond with ring opening.

In the electrocyclic reactions, only one reactant, that is, polyene or in the reverse a cyclo (poly) alkene is involved. Such reactions may be represented as follows:

$$\pi^m + \text{ring opening} \Leftrightarrow \pi^{(m-2)} + \sigma^2 + \text{ring closure}$$

where m = number of π-electrons, in σ^2, 2 stands for the number of electrons in a σ-bond.

Examples:

$$\pi^4 \Leftrightarrow \pi^{4-2} + \sigma^2$$
$$\Leftrightarrow \pi^2 + \sigma^2$$

$$\pi^6 \Leftrightarrow \pi^{6-2} + \sigma^2$$
$$\Leftrightarrow \pi^4 + \sigma^2$$

(iv) Sigmatropic rearrangements

It is a pericyclic reaction in which a σ-bonded group or atom migrates from one end of a π-system to other end in an uncatalyzed intramolecular process.

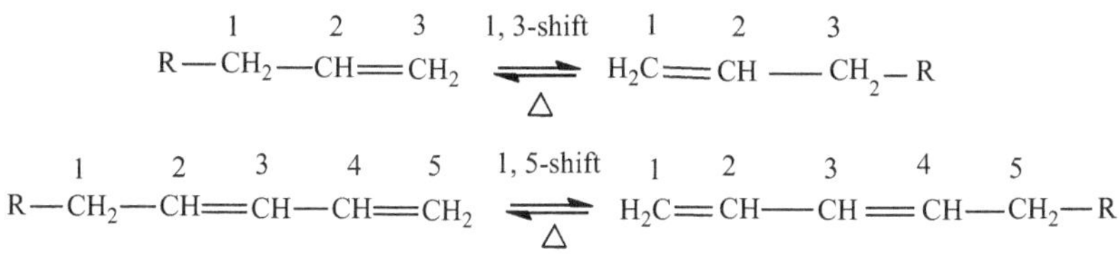

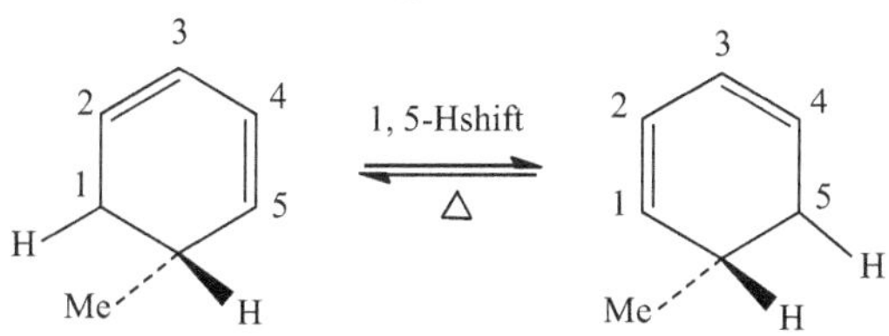

(v) Group transfer reactions

As the name implied, in this type of pericyclic reactions, an atom or a group is transferred from one molecule to another.

The most familiar example of group transfer reaction is the reduction of carbon–carbon double bond with hydrazine in presence of oxidizing agent. The active reducing agent in this reaction is in fact the highly active species di-imide (HN=NH) formed in situ by oxidation of hydrazine.

$$TsNHNH_2 \xrightarrow{\Delta} TsH + HN{=}NH$$

$$NH_2NH_2 \xrightarrow[CuSO_4/EtOH]{O_2} HN{=}NH$$

Ts = Tosyl

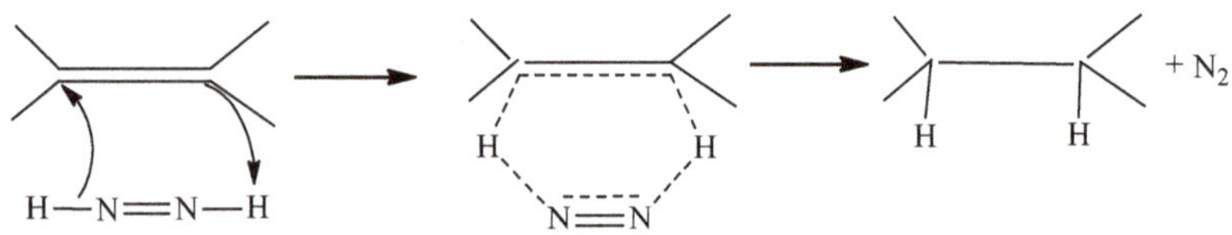

The reaction of diimide with olefin is a concerted "group transfer" reaction proceeding through a six membered cyclic TS. It results symmetrical addition of hydrogen, and is therefore useful for stereospecific introduction of hydrogen on to carbon–carbon double bond with the elimination of N_2.

4.10 Interpretation of pericyclic reactions: different approaches

Since the beginning of investigation by Woodward and Hoffmann for the application of orbital symmetry analyses to pericyclic reaction, three main approaches have so far been developed and these are as follows:

(i) Correlation diagram method
(ii) Frontier molecular orbital (FMO) method and
(iii) Mobius–Hückel method

(i) Correlation diagram method

This approach was due to *Woodward and Hoffmann*, according to which the symmetry properties of the involved MOs of the reactants and products are correlated in a diagram known as *Correlation diagram*. On the basis of such correlation diagram, a set of rules, known as *Selection rules*, were formulated for different pericyclic reactions. According to these rules, reaction paths of certain concerted reactions were described as *symmetry allowed* while some others as *symmetry forbidden*.

(ii) Frontier orbital method

The second approach and perhaps the most popular one is the FMO approach, developed by *Woodward and Fukui* and this is popularly known as HOMO–LUMO method. According to this approach, the symmetry properties of only the HOMO andLUMO are the guiding factors for the interpretation of these concerted reactions.

(iii) Mobius–Hückel method

The third approach was developed by *Mobius and Huckel* and popularized by H. Zimmermann and M. J. S. Dewar. The idea behind this method is that a pericyclic reaction that proceeds through a cyclic TS with aromatic characteristics is energetically favorable and thus allowed. This is known as *Mobius–Hückel method.*

In the present section, mainly *Woodward–Hoffmann* approach of pericyclic reaction will be described and involving some ideas of HOMO and LUMO.

4.11 Woodward–Hoffmann approach

It is now well recognized that molecular transformations involved in chemical reactions are also determined by symmetry of orbitals involved in the reactions. Most of the symmetry rules that explain and predict the course of chemical reaction deal with the changes in electronic structures and vibrational motion in the molecule, which are very strongly symmetry dependent.

Formation of cyclic TSs in concerted pericyclic reaction is necessary for continuous flow of electrons. Also, the cyclic TS must correspond to an arrangement of participation orbitals that can maintain a bonding interaction between the reacting atoms throughout the course of reaction. In subsequent discussion, we will observe that these requirements make pericyclic reactions highly predictable in terms of relative reactivity, stereospecificity and regioselectivity.

According to *Woodward–Hoffmann*, reaction path of such concerted reaction is controlled by symmetry properties of orbitals that are directly involved (HOMO and LUMO) in the reaction. The *Woodward–Hoffmann* approach is applicable when the following two conditions are generally followed:

(i) The reaction must involve a concerted process as its rate determining step, that is, reaction must occur without intermediate.
(ii) During the concerted reaction path, one or more symmetry elements of the entire reacting system must be conserved.

Since the conservation of symmetry properties of participating MOs is of prime importance in pericyclic reactions, it would be desirable to have a brief discussion on symmetry properties of MOs and their determination, and other related issues before applying the *Woodward–Hoffmann* approach in mechanistic interpretation of some pericyclic reactions.

MOs have been found to possess two types of symmetry elements: (i) plane of symmetry or mirror plane (σ) and (ii) twofold axis of symmetry or C_2 axis. Some MOs are symmetric about a mirror plane that bisects the MOs and is perpendicular to the plane of the atoms (σ-frame work). It means if a reflection of MO is carried out on the plane it must result in coincidence with another equivalent MO. The twofold axis of symmetry (C_2) is the symmetry element passing at right angle in the same plane (plane of the atoms) and through the center of the σ-frame work of the atoms forming the MO. If rotation of a MO by 180° about this axis results in an identical MO with original, it is said to have C_2 symmetry or symmetric to C_2.

The following MO has mirror plane of symmetry but has no C_2 symmetry.

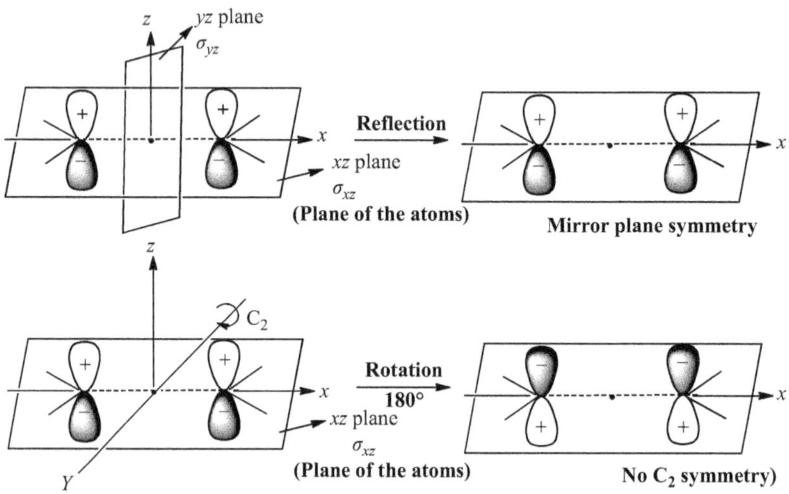

Again, the following MO has no mirror plane symmetry but possesses C_2 symmetry.

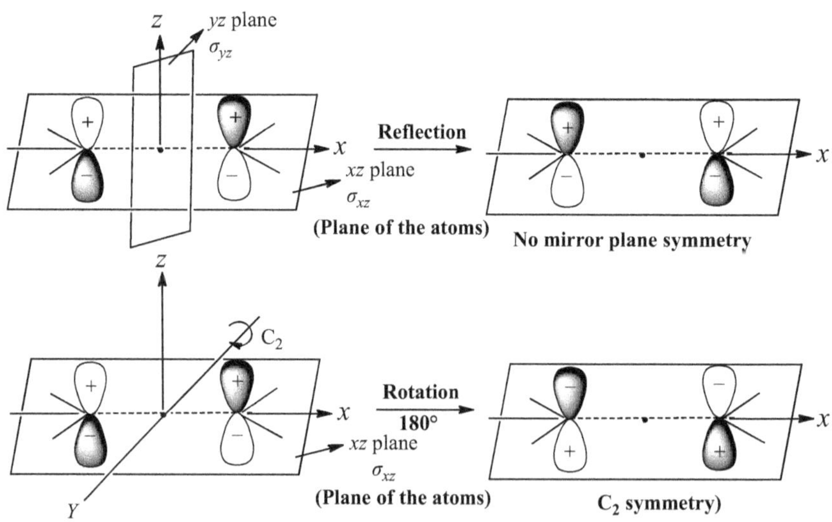

It should be noted that symmetry depends on the symmetry of the interacting MO only and not on the overall symmetry of the molecule as a whole.

Symmetry of MO depends on the number of nodes. For a linear conjugated π-system, the wave function ψ_n will have $(n-1)$ nodes. When $(n-1)$ is zero or an even integer, the MO (ψ_n) will be symmetric with respect to mirror plane (σ) and antisymmetric with respect to C_2. On the other hand, when (n–1) is an odd integer, the MO (ψ_n) will have the symmetry exactly reversed as tabulated below:

Molecular orbital	Node (n–1)	Mirror plane (σ)	C_2
ψ_n	Zero or even	S	A
	Odd	A	S

S = Symmetric
A = Antisymmetric

Conjugated π-systems are of two types: (i) open chain and (ii) planar regular cyclic systems. The number of π-MOs and other details in these two types of Conjugated π-systems are given below:

Conjugated acyclic system	Conjugated planar regular cyclic system
1. Number of bonding π-MOs and antibonding π-MOs are same.	1. Number of bonding π-MOs and antibonding π-MOs are same.
2. MOs follow Aufbau principal and Hund's rule.	2. MOs follow Aufbau principle and Hund's rule.
3. Energy of the MOs is directly proportional to the number of nodal planes. As the nodal plane increases, the energy of the π-MOs increases.	3. Energy of the MOs is directly proportional to the number of nodal planes. As the nodal plane increases, the energy of the π-MOs increases.
4. There will be no degenerate π-MOs.	4. There is always one π-MO of lowest energy followed by a pair of degenerate π-MOs of increasing energy and finally a single π-MO of highest energy.

A conjugated polyene contains $4q\pi$ or $(4q + 2)\pi$ electrons where q = 0, 1, 2, 3 and so on. The π-MOs of a conjugated acyclic molecule may be constructed according to the following generalization:

(i) A π-electron system derived from the interaction of *np* atomic orbitals contain $n\pi$-MOs that differ in energy. Thus, the number of π-MOs are always equal to the number of interacting individual *p*-atomic orbitals.

(ii) Half of the MOs (that is, $n/2$) have lower energy than the isolated *p*-orbitals. These are called bonding molecular orbitals (BMO). The other half ($n/2$) have higher energy than the isolated *p*-orbitals. These are called antibonding molecular orbitals (ABMO).

(iii) For odd number of *p*-atomic orbitals, half will be BMO and the other half ABMO and one π-MO will be nonbonding molecular orbitals (NBMO).

(iv) The BMO of lowest energy has no node. Each MO of increasingly higher energy has one additional node.

(v) The nodes occur between atoms and are arranged symmetrically with respect to the center of p-electron system.
(vi) For MOs, the nodal plane is considered, considering only the interacting *p*-orbitals and node because component of the individual *p*-orbitals are ignored.

A few illustration of π-MOs required in the usual common pericyclic reactions describing all details such as node (*n*), HOMO, LUMO, singly occupied MO (SOMO) may be in the ground state or in the exited state, BMO, ABMO, NBMO, mirror plane [σ_{yz}]–S/A, C_2-S/A (S = symmetric; A = Asymmetric) and electron occupancy are listed below:

(a) **For σ-bond**

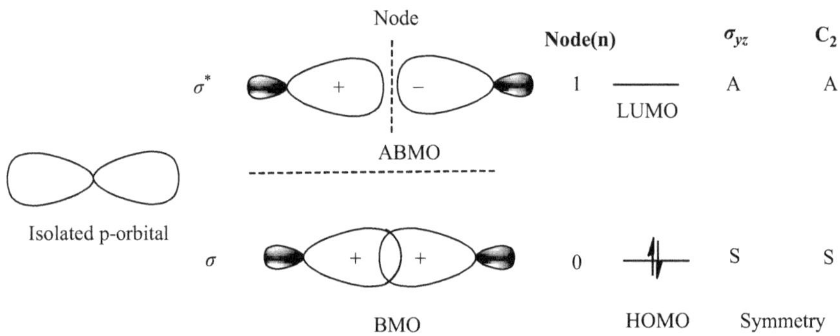

(b) **For ethane**

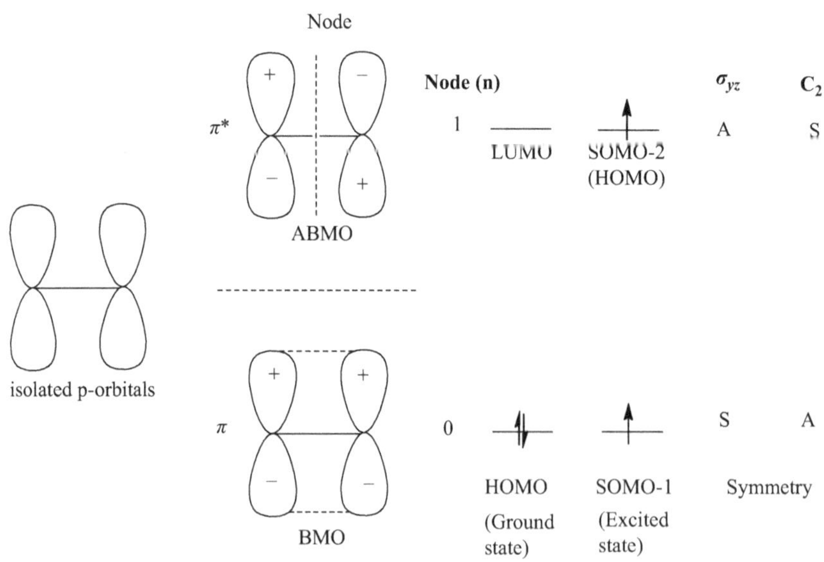

(c) **For allyl system**

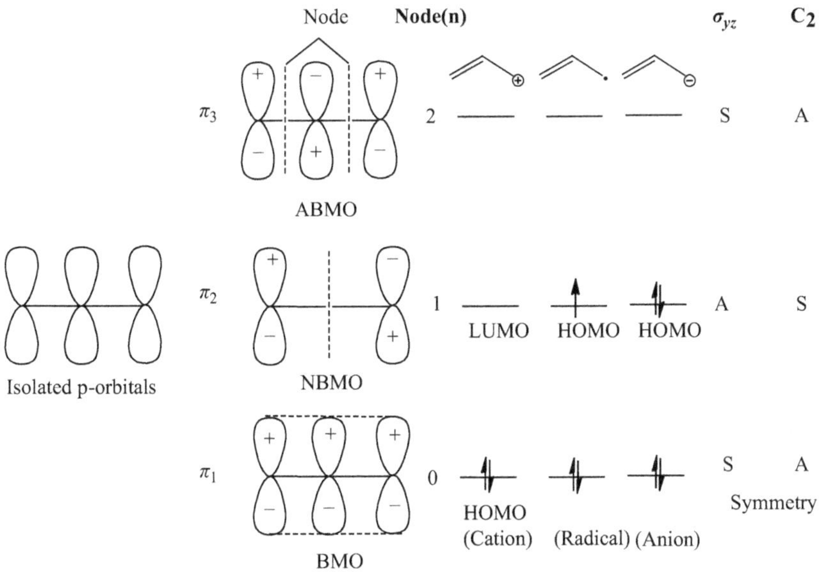

(d) **For 1,3-butadiene system**

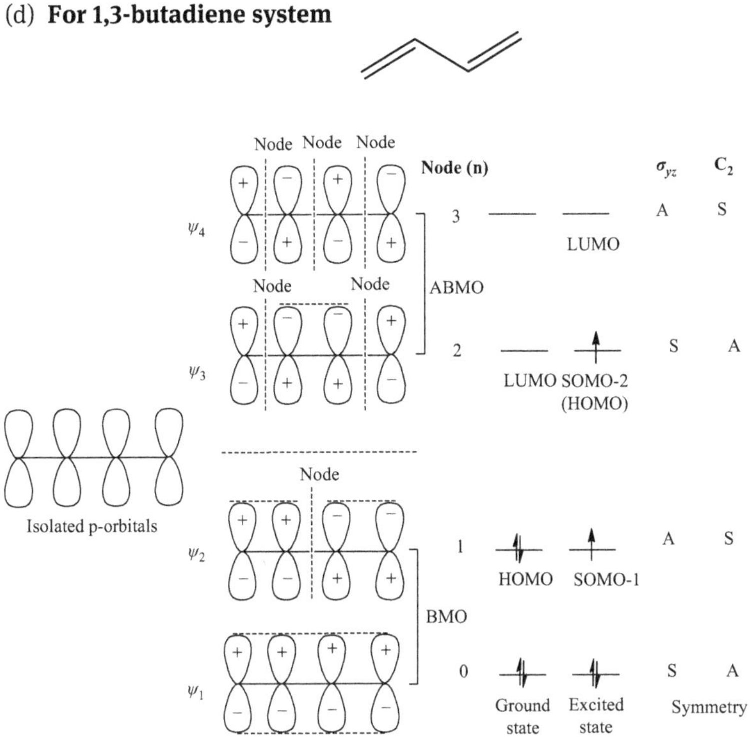

It is notable here that phase situation depends on number of nodes in the ground state and rotation of terminal trigonal carbon or σ-bond in the excited state, whereas HOMO–LUMP depends on number of electron(s) for a particular π-system and ground or excited states.

4.11.1 Symmetry allowed and symmetry forbidden reactions in pericyclic reactions

Pericyclic reactions are exclusively explained on the basis of FMO approach. In this approach, when HOMO and LUMO overlap with the same phase (same sign of the lobe) in the TS, then it is bonding situation and this bonding interaction is called symmetry allowed reaction. But when HOMO–LUMO overlap is not a bonding situation (that is, when opposite sign of lobes interact), then this antibonding situation is called symmetry forbidden reaction. Both the situations are pictorially shown in Fig. 4.19.

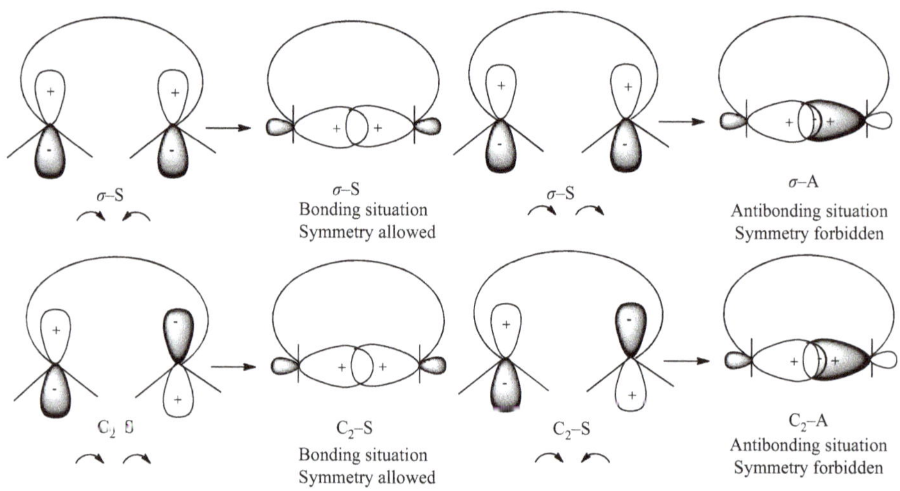

Fig. 4.19: HOMO and LUMO overlapping situation for symmetry allowed and symmetry forbidden reaction.

4.11.2 Conservation of orbital symmetry

In the pericyclic reactions, a set of MOs of the reactant are transformed into a corresponding set of MOs of the product through a concerted process. During the transformation, the symmetry (σ, C_2) of the concerted orbitals are conserved, that is, the orbitals in the TS remain in phase. This maintains some degree of bonding throughout the process giving a concerted character, and the reaction involves a relatively low-energy TS, that is, low activation energy. The reaction so occurred is called symmetry allowed. Contrary to this, if orbital symmetry is destroyed by bringing one or more orbitals out of phase, the TS energy becomes very high (high activation energy) because of antibonding situation and the reaction is called symmetry forbidden.

The overall orbital symmetry of a pericyclic reaction is said to be conserved when each occupied orbital of the reactant transforms smoothly into an occupied orbital of the product with the same symmetry. In order to have appreciable interaction of the HOMO and LUMO, which is the major contributor in the pericyclic reactions, they must be of comparable energies and above all must belong to the same symmetry.

Conservation of orbital symmetry of a pericyclic reaction is usually shown by an orbital correlation diagram, details of which will be given later.

4.11.3 Conrotatory and disrotatory ways of movement in pericyclic reaction

During electrocyclic reaction of an acyclic polyene, each of the terminal trigonal carbon and their substituents must rotate about 90° around the bond to which they are attached to form a σ-bond of the cyclic product. The reverse course is followed in case of ring opening.

If both the terminal trigonal carbons rotate in the same direction, either *clockwise or anticlockwise*, during the cyclization or ring opening is said to be *conrotatory movement*. On the other hand, if both the terminal trigonal carbons rotate in the *opposite direction*, one *lockwise* and the other *anticlockwise*, during the cyclization or ring opening is said to be *disrotatory movement*.

Each type of movement depends upon the following factors:

(i) Thermal or photoinduced reactions.
(ii) Number of π-electrons 4q or (4q + 2) where q = 0, 1, 2, 3. . .
(iii) Stereochemistry of the starting reactant (*cis* or *trans*).
(iv) Steric factors in the TS and the product.
(v) Electronegativity or π-donor capacity of the substituents in the terminal trigonal carbon.
(vi) In case of two disrotatory (inward or outward rotation) and two conrotatory (either clockwise or anticlockwise rotation) possibilities, mode of rotation is controlled by stereoelectronic and steric factors.

Disrotatory and conrotatory movements always give stereoselective products during ring closure and ring opening. Although there are two disrotatory and conrotatory movements possible but they will give same relative stereochemistry in the electrocyclic ring closure reaction for even numbered ring with similar substituents in the terminal trigonal carbons.

The following few examples make it clear.

(i) 4q System

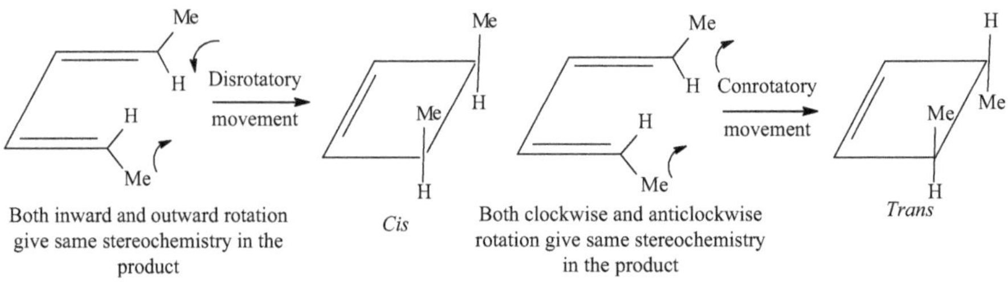

(ii) (4q + 2) System

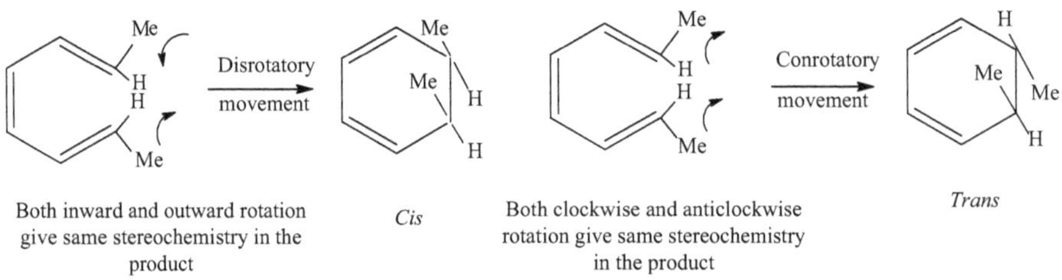

During ring opening, there are also two types of disrotatory and conrotatory movements possible, but the preferred rotation will be controlled by steric factors and π-donor substituents attached to the terminal carbon.

Disrotatory movement

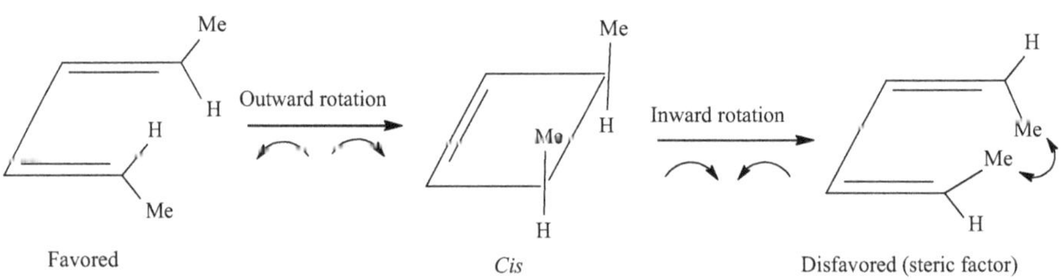

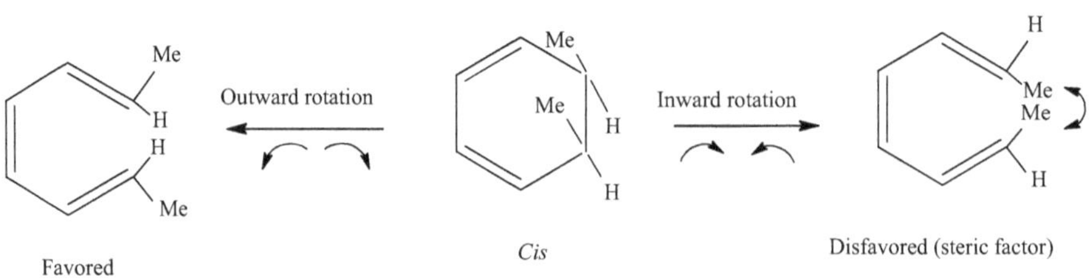

4.11.4 σ, π and ω orbitals and electrons

In pericyclic reactions, the participating electrons are of three types depending upon the nature of orbitals they occupy: σ, π and ω.

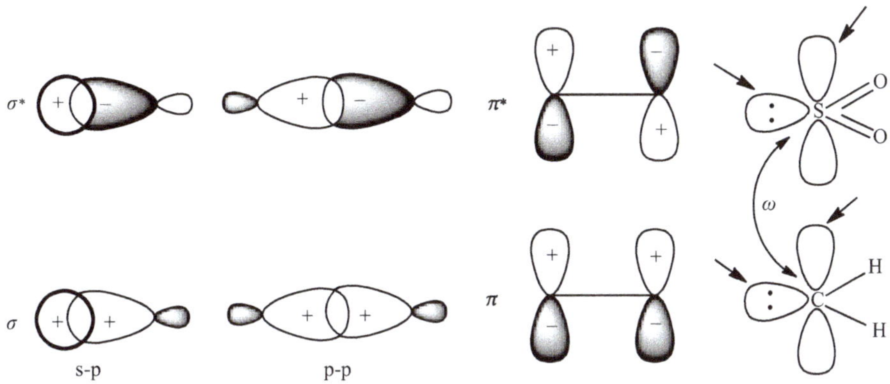

Electrons occupying a single interacting p-orbital is known as ω electrons. The interacting p-orbital may be either filled or vacant, which participate in TS – ω orbitals.

4.11.5 Components in pericyclic reactions

In a pericyclic reaction, a component is a bond or orbital taking part as a single unit. A double bond is π^2 component. The number two (2) is the most important part of this designation and simply refers to the number of electrons. The prefix π tells us the type of electrons. A component may have any number of electrons. For example, butadiene is a π^4 component (4q), hexatriene is a π^6 component (4q + 2), pentadienyl cation is a π^4 component (4q), allyl cation is a π^2 component (4q + 2) and a vacant p-orbital is ω^0 component. A component may not be mixture of π- and σ-electrons. Component usually contains either σ-electrons or contains only π-electrons.

Designation (4q + 2) and (4q) or (4r) simply refer to the number of electrons in the component, where "q" and "r" are integers (0, 1, 2, …etc.). An alkene is a π^2 component and so it is of the (4q + 2) type where q = 0, while diene is a π^4 component and so it is of the 4q or 4r type where q or r = 1.

4.12 Mechanistic interpretation of some pericyclic reactions with symmetry property

(a) Electrocyclic reactions

(i) Cyclobutene–butadiene reversible transformation (4q electron system)

The orbitals of cyclobutene which are directly involved are σ and π and the related antibonding orbitals π^* and σ^*. These orbitals pass on to the four π-MOs of butadiene ψ_1, ψ_2, ψ_3 and ψ_4. All these orbitals are listed in order of increasing energy in the diagram along with their mirror plane symmetry (σ-S/A) and twofold axis of symmetry (C_2-S/A). In the ground state of cyclobutene and butadiene,σ, π and ψ_1, ψ_2 only are filled with electron pair.

The following point is worth considering with regard to cyclobutene–butadiene reversible transformation:

(i) Ring opening and ring closure of cyclobutene ⇔ butadiene may be done either disrotatory or conrotatory movement.

(ii) Ring opening and ring closure of cyclobutene ⇔ butadiene may be done either thermally or photochemically.

(iii) An electrocyclic reaction has been a concerted and cyclic process in which reactant orbitals are converted into product orbitals. It is expected that the TS of such reactions should be intermediate between electronic ground state of the starting material and product. Thus, *the more stable TS will be the one which is able to conserve the symmetry of the reactant orbitals on passing to product orbitals.*

In other words, a symmetric (*S*) orbital in the reactant must get converted in to a same symmetric (*S*) orbital in the product and that an antisymmetric (*A*) orbital in the reactant must get transformed into a same antisymmetric (*A*) orbital in the product. If the symmetries of the reactant and product orbitals do not remain in the same class of symmetry, the reaction will not occur in concerted manner.

In order to judge whether the conservation of symmetry properties essential for pericyclic reactions are being maintained in an electrocyclic reaction, *Woodward and Hoffmann* proposed a set of rules. These rules are derived from the correlation of symmetry characteristics of all the MOs that are directly involved in reaction and such correlation is represented/indicated by a diagram known as *correlation diagram.*

Procedure for construction of correlation diagram

(i) Arrange the reactant orbitals in order of increasing energy from bottom to top in the left hand side of the diagram, including bonding and antibonding MOs.

(ii) Arrange the product orbitals in order of increasing energy from bottom to top in the right hand side of the diagram including bonding and antibonding MOs.

(iii) Place S and A to each reactant and product bonding and antibonding MOs both on the basis of mirror plane symmetry (σ-S) and twofold axis of symmetry (C_2-S).

(iv) Then correlate S or A reactant to the S or A of the product with the same class of symmetry (that is, S reactant to the S product or A reactant to the A product).

(v) If no HOMO/LUMO crossing, the ring opening is thermally allowed (Δ), and if HOMO/LUMO crossing, the ring opening is photochemically allowed ($h\nu$).

Based on the above guidelines, the correlation diagram of cyclobutene ⇔ butadiene is given in Fig. 4.20.

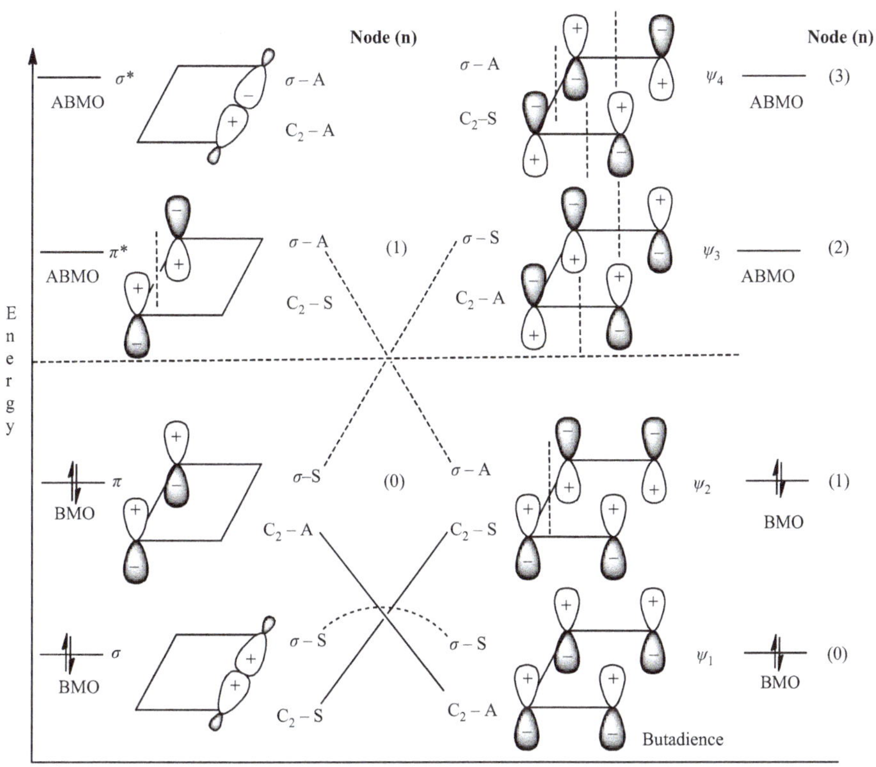

Fig. 4.20: Correlation diagram of cyclobutene ⇔ butadiene. Solid lines, thermally allowed and dashed lines, photochemically allowed reaction process. Nodes are shown by vertical dashed lines.

Disrotatory process: (terminal carbon rotate in the opposite direction)

The σ-orbital is converted into one of the four π-orbitals and this must be either ψ_1 or ψ_3 since the lobes at the terminal atom involved have the same sign. If we follow the unoccupied $\sigma^\star$-orbital similarly in disrotatory opening, it must be one, the ψ_2 or ψ_4 orbital of butadiene with opposite sign lobes at the terminal atoms. On the other hand, the π-orbital of cyclobutene although it does not move can only become ψ_1 or ψ_3 in butadiene, the orbitals with the same sign at those two atoms. Similarly, the $\pi^\star$-orbital will go to ψ_2 or ψ_4 since $\pi^\star$ and ψ_2 or ψ_4 all have lobes of opposite sign at those atoms.

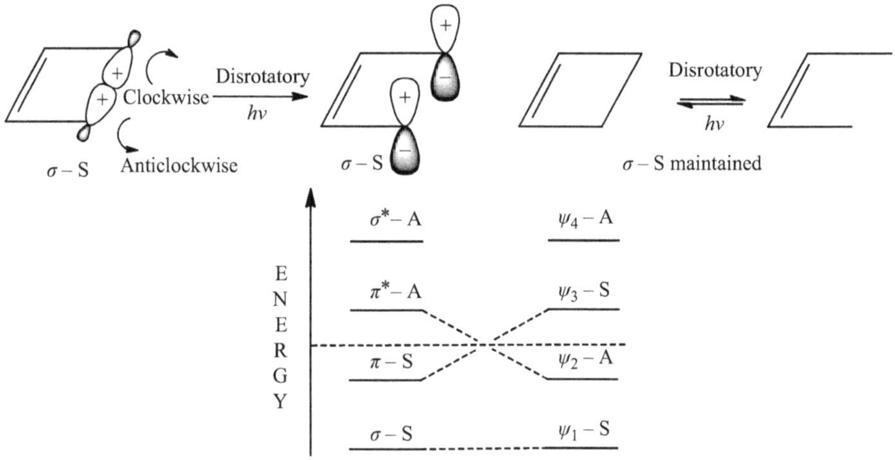

Correlation diagram showing disrotatory interconversion of cyclobutene–butadiene system.

Since the ground state (π-MO) of cyclobutene correlates with the first excited state of butadiene (with mirror plane symmetry, σ–S) and hence disrotatory ring opening (or ring closing) is photochemically a symmetry allowed process (HOMO/LUMO crossing).

$\sigma^2\pi^2$	$\xrightarrow{h\nu}$	$\sigma^2\pi^1\pi^{*1}$	$\rightleftharpoons$	$\psi_1^{\,2}\psi_2^{\,1}\psi_3^{\,1}$	$\xleftarrow{h\nu}$	$\psi_1^{\,2}\psi_2^{\,2}$
Ground state (Cyclobutene)		First excited state		First excited state		Ground state (Butadiene)

Conrotatory process (terminal carbon rotate in the same direction)

Working on similar lines, since the ground state orbitals of cyclobutene are correlated with the ground state orbitals of butadiene (with C_2–S), hence conrotatory ring opening (or ring closing) is thermally a symmetry allowed process (No HOMO/LUMO crossing).

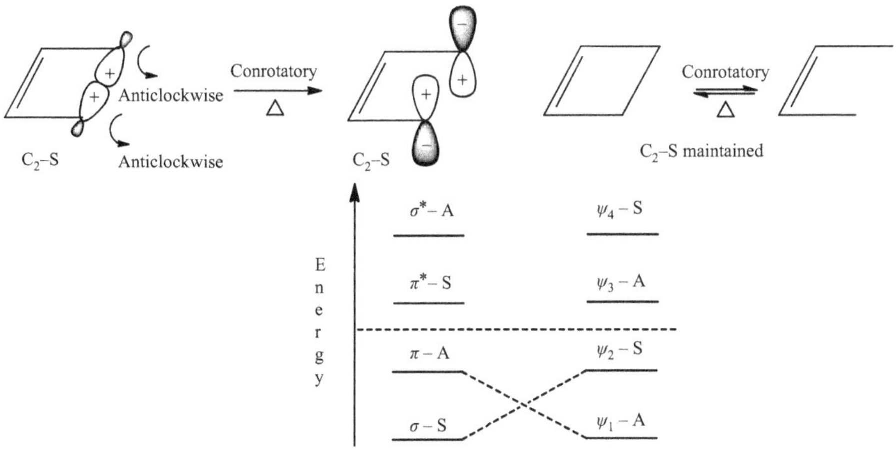

Correlation diagram showing conrotatory interconversion of cyclobutene–butadiene system

$$\sigma^2\,\pi^2 \underset{}{\overset{\Delta}{\rightleftharpoons}} \psi_1^{\,2}\,\psi_2^{\,2}$$

Ground state (Cyclobutene) — Ground state (Butadiene)

(ii) Cyclohexadiene ⇔ Hexatriene (4q + 2 electron system)

Cyclohexadiene — Hexatriene

By following the procedure for construction of correlation diagram, the correlation diagram of the present system is given in Fig. 4.21 along with mirror plane (σ-S/A) symmetry and twofold axis of symmetry (C_2-S/A) and also number of node (s) in each MO.

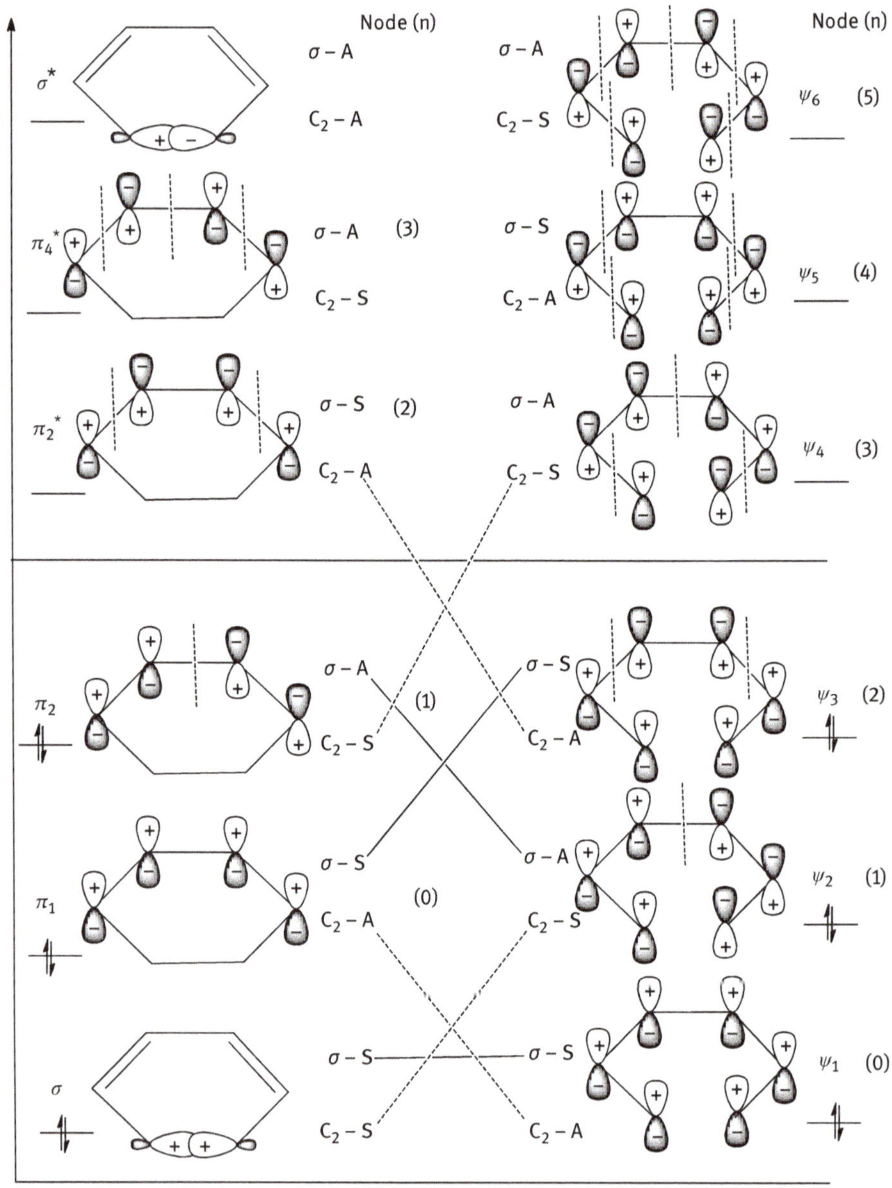

Fig. 4.21: Correlation diagram of **cyclohexadiene** ⇔ hexatriene. Solid lines indicate thermally allowed and dashed lines indicate photochemically allowed reaction process. Nodes are shown by vertical dotted lines.

Disrotatory process (terminal carbons rotate in the opposite direction)

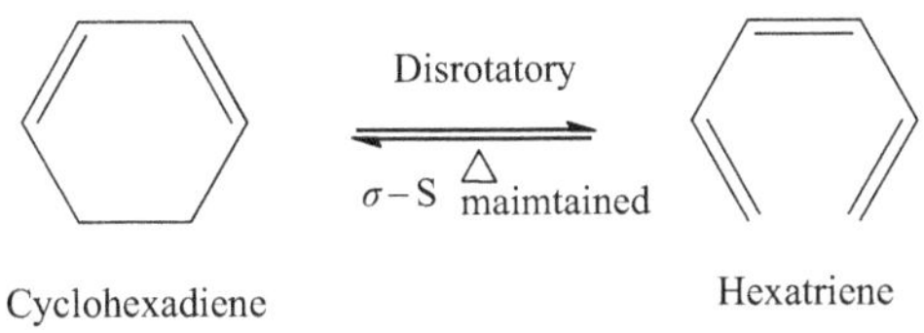

The summary of the correlation diagram showing disrotatory interconversion of cyclohexadiene–hexatriene system is given below:

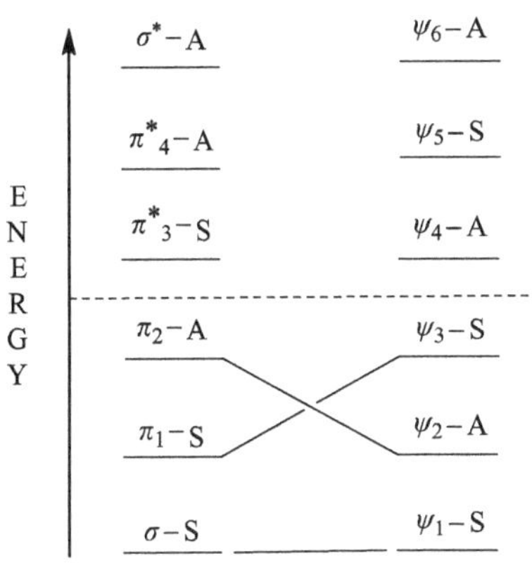

Correlation diagram showing disrotatory interconversion of cyclohexadiene–hexatriene system

Since, the ground state orbitals of cyclohexadiene are correlated with ground state orbitals of hexatriene, disrotatory movement is thermally allowed (no HOMO/LUMO crossing).

$$\underset{\substack{\text{Ground state}\\\text{(Cyclohexadiene)}}}{\sigma^2\pi_1{}^2\pi_2{}^2} \; \underset{\text{Disrotatory}}{\overset{\Delta}{\rightleftharpoons}} \; \underset{\substack{\text{Ground state}\\\text{(Hexatriene)}}}{\psi_1{}^2\psi_2{}^2\psi_2{}^2}$$

Conrotatory process (terminal carbons rotate in the same direction)

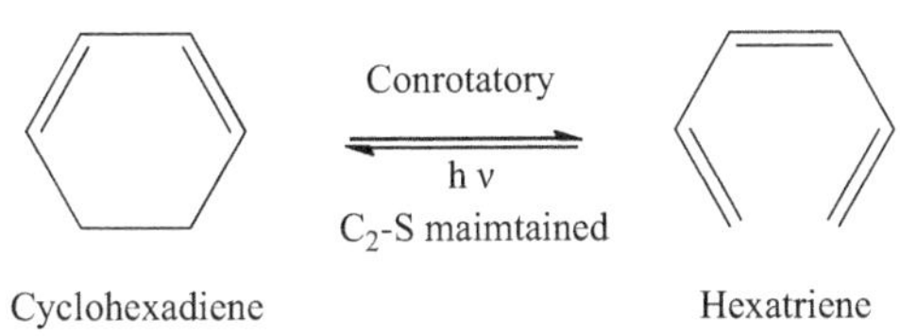

Again, the extract of the correlation diagram showing conrotatory interconversion of cyclohexadiene–hexatriene system is given below:

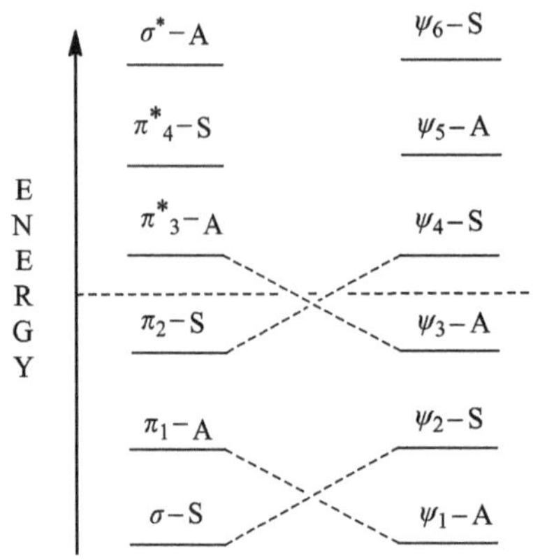

Since ground state (π_2-MO) of cyclohexadiene is correlated with the first excited state of hexatriene (with twofold axis symmetry, C_2–S), hence conrotatory ring opening (or ring closing) is photochemically a symmetry allowed process (HOMO/LUMO crossing).

$$\underset{\text{(cyclohexadiene)}}{\underset{\text{Groundstate}}{\sigma^2\pi_1^2\pi_2^2}} \xrightarrow{h\nu} \underset{\text{state}}{\underset{\text{Firstexcited}}{\sigma^2\pi_1^2\pi_2^1\pi_3^{*1}}} \xrightleftharpoons{h\nu} \underset{\text{state}}{\underset{\text{Firstexcited}}{\psi_1^2\psi_2^2\psi_3^1\psi_4^1}} \xleftarrow{h\nu} \underset{\text{(hexatriene)}}{\underset{\text{Groundstate}}{\psi_1^2\psi_2^2\psi_3^2}}$$

(b) Cycloaddition reactions

It is a special type of pericyclic reaction in which two unsaturated molecules react to form a cyclic compound involving participation of π-electrons in the formation of two new σ-bonds. The reverse reaction, the opening of the cyclic compound with the formation of two π-bonds, is known as reverse or retro cycloaddition. The well-known Diels–Alder reaction, in which a conjugated diene reacts with a substituted alkene, a dienophile to give a cyclohexene derivative is the classical example of cycloaddition reaction.

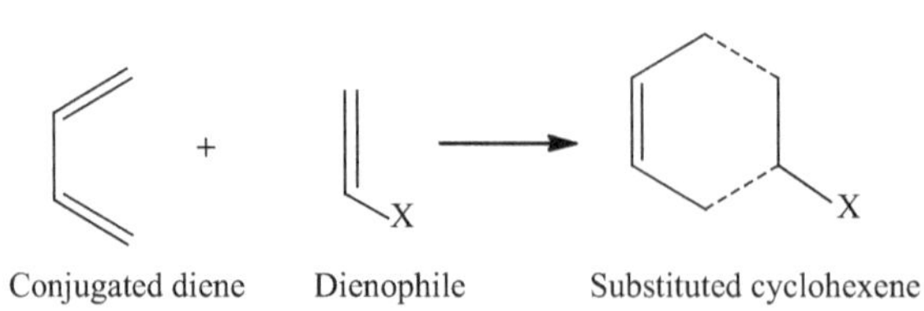

Dimerization of ethylene to cyclobutene is another reaction that has been extensively studied.

Ethylene Cyclobutane

(i) Classification of cycloaddition reactions

Cycloaddition reactions have been classified with respect to two aspects:

(a) Number of π-electrons involved in each of the participation molecule.

(b) Mode of approach of two π-bonded compounds during addition.

When "*m*" number of π-electrons of one molecule and "*n*" number of π-electrons of other molecule is involved, the cycloaddition is expressed as $\pi^m + \pi^n$ and the reaction is called $(m + n)$ addition reaction. The total number of π-electrons of both the molecules, $(m + \text{n})$ may be represented by the expression $4\text{q}\pi$ or by $(4\text{q} + 2)\pi$, where q may be 0, 1, 2,. . ..*n*. Consequently, cycloaddition reactions have been grouped into two categories: (i) cycloaddition of $4\text{q}\pi$ electron system and (ii) the second type is the cycloaddition of $(4\text{q} + 2)\pi$ electron system.

(ii) Mode of approach of two interacting molecules: suprafacial and antarafacial

In cycloaddition process, reaction takes place at the terminals of π-bonds of each molecule and two new σ-bonds are formed by use of π-electrons of the reactants. These two new bonds can either form on the same face of π-systems (suprafacial) or from opposite faces of π-systems (antarafacial).

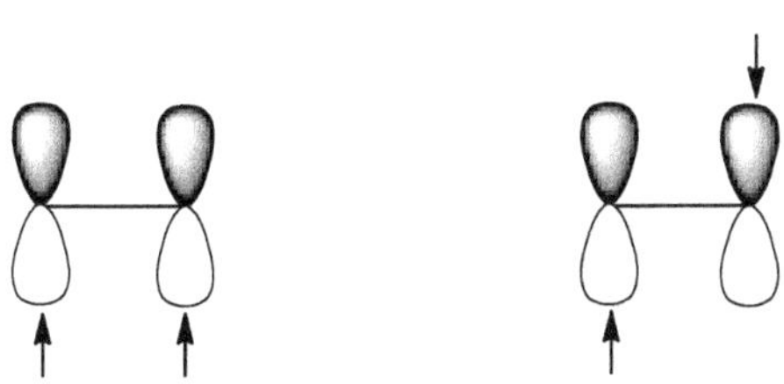

Suprafacial Antarafacial

Suprafacial and antarafacial additions are comparable to *syn* and *anti* addition but unlike *syn* and *anti* addition, suprafacial and antarafacial refer to the actual process and not the net result and will be evident from the following examples.

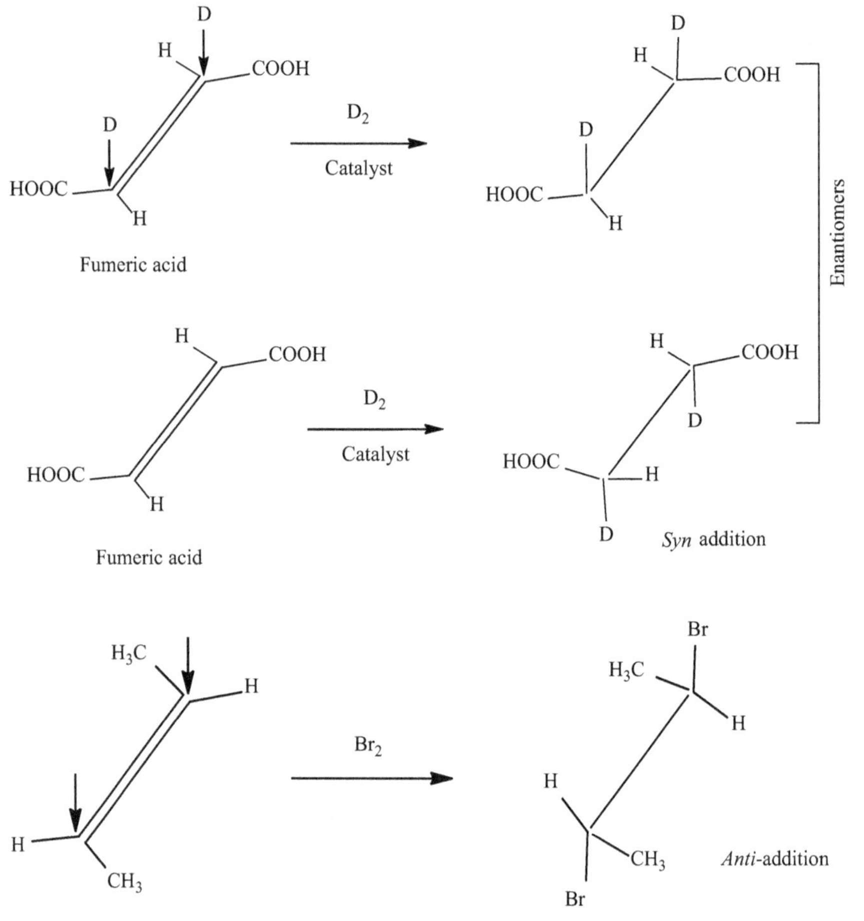

trans-2-Butene

meso-2,3-Dibromobutane

During addition, either both the centers of one component may combine at the two centers of the other unit from the same direction or the combination may take place from the opposite direction. When addition occurs at the same direction, it is known as suprafacial addition and is expressed as "S," but if addition takes place from opposite direction, it is termed as antarafacial usually denoted by "A." Thus, on the basis of mode of approach cycloaddition reactions may be of the following types:

(i) Suprafacial with respect to both the units expressed as (S–S).
(ii) Antarafacial with respect to both the units indicated by (A–A).
(iii) Suprafacial with respect to one and antarafacial with respect to another represented by (S–A).

The above three approaches may be pictorially represented as follows:

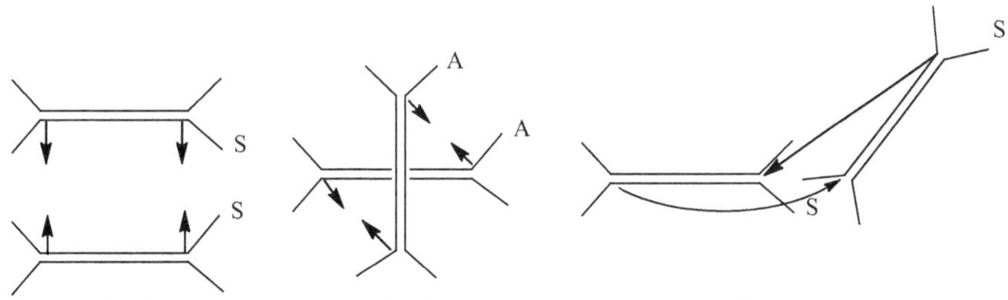

(S-S) Two ethylene molecules in parallel approach

(A-A) Two ethylene molecules in perpendicular approach

(S-A) Two ethylene molecules in inclined perpendicular approach

From the above pictorial representation, it is evident that in the case of (S–S) mode of combination parallel approach of the two interacting units occurs, but in the case of (A–A) and (S–A) mode of combination perpendicular approach of two units is observed.

Among these three modes of addition, (A–A) is very rare and (S–S) is extremely common. Taking into two aspects of classifications, cycloaddition reaction is generally expressed as follows:

$$(\mathrm{a})\pi_{\mathrm{S}}{}^{m}+\pi_{\mathrm{S}}{}^{n}(\mathrm{b})\,\pi_{\mathrm{S}}{}^{m}+\pi_{\mathrm{A}}{}^{n}\text{ and }(\mathrm{c})\,\pi_{\mathrm{A}}{}^{m}+\pi_{\mathrm{A}}{}^{n}$$

(iii) Correlation diagram for cycloaddition reactions

For the purpose of drawing correlation diagram of cycloaddition reactions, the bonding and antibonding MOs of the involved bonds of reacting molecules are identified. With the knowledge of exact mode of approach (parallel or perpendicular) of the participating molecules, the symmetry that is developed in between these interacting molecules is ascertained. Symmetry properties of the reacting systems are tabulated and a correlation diagram is constructed for necessary conclusion.

(iv) Analysis of cycloaddition reaction with symmetry property: Woodward and Hoffmann correlation diagram

(a) Correlation diagram for (2 + 2)π cycloaddition: dimerization of ethylene/ ethene to cyclobutane

In order to understand the control of orbital symmetry on cycloaddition reaction, let us consider the simplest example in which two ethylene molecule approach each other vertically to form a molecule cyclobutene, that is, suprafacial dimerization of two ethylene molecules. Such a system has vertical and horizontal planes of symmetry which are referred to as (σ_v or 1) and (σ_h or 2), respectively. Plane 1 bisects the C–C bond of each ethylene molecule, while plane 2 lies half way between the two molecules of ethylene. These two planes bisect the cyclobutene molecule vertically and horizontally as shown in Fig. 4.22 given below.

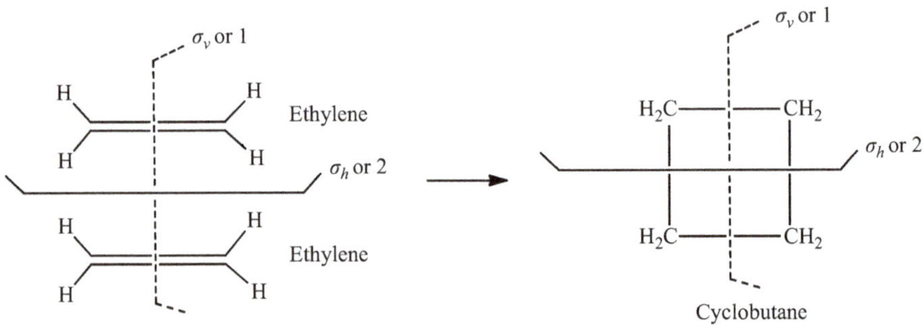

Fig. 4.22: Dimerization of ethylene to cyclobutene.

In this dimerization reaction, we are mainly concerned with four p-orbitals (π-orbitals) of two ethylene and four σ-orbitals of cyclobutene. The shape and symmetry properties of the π- and σ-orbitals with respect to the horizontal plane of symmetry (σ_h) are given first and with respect to the vertical plane of symmetry (σ_v) next. In this reaction, two bonding π-orbitals from each ethylene molecule and their corresponding antibonding orbitals are involved to form two σ-bonding and two antibonding (σ^*) orbitals of cyclobutane. Now, two bonding π-orbitals of two ethylene molecules may approach each other in *two ways* in *parallel condition* forming two σ-MOs, one bonding and one antibonding. Similarly, two antibonding π-orbitals form two σ-MOs, one bonding and another antibonding as shown in Fig. 4.23.

The diagram presented as Fig. 4.23 shows four π-orbitals (two bonding/two antibonding) and four σ-orbitals (two bonding/two antibonding) and correlation of their symmetry properties. It is notable here that more the symmetry in the system less will be energy of the system for π- as well as σ-orbitals. Energy of σ- and π-system follows the increasing order as follows:

$$\text{For } \pi - \text{system: SS} < \text{SA} < \text{AS} < \text{AA}$$

$$\text{For } \sigma - \text{system: SS} < \text{AS} < \text{SA} < \text{AA}$$

A close examination of the diagram leads us to the following two conclusions:

(i) The ground state orbitals of ethylene correlate with an excited state of cyclobutene. Consequently, the combination of two ground state ethylene molecules cannot result in the formation of ground state cyclobutane while conserving the orbital symmetry. Hence, the thermal process is symmetry forbidden.

$$\pi_1^2\pi_2^2 \not\rightleftharpoons \sigma_1^2\sigma_2^2$$

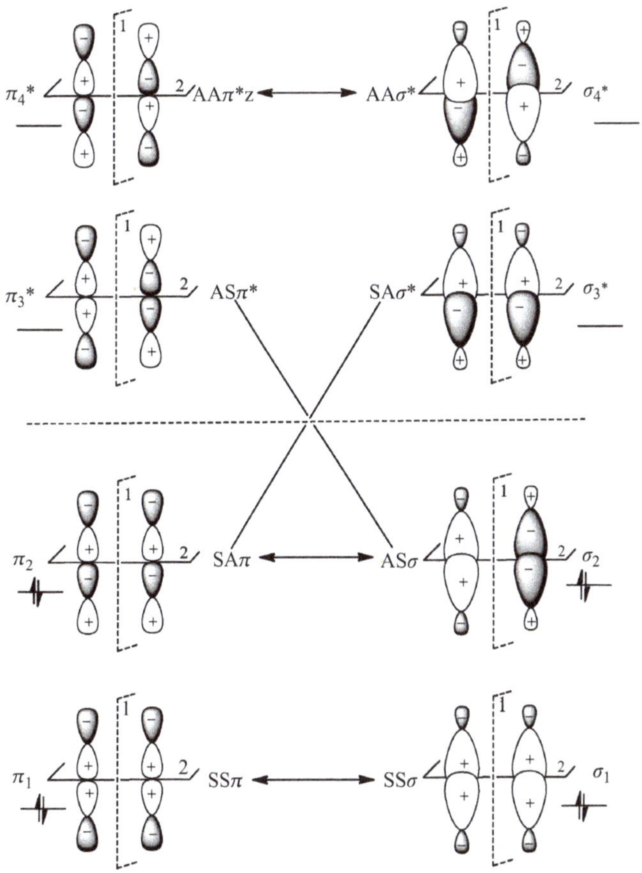

Fig. 4.23: Correlation diagram: Cycloaddition and cycloreversion of ethylene and cyclobutane system.

(ii) As there is correlation between the first excited state of ethylene and the ground state of cyclobutane, the photochemical dimerization of ethylene is symmetry allowed.

$$\pi_1{}^2\pi_2{}^1\pi_3{}^{*1} \rightleftharpoons \sigma_1{}^2\sigma_2{}^1\sigma_3{}^{*1}$$

(b) Correlation diagram for (4 + 2)π cycloaddition: Diels–Alder reactions

The most classical example of (4 + 2)π electron system of cycloaddition is the reversible reaction between butadiene and ethylene system. In this case the reaction is possible if the cycloaddition reaction is of the S–S type, that is, the mode of approach of the two reacting units is parallel to each other as shown below:

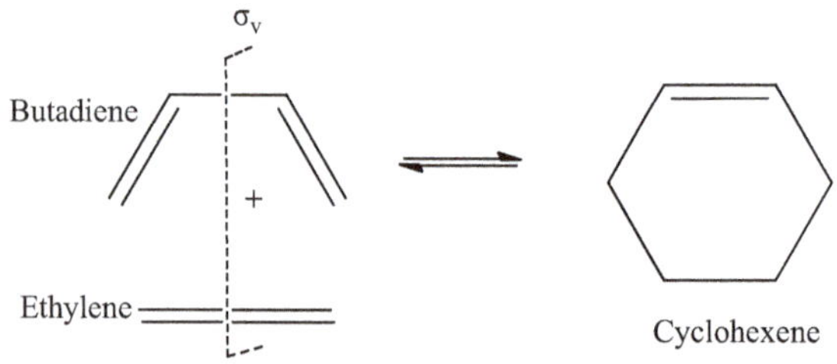

In this mode of approach of reacting molecules, there is only one symmetry element, that is, vertical plane of symmetry (σ_v). This is the developed symmetry which is to be maintained throughout the course of reaction.

In this transformation, we have to take into consideration six orbitals, each of the reactants (butadiene and ethylene) and the product cyclohexene. The ground state MOs of the reactants have been ψ_1 and ψ_2 (of butadiene) and ψ (of ethylene), while ψ_3, $\psi^\star$ and ψ_4 have been the corresponding antibonding orbitals. Similarly, the ground state MOs of cyclohexene may be represented by σ_1, σ_2 and ψ. The remaining three orbitals ($\psi^\star$, $\sigma_3{}^\star$ and $\sigma_4{}^\star$) would be the antibonding.

All these MOs along with their symmetry characteristics with respect to plane of symmetry (σ_v) are listed in order of increasing energy and the orbitals that correlate are connected by line, thereby a correlation diagram (Fig. 4.24) is obtained.

Inspection of correlation diagram showed the following:

(i) The ground state orbitals of butadiene (ψ_1, ψ_2) and ethylene (ψ) correlate with the ground state orbitals of cyclohexene (σ_1, σ_2, ψ) and, therefore, smooth transformation of reactants to the product is a thermally allowed process.

$$\psi_1{}^2\pi^2\psi_2{}^2 \rightleftharpoons \sigma_1{}^2\sigma_1{}^2\pi^2$$

(ii) The photochemical transformation will not be possible because the first excited state of the reactant does not get correlated with the first excited state of the product; rather it gets correlated with the upper excited state of the product. Hence, there exists a symmetry imposed barrier to photochemical reaction of (4 + 2) cycloaddition type.

4.13 Frontier molecular orbital approach: interpretation of pericyclic reactions

(a) Electrocyclic reactions

The important role in deciding the valency on an element or its combining capacity with other elements is played by the electrons in the outermost electron shell.

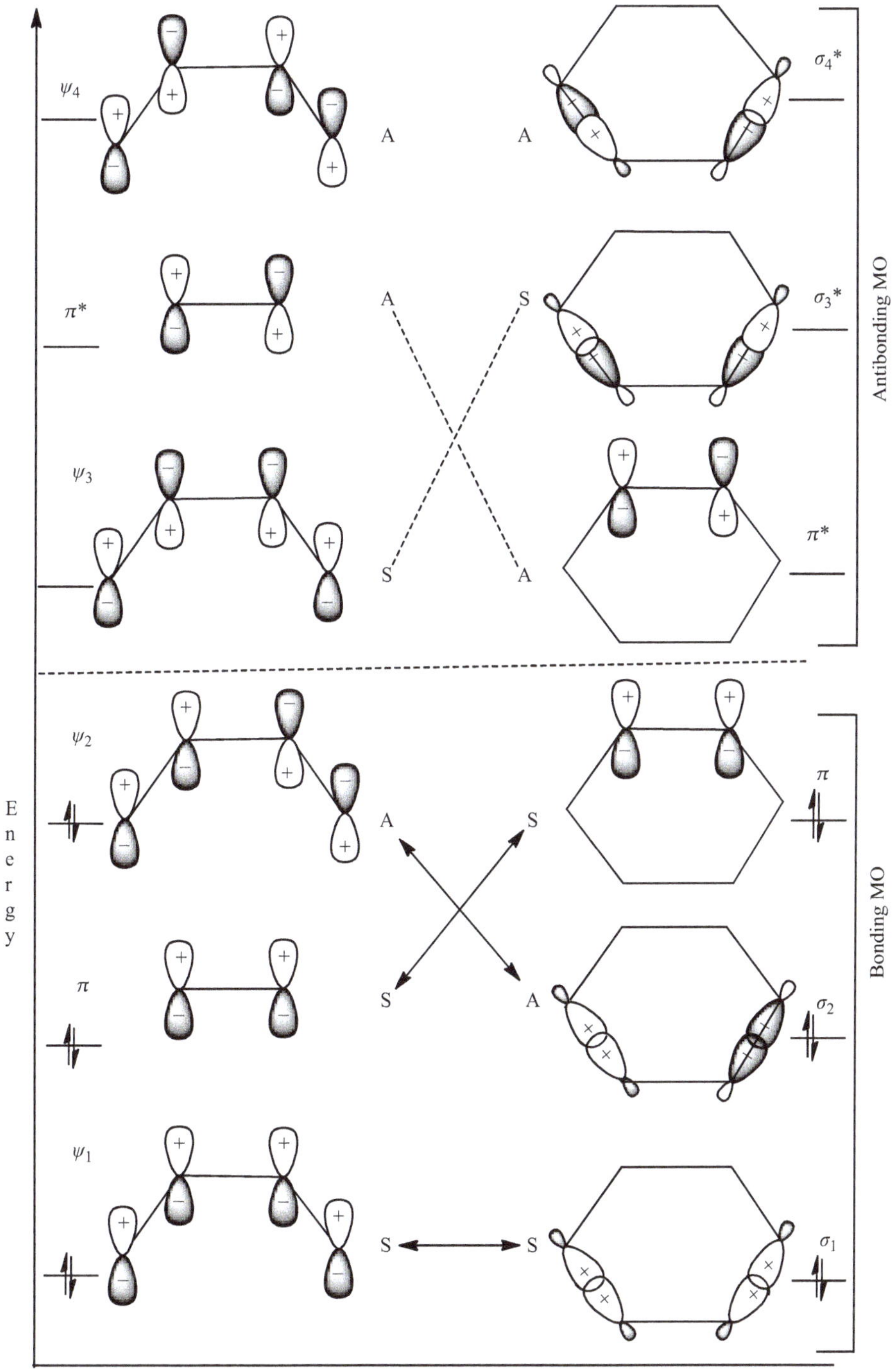

Fig. 4.24: Correlation diagram of Diels–Alder reaction: (4 + 2) cycloaddition.

Similarly, the electrons in the FMOs, HOMO, LUMO and the NBMO that lie in the borderline of the bonding and antibonding should play a very significant role in the reactivities of the molecules. This idea was proposed by Fukui. He introduced the FMO density as an index in deciding the reactivities of different centers within a molecule. In other words, electron densities of HOMO and LUMO in a molecule decide the nucleophilic and electrophilic reactivity, respectively.

The basic theme of FMO approach is that HOMO of one reactant is taken as being similar to the outer valence electrons of an atom. The reaction then involves the overlap between the HOMO, a potential electron donor, with LUMO, a potential electron acceptor, of the other reactant. However, in case of electrolytic reaction, as only one molecule reversibly undergoes ring opening and ring closing, only HOMO needs be considered.

The only guiding principle in this approach has been the symmetry of the HOMO of the open chain partner in an electrocyclic reaction. If this orbital possesses a C_2-symmetry, then the reaction will proceed in conrotatory manner. On the other hand, if this HOMO has σ_v (vertical plane/mirror plane) symmetry, then reaction will occur in disrotatory manner.

Let us consider the case of a $4n\pi$ electron system, such as butadiene ⇔ cyclobutene. Under thermal conditions, the HOMO of the open chain partner in ψ_2 (bonding MO with highest energy). This ψ_2 has C_2 symmetry (see correlation diagram, Fig. 4.20, p.417). Therefore, under thermal conditions, butadiene ⇔ cyclobutene interconversion will occur in conrotatory way. The same conclusion has been drawn from correlation diagram (Fig. 4.20). In photochemical condition the HOMO will be the first excited state. Thus, such HOMO of open chain partner is ψ_3. Here, $(n-1)$ is 2 (even), so its symmetry σ–S (plane of symmetry). So, under this condition the path of the reaction is disrotatory. Similar conclusion has been reached from correlation diagram (Fig. 4.20). The above statement may be explained on the basis of orbital picture of HOMO.

(i) The explanation for this alternative approach has been carried out on the basis of the fact that the overlapping of wave functions of the same sign becomes essential for bond formation. In *thermal condition*, HOMO of ground state of butadiene is ψ_2, and obviously cyclization will become possible only through conrotation.

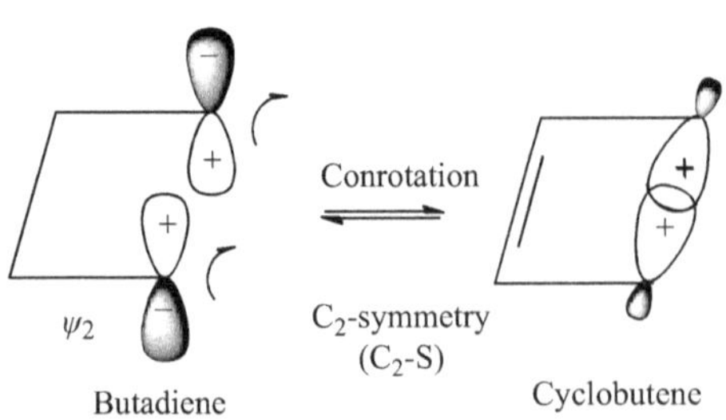

(ii) Irradiation of butadiene is able to promote an electron from ψ_2 to ψ_3, which then become HOMO, and hence bond formation becomes possible only through disrotation.

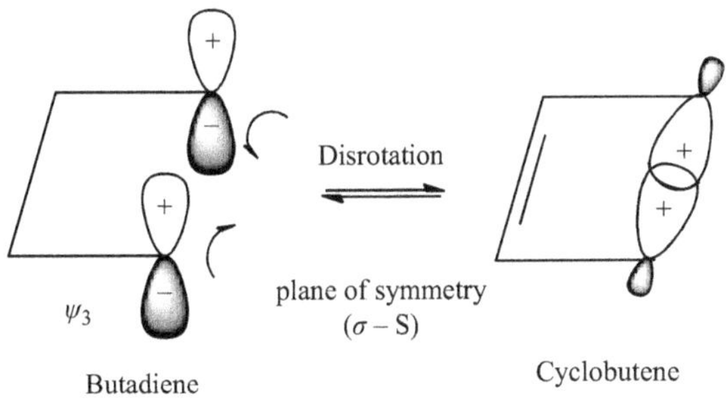

In the case of $(4n + 2)\pi$ electron system, similar explanation may be given. The most classical example is the interconversion of cyclohexadiene into hexatriene.

(i) Under thermal conditions, HOMO of the open chain partner (hexatriene) is ψ_3 (bonding MO of highest energy). Here $(n–1)$ is 2 (even), so it will have plane of symmetry. Therefore, the path of the reaction is disrotation.

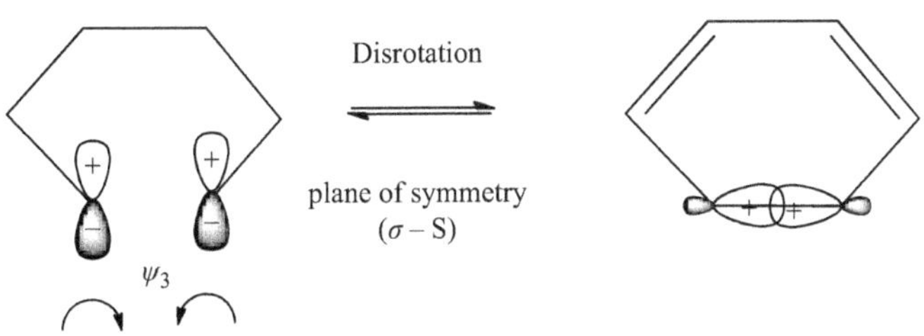

5. In photochemical condition, HOMO of hexatriene is ψ_4 (the first excited state). As $(n–1)$ is 3(odd), it will have C_2-symmetry, so reaction will occur in conrotatory manner.

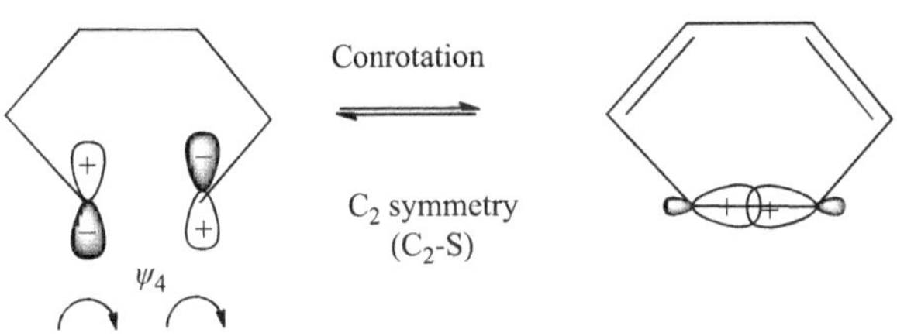

Same conclusions have been drawn from correlation diagram (Fig. 4.21, p.420).

(b) Cycloaddition reactions

Interpretation of $\pi^4{}_s + \pi^2{}_s$ cycloaddition

Cycloaddition reaction may precisely be interpreted in the light of FMO approach and the same conclusion as drawn from Woodward and Hoffmann's correlation diagram can be reached. According to this concept, HOMO of one component and LUMO of another are involved in chemical reaction.

The best known example of $\pi^4{}_s + \pi^2{}_s$ cycloaddition is Diels–Alder reaction, in which a diene reacts with an alkene, generally termed as dienophile, to give a cyclohexene derivative. Let us take butadiene as diene and ethylene as an alkene. In this particular reaction, either HOMO of butadiene and LUMO of ethylene component interact each other or LUMO of butadiene or HOMO of ethylene will participate. But as butadiene component is relatively more electron rich, it will act as an electron donor, that is, HOMO of butadiene system will preferably interact with the LUMO of ethylene.

Based on the above principle, π_2, the HOMO of butadiene component under thermal condition will interact with π^*, the LUMO of ethylene under the same condition (see correlation diagram, Fig. 4.24, p.429). As the signs of lobes at 1, 4 positions of HOMO of butadiene match with the signs of lobes at 1, 2 positions of ethylene, the addition under thermal condition is suprafacially symmetry allowed process.

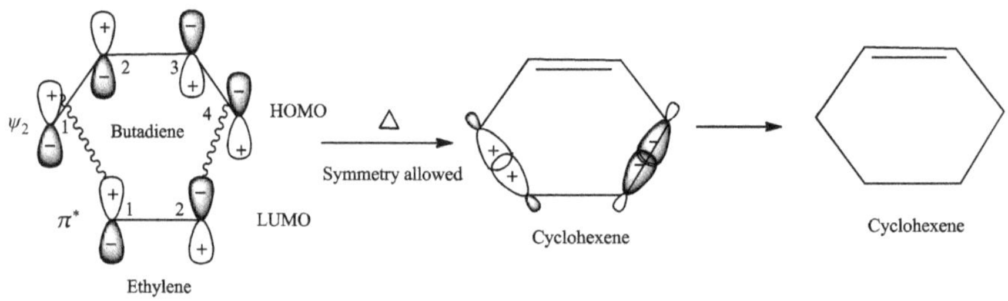

In photochemical conditions, HOMO of butadiene is π_3 and LUMO of ethylene is π^*. These two orbitals will be interacting entities. Inspection of the orbital pictures clearly shows that there is no matching of lobes of same sign in both the directions, and thus the formation of cyclohexene is not possible. Therefore, $\pi^4{}_s + \pi^2{}_s$ cycloaddition is photochemically a symmetry forbidden process.

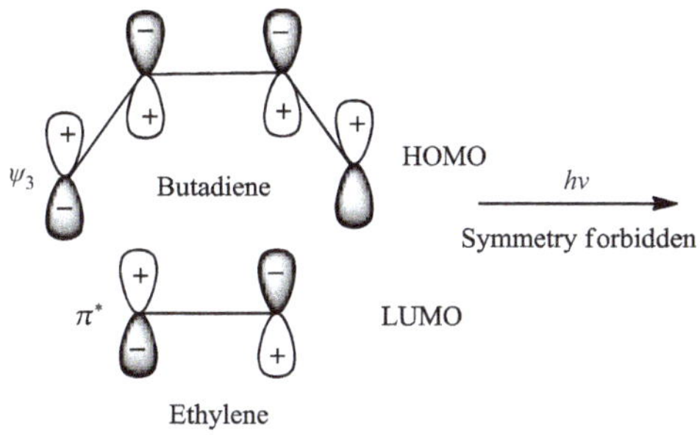

4.14 Woodward and Hoffmann's rules and FMO approach: electrocyclic reactions on the basis of components

(a) Ring closing reactions:

Woodward and Hoffmann proposed two rules for electrocyclic ring closing reactions and these are as follows:

1. A thermal (ground state) electrocyclic reaction is symmetry allowed when total number of $(4q + 2)_s$ and $(4q)_a/(4r)_a$ components is odd.
2. A photochemical (first excited state) electrocyclic reaction is symmetry allowed when total number of $(4q + 2)_a$ and $(4q)_s/(4r)_s$ components is odd.

The applications of these two rules are exemplified here with certain electrocyclic reactions:

(i) 2,4-Hexadiene (4q electron system)
Thermal ring closing

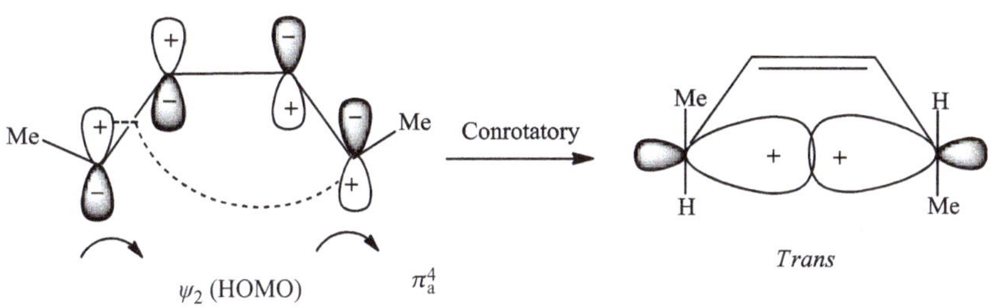

Number of $(4q + 2)_s$ component = 0
Number of $(4q)_a$ component = 1
....
Total components = 1 (odd)

Therefore, ring closing reaction is *thermally symmetry allowed.*

Photochemical ring closing

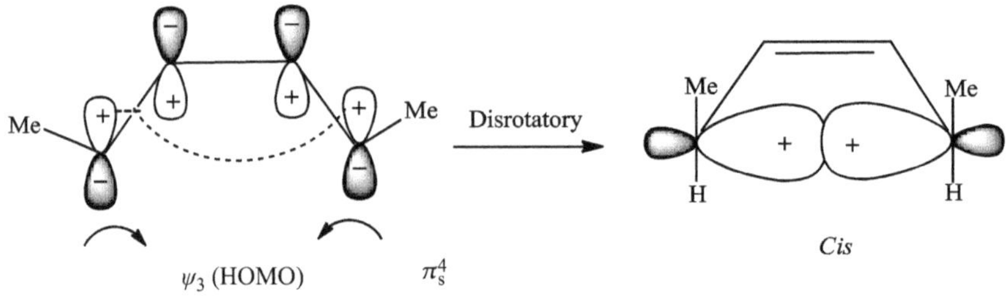

Number of $(4q+2)_a$ component = 0
Number of $(4q)_s$ component = 1
. . . .
Total components = 1 (odd)

So, ring closing reaction is *photochemically symmetry allowed.*

(ii) **2,4, 6-Octatriene [(4q + 2) electron system]**

Thermal ring closing

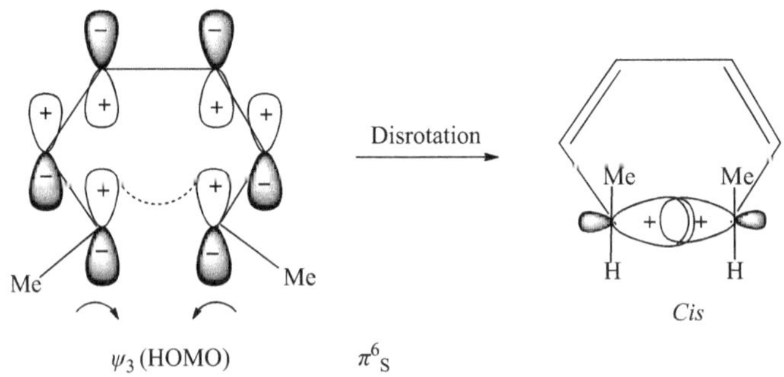

Number of $(4q+2)_S$ component = 1
Number of $(4q)_S$ component = 0
. . . .
Total components = 1 (odd)

So, ring closing reaction is *thermally symmetry allowed.*

Photochemical ring closing

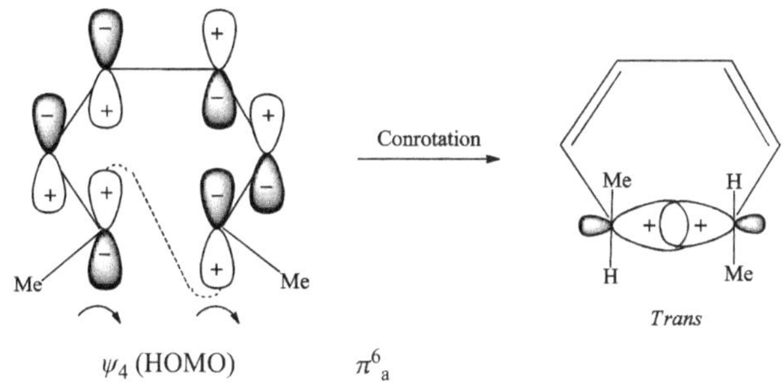

Number of $(4q+2)_a$ component = 1
Number of $(4q)_s$ component = 0
....
Total components = 1 (odd)

So, ring closing reaction is *photochemically symmetry allowed.*

(b) Ring opening reactions

Electrocyclic ring opening reactions by FMO approach may be thought as a two components cycloaddition reaction of σ, π or ω orbitals.

Woodward and Hoffmann proposed two rules for electrocyclic ring opening reactions and these are as follows:

1. A thermal (ground state) electrocyclic reaction is symmetry allowed when total number of $(4q+2)_s$ and $(4q)_a/(4r)_a$ components is odd.
2. A photochemical electrocyclic reaction is symmetry allowed when total number of $(4q+2)_s$ and $(4q)_s/(4r)_s$ components is even.

Examples:

(i) Trans-3,4-dimethyl cyclobutene:
Thermal ring opening

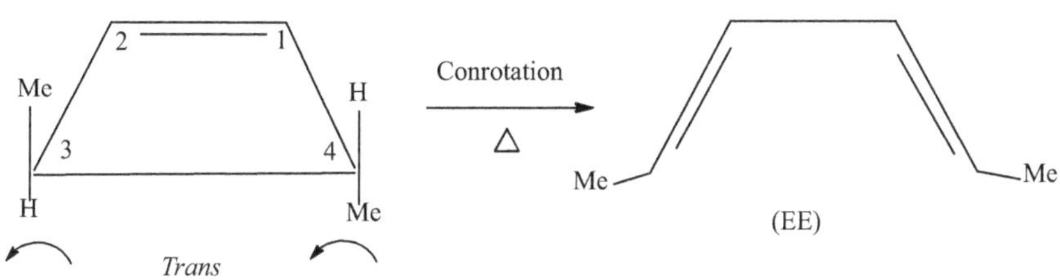

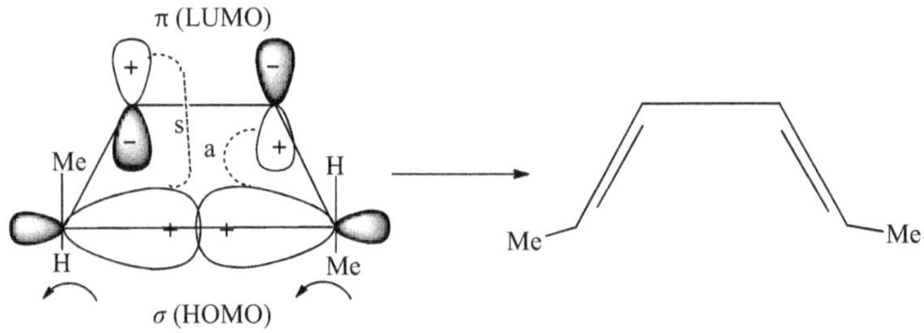

$$(4q+2)_s + (4q)_a$$
$$= \sigma^2 s + \pi^2 a + 0$$
$$= 1 + 0 + 0 = 1 \text{ (odd number)}$$

Therefore *thermal ring* opening process is *symmetry allowed.*

Photochemical ring opening

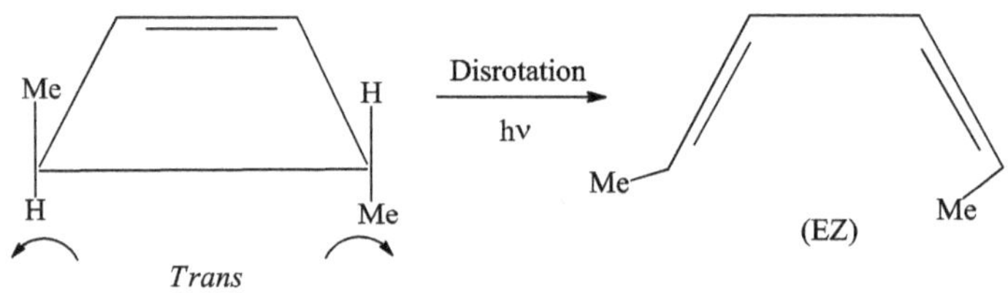

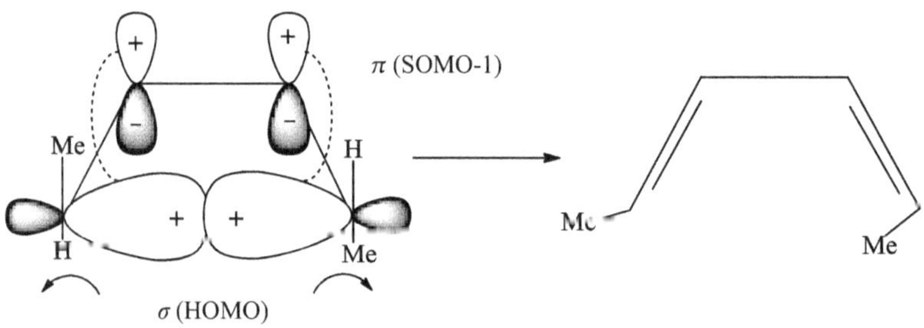

$$(4q+2)_s + (4q)_a$$
$$= \sigma^2 s + \pi^2 a + 0$$
$$= 1 + 1 + 0 = 2 \text{ (even number)}$$

Therefore, *photochemical ring* opening process is *symmetry allowed.*

(ii) Trans-5,6-dimethyl dimethylcyclohexa-1,3-diene thermal ring opening

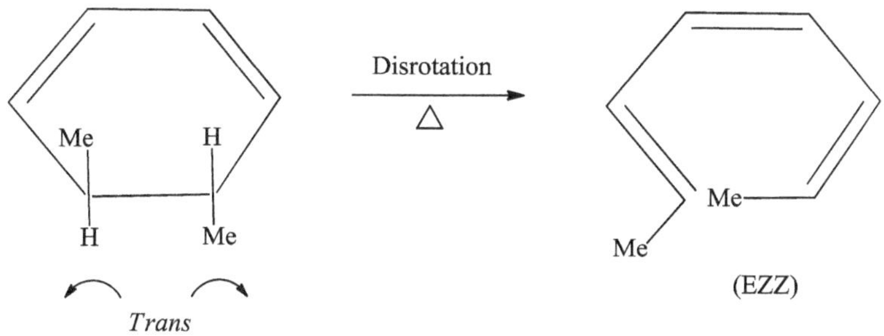

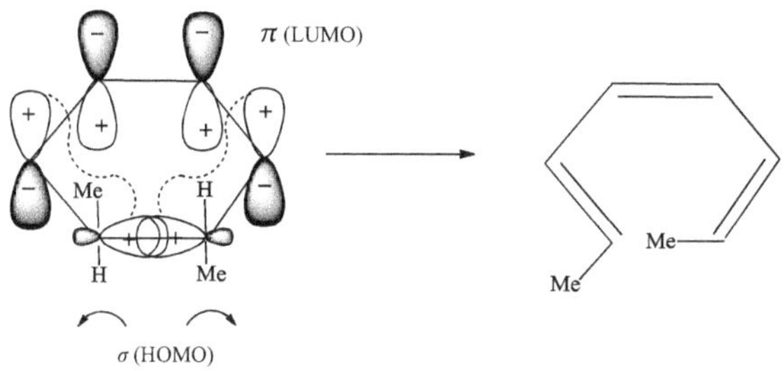

$$(4q+2)_s + (4q)_a$$
$$= \sigma^2 s + \pi^4 s$$
$$= 1 + 0 = 1 \text{(odd number)}$$

Hence, *thermal ring* opening process is *symmetry allowed.*

Photochemical ring opening

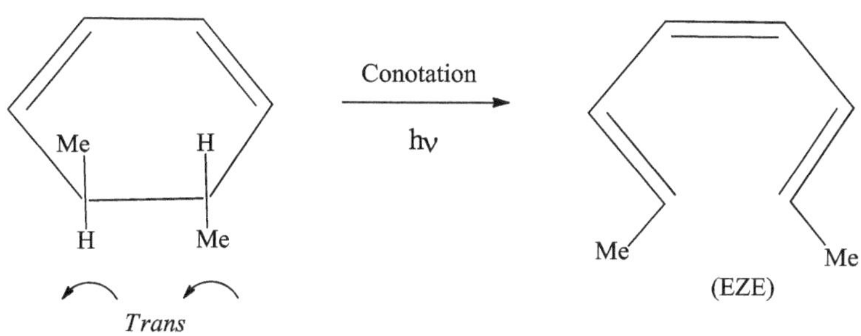

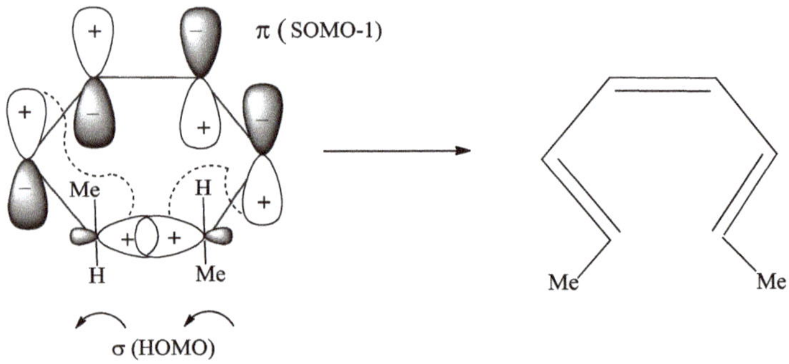

$$(4q+2)s+(4q)a$$
$$=\sigma^2 s+\pi^4 a$$
$$=1+1=2 \text{ (even number)}$$

Hence, *photochemical ring* opening process is *symmetry allowed.*

Exercises

Multiple choice questions/fill in the blank

1. If filled MOs of product are correlated with filled MOs of the reactant, then the reaction is said to be allowed by the following:
   (a) Orbital symmetry (b) Spin symmetry
   (c) Orbital and spin symmetry both (d) None of these

2. The orbitals HOMO and LUMO are coined as “Frontier Orbitals” by:
   (a) Woodward–Hoffmann (b) Pearson
   (c) Fukai (d) Huckel–Mobius

3. The concerted *cis*-addition of H_2 to ethylene is as follows:
   (a) Symmetry allowed (b) Symmetry forbidden
   (c) Cannot be predicted (d) None of these

4. In the nucleophilic displacement reaction, $[Ir^{I}(Cl)(CO)(PPh_3)_2] + CH_3I \rightarrow [[Ir^{III}(Cl)(CO)(PPh_3)_2(CH_3)(I)]$, the nucleophile is as follows:
   (a) Ir(I) HOMO (b) Ir(III) HOMO
   (c) CH_3I LUMO (d) None of these

5. In which one of the following d^n configuration in trigonal bipyramidal symmetry, Berry’s pseudorotation is orbital symmetry allowed?
   (a) d^2 (b) d^3 (c) d^4 (d) d^9

6. Colored compounds are generally more reactive because HOMO–LUMO gas is ....and transition occurs in visible region.
(a) Very large gap (b) Large (c) Small gap (d) All of these

7. For a linear conjugated π-system, the MO (ψ_n) will be symmetric with respect to mirror plane and antisymmetric with respect to C_2 when the number of node is as follows:
(a) 0 (b) 1 (c) 3 (d) 5

8. In cycloaddition reaction when reaction takes place at the terminals of π-bonds of eachmolecule and two new σ-bonds are formed on the same face of the π-systems, theapproach of the two interacting molecules is called.................

Short answer type questions

1. Highlight the various symmetry rules derived from perturbation theory?
2. Explain why the addition to Cl_2 to ethylene is symmetry forbidden.
3. Briefly explain nucleophilic displacement reaction taking a suitable example.
4. What is Berry's pseudorotation? Explain with a suitable example.
5. Just draw the correlation diagram for the conversion of cyclohexadiene to hexatriene.
6. Describe cycloaddition reaction in brief with some suitable examples.
7. What are conrotatory and disrotatory processes in electrocyclic reactions?
8. What do you mean by components in pericyclic reactions? Explain.

Long answer type questions

1. What are symmetry allowed and symmetry forbidden reactions? Explain with reference to some inorganic reactions.
2. Though the direct addition of hydrogen to an olefin is symmetry forbidden, still it is possible to add simultaneously two atoms of hydrogen to unsaturated linkage. Justify this statement with two suitable examples.
3. Based on the concept of HOMO and LUMO overlap, explain nucleophilic displacement reactions with suitable examples.
4. Explain Berry's pseudorotation using the concept of orbital symmetry control taking suitable examples.
5. Based on symmetry rules, predict the stable structure of certain molecules of your choice.
6. What are pericyclic reactions? Present an overview of their classifications.
7. Present the mechanistic interpretation of electrocyclic reactions with symmetry property.

8. Highlight the procedure for construction of correlation diagram. Based on this procedure develop the correlation diagram for cyclobutene ⇔ butadiene and discuss its significance.
9. Present the interpretation of electrocyclic and cycloaddition reactions in the light of FMO approach.

Appendix I

Character tables for chemically important Symmetry groups

1. The nonaxial groups

C_1	E
A	1

C_s	E	σ_h		
A'	1	1	x, y, R_z	x^2, y^2, z^2, xy
A''	1	−1	z, R_x, R_y	yz, xz

C_i	E	i		
A_g	1	1	R_x, R_y, R_z	$x^2, y^2, z^2, xy, xz, yz$
A_u	1	−1	z, y, z	

2. The C_n groups

C_2	E	C_2		
A	1	1	z, R_z	x^2, y^2, z^2, xy
B	1	−1	x, y, R_x, R_y	yz, xz

C_3	E	C_3	C_3^2		
A	1	1	1	z, R_z	x^2+y^2, z^2
E	$\left\{\begin{matrix} 1 \\ 1 \end{matrix}\right.$	$\begin{matrix} \varepsilon \\ \varepsilon^* \end{matrix}$	$\left.\begin{matrix} \varepsilon^* \\ \varepsilon \end{matrix}\right\}$	(x, y) (R_x, R_y)	(x^2+y^2, xy) (yz, xz)

https://doi.org/10.1515/9783110635034-005

C_4	E	C_4	C_2	C_4^3		
A	1	1	1	1	z, R_z	x^2+y^2, z^2
B	1	−1	1	−1		x^2+y^2, xy
E	1	i	−1	−1	(x, y) (R_x, R_y)	(yz, xz)
	1	−i	−1	i		

C_5	E	C_5	C_5^2	C_5^3	C_5^4		$\varepsilon=\exp(2\pi i/5)$
A	1	1	1	1	1	z, R_z	x^2+y^2, z^2
E_1	1	ε	ε^2	ε^{2*}	ε^*	(x, y) (R_x, R_y)	(yz, xz)
	1	ε^*	ε^{2*}	ε^2	ε		
E_2	1	ε^2	ε^*	ε	ε^{2*}		(x^2-y^2, xy)
	1	ε^{2*}	ε	ε^*	ε^2		

C_6	E	C_6	C_3	C_2	C_3^2	C_6^5		$\varepsilon=\exp(2\pi i/6)$
A	1	1	1	1	1	1	z, R_x	x^2+y^2, z^2
B	1	−1	1	−1	1	−1		
E_1	1	ε	$-\varepsilon^*$	−1	$-\varepsilon$	ε^*	(x, y)	(xz, yz)
	1	ε^*	$-\varepsilon$	−1	$-\varepsilon^*$	ε	(R_x, R_y)	
E_2	1	$-\varepsilon^*$	$-\varepsilon$	1	$-\varepsilon^*$	$-\varepsilon$		(x^2-y^2, xy)
	1	$-\varepsilon$	$-\varepsilon^*$	1	$-\varepsilon$	$-\varepsilon^*$		

C_7	E	C_7	C_7^2	C_7^3	C_7^4	C_7^5	C_7^6		$\varepsilon=\exp(2\pi i/7)$
A	1	1	1	1	1	1	1	z, R_x	x^2+y^2, z^2
E_1	1	ε	ε^2	ε^3	ε^{3*}	ε^{2*}	ε^*	(x, y)	(xz, yz)
	1	ε^*	ε^{2*}	ε^{3*}	ε^3	ε^2	ε	(R_x, R_y)	
E_2	1	ε^2	ε^{3*}	ε^*	ε	ε^3	ε^{2*}		(x^2-y^2, xy)
	1	ε^{2*}	ε^3	ε	ε^*	ε^{3*}	ε^2		
E_3	1	ε^3	ε^*	ε^2	ε^{2*}	ε	ε^{3*}		
	1	ε^{3*}	ε	ε^{2*}	ε^2	ε^*	ε^3		

C_8	E	C_8	C_4	C_8^3	C_2	C_8^5	C_4^3	C_8^7		$\varepsilon = \exp(2\pi i/8)$
A	1	1	1	1	1	1	1	1	z, R_x	x^2+y^2, z^2
B	1	−1	1	−1	1	−1	1	−1		
E_1	1	ε	i	$-\varepsilon^*$	−1	$-\varepsilon$	−i	ε^*	(x, y)	(xz, yz)
	1	ε^*	−i	$-\varepsilon$	−1	$-\varepsilon^*$	i	ε	(R_x, R_y)	
E_2	1	i	−1	−i	1	i	−1	−i		(x^2-y^2, xy)
	1	−i	−1	i	1	−i	−1	i		
E_3	1	$-\varepsilon$	i	ε^*	−1	ε	−i	$-\varepsilon^*$		
	1	$-\varepsilon^*$	−i	ε	−1	ε^*	i	$-\varepsilon$		

3. The D_n groups

D_2	E	$C_2(z)$	$C_2(y)$	$C_2(x)$		
A	1	1	1	1		x^2, y^2, z^2
B_1	1	1	−1	−1	z, R_z	xy
B_2	1	−1	1	−1	y, R_y	xz
B_3	1	−1	−1	1	x, R_x	yz

D_3	E	$2C_3$	$3C_2$		
A_1	1	1	1		x^2+y^2, z^2
A_2	1	1	−1	z, R_z	
E	2	−1	0	(x, y) (R_x, R_y)	(x^2-y^2, xy) (xz, yz)

D_4	E	$2C_4$	$C_2(=C_4^2)$	$2C_2'$	$2C_2''$		
A_1	1	1	1	1	1		x^2+y^2, z^2
A_2	1	1	1	−1	−1	z, R_z	
B_1	1	−1	1	1	−1		x^2-y^2
B_2	1	−1	1	−1	1		xy
E	2	0	−2	0	0	(x, y) (R_x, R_y)	(xz, yz)

D_5	E	$2C_5$	$2C_5^2$	$5C_2$		
A_1	1	1	1	1		x^2+y^2, z^2
A_2	1	1	1	−1	z, R_z	
E_1	2	2 cos 72°	2 cos 144°	0	(x, y) (R_x, R_y)	(xz, yz)
E_2	2	2 cos 144°	2 cos 72°	0		(x^2-y^2, xy)

D_6	E	$2C_6$	$2C_3$	C_2	$3C_2'$	$3C_2''$		
A_1	1	1	1	1	1	1		x^2+y^2, z^2
A_2	1	1	1	1	−1	−1	z, R_z	
B_1	1	−1	1	−1	1	−1		
B_2	1	−1	1	−1	−1	1		
E_1	2	1	−1	−2	0	0	(x, y) (R_x, R_y)	(xz, yz)
E_2	2	−1	−1	2	0	0		(x^2-y^2, xy)

4. The C_{nv} groups

C_{2v}	E	C_2	$\sigma_{v(xz)}$	$\sigma_{v(yz)}$		
A_1	1	1	1	1	z	x^2, y^2, z^2
A_2	1	1	−1	−1	R_z	xy
B_1	1	−1	1	−1	x, R_y	xz
B_2	1	−1	−1	1	y, R_x	yz

C_{3v}	E	$2C_3$	$3\sigma_v$		
A_1	1	1	1	z	x^2+y^2, z^2
A_2	1	1	−1	R_z	
E	2	−1	0	(x, y) (R_x, R_y)	(x^2-y^2, xy) (xz, yz)

C_{4v}	E	$2C_4$	C_2	$2\sigma_v$	$2\sigma_d$		
A_1	1	1	1	1	1	z	x^2+y^2, z^2
A_2	1	1	1	−1	−1	R_z	
B_1	1	−1	1	1	−1		x^2-y^2
B_2	1	−1	1	−1	1		xy
E	2	0	−2	0	0	(x, y) (R_x, R_y)	(xz, yz)

C_{5v}	E	$2C_5$	$2C_5^2$	$5\sigma_v$		
A_1	1	1	1	1	z	x^2+y^2, z^2
A_2	1	1	1	−1	R_z	
E_1	2	2 cos 72°	2 cos 144°	0	(x, y) (R_x, R_y)	(xz, yz)
E_2	2	2 cos 144°	2 cos 72°	0		(x^2-y^2, xy)

C_{6v}	E	$2C_6$	$2C_3$	C_2	$3\sigma_v$	$3\sigma_d$		
A_1	1	1	1	1	1	1	z	x^2+y^2, z^2
A_2	1	1	1	1	−1	−1	R_z	
B_1	1	−1	1	−1	1	−1		
B_2	1	−1	1	−1	−1	1		
E_1	2	1	−1	−2	0	0	(x, y) (R_x, R_y)	(xz, yz)
E_2	2	−1	−1	2	0	0		(x^2-y^2, xy)

5. The C_{nh} groups

C_{2h}	E	C_2	i	σ_h		
A_g	1	1	1	1	R_z	x^2, y^2, z^2, xy
B_g	1	−1	1	−1	R_x, R_y	xz, yz
A_u	1	1	−1	−1	z	
B_u	1	−1	−1	1	x, y	

C_{3h}	E	C_3	C_3^2	σ_h	S_3	S_6^5		$\varepsilon = \exp(2\pi i/3)$
A'	1	1	1	1	1	1	R_z	x^2+y^2, z^2
E'	1	ε	ε^*	1	ε	ε^*	(x, y)	(x^2-y^2, xy)
	1	ε^*	ε	−1	ε^*	ε		
A''	1	1	1	−1	−1	−1	z	
E''	1	ε	ε^*	−1	$-\varepsilon$	$-\varepsilon^*$	(R_x, R_y)	(xz, yz)
	1	ε^*	ε	−1	$-\varepsilon^*$	$-\varepsilon$		

C_{4h}	E	C_4	C_2	C_4^3	i	S_4^3	σ_h	S_4		
A_g	1	1	1	1	1	1	1	1	R_z	x^2+y^2, z^2
B_g	1	−1	1	−1	1	−1	1	−1		(x^2-y^2, xy)
E_u	1	i	−1	−i	1	i	−1	−i	R_x, R_y	(xz, yz)
	1	−i	−1	i	1	−i	−1	i		
A_u	1	1	1	1	−1	−1	−1	−1	z	
B_u	1	−1	1	−1	−1	1	−1	1		
E_u	1	i	−1	−i	−1	−i	1	i	(x, y)	
	1	−i	−1	i	−1	i	1	−i		

C_{5h}	E	C_5	C_5^2	C_5^3	C_5^4	σ_h	S_5	S_5^7	S_5^3	S_5^9		$\varepsilon = \exp(2\pi i/5)$
A'	1	1	1	1	1	1	1	1	1	1	R_z	x^2+y^2, z^2
E_1'	1	ε	ε^2	ε^{2*}	ε^*	1	ε	ε^2	ε^{2*}	ε^{9*}	(x, y)	
	1	ε^*	ε^{2*}	ε^2	ε	1	ε^*	ε^{2*}	ε^2	ε		
E_2'	1	ε^2	ε^*	ε	ε^{2*}	1	ε^2	ε^*	ε	ε^{2*}		(x^2-y^2, xy)
	1	ε^{2*}	ε	ε^*	ε^2	1	ε^{2*}	ε	ε^*	ε^2		
A''	1	1	1	1	1	−1	−1	−1	−1	−1	z	
E_1''	1	ε	ε^2	ε^{2*}	ε^*	−1	$-\varepsilon$	$-\varepsilon^2$	$-\varepsilon^{2*}$	$-\varepsilon^*$	(R_x, R_y)	(xz, yz)
	1	ε^*	ε^{2*}	ε^2	ε	−1	$-\varepsilon^*$	$-\varepsilon^{2*}$	$-\varepsilon^2$	$-\varepsilon$		
E_2''	1	ε^2	ε^*	ε	ε	−1	$-\varepsilon^2$	$-\varepsilon^*$	$-\varepsilon$	$-\varepsilon^{2*}$		
	1	ε^{2*}	ε	ε^*	ε^2	−1	ε^{2*}	$-\varepsilon$	$-\varepsilon^*$	$-\varepsilon^2$		

C_{6h}	E	C_6	C_3	C_2	C_3^2	C_6^5	i	S_3^5	S_6^5	σ_h	S_6	S_3		$\varepsilon = \exp(2\pi i/6)$
A_g	1	1	1	1	1	1	1	1	1	1	1	1	R_z	x^2+y^2, z^2
B_g	1	−1	1	−1	1	−1	1	−1	1	−1	1	−1		
E_{1g}	1	ε	$-\varepsilon^*$	−1	$-\varepsilon$	ε^*	1	ε	$-\varepsilon^*$	−1	$-\varepsilon$	ε^*	(R_x, R_y)	(xz, yz)
	1	ε^*	$-\varepsilon$	−1	$-\varepsilon^*$	ε	1	ε^*	$-\varepsilon$	−1	$-\varepsilon^*$	ε		
E_{2g}	1	$-\varepsilon^*$	$-\varepsilon$	1	$-\varepsilon^*$	$-\varepsilon$	1	$-\varepsilon^*$	$-\varepsilon$	1	$-\varepsilon^*$	$-\varepsilon$		(x^2-y^2, xy)
	1	$-\varepsilon$	$-\varepsilon^*$	1	$-\varepsilon$	$-\varepsilon^*$	1	$-\varepsilon$	$-\varepsilon^*$	1	$-\varepsilon$	$-\varepsilon^*$		
A_u	1	1	1	1	1	1	−1	−1	−1	−1	−1	−1	z	
B_u	1	−1	1	−1	1	−1	−1	1	−1	1	−1	1		
E_{1u}	1	ε	$-\varepsilon^*$	−1	$-\varepsilon$	ε^*	−1	$-\varepsilon$	ε^*	1	ε	$-\varepsilon^*$	(x, y)	
	1	ε^*	$-\varepsilon$	−1	$-\varepsilon^*$	ε	−1	$-\varepsilon^*$	ε	1	ε^*	$-\varepsilon$		
E_{2u}	1	$-\varepsilon^*$	ε	1	$-\varepsilon^*$	$-\varepsilon$	−1	ε^*	ε	−1	ε^*	ε		
	1	$-\varepsilon$	$-\varepsilon^*$	1	$-\varepsilon$	$-\varepsilon^*$	−1	ε	ε^*	−1	ε	ε^*		

6. The D_{nh} groups

D_{2h}	E	$C_2(z)$	$C_2(y)$	$C_2(x)$	i	$\sigma(xy)$	$\sigma(xz)$	$\sigma(yz)$		
A_g	1	1	1	1	1	1	1	1		x^2, y^2, z^2
B_{1g}	1	1	−1	−1	1	1	−1	−1	R_z	xy
B_{2g}	1	−1	1	−1	1	−1	1	−1	R_y	xz
B_{3g}	1	−1	−1	1	1	−1	−1	1	R_x	yz
A_u	1	1	1	1	−1	−1	−1	−1		
B_{1u}	1	1	−1	−1	−1	−1	1	1	z	
B_{2u}	1	−1	1	−1	−1	1	−1	1	y	
B_{3u}	1	−1	−1	1	−1	1	1	−1	x	

D_{3h}	E	$2C_3$	$3C_2$	σ_h	$2S_3$	$3\sigma_h$		
A_1'	1	1	1	1	1	1		x^2+y^2, z^2
A_2'	1	1	−1	1	1	−1	R_z	
E'	2	−1	0	2	−1	0	(x, y)	(x^2-y^2, xy)
A_1''	1	1	1	−1	−1	−1		
A_2''	1	1	−1	−1	−1	1	z	
E''	2	−1	0	−2	1	0	(R_x, R_y)	(xz, yz)

D_{4h}	E	$2C_4$	C_2	$2C_2'$	$2C_2''$	i	$2S_4$	σ_h	$2\sigma_v$	$2\sigma_d$		
A_{1g}	1	1	1	1	1	1	1	1	1	1		x^2+y^2, z^2
A_{2g}	1	1	1	−1	−1	1	1	1	−1	−1	R_z	
B_{1g}	1	−1	1	1	−1	1	−1	1	1	−1		(x^2-y^2)
B_{2g}	1	−1	1	−1	1	1	−1	1	−1	1		xy
E_g	2	0	−2	0	0	2	0	−2	0	0	(R_x, R_y)	(xz, yz)
A_{1u}	1	1	1	1	1	−1	−1	−1	−1	−1		
A_{2u}	1	1	1	−1	−1	−1	−1	−1	1	1	z	
B_{1u}	1	−1	1	1	−1	−1	1	−1	−1	1		
B_{2u}	1	−1	1	−1	1	−1	1	−1	1	−1		
E_u	2	0	−2	0	0	−2	0	2	0	0	(x, y)	

D_{5h}	E	$2C_5$	$2C_5^2$	$5C_2$	σ_h	$2S_5$	$2S_5^3$	$5\sigma_v$		
A_1'	1	1	1	1	1	1	1	1		(x^2+y^2, z^2)
A_2'	1	1	1	−1	1	1	1	−1	R_z	
E_1'	2	2cos72°	2cos144°	0	2	2cos72°	2cos144°	0	(x, y)	
E_2'	2	2cos144°	2cos72°	0	2	2cos144°	2cos72°	0		(x^2-y^2, xy)
A_1''	1	1	1	1	−1	−1	−1	−1		
A_2''	1	1	1	−1	−1	−1	−1	1	z	
E_1''	2	2cos72°	2cos144°	0	−2	−2cos72°	−2cos144°	0	(R_x, R_y)	(xz, yz)
E_2''	2	2cos144°	2cos72°	0	−2	−2cos144°	−2cos72°	0		

D_{6h}	E	$2C_6$	$2C_3$	C_2	$3C_2'$	$3C_2''$	i	$2S_3$	$2S_6$	σ_h	$3\sigma_d$	$3\sigma_v$		
A_{1g}	1	1	1	1	1	1	1	1	1	1	1	1		x^2+y^2, z^2
A_{2g}	1	1	1	1	−1	−1	1	1	1	1	−1	−1	R_z	
B_{1g}	1	−1	1	−1	1	−1	1	−1	1	−1	1	−1		
B_{2g}	1	−1	1	−1	−1	1	1	−1	1	−1	−1	1		
E_{1g}	2	1	−1	−2	0	0	2	1	−1	−2	0	0	(R_x, R_y)	(xz, yz)
E_{2g}	2	−1	−1	2	0	0	2	−1	−1	2	0	0		(x^2-y^2, xy)
A_{1u}	1	1	1	1	1	1	−1	−1	−1	−1	−1	−1		
A_{2u}	1	1	1	1	−1	−1	−1	−1	−1	−1	1	1	z	
B_{1u}	1	−1	1	−1	1	−1	−1	1	−1	1	−1	1		
B_{2u}	1	−1	1	−1	−1	1	−1	1	−1	1	1	−1		
E_{1u}	2	1	−1	−2	0	0	−2	−1	1	2	0	0	(x, y)	
E_{2u}	2	−1	−1	2	0	0	−2	1	1	−2	0	0		

D_{8h}	E	$2C_8$	$2C_8^3$	$2C_4$	C_2	$4C_2'$	$4C_2''$	i	$2S_8$	$2S_8^3$	$2S_4$	σ_h	$4\sigma_d$	$4\sigma_v$		
A_{1g}	1	1	1	1	1	1	1	1	1	1	1	1	1	1		x^2+y^2, z^2
A_{2g}	1	1	1	1	1	−1	−1	1	1	1	1	1	−1	−1	R_z	
B_{1g}	1	−1	−1	1	1	1	−1	1	−1	−1	1	1	1	−1		
B_{2g}	1	−1	−1	1	1	−1	1	1	−1	−1	1	1	−1	1		
E_{1g}	2	$\sqrt{2}$	$-\sqrt{2}$	0	−2	0	0	2	$\sqrt{2}$	$-\sqrt{2}$	0	−2	0	0	(R_x, R_y)	(xz, yz)
E_{2g}	2	0	0	−2	2	0	0	2	0	0	−2	2	0	0		(x^2-y^2, xy)
E_{3g}	2	$-\sqrt{2}$	$\sqrt{2}$	0	−2	0	0	2	$-\sqrt{2}$	$\sqrt{2}$	0	−2	0	0		
A_{1u}	1	1	1	1	1	1	1	−1	−1	−1	−1	−1	−1	−1		
A_{2u}	1	1	1	1	1	−1	−1	−1	−1	−1	−1	−1	1	1	z	
B_{1u}	1	−1	−1	1	1	1	−1	−1	1	1	−1	−1	−1	1		
B_{2u}	1	−1	−1	1	1	−1	1	−1	1	1	−1	−1	1	−1		
E_{1u}	2	$\sqrt{2}$	$-\sqrt{2}$	0	−2	0	0	−2	$-\sqrt{2}$	$\sqrt{2}$	0	2	0	0	(x, y)	
E_{2u}	2	0	0	−2	2	0	0	−2	0	0	2	−2	0	0		
E_{3u}	2	$-\sqrt{2}$	$\sqrt{2}$	0	−2	0	0	−2	$\sqrt{2}$	$-\sqrt{2}$	0	2	0	0		

7. The D_{nd} groups

D_{2d}	E	$2S_4$	C_2	$2C_2'$	$2\sigma_d$		
A_1	1	1	1	1	1		x^2+y^2, z^2
A_2	1	1	1	−1	−1	R_z	x^2-y^2
B_1	1	−1	1	1	−1		xy
B_2	1	−1	1	−1	1	z	
E_1	2	0	−2	0	0	(x, y); (R_x, R_y)	(xz, yz)

D_{3d}	E	$2C_3$	$3C_2$	i	$2S_6$	$3\sigma_d$		
A_{1g}	1	1	1	1	1	1		x^2+y^2, z^2
A_{2g}	1	1	−1	1	1	−1	R_z	
E_g	2	−1	0	2	−1	0	(R_x, R_y)	(x^2-y^2, xy), (xz, yz)
A_{1u}	1	1	1	−1	−1	−1		
A_{2u}	1	1	−1	−1	−1	1	z	
E_u	2	−1	0	−2	1	0	(x, y)	

D_{4d}	E	$2S_8$	$2C_4$	$2S_8^3$	C_2	$4C_2'$	$4\sigma_d$		
A_1	1	1	1	1	1	1	1		x^2+y^2, z^2
A_2	1	1	1	1	1	−1	−1	R_z	
B_1	1	−1	1	−1	1	1	−1		
B_2	1	−1	1	−1	1	−1	1	z	
E_1	2	$\sqrt{2}$	0	$-\sqrt{2}$	−2	0	0	(x, y)	
E_2	2	0	−2	0	2	0	0		(x^2-y^2, xy)
E_3	2	$-\sqrt{2}$	0	$\sqrt{2}$	−2	0	0	(R_x, R_y)	(xz, yz)

D_{5d}	E	$2C_5$	$2C_5^2$	$5C_2$	i	$2S_{10}^3$	$2S_{10}$	$5\sigma_d$		
A_{1g}	1	1	1	1	1	1	1	1		(x^2+y^2, z^2)
A_{2g}	1	1	1	−1	1	1	1	−1	R_z	
E_{1g}	2	2cos72°	2cos144°	0	2	2cos72°	2cos144°	0	(R_x, R_y)	(xz, yz)
E_{2g}	2	2cos144°	2cos72°	0	2	2cos144°	2cos72°	0		(x^2-y^2, xy)
A_{1u}	1	1	1	1	−1	−1	−1	−1		
A_{2u}	1	1	1	−1	−1	−1	−1	1	z	
E_{1u}	2	2cos72°	2cos144°	0	−2	−2cos72°	−2cos144°	0	(x, y)	
E_{2u}	2	2cos144°	2cos72°	0	−2	−2cos144°	−2cos72°	0		

D_{6d}	E	$2S_{12}$	$2C_6$	$2S_4$	$2C_3$	$2S_{12}^5$	C_2	$6C_2'$	$6\sigma_d$		
A_1	1	1	1	1	1	1	1	1	1		x^2+y^2, z^2
A_2	1	1	1	1	1	1	1	−1	−1	R_z	
B_1	1	−1	1	−1	1	−1	1	1	−1		
B_2	1	−1	1	−1	1	−1	1	−1	1	z	
E_1	2	$\sqrt{3}$	1	0	−1	$-\sqrt{3}$	−2	0	0	(x, y)	
E_2	2	1	−1	−2	−1	1	2	0	0		(x^2-y^2, xy)
E_3	2	0	−2	0	2	0	−2	0	0		
E_4	2	−1	−1	2	−1	−1	2	0	0		
E_5	2	$-\sqrt{3}$	1	0	−1	$\sqrt{3}$	−2	0	0	(R_x, R_y)	(xz, yz)

8. The S_n groups

S_4	E	S_4	C_2	S_4^3		
A	1	1	1	1	R_z	x^2+y^2, z^2
B	1	−1	1	−1	z	(x^2-y^2, xy)
E	1	i	−1	−i	(x,y); (R_x, R_y)	(xz, yz)
	1	−i	−1	i		

S_6	E	C_3	C_3^2	i	S_6^5	S_6		$\varepsilon = \exp(2\pi i/6)$
A_g	1	1	1	1	1	1	R_z	x^2+y^2, z^2
E_g	1	ε	ε^*	1	ε	ε^*	(R_x, R_y)	(x^2-y^2, xy); (xz, yz)
	1	ε^*	ε	1	ε^*	ε		
A_u	1	1	1	−1	−1	−1	z	
E_u	1	ε	ε^*	−1	$-\varepsilon$	$-\varepsilon^*$	(x,y)	
	1	ε^*	ε	−1	$-\varepsilon^*$	$-\varepsilon$		

S_8	E	S_8	C_4	S_8^3	C_2	S_8^5	C_4^3	S_8^7		$\varepsilon = \exp(2\pi i/8$
A	1	1	1	1	1	1	1	1	R_z	x^2+y^2, z^2
B	1	−1	1	−1	1	−1	1	−1	z	
E_1	1	ε	i	$-\varepsilon^*$	−1	$-\varepsilon$	−i	ε^*	(x, y); (R_x, R_y)	
	1	ε^*	−i	$-\varepsilon$	−1	$-\varepsilon^*$	i	ε		
E_2	1	i	−1	−i	1	i	−1	−i		(x^2-y^2, xy)
	1	−i	−1	i	1	−i	−1	i		
E_3	1	$-\varepsilon^*$	−i	ε	−1	ε^*	i	$-\varepsilon$		(xz, yz)
	1	$-\varepsilon$	i	ε^*	−1	ε	i	$-\varepsilon^*$		

9. The cubic groups

T	E	$4C_3$	$4C_3^2$	$3C_2$		$\varepsilon = \exp(2\pi i/3)$
A	1	1	1	1		$(x^2+y^2+z^2)$
E	1	ε	ε^*	1		$(2z^2-x^2-y^2, x^2-y^2)$
	1	ε^*	ε	1		
T	3	0	0	1	(R_x, R_y, R_z); (x, y, z)	(xy, xz, yz)

T_h	E	$4C_3$	$4C_3^2$	$3C_2$	i	$4S_6$	$4S_6^5$	$3\sigma_h$		$\varepsilon = \exp(2\pi i/3)$
A_g	1	1	1	1	1	1	1	1		$(x^2+y^2+z^2)$
A_u	1	1	1	1	−1	−1	−1	−1		
E_g	1	ε	ε^*	1	1	ε	ε^*	1		$(2z^2-x^2-y^2,$
	1	ε^*	ε	1	1	ε^*	ε	1		$x^2-y^2)$
E_u	1	ε	ε^*	1	−1	$-\varepsilon$	$-\varepsilon^*$	−1		
	1	ε^*	ε	1	−1	$-\varepsilon^*$	$-\varepsilon$	−1		
T_g	3	0	0	−1	−3	0	0	−1	(R_x, R_y, R_z)	(xy, xz, yz)
T_u	3	0	0	−1	−3	0	0	1	(x, y, z)	

T_d	E	$8C_3$	$3C_2$	$6S_4$	$6\sigma_d$		
A_1	1	1	1	1	1		$x^2+y^2+z^2$
A_2	1	1	1	−1	−1		
E	2	−1	2	0	0		$(2z^2-x^2-y^2, x^2-y^2)$
T_1	3	0	−1	1	−1	(R_x, R_y, R_z)	
T_2	3	0	−1	−1	1	(x, y, z)	(xy, xz, yz)

O	E	$6C_4$	$3C_2(=C_4^2)$	$8C_3$	$6C_2$		
A_1	1	1	1	1	1		$x^2+y^2+z^2$
A_2	1	−1	1	1	−1		
E	2	0	2	−1	0		$(2z^2-x^2-y^2, x^2-y^2)$
T_1	3	1	−1	0	−1	(Rx, Ry, Rz); (x, y, z)	
T_2	3	−1	−1	0	1		(xy, xz, yz)

O_h	E	$8C_3$	$3C_2$	$6C_4$	$6C_2$	i	$6S_4$	$8S_6$	$3\sigma_h$	$6\sigma_d$		
A_{1g}	1	1	1	1	1	1	1	1	1	1		$x^2+y^2+z^2$
A_{2g}	1	1	−1	−1	1	1	1	1	−1	−1		
E_g	2	−1	2	0	0	2	−1	2	0	0		$(2z^2-x^2-y^2, x^2-y^2)$
T_{1g}	3	0	−1	1	−1	3	0	−1	1	−1	(R_x, R_y, R_z)	
T_{2g}	3	0	−1	−1	1	3	0	−1	−1	1		(xy, xz, yz)
A_{1u}	1	1	1	1	1	−1	−1	−1	−1	−1		
A_{2u}	1	1	1	−1	−1	−1	−1	−1	1	1		
E_u	2	−1	2	0	0	−2	1	−2	0	0		
T_{1u}	3	0	−1	1	−1	−3	0	1	−1	1	(x, y, z)	
T_{2u}	3	0	−1	−1	1	−3	0	1	1	−1		

10. **The groups $C_{\infty v}$ and $D_{\infty h}$ for linear molecules**

$C_{\infty v}$	E	$2C_\infty^\phi$	...	$\infty\sigma_v$		
$A_1 \equiv \Sigma^+$	1	1	...	1	z	x^2+y^2, z^2
$A_2 \equiv \Sigma^-$	1	1	...	−1	R_z	
$E_1 \equiv \Pi$	2	$2\cos\phi$	...	0	$(x, y); (R_x, R_y)$	(xz, yz)
$E_2 \equiv \Delta$	2	$2\cos 2\phi$	...	0		(x^2-y^2, xy)
$E_3 \equiv \Phi$	2	$2\cos 3\phi$	...	0		
...	...	...	...	...		

$D_{\infty h}$	E	$2C_\infty^\phi$	...	$\infty\sigma_v$	i	$2S_\infty^\phi$	...	∞C_2		
Σ_g^+	1	1	...	1	1	1	...	1		x^2+y^2, z^2
Σ_g^-	1	1	...	−1	1	1	...	−1	R_z	
Π_g	2	$2\cos\phi$	...	0	2	$-2\cos\phi$	...	0	(R_x, R_y)	(xz, yz)
Δ_g	2	$2\cos 2\phi$	...	0	2	$2\cos 2\phi$	...	0		(x^2-y^2, xy)
...	...	...	...	...	...	...	...	...		
Σ_u^+	1	1	...	1	−1	−1	...	−1	z	
Σ_u^-	1	1	...	−1	−1	−1	...	1		
Π_u	2	$2\cos\phi$	...	0	−2	$2\cos\phi$	...	0	(x, y)	
Δ_u	2	$2\cos 2\phi$	...	0	−2	$-2\cos 2\phi$	...	0		
...	...	...	...	...	...	...	...	...		

11. The Icosahedral groups*

I_h	E	$12C_5$	$12C_5^3$	$20C_3$	$15C_2$	i	$12S_{10}$	$12S_{10}^3$	$20S_6$	15σ		
A_g	1	1	1	1	1	1	1	1	1	1		$x^2+y^2+z^2$
T_{1g}	3	$\frac{1}{2}(1+\sqrt{5})$	$\frac{1}{2}(1-\sqrt{5})$	0	−1	3	$\frac{1}{2}(1-\sqrt{5})$	$\frac{1}{2}(1+\sqrt{5})$	0	−1		
T_{2g}	3	$\frac{1}{2}(1-\sqrt{5})$	$\frac{1}{2}(1+\sqrt{5})$	0	−1	3	$\frac{1}{2}(1+\sqrt{5})$	$\frac{1}{2}(1-\sqrt{5})$	0	−1	(R_x, R_y, R_z)	
G_g	4	−1	−1	1	0	4	−1	−1	1	0		
H_g	5	0	0	−1	1	5	0	0	−1	1		$(2z^2-x^2-y^2, x^2-y^2, xy, yz, xz)$
A_u	1	1	1	1	1	−1	−1	−1	−1	−1		
T_{1u}	3	$\frac{1}{2}(1+\sqrt{5})$	$\frac{1}{2}(1-\sqrt{5})$	0	−1	−3	$-\frac{1}{2}(1-\sqrt{5})$	$-\frac{1}{2}(1+\sqrt{5})$	0	1	(x, y, z)	
T_{2u}	3	$\frac{1}{2}(1-\sqrt{5})$		0	−1	−3	$-\frac{1}{2}(1+\sqrt{5})$	$-\frac{1}{2}(1-\sqrt{5})$	0	1		
G_u	4	−1	−1	1	0	−4	1	1	−1	0		
H_u	5	0	0	−1	1	−5	0	0	1	−1		

*For the pure rotation group I, the outlined section in the upper left is the character table; the g superscripts should, of course, be dropped and (x, y, z) assigned for the T_1 representation.

Appendix II

Correlation table for O_h and T_d groups

The following tables show how the representations of groups O_h and T_d are changed or decomposed into those of its subgroups when the symmetry is altered or lowered. These tables cover only representations of use in dealing with the more common symmetries of complexes. A complete collection of correlation tables will be found as Appendix B in "Molecular Symmetry and Group Theory" by Robert L. Carter, John Wiley Sons, Inc. 1998.

Correlation table for O_h

O_h	O	T_d	D_{4h}	D_{2d}	C_{4v}	C_{2v}	D_{3d}	D	C_{2h}
A_{1g}	A_1	A_1	A_{1g}	A_1	A_1	A_1	A_{1g}	A_1	A_g
A_{2g}	A_2	A_2	B_{1g}	B_1	B_1	A_2	A_{2g}	A_2	B_g
E_g	E	E	$A_{1g} + B_{1g}$	$A_1 + B_1$	$A_1 + B_1$	$A_1 + A_2$	E_g	E	$A_g + B_g$
T_{1g}	T_1	T_1	$A_{2g} + E_g$	$A_2 + E$	$A_2 + E$	$A_2 + B_1 + B_2$	$A_{2g} + E_g$	$A_1 + E$	$2A_g + B_g$
T_{2g}	T_2	T_2	$B_{2g} + E_g$	$B_2 + E$	$B_2 + E$	$A_1 + B_1 + B_2$	$A_{1g} + E_g$	$A_1 + E$	$2A_g + B_g$
A_{1u}	A_1	A_2	A_{1u}	B_1	A_2	A_2	A_{1u}	A_1	A_u
A_{2u}	A_2	A_1	B_{1u}	A_1	B_2	A_1	A_{2u}	A_2	B_u
E_u	E	E	$A_{1u} + B_{1u}$	$A_1 + B_1$	$B_2 + B_2$	$A_1 + A_2$	E_u	E	$A_u + B_u$
T_{1u}	T_1	T_2	$A_{2u} + E_u$	$B_2 + E$	$A_1 + E$	$A_1 + B_1 + B_2$	$A_{2u} + E_u$	$A_2 + E$	$A_u + 2B_u$
T_{2u}	T_2	T_1	$B_{2u} + E_u$	$A_2 + E$	$B_1 + E$	$A_2 + B_1 + B_2$	$A_{1u} + E_u$	$A_1 + E$	$2A_u + B_u$

Correlation table for T_d

T_d	D_{2d}	C_{3v}	S_4	D_2	C_{2v}	C_3	C_2	C_s
A_1	A_1	A_1	A	A	A_1	A	A	A′
A_2	B_1	A_2	B	A	A_2	A	A	A″
E	$A_1 + B_1$	E	A + B	2A	$A_1 + A_2$	E	2A	A′ + A″
T_1	$A_2 + E$	$A_2 + E$	A + E	$B_1 + B_2 + B_3$	$A_2 + B_1 + B_2$	A + E	A + 2B	A′ + 2A″
T_2	$B_2 + E$	$A_1 + E$	B + E	$B_1 + B_2 + B_3$	$A_1 + B_1 + B_2$	A + E	A + 2B	2A′ + A″

Appendix III

Some useful trigonometric relations used in text

1.	sin $(-\theta)$	$= -\sin\theta$
2.	cos $(-\theta)$	$= \cos\theta$
3.	sin $(90 + \theta)$	$= \cos\theta$
4.	cos $(90 + \theta)$	$= -\sin\theta$
5.	sin $(90 - \theta)$	$= \cos\theta$
6.	cos $(90 -\theta)$	$= \sin\theta$
7.	sin $(180 + \theta)$	$= -\sin\theta$
8.	cos $(180 + \theta)$	$= -\cos\theta$
9.	sin $(180 - \theta)$	$= \sin\theta$
10.	cos $(180 - \theta)$	$= -\cos\theta$
11.	sin $(360 + \theta)$	$= \sin\theta$
12.	cos $(360 + \theta)$	$= \cos\theta$
13.	sin $(360 - \theta)$	$= -\sin\theta$
14.	cos $(360 - \theta)$	$= \cos\theta$

$$\sin^2\theta + \cos^2\theta = 1$$

$$\tan\theta = \frac{\sin\theta}{\mathrm{cin}\,\theta}$$

$$e^{i\theta} - e^{-i\theta} = 2i\ \sin\theta$$

$$\sin\ (A + B) = \sin A \cos B + \cos A \sin B$$

$$\sin\ (A - B) = \sin A \cos B - \cos A \sin B$$

$$\cos\ (A + B) = \cos A \cos B - \sin A \sin B$$

$$\cos\ (A - B) = \cos A \cos B + \sin A \sin B$$

$$\cos^2\theta - \sin^2\theta = \cos\ 2\theta$$

Value of trigonometric functions for some common angles:

θ	0°	30°	45°	60°	90°	180°	270°	360°
sin	0	1/2	1/√2	√3/2	1	0	−1	0
cos	1	√3/2	1/√2	1/2	0	−1	0	1
tan	0	1/√2	1/2	√3	–	0	–	0

θ	0°	30°	45°	60°	90°	120°	135°	150°	180°	210°	225°
sin	0	1/2	1/√2	√3/2	1	√3/2	1/√2	1/2	0	−1/2	−1/√2
cos	1	√3/2	1/√2	1/2	0	−1/2	−1/√2	−√3/2	−1	−√3/2	−1/√2

θ	240°	270°	300°	315°	330°	360°
sin	−√3/2	−1	−√3/2	−1/√2	−1/2	0
cos	−1/2	0	1/2	1/√2	√3/2	1

Bibliography

The following books/Research papers are recommended for further reading:

[1] F. A. Cotton, "*Chemical Applications of Group Theory*", 3rd Edition, Wiley Eastern Pvt. Ltd., New Delhi (1999).
[2] Y. Tanabe and S. Sugano, *On the Absorption Spectra of Complexe ions*, J. Phys. Soc., Japan, 9 (1954) 753–766.
[3] D. S. Schonland, "*Molecular Symmetry, an Introduction to Group Theory and its Uses in Chemistry*", Van Nostrand Reinhold Company Ltd., New York (1965).
[4] B. E. Douglas and C. A. Hollingsworth, "*Symmetry in Bonding and Spectra, an Introduction*", Academic Press, Inc., New York (1985).
[5] L. H. Hall, "*Group Theory and Symmetry in Chemistry*", McGraw-Hill Book Company, New York (1969).
[6] J. M. Hollas, *Symmetry in Chemistry*, Chapman and Hall Ltd., New York (1972).
[7] I. Hargittai and M. Hargittai, "*Symmetry through the Eyes of a Chemist*", Verlagsgesellschaft, Germany (1986).
[8] T. M. Dunn, D. S. McClure and R. G. Pearson, "*Crystal Field Theory*", Harper and Row, New York (1965).
[9] J. R. Ferraro, "*Introductory Group Theory and its Application to Molecular structure*", Plenum Press, New York (1969).
[10] H. H. Jaffe and Milton Orchin, "*Symmetry in Chemistry*", Wiley Eastern Pvt. Ltd., New Delhi (1971).
[11] I. S. Dmitriev, "*Symmetry in the World of Molecules*", Mir Publishers, Moscow (1979), Harmondsworth, Penguin, (1970).
[12] F. A. Cotton, G. Wilkinson, C. A Murillo and M. Bochmann "*Advanced Inorganic Chemistry*", 6th Edition, John Wiley & Sons, Inc., New York (1999).
[13] Manas Chanda, "*Atomic Structure and Chemical bond*", Tata McGraw Hill Publishing Co. Ltd., New Delhi (1979).
[14] John D. Roberts, "*Molecular Orbital Calculations*", W. A. Benjamin, Inc., New York (1962).
[15] Russel S. Drago, "*Physical Methods in Inorganic Chemistry*", Affiliated East West Pvt. Ltd., New Delhi (1968).
[16] Alan Vincent, "*Molecular Symmetry and Group Theory- Programmed Introduction to Chemical Applications*, Wiley, New York (1977).
[17] M. S. Gopinathan and V. Ramakrishnan, "Group Theory in Chemistry", Vishal Publications, Educational Publishers, Jalandhar (1998).
[18] K. V. Raman, "*Group Theory and its applications to Chemistry*", Tata McGraw Hill Publishing Co. Ltd., New Delhi (1990).
[19] A. Salahuddin Kunju and G. Krishnan, "*Group Theory and its Applications in Chemistry*", PHI Learning private Limited, New Delhi (2010).
[20] K. Veera Reddy, "*Symmetry and Spectroscopy of Molecules*", New Age International Publishers, New Delhi (1998).
[21] P. K. Bhattacharya, "*Group Theory and its Chemical Applications*", Himalaya Publishing House, New Delhi (2011).
[22] B. E. Douglas, D. H. McDaniel and J. J. Alexander, "*Concepts and Models of Inorganic Chemistry*", 2nd Edition, John Wiley & Sons, Inc., New York (1999).
[23] James E. Huheey, "*Inorganic Chemistry: Principles of Structure and Reactivity*", 3rd Edition, Harper International SI Edition, London (1983).
[24] D. N. Sathyanarayana, "*Electronic Absorption Spectroscopy and Related Techniques*", University Press (India) Limited, Hyderabad (2001).

https://doi.org/10.1515/9783110635034-006

[25] B. N. Figgis, "*Introduction to Ligand Fields*", First Wiley Eastern Reprint, Wiley Eastern Limited, New Delhi (1976).

[26] B. S. Garg, "*Chemical Applications of Molecular symmetry and group Theory*", First Edition, Macmillan Publishers India Ltd., Delhi (2012).

[27] D. Bonchev and D. H. Rouvray (Editors), "*Chemical Group Theory: techniques and Applications*", Australia (1995).

[28] Mark Ladd, "*Symmetry and Group Theory in Chemistry*", First Edition, Harwood Publishing Limited, England (1998).

[29] Philip R. Bunker, "*Molecular Symmetry and Spectroscopy*" Academic Press, Inc., New York (1979).

[30] M. Orchin and M. M. Jaffe, *Symmetry, Point groups and Character tabales Part I: Symmetry operations and their importance for Chemical problems, J. Chem. Educ.*, 47 (1970) 246.

[31] P. B. Dorain, "*Symmetry in Inorganic Chemistry*", Addison-Wesley, Reading, Massachusetts (1965).

[32] J. P. Fackler, Jr., "*Symmetry in Coordination Chemistry*", Academic Press,New York, 1971.

[33] R. C. Maurya, *Molecular Symmetry and Group Theoretical Approach of Chemical bonding*, Lap Lambert Academic Publishing, Mauritius (2017).

[34] J. A. Greathouse, Group Theory Calculations Involving Linear Molecules, The Chemical Educator, Springer-verlag, New york, Inc. (1996).

[35] T. David Westmoreland, *Symmetry control of chemical reactions: Applications to the Berry pseudorotation of five-coordinate transition metal complexes*, Inorg. Chim. Acta, 361 (2008) 1187–1191.

[36] R. G. Pearson, *Symmetry Rules for Chemical Reactions*, Account Chem. Res., 4 (1971) 152–160.

[37] H. A. Jahn and E. Teller, *Stability of Polyatomic Molecules in Degenerate Electronic States. I. Orbital Degeneracy*, Proc. Roy. Soc., A161 (1397) 220–235.

[38] T. Chatterji and T. C. Hansen, *Magatoelastic effects in Jahn-Teller distorted CrF_2 and CuF_2 studied by neutron powder diffraction*, J. Phys.: Condensd matter, 23 (27) (2011) 1–10.

[39] C. J. Ballhausen, Introduction to Ligand Field Theory, McGraw-Hill Book Company, Inc, New York, 1971.

Index

https://doi.org/10.1515/9783110635034-007

www.ingramcontent.com/pod-product-compliance
Lightning Source LLC
Chambersburg PA
CBHW061152240926
51846CB00006B/995
* 9 7 8 3 1 1 0 6 3 4 9 6 9 *